Lecture Notes in Artificial Intelligence   8536

Subseries of Lecture Notes in Computer Science

Lecture Notes in Artificial Intelligence 8536

Subseries of Lecture Notes in Computer Science

LNAI Series Editors

Randy Goebel
University of Alberta, Edmonton, Canada
Yuzuru Tanaka
Hokkaido University, Sapporo, Japan
Wolfgang Wahlster
DFKI and Saarland University, Saarbrücken, Germany

LNAI Founding Series Editor

Jörg Siekmann
DFKI and Saarland University, Saarbrücken, Germany

Chris Cornelis   Marzena Kryszkiewicz
Dominik Ślęzak
Ernestina Menasalvas Ruiz   Rafael Bello
Lin Shang (Eds.)

Rough Sets and Current Trends in Soft Computing

9th International Conference, RSCTC 2014
Granada and Madrid, Spain, July 9-13, 2014
Proceedings

 Springer

Volume Editors

Chris Cornelis
University of Granada, Spain
E-mail: chris.cornelis@decsai.ugr.es

Marzena Kryszkiewicz
Warsaw University of Technology, Poland
E-mail: mkr@ii.pw.edu.pl

Dominik Ślęzak
University of Warsaw, Poland
E-mail: slezak@mimuw.edu.pl

Ernestina Menasalvas Ruiz
Polytechnic University of Madrid
E-mail: ernestina.menasalvas@upm.es

Rafael Bello
Central University of Las Villas, Cuba
E-mail: rbellop@uclv.edu.cu

Lin Shang
Nanjing University, China
E-mail: shanglin@nju.edu.cn

ISSN 0302-9743   e-ISSN 1611-3349
ISBN 978-3-319-08643-9   e-ISBN 978-3-319-08644-6
DOI 10.1007/978-3-319-08644-6
Springer Cham Heidelberg New York Dordrecht London

Library of Congress Control Number: 2014941828

LNCS Sublibrary: SL 7 – Artificial Intelligence

Typesetting: Camera-ready by author, data conversion by Scientific Publishing Services, Chennai, India

Printed on acid-free paper

Springer is part of Springer Science+Business Media (www.springer.com)

Preface

This volume contains the papers accepted for presentation at the 9th International Conference on Rough Sets and Current Trends in Computing (RSCTC 2014), which, along with the Second International Conference on Rough Sets and Emerging Intelligent Systems Paradigms (RSEISP 2014), was held as a major part of the 2014 Joint Rough Set Symposium (JRS 2014), during July 9–13, 2014, in Granada and Madrid, Spain. In addition, JRS 2014 also hosted the workshop on Rough Sets: Theory and Applications (RST&A), held in Granada on July 9, 2014.

JRS was organized for the first time in 2007 and was re-established in 2012 as the major flagship IRSS-sponsored event gathering different rough-set-related conferences and workshops every year. This year, it provided a forum for researchers and practitioners interested in rough sets, fuzzy sets, intelligent systems, and complex data analysis.

RSCTC is a biannual international conference that showcases the state of the art in rough set theory, current computing methods and their applications. It provides researchers and practitioners interested in emerging information technologies an opportunity to highlight innovative research directions, novel applications, and to emphasize relationships between rough sets and related areas. RSCTC is an outcome of a series of annual International Workshops devoted to the subject of rough sets, started in Poznań, Poland, in 1992.

JRS 2014 received 120 submissions that were carefully reviewed by three or more Program Committee members or external reviewers. Papers submitted to special sessions were subject to the same reviewing procedure as those submitted to regular sessions. After the rigorous reviewing process, 40 regular papers (acceptance rate 33.3%) and 37 short papers were accepted for presentation at the symposium and publication in two volumes of the JRS proceedings. This volume contains the papers accepted for the conference RSCTC 2014.

It is truly a pleasure to thank all those people who helped this volume to come into being and to turn JRS 2014 into a successful and exciting event. In particular, we would like to express our appreciation for the work of the JRS 2014 Program Committee members who helped assure the high standards of accepted papers. Also, we are grateful to the organizers of special sessions: Enrique Herrera-Viedma, Francisco Javier Cabrerizo, Ignacio Javier Pérez, Lluis Godo, Thomas Vetterlein, Manuel Ojeda-Aciego, Sergei O. Kuznetsov, Pablo Cordero, Isaac Triguero, Salvador García, Robert Bembenik, Dariusz Gotlib, Grzegorz Protaziuk, Bing Zhou, Hong Yu, and Huaxiong Li. Furthermore, we want to thank the RST&A 2014 workshop chairs (Piotr Artiemjew and Nele Verbiest), and we also gratefully acknowledge the generous help of the remaining JRS 2014 chairs- Hiroshi Motoda, Zbigniew W. Raś, Davide Ciucci, Jesús Medina-Moreno, Consuelo Gonzalo, and Salvador García, as well as of the Steering Committee members Henryk Rybiński, Roman Słowiński, Guoyin Wang, and Yiyu Yao, for their hard work and valuable suggestions with respect to preparation of the proceedings and conference organization. Also, we want to pay tribute to our honorary chairs, Lotfi Zadeh and Andrzej Skowron, whom we deeply respect for their countless contributions to the field. We also wish to express our thanks to Bernard De Baets, Francisco Herrera, and Jerzy Stefanowski for accepting to be plenary speakers of JRS 2014. Last but not least, we would like to thank all the authors of JRS 2014, without whose high-quality contributions it would not have been possible to organize the symposium.

We also want to take this opportunity to thank our sponsors, in particular, Infobright Inc. for being the industry sponsor of the entire event, and Springer for contributing the best paper award.

We are very thankful to Alfred Hofmann and the excellent LNCS team at Springer for their help and co-operation. We would also like to acknowledge the use of EasyChair, a great conference management system.

Finally, it is our sincere hope that the papers in the proceedings may be of interest to the readers and inspire them in their scientific activities.

July 2014

Chris Cornelis
Marzena Kryszkiewicz
Dominik Ślęzak
Ernestina Menasalvas Ruiz
Rafael Bello
Lin Shang

JRS 2014 Organization

General Chairs

Ernestina Menasalvas Ruiz   Polytechnic University of Madrid, Spain
Dominik Ślęzak   University of Warsaw, Poland & Infobright Inc.,
    Poland

Steering Committee

Henryk Rybiński   Warsaw University of Technology, Poland
Roman Słowiński   Poznań University of Technology, Poland
Guoyin Wang   Chongqing University of Posts and
    Telecommunications, China
Yiyu Yao   University of Regina, Canada

Program Chairs

Chris Cornelis   University of Granada, Spain (RSCTC 2014)
Marzena Kryszkiewicz   Warsaw University of Technology, Poland
    (RSEISP 2014)

Program Co-chairs for RSCTC 2014

Rafael Bello   Central University of Las Villas, Cuba
Lin Shang   Nanjing University, China

Program Co-chairs for RSEISP 2014

Hiroshi Motoda   Osaka University, Japan
Zbigniew W. Raś   University of North Carolina at Charlotte, USA
    & Warsaw University of Technology, Poland

Special Session and Tutorial Chairs

Davide Ciucci   University of Milano-Bicocca, Italy
Jesús Medina-Moreno   University of Cádiz, Spain

RST&A Workshop Chairs

Piotr Artiemjew   University of Warmia and Masuria in Olsztyn,
    Poland
Nele Verbiest   Ghent University, Belgium

Local Chairs

Salvador García   University of Jaén, Spain (Granada)
Consuelo Gonzalo   Polytechnic University of Madrid, Spain
                                   (Madrid)

Program Committee

Arun Agarwal   Joao Gomes
Piotr Andruszkiewicz   Anna Gomolińska
Piotr Artiemjew   Santiago Gonzalez Tortosa
S. Asharaf   Consuelo Gonzalo Martín
Sanghamitra Bandyopadhyay   Dariusz Gotlib
Mohua Banerjee   Salvatore Greco
Andrzej Bargiela   Jerzy Grzymała-Busse
Luis Baumela Molina   Jonatan Gómez
Jan G. Bazan   Jun He
Robert Bembenik   Francisco Herrera
José Manuel Benítez   Enrique Herrera-Viedma
Jerzy Błaszczyński   Pilar Herrero
Nizar Bouguila   Chris Hinde
Humberto Bustince   Shoji Hirano
Francisco Javier Cabrerizo   Władyslaw Homenda
Yongzhi Cao   Qinghua Hu
Mihir K. Chakraborty   Masahiro Inuiguchi
Chien-Chung Chan   Ryszard Janicki
Pablo Cordero   Andrzej Janusz
Oscar Cordón   Richard Jensen
Zoltán Csajbók   Jouni Järvinen
Bijan Davvaz   Janusz Kacprzyk
Martine De Cock   Etienne Kerre
Antonio de La Torre   Beata Konikowska
Dayong Deng   Jacek Koronacki
Lipika Dey   Vladik Kreinovich
Fernando Diaz   Yasuo Kudo
Maria Do Carmo Nicoletti   Sergei O. Kuznetsov
Didier Dubois   Huaxiong Li
Ivo Düntsch   Tianrui Li
Zied Elouedi   Jiye Liang
Javier Fernandez   Churn-Jung Liau
Wojciech Froelich   Antoni Ligęza
Salvador García   Mario Lillo Saavedra
Piotr Gawrysiak   Pawan Lingras
Guenther Gediga   Dun Liu
Lluis Godo   Dickson Lukose

Neil Mac Parthalain	Wojciech Rząsa
Luis Magdalena	Yvan Saeys
Victor Marek	Hiroshi Sakai
María José Martín	Abdel-Badeeh Salem
Benedetto Matarazzo	Miguel Ángel Sanz-Bobi
Luis Mengual	Gerald Schaefer
Jusheng Mi	Steven Schockaert
Duoqian Miao	Jesús Serrano-Guerrero
Tamás Mihálydeák	B. Uma Shankar
Evangelos Milios	Qiang Shen
Fan Min	Marek Sikora
Sadaaki Miyamoto	Arul Siromoney
Javier Montero	Myra Spiliopoulou
Jesús Montes	Jerzy Stefanowski
Mikhail Moshkov	John G. Stell
Hiroshi Motoda	Jarosław Stepaniuk
Santiago Muelas Pascual	Zbigniew Suraj
Tetsuya Murai	Piotr Synak
Michinori Nakata	Andrzej Szałas
Amedeo Napoli	Marcin Szczuka
Hung Son Nguyen	Domenico Talia
Sinh Hoa Nguyen	Vicenç Torra
Vilem Novak	Isaac Triguero
Hannu Nurmi	Li-Shiang Tsay
Manuel Ojeda-Aciego	Shusaku Tsumoto
Jose Angel Olivas Varela	Athina Vakali
Piero Pagliani	Nam Van Huynh
Krzysztof Pancerz	Nele Verbiest
Gabriella Pasi	José Luis Verdegay
Witold Pedrycz	Thomas Vetterlein
Georg Peters	Krzysztof Walczak
James F. Peters	Xin Wang
Frederick Petry	Junzo Watada
Jose María Peña	Geert Wets
Jonas Poelmans	Alicja Wieczorkowska
Lech Polkowski	Arkadiusz Wojna
Henri Prade	Marcin Wolski
Grzegorz Protaziuk	Michał Woźniak
Ignacio Javier Pérez	Jakub Wróblewski
Keyun Qin	Wei-Zhi Wu
Anna Maria Radzikowska	Yong Yang
Vijay V. Raghavan	Jingtao Yao
Sheela Ramanna	Hong Yu
Francisco P. Romero	Sławomir Zadrożny
Dominik Ryżko	Danuta Zakrzewska

Yan-Ping Zhang   William Zhu
Bing Zhou   Wojciech Ziarko

External Reviewers

Ola Amayri   Seyednaser Nourashrafeddin
Fernando Bobillo   Robert Olszewski
María Eugenia Cornejo Piñero   Stanisław Oszczak
Lynn D'eer   Wojciech Pachelski
Subasish Das   Emanuele Panzeri
Glad Deschrijver   Sandra Sandri
Waldemar Koczkodaj   Andrzej Stateczny
Adam Lenarcic   Zbigniew Szymański
Tetsuya Murai   Peter Vojtáš
Petra Murinova   Xiangru Wang
Julian Myrcha

Table of Contents

Three-Way Decisions and Probabilistic Rough Sets

New Trends in Formal Concept Analysis and Related Methods

Fuzzy Decision Making and Consensus

Soft Computing for Learning from Data

Web Information Systems and Decision Making

Image Processing and Intelligent Systems

Orthopairs in the 1960s:
Historical Remarks and New Ideas

Davide Ciucci

Dipartimento di Informatica, Sistemistica e Comunicazione
Università di Milano – Bicocca
Viale Sarca 336 – U14, I-20126 Milano, Italia
ciucci@disco.unimib.it

Abstract. Before the advent of fuzzy and rough sets, some authors in the 1960s studied three-valued logics and pairs of sets with a meaning similar to those we can encounter nowadays in modern theories such as rough sets, decision theory and granular computing. We revise these studies using the modern terminology and making reference to the present literature. Finally, we put forward some future directions of investigation.

1 Introduction

An orthopair on a universe U is a pair of subsets $A, B \subseteq U$ such that $A \cap B = \emptyset$. From a purely set-theoretical standpoint an orthopair is equivalent to a pair of nested sets $A \subseteq C$ once defined $C := B^c$ ($\cdot^c$ denoting the set complement with respect to the universe U). Clearly, an orthopair partitions the universe in three sets: A, B and $(A \cup B)^c$. So, a bijection between orthopairs and three-valued sets can be established. Given an orthopair (A, B), a three-valued set $f : U \mapsto \{0, \frac{1}{2}, 1\}$ can be defined as $f(x) = 1$ if $x \in A$, $f(x) = 0$ if $x \in B$ and $f(x) = \frac{1}{2}$ otherwise[1]. Vice versa, from a three-valued set f, an orthopair can be defined as the inverse of the previous mapping. It follows that we can equivalently study orthopairs, nested pairs or three-valued sets and in the following we will mix these three approaches keeping in mind their syntactical equivalence. Let us also notice that this tri-partition directly points to three-way decision [33] whose aim is to partition in three a universe and to the theory of opposition, in particular to a hexagon of oppositions, that can be naturally defined from a tri-partition [12,17].

Several interpretations can be attached to the two sets A, B, here is some example:

- They can be a set of true and false propositional variables;
- They can be a set of examples and counterexamples;
- A can represent the elements surely belonging to a given concept and B those surely not belonging;

---

[1] This translation reflects somehow the semantics usually assigned to orthopairs. Without giving any semantics to the two sets A, B and to the three-valued sets, any permutation in the assigned values defines a bijection.

C. Cornelis et al. (eds.): RSCTC 2014, LNAI 8536, pp. 1–12, 2014.
© Springer International Publishing Switzerland 2014

- A can represent the elements surely belonging to a given concept and B can represent a region of uncertainty: the elements about which either we cannot decide on their (Boolean) belongingness or that belong to a certain extent.

The last point emphasize a difference in the interpretation of the no-certainty zone: *unknown*, an epistemic notion, and *borderline*, an ontic notion. This difference is reflected to a greater extent by the third truth value ($\frac{1}{2}$) meaning: under the epistemic view it can stand for unknown, possible, inconsistent, etc. and under the ontic one, it can be interpreted as borderline (or half-true), undefined (in the sense of Kleene [23], that is outside the definition domain of a given function), irrelevant, etc. For reasoning purposes, it can be argued that three-valued truth-functional logics are suited in the ontic case whereas in the epistemic one non-truth functional, such as modal logic or possibility theory, are better placed [15].

Orthopairs (and nested pairs) appear in different contexts such as rough sets and ill-known sets, three-way decision, three-valued logic, shadowed sets, etc... They have been studied from a logical-algebraic standpoint by several authors and also in general contexts where the underlying structure is weaker than a Boolean algebra (for an overview see [8] and further historical remarks can be found in [28, Frame 10.11]). In Section 2, we will revise some of the operations that can be defined on these structures. For a more complete study on operations we refer to [8,24,11] for orthopairs, an overview on three values is given in [10] and about the relationship between orthopairs and three-valued operations see [13].

The aim of the paper is two-fold, as the title suggests. From one side, it presents an overview of some papers published before the coming of fuzzy sets and rough sets, which contain several ideas about concepts that will be developed in the following years. Then, we show that some ideas contained in those papers are still innovative and can give some insight to the paradigms connected with orthopairs. In particular, we will take into account the works by Fadini [19,18], Andreoli [1,2,3] and the work by Gentilhomme [21]. Only the last one is to some extent known in the fuzzy logic community and cited by some authors[2], the first two are almost unknown. On the other hand, we remark that it is out of the scope of the present work to survey all the operations existing in rough set theory, for this aspect we refer the interested reader to [4,5,13].

2   Preliminary Notions

As said in the introduction, several interpretations to three-valued sets can be given and this reflects also on the different operations that we are entitled to introduce. The same situation applies to orthopairs and nested pairs. As we will see in Section 4, according to which interpretation or representation we use, different operations naturally arise.

---

[2] The paper [21] has 95 citations on Google Scholar at 23 April 2014.

Indeed, let us consider a concept A that we want to describe and a pair of sets (A_1, A_2) that we use to effectively represent it. Usually, A_1 represents the elements that *surely belong* to A. This corresponds in rough set theory to the lower approximation and so, from now on, we will denote it as A_l. The second set can have three (mutually definable) interpretations that we are going to use through the paper:

- A_2 contains the objects *possibly belonging* to A, that is, it correponds to the upper approximation and we denote it as A_u.
- A_2 contains the objects *surely not belonging* to concept A, that is, the exterior region. Thus, we are going to denote it as A_e.
- A_2 represents the objects on which we are *undecided* about their belongingness to A, this is the boundary region and it will be named A_{bnd}.

Needless to stress (A_l, A_u) is a nested pair, whereas (A_l, A_e) and (A_l, A_{bnd}) are orthopairs. We are not going to consider here the case where the first element A_1 is different from A_l, which will generate other orthopairs, such as (A_u, A_e) and (A_e, A_{Bnd}). From now on, A and B will stand for orthopairs $(A_l, A_e), (B_l, B_e)$ or nested pairs $(A_l, A_u), (B_l, B_u)$.

We now introduce some operations and order relations that have been introduced on ortho (nested) pairs. This is not an exhaustive list, but a presentation of some operations already known in literature that are helpful for the following discussion. Other operations can be found in [8,25,24,11,10,14].

Let us first consider unary operations, usually meant to model a negation. We have on three values the involutive, the intuitionistic and the paraconsistent negations which extend the Boolean negation and thus differ only on the negation of the third-truth value, respectively defined as $\frac{1}{2}' = \frac{1}{2}$, $\sim \frac{1}{2} = 0$ and $-\frac{1}{2} = 1$. Once translated to orthopairs and nested pairs these negations are defined as follows:

$$(A_l, A_e)' := (A_e, A_l) \qquad\qquad (A_l, A_u)' := (A_u^c, A_l^c) \qquad (1)$$
$$\sim (A_l, A_e) := (A_e, A_e^c) \qquad\qquad \sim (A_l, A_u) := (A_u^c, A_u^c) \qquad (2)$$
$$-(A_l, A_e) := (A_l^c, A_l) \qquad\qquad -(A_l, A_u) := (A_l^c, A_l^c) \qquad (3)$$

As far as binary operations are concerned, we are mainly interested in conjunction and disjunction, as well as related order relations. So, the basic meet and join on three-values are the min and max (Kleene conjunction and disjunction [23]) corresponding to the usual ordering on numbers: $0 < \frac{1}{2} \leq 1$. On orthopairs A and B, this ordering is known as the *truth ordering* [6] and it reads $A_l \subseteq B_l$ and $B_e \subseteq A_e$ or equivalently $A_u \subseteq B_u$ on nested pairs. The meet and join operations are respectively defined on orthopairs as

$$(A_l, A_e) \sqcap (B_l, B_e) := (A_l \cap B_l, A_e \cup B_e) \qquad (4)$$
$$(A_l, A_e) \sqcup (B_l, B_e) := (A_l \cup B_l, A_e \cap B_e) \qquad (5)$$

The other usually considered order relation on orthopairs is the *knowledge ordering* [6,32], also known as *semantic precision* [24]: $A \preceq_k B$ if $A_l \subseteq B_l$ and

$A_e \subseteq B_e$ (equiv., $B_u \subseteq A_u$). As its name reflects, this ordering means that A is less informative than B. It is a just a partial order which corresponds on three values to $\frac{1}{2} \leq \{0, 1\}$. Thus, it does not generate a join operator but only a meet one, that is the min with respect to this ordering, and it corresponds on orthopairs to the *optimistic combination operator* [24].

However, by transforming this order relation into a total one, two different orderings and two different conjunctions and disjunctions are generated. These orderings read on orthopairs as:

$$(A_l, A_e) \preceq_e (B_l, B_e) \quad \text{iff} \quad A_e \subseteq B_e \text{ and } A_l \cup A_e \subseteq B_l \cup B_e, \qquad (6)$$

$$(A_l, A_e) \preceq_l (B_l, B_e) \quad \text{iff} \quad A_l \subseteq B_l, \text{ and } A_l \cup A_e \subseteq B_l \cup B_e, \qquad (7)$$

So, in both cases A is less informative than B. Let us notice, indeed, that $A_l \cup A_e = A^c_{Bnd}$ and so the second condition of both relations can be equivalently stated as $B_{bnd} \subseteq A_{Bnd}$. In the first case A is at least as negative as B, whereas the second ordering means that A is at least as positive as B. From an information point of view these two orderings are, thus, less demanding than the information ordering and we have that $A \preceq_k B$ iff $A \preceq_e B$ and $A \preceq_l B$.

Other orders on three values are respectively $\frac{1}{2} \leq_e 1 \leq_e 0$ and $\frac{1}{2} \leq_l 0 \leq_l 1$ (we are using the symbol $\leq$ on numbers and $\preceq$ on pairs). The interpretation of these orderings on three values is not interesting per se, but they generate two important pairs of conjunction and disjunction:

- the Sobociński operations [31], corresponding to uninorms with neutral element $\frac{1}{2}$ [22]. The conjunction corresponds to the max with respect to the order $\leq_e$ and the disjunction is the max with respect to $\leq_l$. They are used in conditional events to fuse conditionals [16]. Their definition on orthopairs is given in equations 8.

$$(A_l, A_e) \sqcap_S (B_l, B_e) := (A_l, A_e) \sqcup_e (B_l, B_e) = (A_l \backslash B_e \cup B_l \backslash A_e, A_e \cup B_e)$$
$$(8a)$$

$$(A_l, A_e) \sqcup_S (B_l, B_e) := (A_l, A_e) \sqcup_l (B_l, B_e) = (A_l \cup B_l, A_e \backslash B_l \cup B_e \backslash A_l)$$
$$(8b)$$

- the weak Kleene meet and join [23], respectively corresponding to the min with respect to $\leq_l$ and $\leq_e$. In this case, the third value is interpreted as undefined. On orthopairs they generate the operations in equations 9.

$$(A_l, A_e) \sqcap_K (B_l, B_e) := (A_l, A_e) \sqcap_e (B_l, B_e) \qquad (9a)$$
$$:= ((A_l \cap B_l) \cup [(A_l \cap B_e) \cup (B_l \cap A_e)], A_e \cap B_e))$$
$$(A_l, A_e) \sqcup_K (B_l, B_e) := (A_l, A_e) \sqcap_l (B_l, B_e) \qquad (9b)$$
$$:= (A_l \cap B_l, (A_e \cap B_e) \cup [(A_e \cap B_l) \cup (B_e \cap A_l)])$$

Finally, we can introduce six different negations by using these two orderings similarly as was done in equations (1)–(3). Some of these negations will be discussed in Section 4.2.

3   Interpretation

We start now to analyze how the three authors under investigation approach orthopairs. Here, we consider the interpretation attached to a pair of sets and in the following section, we see which operations are defined on them.

Orthopairs are introduced by Fadini [18,19] under the name *complex classes*. They are defined as pairs of the kind (A_l, A_{Bnd}), so with the meaning lower-boundary. Using Fadini's terminology: A_l is the extension of a class and A_{Bnd} contains "all the elements whose belonging to the class has the third truth value"[3] [18]. The union $A_l \oplus A_{Bnd}$ is the *complex extension* of a class, and to distinguish between the two different parts of the extension a new unity i is introduced as in complex numbers and a complex class C is thus denoted as $C = A_l \cup i A_{Bnd}$.

Fadini is aware that different meanings can be attached to the third truth value and that operations have to be chosen accordingly to the interpretation. He gives the interpretation of *indeterminate* whose meaning is "clearly different from *unknown*", and for which the tertium non datur principle should not hold (contrary to the unknown case). The term "indeterminate" is taken from Reichenbach [29], that is, from quantum mechanics. So, it seems that the third value has an ontic nature and it is not a knowledge flaw. This point of view is clarified by Fadini himself in his later book on fuzzy sets, where he says that indeterminate stands for a third truth value which is neither true nor false and it is different from *unknown* or *unknowable* [20]. Moreover, in this book he also studies the case of *unknown* which represents "the indecision between true and false and so it is not a real third truth value". In order to manage this case he refers to a doxastic logic [26].

Andreoli [2,3] studies the generalizations of Boolean algebras and Boolean sets in two directions. The first one, is by extending the set of truth values (the membership function in case of sets). So, in the case of three values, he classifies objects as "interior", "exterior", "boundary" or also as "accepted", "rejected", "undecided". So, we can see that the former terminology coincide with the standard rough-set theory one whereas the second interpretation echoes the terminology of three-way decision theory [33]. The second direction is what he calls "levels", whose motivations arise from genetics. Indeed, he gives the example of a gene with a dominant allele A and a recessive one a. Then, we can have three different kinds of pair: AA, $Aa \equiv aA$ and aa with the order $AA > Aa > aa$ or if we want to distinguish between Aa and aA with a Boolean lattice structure (see Figure 1). We notice, however, that an operation of Boolean operators in this context is not provided. Perhaps, we could hazard that the min/max can give the combination with the minor/major number of recessive alleles possible.

This idea of levels reminds the granular computing approach to represent knowledge. This similarity is also supported by Andreoli's idea that two oper-

---

[3] All translations from Italian (and in case of Gentilhomme, from French) are under our complete responsibility.

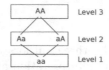

Fig. 1. Three levels of a Boolean algebra

ations named "refinement" and "attenuation" can be introduced on levels [3]. The first operation introduces a new level, that is, we can move from two to three by splitting one level into other two. The second operation acts in the opposite direction by fusing together two different levels. These operations can be encountered nowadays in granular computing acting on levels or on concepts, for instance in [7] we defined a refinement operation and an elimination one on an ontology's concepts. He also gives some hints on the fact that these two operations can be generalized to more generic structures, not only to three values. We postpone the investigation of this idea to a future work. We also notice that in [1] a general study on the algebraic structure of pairs from a Boolean algebra is given. These pairs, however, are not necessarily disjoint and an interpretation is not provided.

Finally, Gentilhomme pairs are in the form lower-upper (A_l, A_u) and they are named fuzzy ("flou") sets. The lower set is the "certainty zone", the upper the "maximal extension" and the boundary, the fuzzy zone. He also notices that a flou set can be equivalently expressed as the orthopair lower-exterior.

The interpretation of the boundary (equivalently, the third value) has an ontic nature (the value $\frac{1}{2}$ is named "maybe"). Indeed, it refers to linguistic problems where we are unable to correctly classify some linguistic object and this "hesitation" is a "matter of the language" [21]. So, the problem is intrinsic to the object under study, not to the observer. However, in some passages Gentilhomme also comments on the causes of this "partial failure" in classifying objects with certainty, opening the door to some problems also in the observers. Indeed, in the list of causes given by Gentilhomme, we have:

- the fact that different agents have different opinions. This could lead to an interpretation of the third truth value as *inconsistent* and thus to paraconsistent logic where two (or more) agents can be in accordance or not on the Boolean truth value to assign.
- the fact that data cannot be analyzed in a certain way or that we cannot apply some given criteria to the data.

Finally, we notice that he also uses the terminology of complex numbers when generalizing to not nested pairs (A, B). Analogously to the numerical case, the sense is that "all the symbols do not have an immediate interpretation [...] but they obey to similar formal rules". He also notices that such pairs can be defined by a set "completely" fuzzy $(\emptyset, B)$ and a set "totally imaginary" $(A, \emptyset)$ as $(A, B) = (A, \emptyset) \sqcup (\emptyset, B)$.

4 Operations

In this section, we look at the operations introduced by the three authors in their works making reference, when possible, to those introduced in Section 2 and pointing out which operations are new and the relations with other theories. Of course, these operations are defined in an abstract setting and they should be adapted to the context where they are used. For instance, in rough set theory, not all orthopairs are representable as rough sets, due to the partition generated by the equivalence relation; this entails a non truth-functional behaviour of rough sets [13].

4.1 Intersection, Union and Difference

Andreoli is interested in a general study on orthopairs (sometimes only pairs as in [1]) and he considers as plausible all the binary operations that are associative, commutative and idempotent. There are six operations of this kind and they are pairwise linked by a de Morgan property (using the standard involutive negation (1)). They are the Kleene min and max, Sobocinski conjunction and disjunction (equations 8) and weak Kleene conjunction and disjunction (equations 9). He also notices that they correspond to the min or max with respect to different orderings on the three values. So, for instance Sobocinski conjunction is the max with respect to the order $\leq_e$ (equiv., the min with respect to the opposite order $0 \leq 1 \leq \frac{1}{2}$).

On the other hand, both Fadini and Gentilhomme consider as intersection and union the standard Kleene operations (min and max on three values). This is coherent with respect to the ontic interpretation they have in mind as explained in the previous sections.

Besides conjunction and disjunction, Gentilhomme also devotes some efforts to define a difference between two nested pairs. He considers as the correct definition of a difference the following one ($\cdot'$ is the involutive negation on pairs as defined in equation (1)):

$$A \setminus B := A \sqcap B' = (A_l \setminus B_u, A_u \setminus B_l) \tag{10}$$

that corresponds to an "experimental reality" where the certainty zone is exactly $A_l \setminus B_u$. However, he also introduces two other differences: the greatest, that consists in accepting the maximum of the risk, and the smallest, that is accepting no risk. They are respectively defined as

$$A \setminus_g B := (A_l \setminus B_l, A_u \setminus B_l) \tag{11}$$

$$A \setminus_l B := (A_l \setminus B_u, A_u \setminus B_u) \tag{12}$$

It is interesting to notice that these difference operators can be obtained as in equation (10) but using the intuitionistic and paraconsistent negations of equations (2) and (3), that is, we have $A \setminus_g B = A \sqcap -B$ and $A \setminus_l B = A \sqcap \sim B$.

Of course, this reference to the risk occurring to using one operation instead of another points to decision theory and so to the possibility to introduce these (and other) operations on three-way decision theory. This aspect deserves a further investigation in the future.

All these differences are not equal to the difference on orthopairs defined in [25] as $(A_l \setminus B_e, A_e \setminus B_l)$. This last corresponds to making the orthopair "consistent", that is removing from the positive/negative part of A the negative/positive part of B so as to avoid conflict between A and B. The corresponding operation on nested pairs reads $(A_l \cap B_u, A_u \cup B_u^c)$ whose interpretation is not so clear. Thus, in the difference case, it is evident that even though mathematically equivalent, nested and orthopairs give birth to different operations which make sense more in one case than in the other.

4.2 Negations

All the three authors consider the involutive negation as one possibility. Gentilhomme just considers this one, whereas Andreoli and Fadini give more solutions.

Indeed, in order to obtain a negation, Fadini considers what happens in negating the Boolean part, the complex part or both. As a result, not all these operations are an extension of the Boolean negation on two values. By the negation of the Boolean part we get the standard involutive negation. By negating the complex part, we obtain a swapping of 0 with $\frac{1}{2}$: on nested pairs, $(A_l, A_u)^* := (A_l, A_l \cup A_e)$ and on orthopairs, $(A_l, A_e)^* := (A_l, A_{Bnd})$. That is, in the case of orthopairs the negation consists in a switch of the interpretation from necessity-impossibility to necessity-unknown. Let us notice that it corresponds to an involutive negation based on the order $\leq_e$ (i.e., the "middle" value is 1).

Finally, by negating both the Boolean and the complex part we get two negations: $\overline{(A_l, A_u)} := (A_l^c, U)$ and $\underline{(A_l, A_u)} := (A_{Bnd}, U)$. On three values, the first negation is such that $\overline{1} = \frac{1}{2}$, $\overline{0} = \overline{\frac{1}{2}} = 1$, which is the *complete negation* introduced by Reichenbach and has its reasons[4] in quantum mechanics [29]. We remark that it can be obtained by a paraconsistent-like negation based on the order $\leq_l$. On the same ordering, the intuitionistic-like negation corresponds to the second negation by Fadini, and it is defined as $\overline{1} = \overline{0} = \frac{1}{2}$, $\overline{\frac{1}{2}} = 1$. In both cases, nothing is false, the difference lies in what is true: in the first case it corresponds to what was not true (that is, false or unknown), in the second to what was not known (i.e., in the boundary).

Moreover, he also considers the possibility that a negation of a Boolean or complex part is allowed to contain *only* elements outside the class itself instead of *all and only* the elements. In this way, a further negation is introduced by swapping the Boolean and complex part. In terms of nested pairs: $(A_l, A_u)^\circ = (A_u \setminus A_l, A_u)$ and of three-values $0^\circ = 0$, $\frac{1}{2}^\circ = 1$ and $1^\circ = \frac{1}{2}$. This corresponds to an involutive negation based on the order $\leq_l$ ($\frac{1}{2} \leq_l 0 \leq_l 1$).

Andreoli, apart from the standard involutive negation, introduces on three values an interesting approach, which is different from what we have seen until

---

[4] Reichenbach justifies the name *complete* by the fact that $a \vee \overline{a}$ is a tautology and this form of excluded middle is required by quantum mechanics.

now. In a certain sense, it can be seen as a Boolean negation on the set of truth values. Indeed, if $t \in \{0, \frac{1}{2}, 1\}$, then the negation of $\{t\}$ is $\{t\}^c$. This is justified by the interpretation of three values as *chosen, rejected, undecided*. So, for instance, the negation of undecided is *decided* which can mean either chosen or refused. With respect to decision theory he then suggests that one can arrive at a final decision with a two steps "Boolean" procedure. That is, at a first step there is the division of the world in decided and undecided and in the second step decided is further classified as rejected or chosen. This is a sort of sequential reasoning which however works in the opposite direction with respect to the three-way sequential decision theory [34] (see Figure 2).

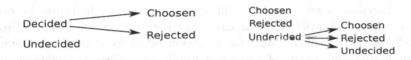

Fig. 2. Sequential decision making (Andreoli left, Yao right)

Indeed, this last consists in developing at the second step what is left undecided in the first one. In this way, through a sequence of three-way decision it is possible to arrive at a Boolean decision, classifying all objects as accepted or rejected. This strategy is pointed out (but not developed) by Gentilhomme with respect to classification: "we classify what we are able to classify and we devote a "cesspool" to the rebel elements" [21].

Going back to the negation, when applied to orthopairs we get that the negation of (A_l, A_e) is $((A_l)^c, A_e^c) = ((A^c)_u, A_u)$ which is no more an orthopair and the intersection of A_l^c and A_u is the boundary A_{Bnd}. Pairs of this kind make sense when we want to model conflicting information, such as in Belnap or paraconsistent logics, so that an object can be in both sets of the pair [11,14]. The role of this negation in decision theory and conflicting information should be better investigated.

Finally, let us notice that the fact that at a first step we do not take a definite position but just exclude one possibility among three, that is we wonder if a proposition is false (either true) or unknown has been also discussed in terms of orthopairs arising from formal contexts in [27]. Similarly, in the translation from three-valued to epistemic logics [9] we admit the possibility to represent that a valuation of a formula is $\leq \frac{1}{2}$ (false or unknown) or $\geq \frac{1}{2}$ (unknown or true).

4.3 Inclusion

According to Fadini [19], the main property to define an inclusion is based on the two parts (the Boolean and boundary ones) of his complex-class notion. Indeed, the inclusion between two complex classes $A_l \oplus iA_{Bnd}$ and $B_l \oplus iB_{Bnd}$ holds if it holds for the union of the two parts $A_l \cup A_{Bnd} \subseteq B_l \cup B_{Bnd}$. This, on nested pairs, reduces to $A_u \subseteq B_u$. Fadini outlined five ways to fulfill this condition, namely (sub1) $A_l \subseteq B_l$, $A_u \subseteq B_u$ it is named *proper inclusion* since it is the only one to be an order relation. We can see that it is the standard inclusion relation on nested pairs, corresponding to the standard order relation on three-values;

(sub2) $A_u \subseteq B_l$ named *total inclusion* for the intuitive reason that the concept A it totally contained in B, both in its certainty and possibility parts;

(sub3) $A_{bnd} \subseteq B_l$, $A_l \subseteq B_{bnd}$ it is called *improper* or *inverted inclusion*;

(sub4) $A_u \subseteq B_{Bnd}$, the *total improper inclusion*;

(sub5) A_{bnd} and A_l are both contained in B_u and they have a non-empty intersection with both B_l and B_{bnd}, this is the *mixed inclusion*.

As can be seen his notions of inclusion are greatly influenced by his interpretation of the orthopair as lower-boundary. This view generates original (pseudo) order relations that not always can find an easy interpretation outside this framework.

Gentilhomme presents six different inclusion relations, defined according to different mutual behaviours of lower and upper sets. These six relations form a lattice where the order relation is given by "implies", that is an inclusion is smaller than another if the second one implies the first[5]. Only three of these relations are considered meaningful, named *normal*, since "it is best suited for calculations": $A_l \subseteq B_l$ and $A_u \subseteq B_u$ (again the standard inclusion on orthopairs); *strong* and *weak* defined, respectively, as $A_u \subseteq B_l$ and $A_l \subseteq B_u$, the justification of the name being intuitive.

Finally, Andreoli does not study directly the inclusion relations on three values/orthopairs but just on four values in [2]. On three values, they are indirectly considered when studying the join and meet operations (see section 4.1). As can be seen no one considered the knowledge ordering but they have defined order relations never encountered before.

With respect to the relationship with implications, the only reference is by Fadini in [18]. He does not consider his inclusion operations to be the corresponding of an inference but of a conditional *if a, then b* since they hold only when a is true. Oddly, no further discussion on implication is present in other Fadini's works nor in Andreoli and Gentilhomme.

5   Conclusion

In the present paper, we reviewed some (old) works on orthopairs, related them to modern theories such as rough sets, decision theory and granular computing. We saw how different interpretations of an orthopair can influence the definition of operations. Some new operations (with respect to what is usually considered nowadays) have been found: the difference operations in Gentilhomme and the negations in Fadini and Andreoli. In particular these operations are often given an interpretation in decision theory. So, as a future work, it is worth considering the possibility to study orthopair operations in three-way decision. More generally, the role of the negation in orthopairs and conflicting information needs

---

[5] The relation of this lattice with implication lattices in rough sets [30] should be studied in the future.

a thorough understanding. Further, the role of levels in generalized structures outlined by Andreoli could be of some interest in granular computing. Finally, an algebraic study of these new operations deserves some attention and an ad hoc study.

References

1. Andreoli, G.: Struttura delle algebre di Boole e loro estensione quale calcolo delle classi. Giornale di Matematiche di Battaglini LXXXV, 141–171 (1957)
2. Andreoli, G.: Algebre di Boole - algebre di insieme - algebre di livelli. Giornale di Matematiche di Battaglini LXXXVII, 3–22 (1959)
3. Andreoli, G.: Dicotomie e tricotomie (anelli booleani triadici ed algebre booleane a tre valori). La Ricerca, pp. 1–10 (1961)
4. Banerjee, M., Chakraborty, K.: Algebras from rough sets. In: Pal, S., Skowron, A., Polkowski, L. (eds.) Rough-Neural Computing, pp. 157–188. Springer (2004)
5. Banerjee, M., Khan, M. A.: Propositional logics from rough set theory. In: Peters, J.F., Skowron, A., Düntsch, I., Grzymała-Busse, J.W., Orłowska, E., Polkowski, L. (eds.) Transactions on Rough Sets VI. LNCS, vol. 4374, pp. 1–25. Springer, Heidelberg (2007)
6. Belnap, N.D.: A useful four-valued logic. In: Dunn, J.M., Epstein, G. (eds.) Modern Uses of Multiple-Valued Logic, pp. 8–37. D. Reidel Publishing Company (1977)
7. Calegari, S., Ciucci, D.: Granular computing applied to ontologies. International Journal of Approximate Reasoning 51(4), 391–409 (2010)
8. Ciucci, D.: Orthopairs: A simple and widely used way to model uncertainty. Fundam. Inform. 108(3-4), 287–304 (2011)
9. Ciucci, D., Dubois, D.: A modal theorem-preserving translation of a class of three-valued logics of incomplete information. Journal of Applied Non-Classical Logics 23(4), 321–352 (2013)
10. Ciucci, D., Dubois, D.: A map of dependencies among three-valued logics. Information Sciences 250, 162–177 (2013), Corrigendum: Information Sciences 256, 234–235 (2014)
11. Ciucci, D., Dubois, D., Lawry, J.: Borderline vs. unknown: a comparison between three-valued valuations, partial models, and possibility distributions. In: Proceedings of the Workshop Weighted Logics for AI at ECAI 2012, pp. 83–90 (2012), http://www2.lirmm.fr/ecai2012/images/stories/ecai_doc/pdf/workshop/W34_wl4ai-working-notes.pdf
12. Ciucci, D., Dubois, D., Prade, H.: Oppositions in rough set theory. In: Li, T., Nguyen, H.S., Wang, G., Grzymala-Busse, J., Janicki, R., Hassanien, A.E., Yu, H. (eds.) RSKT 2012. LNCS, vol. 7414, pp. 504–513. Springer, Heidelberg (2012)
13. Ciucci, D., Dubois, D.: Three-valued logics, uncertainty management and rough sets. In: Peters, J.F., Skowron, A. (eds.) Transactions on Rough Sets XVII. LNCS, vol. 8375, pp. 1–32. Springer, Heidelberg (2014)
14. Ciucci, D., Dubois, D., Lawry, J.: Borderline vs. unknown - comparing three-valued representation of imperfect information. International Journal of Approximate Reasoning (2013) (submitted)
15. Dubois, D.: Reasoning about ignorance and contradiction: many-valued logics versus epistemic logic. Soft Comput. 16(11), 1817–1831 (2012)
16. Dubois, D., Prade, H.: Conditional objects as nonmonotonic consequence relationships. IEEE Transaction of Sysyems, Man, and Cybernetics 24(12), 1724–1740 (1994)

17. Dubois, D., Prade, H.: From Blanché's hexagonal organization of concepts to formal concept analysis and possibility theory. Logica Universalis 6(1-2), 149–169 (2012)
18. Fadini, A.: Il calcolo delle classi in una logica a tre valori di verità. Giornale di Matematiche di Battaglini XC, 72–91 (1962)
19. Fadini, A.: Teoria degli elementi complessi nelle Algebre di Boole. Annali del Pontificio Istituto Superiore di Scienze e Lettere "S. Chiara" 12, 223–243 (1962)
20. Fadini, A.: Introduzione alla Teoria degli Insiemi Sfocati. Liguori Editore, Napoli (1979)
21. Gentilhomme, M.Y.: Les ensembles flous en linguistique. Cahiers de linguistique théorique et appliquée. Bucarest 47, 47–65 (1968)
22. Grabisch, M., Marichal, J.L., Mesiar, R., Pap, E.: Aggregation Functions. Cambridge University Press (2009)
23. Kleene, S.C.: Introduction to metamathematics. North–Holland Pub. Co., Amsterdam (1952)
24. Lawry, J., Dubois, D.: A bipolar framework for combining beliefs about vague propositions. In: Brewka, G., Eiter, T., McIlraith, S.A. (eds.) Principles of Knowledge Representation and Reasoning: Proceedings of the Thirteenth International Conference, pp. 530–540. AAAI Press (2012)
25. Lawry, J., Tang, Y.: On truth-gaps, bipolar belief and the assertability of vague propositions. Artif. Intell. 191-192, 20–41 (2012)
26. Nola, A.D., Gerla, G.: A three-valued doxastic logic. La Ricerca XXXII, 19–33 (1981)
27. Pagliani, P.: Information gaps as communication needs: A new semantic foundation for some non-classical logics. Journal of Logic, Language and Information 6(1), 63–99 (1997)
28. Pagliani, P., Chakraborty, M.: A Geometry of Approximation. Springer (2008)
29. Reichenbach, H.: Philosophic Foundations of Quantum Mechanics. University of California Press, Berkeley (1954)
30. Samanta, P., Chakraborty, M.K.: Generalized rough sets and implication lattices. In: Peters, J.F., Skowron, A., Sakai, H., Chakraborty, M.K., Slezak, D., Hassanien, A.E., Zhu, W. (eds.) Transactions on Rough Sets XIV. LNCS, vol. 6600, pp. 183–201. Springer, Heidelberg (2011)
31. Sobociński, B.: Axiomatization of a partial system of three-value calculus of propositions. J. of Computing Systems 1, 23–55 (1952)
32. Yao, Y.: Interval sets and interval-set algebras. In: Proceedings of the 8th IEEE International Conference on Cognitive Informatics, pp. 307–314 (2009)
33. Yao, Y.: An Outline of a Theory of Three-way Decisions. In: Yao, J., Yang, Y., Słowiński, R., Greco, S., Li, H., Mitra, S., Polkowski, L. (eds.) RSCTC 2012. LNCS, vol. 7413, pp. 1–17. Springer, Heidelberg (2012)
34. Yao, Y.: Granular computing and sequential three-way decisions. In: Lingras, P., Wolski, M., Cornelis, C., Mitra, S., Wasilewski, P. (eds.) RSKT 2013. LNCS, vol. 8171, pp. 16–27. Springer, Heidelberg (2013)

General Negations for Residuated Fuzzy Logics*

María Eugenia Cornejo Piñero[1], Jesús Medina-Moreno[2],
and Eloisa Ramírez-Poussa[1]

[1] Department of Statistic and O.R., University of Cádiz, Spain
{mariaeugenia.cornejo,eloisa.ramirez}@uca.es
[2] Department of Mathematics, University of Cádiz, Spain
jesus.medina@uca.es

Abstract. Involutive residuated negations are usually considered in residuated fuzzy logics and they are also based on continuous triangular norms. This paper introduces a generalization of these negations using flexible conjunctors, several properties of them and the corresponding disjunctive dual operators associated with the conjunctor.

Keywords: Residuated fuzzy logics, residuated negations, adjoint triples.

1 Introduction

Adjoint triples were firstly used considering the adjoint conjunctor and only one implication in Logic Programming [16,20]. Later they were assumed in Logic Programming [19] and in other frameworks, such as in general substructural logics [1], Fuzzy Formal Concept Analysis [18], Fuzzy Relation Equations [9] and Rough Set Theory [7], with the goal of providing flexible settings in order to increase the range of possible applications of these frameworks. Moreover, several properties of them and comparisons have been given in several papers [4,5].

On the other hand, negation operators are widely studied [8,11,22,23] and they are usually considered in fuzzy logics [3,12,14] to simulate the ability of a human brain in order to make decisions. For instance, the use of negations in logic programming allows the definition of nonmonotonic logic programs, but require an extra effort to obtain (stable) models [17]. To the best of our knowledge, the most general negations were introduced by Georgescu and Popescu in [15] and are called pairs of weak negations.

One class of useful negations are the residuated negations, which are defined from residuated implications of a t-norm [2,13,21]. This paper considers adjoint triples in order to present a generalization of residuated negations. Moreover, a comparison with the pairs of weak negations is introduced, which shows that adjoint negations are more general. Indeed, we have proven that every pair of weak negations can be obtained from an adjoint triple. As a consequence, a very interesting result is obtained, that is, every pair of weak negation can be derived from the implications of an adjoint triple.

* Partially supported by the Spanish Science Ministry project TIN2012-39353-C04-04 and by Junta de Andalucía project P09-FQM-5233.

C. Cornelis et al. (eds.): RSCTC 2014, LNAI 8536, pp. 13–22, 2014.

The disjunction dual to an adjoint conjunctor is also defined and several properties are presented. In the case of strong adjoint negations, two implications associated with the disjunction are defined and they satisfy, together with the adjoint conjunctor, the dual adjoint property.

2   Preliminaries

In this section, we will set the basic notions used in the paper. Firstly, we recall the definition of adjoint triple and some interesting properties obtained from this notion.

Adjoint triples are general operators which have been developed to increase the flexibility of the framework in which they are used, since, for example, the conjunctors in adjoint triples are neither required to be commutative nor associative.

As the commutativity is not assumed, we have two different ways of generalizing the well-known adjoint property between a t-norm and its residuated implications, depending on which argument is fixed.

Definition 1. Let $(P_1, \leq_1)$, $(P_2, \leq_2)$, $(P_3, \leq_3)$ be posets and $\& \colon P_1 \times P_2 \to P_3$, $\swarrow \colon P_3 \times P_2 \to P_1$, $\nwarrow \colon P_3 \times P_1 \to P_2$ be mappings, then $(\&, \swarrow, \nwarrow)$ is an adjoint triple with respect to P_1, P_2, P_3 if $\&$, $\swarrow, \nwarrow$ satisfy the adjoint property:

$$x \leq_1 z \swarrow y \quad \text{iff} \quad x \& y \leq_3 z \quad \text{iff} \quad y \leq_2 z \nwarrow x$$

where $x \in P_1$, $y \in P_2$ and $z \in P_3$.

From the adjoint property we obtain the following results.

Proposition 1. If $(\&, \swarrow, \nwarrow)$ is an adjoint triple w.r.t. the posets $(P_1, \leq_1)$, $(P_2, \leq_2)$ and $(P_3, \leq_3)$, then

1. $\&$ is order-preserving on both arguments.
2. $\swarrow, \nwarrow$ are order-preserving on the first argument and order-reversing on the second argument.

Proposition 2. Given an adjoint triple $(\&, \swarrow, \nwarrow)$ with respect to the bounded partially ordered sets $(P_1, \leq_1, \bot_1, \top_1)$, $(P_2, \leq_2, \bot_2, \top_2)$ and $(P_3, \leq_3, \bot_3, \top_3)$, the following boundary conditions hold

1. $\bot_1 \& y = \bot_3$ and $x \& \bot_2 = \bot_3$, for all $x \in P_1$, $y \in P_2$.
2. $z \nwarrow \bot_1 = \top_2$ and $z \swarrow \bot_2 = \top_1$, for all $z \in P_3$.

The result below shows that given a conjunctor, if there exist its residuated implications, they are unique.

Proposition 3. Given the conjunctor $\&$ of an adjoint triple, its residuated implications $\swarrow$ and $\nwarrow$ are unique.

Proof. For the implication $\swarrow$, let us suppose there are $\swarrow^1$ and $\swarrow^2$ satisfying the adjoint property with respect to the same conjunctor $\&$. From inequalities $z \swarrow^1 y \leq_1 z \swarrow^1 y$ and $z \swarrow^2 y \leq_1 z \swarrow^2 y$ and applying the adjoint property we obtain the result.

Similarly for the implication $\nwarrow$. □

Example of adjoint triples are the Gödel, product and Łukasiewicz t-norms together with their residuated implications. Note that, these t-norms are commutative and so, the residuated implications satisfy that $\swarrow^G = \nwarrow_G$, $\swarrow^P = \nwarrow_P$ and $\swarrow^L = \nwarrow_L$. These adjoint triples are defined on $[0,1]$ as:

$$\&_G(x,y) = \min(x,y) \qquad z \nwarrow_G x = \begin{cases} 1 & \text{if } x \leq z \\ z & \text{otherwise} \end{cases}$$

$$\&_P(x,y) = x \cdot y \qquad z \nwarrow_P x = \min(1, z/x)$$

$$\&_L(x,y) = \max(0, x+y-1) \qquad z \nwarrow_L x = \min(1, 1-x+z)$$

Example 1. Given $m \in \mathbb{N}$, the set $[0,1]_m$ is a regular partition of $[0,1]$ in m pieces, for example $[0,1]_2 = \{0, 0.5, 1\}$ divide the unit interval in two pieces.

A discretization of the product t-norm is the operator $\&_P^* \colon [0,1]_{20} \times [0,1]_8 \to [0,1]_{100}$ defined, for each $x \in [0,1]_{20}$ and $y \in [0,1]_8$ as:

$$x \,\&_P^*\, y = \frac{\lceil 100 \cdot x \cdot y \rceil}{100}$$

whose residuated implications $\swarrow_P^* \colon [0,1]_{100} \times [0,1]_8 \to [0,1]_{20}$, $\nwarrow_P^* \colon [0,1]_{100} \times [0,1]_{20} \to [0,1]_8$ are defined as:

$$b \swarrow_P^* a = \frac{\lfloor 20 \cdot \min\{1, b/a\} \rfloor}{20} \qquad b \nwarrow_P^* c = \frac{\lfloor 8 \cdot \min\{1, b/c\} \rfloor}{8}$$

where $\lceil \_ \rceil$ and $\lfloor \_ \rfloor$ are the ceiling and the floor functions, respectively.

Hence, the triple $(\&_P^*, \swarrow_P^*, \nwarrow_P^*)$ is an adjoint triple and the operator $\&_P^*$ is neither commutative nor associative. Similar adjoint triples can be obtained from the Gödel and Łukasiewicz t-norms.

In order to make this contribution self-contained, the formal definition of closure operator and Galois connection are recalled.

Definition 2. *Let $(P, \leq)$ be poset. A mapping $c \colon P \times P \to P$ is called a* closure operator *on P if, for all $x, y \in P$:*

1. $x \leq c(x)$,
2. *If $x \leq y$ then $c(x) \leq c(y)$,*
3. $c(c(x)) = c(x)$.

Definition 3. *Let $(P_1, \leq_1)$ and $(P_2, \leq_2)$ be posets, and $^\downarrow : P_1 \to P_2$, $^\uparrow : P_2 \to P_1$ mappings, the pair $(^\uparrow, ^\downarrow)$ forms a* Galois connection *between P_1 and P_2 if, for all $x \in P_1$ and $y \in P_2$, the following equivalence holds:*

$$x \leq y^\uparrow \quad \text{if and only if} \quad y \leq x^\downarrow$$

Regarding the residuated implications $\swarrow, \nwarrow$ of a conjunctor, it is easy to check that the operators $_z \swarrow : P_2 \to P_1$ and $^z \nwarrow : P_1 \to P_2$ defined as $_z \swarrow (y) = z \swarrow y, ^z \nwarrow (x) = z \nwarrow x$, for all $x \in P_1, y \in P_2$, respectively, form a Galois connection [10].

3   Adjoint Negations

In the following, the notion of adjoint negations is introduced. These operators are defined from an adjoint triple and are a generalization of the residuated negations [2,13,21]. Moreover, the relation between adjoint negations and pairs of weak negations, introduced by Georgescu and Popescu in [15], is studied. The most important result of this section shows that every pair of weak negations may be obtained from an adjoint triple, which leads us to conclude that pairs of weak negations are a particular case of adjoint negations.

First of all, the notion of adjoint negations is presented.

Definition 4. *Given an adjoint triple $(\&, \swarrow, \nwarrow)$ with respect to a lower bounded poset $(P, \leq, \bot)$, the mappings $n_s, n_n : P \to P$ defined, for all $x, y \in P$, as*

$$n_s(y) = \bot \swarrow y \qquad n_n(x) = \bot \nwarrow x$$

are called adjoint negations *on P.*

The operators n_s and n_n satisfying that $x = n_n(n_s(x)) = n_s(n_n(x))$, for all $x \in P$, are called strong adjoint negations on P.

Since $(_z\swarrow, {}^z\nwarrow)$ is a Galois connection, the following properties are straightforwardly obtained.

Proposition 4. *Let $(P, \leq, \bot, \top)$ be a bounded poset and n_s, n_n adjoint negations on P. The following statements hold:*

1. *$n_s(\bot) = n_n(\bot) = \top$;*
2. *$n_s n_n n_s = n_s$ and $n_n n_s n_n = n_n$;*
3. *$n_s n_n$ and $n_n n_s$ are closure operators;*
4. *$x \leq n_s(y)$ iff $y \leq n_n(x)$, for all $x, y \in P$;*
5. *When the supremum and the infimum exist, for any $X, Y \subseteq P$,*
   (a) $n_s(\bigvee_{y \in Y} y) = \bigwedge_{y \in Y} n_s(y)$,
   (b) $n_n(\bigvee_{x \in X} x) = \bigwedge_{x \in X} n_n(x)$.

Now, we will prove that these negations generalize pairs of weak negations introduced in [15]. First of all, the definition of pair of weak negations will be recalled.

Definition 5 ([15]). *Let* $(P, \leq, \bot, \top)$ *be a bounded partially ordered set and two functions* $n_1 \colon P \to P$, $n_2 \colon P \to P$, *the pair* (n_1, n_2) *is said to be a pair of weak negations on* P, *if the following conditions hold, for all* $x \in P$:

1. $n_1(\top) = n_2(\top) = \bot$;
2. n_1 *and* n_2 *are antitone;*
3. $x \leq n_2 n_1(x)$ *and* $x \leq n_1 n_2(x)$.

Theorem 1. *Given a pair of weak negations* (n_1, n_2) *on* P, *there exists an adjoint triple* $(\&, \swarrow, \nwarrow)$ *with respect to* P *such that* $n_1 = n_s$ *and* $n_2 - n_n$.

As a consequence of the theorem above, the definition of adjoint negations is more general than pairs of weak negations.

The next example shows an idea in order to prove the previous theorem.

Example 2. Regarding a pair of weak negations (n_1, n_2) on P, the operators $\&_1, \swarrow$ and $\nwarrow$ defined as:

$$x \&_1 y = \begin{cases} \top & \text{if } x \nleq n_1(y) \\ \bot & \text{if } x \leq n_1(y) \end{cases}$$

$$z \swarrow y = \begin{cases} n_1(y) & \text{if } z \neq \top \\ \top & \text{if } z = \top \end{cases}$$

$$z \nwarrow x = \begin{cases} n_2(x) & \text{if } z \neq \top \\ \top & \text{if } z = \top \end{cases}$$

form an adjoint triple $(\&_1, \swarrow, \nwarrow)$ with respect to P satisfying that $n_1 = n_s$ and $n_2 = n_n$.

Note that the adjoint conjunctor may be defined as:

$$x \&_2 y = \begin{cases} \top & \text{if } y \nleq n_2(x) \\ \bot & \text{if } y \leq n_2(x) \end{cases}$$

and both definitions are equivalent applying Proposition 4(4), since (n_1, n_2) is a Galois connection. Therefore, $\&_1 = \&_2$. It is easily verified that last equality does not mean that $n_1 = n_2$.

However, the converse is not true since, for instance, the boundary conditions of pair of weak negations $n_s(\top) = n_n(\top) = \bot$ need not be satisfied.

Let $(L, \preceq)$ be a complete lattice, where $L = \{\bot, x_1, t, x_2, \top\}$ and $\bot \leq x_1 \leq t \leq x_2 \leq \top$, the conjunctor $\& \colon L \times L \to L$, defined from Table 1, and the residuated implications $\swarrow, \nwarrow \colon L \times L \to L$ of the operator $\&$, defined from Table 2.

Table 1. Definition of &

&	$\perp$	x_1	t	x_2	$\top$
$\perp$	$\perp$	$\perp$	$\perp$	$\perp$	$\perp$
x_1	$\perp$	$\perp$	$\perp$	$\perp$	$\perp$
t	$\perp$	$\perp$	t	t	t
x_2	$\perp$	x_1	t	t	x_2
$\top$	$\perp$	x_1	t	x_2	$\top$

Table 2. Definition of $\swarrow$ and $\nwarrow$

$\swarrow$	$\perp$	x_1	t	x_2	$\top$
$\perp$	$\top$	t	x_1	x_1	x_1
x_1	$\top$	$\top$	x_1	x_1	x_1
t	$\top$	$\top$	$\top$	x_2	t
x_2	$\top$	$\top$	$\top$	$\top$	x_2
$\top$	$\top$	$\top$	$\top$	$\top$	$\top$

$\nwarrow$	$\perp$	x_1	t	x_2	$\top$
$\perp$	$\top$	$\top$	x_1	$\perp$	$\perp$
x_1	$\top$	$\top$	x_1	x_1	x_1
t	$\top$	$\top$	$\top$	x_2	t
x_2	$\top$	$\top$	$\top$	$\top$	x_2
$\top$	$\top$	$\top$	$\top$	$\top$	$\top$

Taking into account Definition 1, the triple $(\&, \swarrow, \nwarrow)$ is an adjoint triple w.r.t. the complete lattice L. Therefore, the adjoint negations n_s and n_n can be defined from this adjoint triple.

Straightforwardly, by definition, the operators n_s and n_n are order-reversing and, by Proposition 4, we obtain that $n_s n_n$ and $n_n n_s$ are closure operators. Hence, n_s and n_n satisfy Conditions (2) and (3) in Definition 5.

However, it is easy to check that $n_s(\top) = \perp \swarrow \top = x_1 \neq \perp$. Therefore, the boundary condition assumed in the definition of pair of weak negations $n_s(\top) = \perp$ is not satisfied by this adjoint negation.

Therefore, adjoint negations generalize pairs of weak negations and so, these operators can be considered in more flexible frameworks. Moreover, the properties introduced in [15] can be applied to the adjoint negations.

Furthermore, we need to note that, although pairs of weak negations are defined in a general poset $(P, \leq)$, the properties presented in [15] are given for a bounded chain, which reduces the applicability of these negations. Indeed, the property $n_1(x \wedge y) = n_1(x) \vee n_1(y)$, for all $x, y \in P$, is considered in Proposition 3.1 of [15], nevertheless this is not true in a general lattice, as the following example shows.

Example 3. Let $(L, \preceq)$ be the lattice showed in Figure 1 and $(\&_G, \swarrow^G, \nwarrow_G)$ be an adjoint triple with respect to L, which is obtained from the Gödel conjunctor together with its residuated implications, which satisfy that $\swarrow^G = \nwarrow_G$, defined as:

$$x \&_G y = \inf\{x, y\} \qquad z \swarrow^G y = \begin{cases} \top & \text{if } y \leq z \\ z & \text{otherwise} \end{cases}$$

In the following, we will check that there exist two elements in L such that $n_s(x \wedge y) \neq n_s(x) \vee n_s(y)$:

$$n_s(x \wedge y) = n_s(\bot) = \top$$
$$n_s(x) \vee n_s(y) = (\bot \nearrow^G x) \vee (\bot \nearrow^G y) = \bot$$

Fig. 1. $(L, \preceq)$

4   Disjunction Dual to the Conjunctor

In this section, the definition of disjunction dual to an adjoint conjunctor of an adjoint triple is presented. This definition is given from two adjoint negations obtained by an adjoint triple with respect to a lower bounded lattice.

Several properties will be shown with the purpose of proving that, given a disjunction dual to an adjoint conjunctor, there exist two residuated implications associated with the disjunction which satisfy a dual equivalence to the adjoint property. First of all, the disjunction dual to the conjunctor is introduced.

Definition 6. *Given an adjoint triple* $(\&, \nearrow, \nwarrow)$ *with respect to a lower bounded poset* $(P, \leq, \bot)$, *a disjunction dual to the conjunctor* $\oplus \colon P \times P \to P$, *is defined, using the negation operators* $n_s \colon P \to P$ *and* $n_n \colon P \to P$, *as follows:*

$$x \oplus y = n_n(n_s(x) \& n_s(y))$$

Note that, the disjunction dual to the conjunctor may be defined using any combination of negations in each place of the previous expression.

The following proposition shows interesting results satisfied by $\oplus$, which are obtained straightforwardly from definitions and properties of the operators $\&$, n_s and n_n. For,

Proposition 5. *Let* $(\&, \nearrow, \nwarrow)$ *be an adjoint triple with respect to a lower bounded poset* $(P, \leq, \bot)$, *the disjunction dual to the conjunctor* $\oplus$ *satisfies the following properties:*

1. $\oplus$ *is order-preserving on both arguments, i.e. if* $x_1, x_2, x, y_1, y_2, y \in P$ *such that* $x_1 \leq x_2$, $y_1 \leq y_2$, *then* $(x_1 \oplus y) \leq (x_2 \oplus y)$ *and* $(x \oplus y_1) \leq (x \oplus y_2)$.
2. *If* $(P, \leq, \bot, \top)$ *is a bounded poset, then* $\top \oplus y = \top$ *and* $x \oplus \top = \top$, *for all* $x, y \in P$.

Since it is interesting that the disjunction satisfies the previous boundary conditions, from now on, we will assume a bounded poset $(P, \leq, \perp, \top)$.

The following technical result states the conditions that two adjoint negations n_s, n_n and the conjunctor of an adjoint triple $(\&, \swarrow, \nwarrow)$ must satisfy, in order to ensure that $\perp$ is the left identity element for $\oplus$.

Proposition 6. *Let* $(\&, \swarrow, \nwarrow)$ *be an adjoint triple with respect to P such that their corresponding n_s and n_n are strong adjoint negations on P. If $\top$ is the left identity element for $\&$, that is $\top \& y = y$, for all $y \in P$, then the equality $\perp \oplus y = y$ holds, for all $y \in P$.*

A similar result is obtained if the boundary condition $x \& \top = x$, for all $x \in P$, is considered.

Since the disjunction dual $\oplus$ is defined from a non commutative conjunctor $\&$, two different residuated implications can be defined.

Definition 7. *Given an adjoint triple* $(\&, \swarrow, \nwarrow)$ *with respect to P such that their induced negations n_s and n_n are strong adjoint negations on P, the following two operators are defined on P.*

$$z \swarrow^{\oplus} y = n_n(n_s(z) \swarrow n_s(y))$$
$$z \nwarrow_{\oplus} x = n_n(n_s(z) \nwarrow n_s(x))$$

for all $x, y \in P$.

The next proposition shows that the previous operators are indeed the residuated implications associated with the disjunction dual to the conjunctor of an adjoint triple.

Proposition 7. *Given an adjoint triple* $(\&, \swarrow, \nwarrow)$ *with respect to P such that their induced negations n_s and n_n are strong adjoint negations on P.*

Then, there exist two mappings $\swarrow^{\oplus}, \nwarrow_{\oplus} : P \times P \to P$ satisfying the equivalence

$$z \swarrow^{\oplus} y \leq x \quad \text{iff} \quad z \leq x \oplus y \quad \text{iff} \quad z \nwarrow_{\oplus} x \leq y \tag{1}$$

for all $x, y, z \in P$.

Similarly to Proposition 3 referring to adjoint triples, the following result shows that the implications $\swarrow^{\oplus}$ and $\nwarrow_{\oplus}$ of the triple $(\oplus, \swarrow^{\oplus}, \nwarrow_{\oplus})$, are unique, considering Equivalence (1).

Proposition 8. *Given the disjunction dual to an adjoint conjunctor $\&$, and a triple $(\oplus, \swarrow^{\oplus}, \nwarrow_{\oplus})$ satisfying Equivalence (1), the operators $\swarrow^{\oplus}$ and $\nwarrow_{\oplus}$ are unique.*

5   Conclusions and Future Work

The use of residuated negations is very useful in fuzzy logic. This paper has considered the general setting of adjoint triples in order to introduce a generalization of residuated negations. Indeed, we have proven that the introduced operators are more general than pairs of weak negations presented by Georgescu and Popescu in [15]. Although the former are defined in a poset, the properties are proven in a boundary chain and, therefore, the spectrum of possible logics in which they can be considered is reduced, as we have shown in the paper. Moreover, we have proven that every pair of weak negations can be obtained from an adjoint triple. Therefore, since every classical residuated negation is a particular case of a pair of weak negations, they can be derived from the implications of an adjoint triple.

Furthermore, the disjunction dual to an adjoint conjunctor is defined and several properties are presented. For instance, when strong adjoint negations are considered, two dual adjoint implications are obtained satisfying the dual adjoint property.

In the future, the definition of adjoint negations will be generalized considering an adjoint triple defined in three differents posets and the proposed negations will be compared with other ones, such as the negations introduced in the framework of the extended-order algebras [8]. In addition, it would be interesting to see if the representation theorems of negations given in [6] can be extended to our framework.

References

1. Cintula, P., Horčík, R., Noguera, C.: Non-associative substructural logics and their semilinear extensions: axiomatization and completeness properties. Review of Symbolic Logic 6(3), 394–423 (2013)
2. Cintula, P., Klement, E.P., Mesiar, R., Navara, M.: Residuated logics based on strict triangular norms with an involutive negation. Mathematical Logic Quarterly 52(3), 269–282 (2006)
3. Cintula, P., Klement, E.P., Mesiar, R., Navara, M.: Fuzzy logics with an additional involutive negation. Fuzzy Sets and Systems 161(3), 390–411 (2010), Fuzzy Logics and Related Structures
4. Cornejo, M., Medina, J., Ramírez, E.: A comparative study of adjoint triples. Fuzzy Sets and Systems 211, 1–14 (2013)
5. Cornejo, M.E., Medina, J., Ramírez, E.: Implication triples versus adjoint triples. In: Cabestany, J., Rojas, I., Joya, G. (eds.) IWANN 2011, Part II. LNCS, vol. 6692, pp. 453–460. Springer, Heidelberg (2011)
6. Cornelis, C., Arieli, O., Deschrijver, G., Kerre, E.E.: Uncertainty modeling by bilattice-based squares and triangles. IEEE Transactions on Fuzzy Systems 15(2), 161–175 (2007)
7. Cornelis, C., Medina, J., Verbiest, N.: Multi-adjoint fuzzy rough sets: Definition, properties and attribute selection. International Journal of Approximate Reasoning 55, 412–426 (2014)
8. Della Stella, M.E., Guido, C.: Associativity, commutativity and symmetry in residuated structures. Springer Science+Business Media 30(2), 363–401 (2013)
9. Díaz, J.C., Medina, J.: Multi-adjoint relation equations: Definition, properties and solutions using concept lattices. Information Sciences 253, 100–109 (2013)
10. Díaz-Moreno, J., Medina, J., Ojeda-Aciego, M.: On basic conditions to generate multi-adjoint concept lattices via galois connections. International Journal of General Systems 43(2), 149–161 (2014)

11. Esteva, F.: Negaciones en retículos completos. Stochastica I, 49–66 (1975)
12. Esteva, F., Godo, L.: Monoidal t-norm based logic: towards a logic for left-continuous t-norms. Fuzzy Sets and Systems 124, 271–288 (2001)
13. Esteva, F., Godo, L., Hájek, P., Navara, M.: Residuated fuzzy logics with an involutive negation. Archive for Mathematical Logic 39(2), 103–124 (2000)
14. Fodor, J.: Nilpotent minimum and related connectives for fuzzy logic. In: Proc. FUZZ-IEEE 1995, pp. 2077–2082 (1995)
15. Georgescu, G., Popescu, A.: Non-commutative fuzzy structures and pairs of weak negations. Fuzzy Sets and Systems 143, 129–155 (2004)
16. Julian, P., Moreno, G., Penabad, J.: On fuzzy unfolding: A multi-adjoint approach. Fuzzy Sets and Systems 154(1), 16–33 (2005)
17. Madrid, N., Ojeda-Aciego, M.: Measuring inconsistency in fuzzy answer set semantics. IEEE Transactions on Fuzzy Systems 19(4), 605–622 (2011)
18. Medina, J., Ojeda-Aciego, M., Ruiz-Calviño, J.: Formal concept analysis via multi-adjoint concept lattices. Fuzzy Sets and Systems 160(2), 130–144 (2009)
19. Medina, J., Ojeda-Aciego, M., Valverde, A., Vojtáš, P.: Towards biresiduated multi-adjoint logic programming. In: Conejo, R., Urretavizcaya, M., Pérez-de-la-Cruz, J.-L. (eds.) CAEPIA/TTIA 2003. LNCS (LNAI), vol. 3040, pp. 608–617. Springer, Heidelberg (2004)
20. Medina, J., Ojeda-Aciego, M., Vojtáš, P.: Multi-adjoint logic programming with continuous semantics. In: Eiter, T., Faber, W., Truszczyński, M. (eds.) LPNMR 2001. LNCS (LNAI), vol. 2173, pp. 351–364. Springer, Heidelberg (2001)
21. San-Min, W.: Logics for residuated pseudo-uninorms and their residua. Fuzzy Sets and Systems 218, 24–31 (2013), Theme: Logic and Algebra
22. Trillas, E.: Sobre negaciones en la teoría de conjuntos difusos. Stochastica III, 47–60 (1979)
23. Vetterlein, T., Ciabattoni, A.: On the (fuzzy) logical content of cadiag-2. Fuzzy Sets and Systems 161(14), 1941–1958 (2010), Theme: Fuzzy and Uncertainty Logics

Improving the β-Precision and OWA Based Fuzzy Rough Set Models: Definitions, Properties and Robustness Analysis

Lynn D'eer and Nele Verbiest

Department of Applied Mathematics, Computer Science and Statistics,
Ghent University, Krijgslaan 281 (S9), B-9000 Gent, Belgium
{Lynn.Deer,Nele.Verbiest}@UGent.be

Abstract. Since the early 1990s, many authors have studied fuzzy rough set models and their application in machine learning and data reduction. In this work, we adjust the β-precision and the ordered weighted average based fuzzy rough set models in such a way that the number of theoretical properties increases. Furthermore, we evaluate the robustness of the new models a-β-PREC and a-OWA to noisy data and compare them to a general implicator-conjunctor-based fuzzy rough set model.

Keywords: fuzzy sets, rough sets, hybridization, lower and upper approximation, implication, conjunction, beta-precision, OWA, robustness.

1 Introduction

Rough set theory (Pawlak [1], 1982) characterizes uncertainty due to incomplete information, by dividing a set of objects according to their indiscernibility towards each other, modeled by an equivalence relation. In particular, the lower and upper approximation of a set are constructed. The former includes the objects certainly belonging to the set, while the latter excludes the objects certainly not belonging to the set.

Furthermore, fuzzy set theory (Zadeh [2], 1965) extends classical or crisp sets in the sense that intermediary membership degrees, mostly between 0 and 1, can be obtained. This theory is used when dealing with gradual information or vague concepts.

Hybridization of both theories has its origin in the early 1990s, when Dubois and Prade [3] presented the first fuzzy rough set model. From then on, research on fuzzy rough set models grows, mainly due to its proven application in machine learning and, in particular, in feature selection.

Many fuzzy rough set models are intuitively constructed by substituting the Boolean implication and conjunction in Pawlak's model by fuzzy implicators and conjunctors, as well as the universal and existential quantifier by the infimum and supremum operator. In addition, approximations by general fuzzy relations are studied, instead of considering fuzzy equivalence relations. All these studies can be covered by a general implicator-conjunctor-based fuzzy rough set model (IC-model), discussed in [4].

C. Cornelis et al. (eds.): RSCTC 2014, LNAI 8536, pp. 23–34, 2014.
© Springer International Publishing Switzerland 2014

However, the main disadvantage of this model is its use of the infimum and supremum operator. Both operators are very sensitive to noise in the data and/or outlying samples. To overcome this problem, authors have studied robust fuzzy rough set models. We focus on the following two models: the β-precision fuzzy rough set model introduced by Fernández-Salido and Murakami in 2003 [5,6] and the ordered weighted average (OWA) based fuzzy rough set model introduced by Cornelis et al. in 2010 [7]. Both models use aggregation operators instead of the inf- and sup-operator and preliminary work showed that they have interesting theoretical and practical assets. Unfortunately, they do not satisfy the inclusion property, which is required if we want the approximations to be on both sides of the set to be approximated, and is important for feature selection [8].

In this work, we overcome this drawback by adjusting the two models. Inspiration for this adjustment was given by Inuiguchi [9]. We present the adapted β-precision and OWA based fuzzy rough set models, shortly called the a-β-PREC- and a-OWA-model. Moreover, we discuss their properties and analyze their robustness in comparison with the IC-model and the original robust models.

The remainder of this paper is as follows: in Section 2, we briefly recall the IC-model. In Section 3, we recall the β-precision and OWA based fuzzy rough set models and propose adaptations. Furthermore, we discuss which properties of the IC-model are maintained by the new models. In Section 4, we compare the five models with respect to their robustness to noisy data. Finally, we conclude and state future work in Section 5.

2   The IC-Model

Consider a *fuzzy approximation space*, i.e., a couple (U, R) consisting of a non-empty set U and a binary fuzzy relation R in U, and a general format for approximation operators using implicators[1] and conjunctors[2]:

Definition 1. *[4] Let (U, R) be a fuzzy approximation space, A a fuzzy set in U, $\mathcal{I}$ an implicator and $\mathcal{C}$ a conjunctor. The $(\mathcal{I}, \mathcal{C})$-fuzzy rough approximation of A by R is the pair of fuzzy sets $(R\downarrow_{\mathcal{I}} A, R\uparrow_{\mathcal{C}} A)$ defined by, for $x \in U$,*

$$(R\downarrow_{\mathcal{I}} A)(x) = \inf_{y \in U} \mathcal{I}(R(y, x), A(y)),$$

$$(R\uparrow_{\mathcal{C}} A)(x) = \sup_{y \in U} \mathcal{C}(R(y, x), A(y)).$$

A pair (A_1, A_2) of fuzzy sets in U is called a fuzzy rough set *in (U, R) if there exists a fuzzy set A in U such that $A_1 = R\downarrow_{\mathcal{I}} A$ and $A_2 = R\uparrow_{\mathcal{C}} A$.*

---

[1] An *implicator* $\mathcal{I}$ is a mapping $\mathcal{I} \colon [0, 1]^2 \to [0, 1]$ satisfying $\mathcal{I}(1, 0) = 0$, $\mathcal{I}(1, 1) = \mathcal{I}(0, 1) = \mathcal{I}(0, 0) = 1$ which is decreasing in the first and increasing in the second argument.

[2] A *conjunctor* $\mathcal{C}$ is a mapping $\mathcal{C} \colon [0, 1]^2 \to [0, 1]$ which is increasing in both arguments and which satisfies $\mathcal{C}(0, 0) = \mathcal{C}(0, 1) = \mathcal{C}(1, 0) = 0$ and $\mathcal{C}(1, 1) = 1$. It is called a *border conjunctor* if it satisfies $\mathcal{C}(1, x) = x$ for all x in $[0, 1]$. A commutative and associative border conjunctor $\mathcal{T}$ is called a *t-norm*.

In Table 1, the extensions of the classical rough set properties to a fuzzy approximation space are shown; (U, R), (U, R_1) and (U, R_2) are fuzzy approximation spaces, A, B and $\hat{\alpha}^3$ are fuzzy sets in U, $\mathcal{I}$ is an implicator, $\mathcal{C}$ a conjunctor, $\mathcal{N}$ an involutive negator[4] and R' the inverse relation of R, defined by, for $x, y \in U$, $R'(y, x) = R(x, y)$.

The following proposition summarizes under which conditions all properties in Table 1 hold.

Proposition 1. *[4] Let $\mathcal{C}$ be a left-continuous t-norm $\mathcal{T}$ and $\mathcal{I}$ its R-implicator, i.e., for x, y in $[0, 1]$, $\mathcal{I}_\mathcal{T}(x, y) = \sup\{\gamma \in [0, 1] \mid \mathcal{T}(x, \gamma) \leq y\}$. If the fuzzy relation R is reflexive, i.e., for x in U, $R(x, x) = 1$, and $\mathcal{T}$-transitive, i.e., for x, y, z in U, $\mathcal{T}(R(x, y), R(y, z)) \leq R(x, z)$, then all properties in Table 1 hold.*

Table 1. Properties in a fuzzy approximation space

(D) Duality	$R{\downarrow}_\mathcal{I} A = (R{\uparrow}_\mathcal{C}(A^\mathcal{N}))^\mathcal{N}$
(A) Adjointness	$A \subseteq R{\uparrow}_\mathcal{C} B \Leftrightarrow R'{\downarrow}_\mathcal{I} A \subseteq B$
(INC) Inclusion	$R{\downarrow}_\mathcal{I} A \subseteq A \subseteq R{\uparrow}_\mathcal{C} A$
(SM) Set monotonicity	$A \subseteq B \Rightarrow \begin{cases} R{\downarrow}_\mathcal{I} A \subseteq R{\downarrow}_\mathcal{I} B \\ R{\uparrow}_\mathcal{C} A \subseteq R{\uparrow}_\mathcal{C} B \end{cases}$
(RM) Relation monotonicity	$R_1 \subseteq R_2 \Rightarrow \begin{cases} R_2{\downarrow}_\mathcal{I} A \subseteq R_1{\downarrow}_\mathcal{I} A \\ R_1{\uparrow}_\mathcal{C} A \subseteq R_2{\uparrow}_\mathcal{C} A \end{cases}$
(IU) Intersection and Union	$R{\downarrow}_\mathcal{I}(A \cap B) = R{\downarrow}_\mathcal{I} A \cap R{\downarrow}_\mathcal{I} B$ $R{\uparrow}_\mathcal{C}(A \cup B) = R{\uparrow}_\mathcal{C} A \cup R{\uparrow}_\mathcal{C} B$
(ID) Idempotence	$R{\downarrow}_\mathcal{I}(R{\downarrow}_\mathcal{I} A) = R{\downarrow}_\mathcal{I} A$ $R{\uparrow}_\mathcal{C}(R{\uparrow}_\mathcal{C} A) = R{\uparrow}_\mathcal{C} A$
(CS) Constant sets	$R{\downarrow}_\mathcal{I} \hat{\alpha} = \hat{\alpha}$ $R{\uparrow}_\mathcal{C} \hat{\alpha} = \hat{\alpha}$

3   Robust Fuzzy Rough Set Models

A disadvantage of the IC-model is that the infimum and supremum operator are very sensitive to noise in the data: a small error in the data set can change the outcome of the model drastically. To overcome this problem, robust fuzzy rough set models are studied.

In literature, many robust fuzzy rough set models are defined. A first group of robust models is based on frequency: these models only take a subset of U into account when computing the lower and upper approximation [10,11,12]. Furthermore, there are robust models that use vague quantifiers to compute the

---

[3] $\forall \alpha \in [0, 1]$, we denote with $\hat{\alpha}$ the constant α-set in U, i.e., $\forall x \in U : \hat{\alpha}(x) = \alpha$.

[4] A *negator* $\mathcal{N}$ is a decreasing mapping $\mathcal{N} : [0, 1] \to [0, 1]$ which satisfies $\mathcal{N}(0) = 1$ and $\mathcal{N}(1) = 0$. It is *involutive* if for all $x \in [0, 1]$, $\mathcal{N}(\mathcal{N}(x)) = x$. The standard negator $\mathcal{N}_S$ is defined by, for x in $[0, 1]$, $\mathcal{N}_S(x) = 1 - x$.

approximation operators [13] or that modify the fuzzy set A which is approximated [14].

In this work, we focus on robust models that replace the inf- and sup-operator by aggregation operators. It is known that they are highly robust against noise and that they satisfy the properties (SM) and (RM), which are important in feature selection where the goal is to find a minimal subset of features [15]. More specifically, we concentrate on the β-precision (β-PREC, [5,6]) and OWA based (OWA, [7]) fuzzy rough set models. We adapt both models in such a way that the inclusion property is guaranteed, and hence the original idea of Pawlak. In addition, we compare the original and adapted models to the IC-model from both theoretical and practical view.

In the remainder of this article, we assume the universe U to be finite. This is not a limitation for practical purposes, since data sets in real applications are always finite.

3.1 The Original β-Precision and OWA Based Fuzzy Rough Set Models

First, we recall the fuzzy rough set model based on β-precision quasi-t-norms and quasi-t-conorms:

Definition 2. *[5,6] Given a t-norm $\mathcal{T}$, a t-conorm[5] $\mathcal{S}$, $\beta \in [0,1]$ and $n \in \mathbb{N} \setminus \{0,1\}$, the corresponding β-precision quasi-t-norm $\mathcal{T}_\beta$ and β-precision quasi-t-conorm $\mathcal{S}_\beta$ of order n are $[0,1]^n \to [0,1]$ mappings such that for all $x = (x_1, \ldots, x_n)$ in $[0,1]^n$,*

$$\mathcal{T}_\beta(x) = \mathcal{T}(y_1, \ldots, y_{n-m}),$$
$$\mathcal{S}_\beta(x) = \mathcal{S}(z_1, \ldots, z_{n-p}),$$

where y_i is the ith greatest element of x and z_i is the ith smallest element of x, and

$$m = \max\left\{ i \in \{0, \ldots, n\} \mid i \leq (1-\beta)\sum_{j=1}^{n} x_j \right\},$$

$$p = \max\left\{ i \in \{0, \ldots, n\} \mid i \leq (1-\beta)\sum_{j=1}^{n} (1-x_j) \right\}.$$

Note that for $\beta = 1$ the original t-norm $\mathcal{T}$ and t-conorm $\mathcal{S}$ are retrieved.

The β-precision fuzzy rough set model (shortly, β-PREC-model) is defined as follows:

---

[5] A *t-conorm* $\mathcal{S}$ is a mapping $\mathcal{S} \colon [0,1]^2 \to [0,1]$ that is increasing in both arguments, commutative, associative and satisfies for x in $[0,1]$, $\mathcal{S}(x,0) = x$.

Definition 3. *[6] Let $\mathcal{T}$ be a t-norm, $\mathcal{S}$ a t-conorm and $\beta \in [0,1]$. Given an implicator $\mathcal{I}$ and a conjunctor $\mathcal{C}$, the β-precision fuzzy rough approximation of A by R is the pair of fuzzy sets $(R{\downarrow}_{\mathcal{I},\mathcal{T}_\beta}A, R{\uparrow}_{\mathcal{C},\mathcal{S}_\beta}A)$, defined by, for $x \in U$:*

$$(R{\downarrow}_{\mathcal{I},\mathcal{T}_\beta}A)(x) = \mathcal{T}_\beta^{y \in U} \langle \mathcal{I}(R(y,x), A(y)) \rangle,$$

$$(R{\uparrow}_{\mathcal{C},\mathcal{S}_\beta}A)(x) = \mathcal{S}_\beta^{y \in U} \langle \mathcal{C}(R(y,x), A(y)) \rangle.$$

For $\mathcal{T} = \min$ and $\mathcal{S} = \max$ the following hold: if $\beta = 1$, the IC-model is obtained, and if $\beta < 1$, the approximation operators of the β-PREC-model satisfy $R{\downarrow}_{\mathcal{I}}A \subseteq R{\downarrow}_{\mathcal{I},\mathcal{T}_\beta}A$ and $R{\uparrow}_{\mathcal{T},\mathcal{S}_\beta}A \subseteq R{\uparrow}_{\mathcal{T}}A$.

The following properties hold for the β-PREC-model:

Proposition 2. *Let $\mathcal{T}$ be a t-norm and $\mathcal{S}$ its $\mathcal{N}_S$-dual t-conorm, i.e., for x, y in $[0,1]$, $\mathcal{S}(x,y) = 1 - \mathcal{T}(1-x, 1-y)$. Let $\beta \in [0,1]$. If the pair $(\mathcal{I}, \mathcal{C})$ consists of an implicator $\mathcal{I}$ and the conjunctor induced by $\mathcal{I}$ and $\mathcal{N}_S$, i.e., for x, y in $[0,1]$, $\mathcal{C}(x,y) = 1 - \mathcal{I}(x, 1-y)$, then (D) w.r.t. $\mathcal{N}_S$ holds for the β-PREC-model.*

Proposition 3. *The β-PREC-model satisfies (SM) and (RM).*

Secondly, we recall the fuzzy rough set model based on OWA operators:

Definition 4. *[16] Given a sequence D of n scalar values and a weight vector $W = \langle w_1, \ldots, w_n \rangle$ of length n, such that for all $i \in \{1, \ldots, n\}$, $w_i \in [0,1]$, and $\sum_{i=1}^{n} w_i = 1$ (an OWA weight vector of length n). Let σ be the permutation on $\{1, \ldots, n\}$ such that $d_{\sigma(i)}$ is the ith largest value of D. The OWA operator acting on D yields the value $\mathrm{OWA}_W(D) = \sum_{i=1}^{n} w_i d_{\sigma(i)}$.*

The OWA operator allows to consider a wide variety of aggregation strategies. For instance, the maximum and minimum are represented by the weight vectors $W_{\max} = \langle 1, 0, \ldots, 0 \rangle$ and $W_{\min} = \langle 0, \ldots, 0, 1 \rangle$, respectively. To measure how similar an OWA operator is to the maximum and minimum, the orness and andness degree are used. Let W be an OWA weight vector of length n, the *orness* and *andness degree* of W are defined by

$$\mathrm{orness}(W) = \frac{1}{n-1} \sum_{i=1}^{n} ((n-i) \cdot w_i),$$

$$\mathrm{andness}(W) = 1 - \mathrm{orness}(W).$$

The OWA based fuzzy rough set model (shortly OWA-model) is defined as follows:

Definition 5. *[7] Given an implicator $\mathcal{I}$, a conjunctor $\mathcal{C}$ and OWA weight vectors W_1 and W_2 of length n, with $n = |U|$, and such that $\mathrm{andness}(W_1) > 0.5$*

and orness$(W_2) > 0.5$. *The* (W_1, W_2)-*fuzzy rough approximation of* A *by* R *is the pair of fuzzy sets* $(R\downarrow_{\mathcal{I},W_1} A, R\uparrow_{\mathcal{C},W_2} A)$ *defined by, for* $x \in U$:

$$(R\downarrow_{\mathcal{I},W_1} A)(x) = \underset{y \in U}{\text{OWA}_{W_1}} \langle \mathcal{I}(R(y,x), A(y)) \rangle,$$

$$(R\uparrow_{\mathcal{C},W_2} A)(x) = \underset{y \in U}{\text{OWA}_{W_2}} \langle \mathcal{C}(R(y,x), A(y)) \rangle.$$

By varying the OWA weight vectors, different fuzzy rough set models can be obtained. For the weight vectors $W_1 = W_{\min}$ and $W_2 = W_{\max}$, we obtain the IC-model. If other OWA weight vectors are used, more weight will be given to higher, resp. lower values, so it always holds that $R\downarrow_{\mathcal{I}} A \subseteq R\downarrow_{\mathcal{I},W_1} A$ and $R\uparrow_{\mathcal{C},W_2} A \subseteq R\uparrow_{\mathcal{C}} A$.

The following properties hold for the OWA-model:

Proposition 4. *Let* W_1 *be a weight vector such that* andness$(W_1) > 0.5$ *and let* W_2 *be of the same length* n *such that* $(W_2)_i = (W_1)_{n-i+1}$ *for* $i \in \{1, \dots, n\}$. *Let* $(\mathcal{I}, \mathcal{C})$ *be a pair consisting of an implicator* $\mathcal{I}$ *and the conjunctor induced by* $\mathcal{I}$ *and* $\mathcal{N}_S$, *then* (D) *w.r.t.* $\mathcal{N}_S$ *holds for the OWA-model.*

Proposition 5. *The OWA-model satisfies* (SM) *and* (RM)*.*

3.2   The Adapted β-Precision and OWA Based Fuzzy Rough Set Models

A drawback of both models is that they do not satisfy the inclusion property. It means that the lower approximation is not necessarily contained in the approximated set. This is something we want to avoid in feature selection, where the goal is to find a smaller set of features. However, by adjusting the models, we can force the inclusion property to hold.

We begin with adapting the β-precision fuzzy rough set model:

Definition 6. *Let* $\mathcal{T}$ *be a t-norm,* $\mathcal{S}$ *a t-conorm and* $\beta \in [0,1]$. *Given an implicator* $\mathcal{I}$ *and a conjunctor* $\mathcal{C}$, *the adapted* β-*precision fuzzy rough approximation of* A *by* R *is the pair of fuzzy sets* $(R\downarrow^a_{\mathcal{I},\mathcal{T}_\beta} A, R\uparrow^a_{\mathcal{C},\mathcal{S}_\beta} A)$, *defined by, for* $x \in U$:

$$(R\downarrow^a_{\mathcal{I},\mathcal{T}_\beta} A)(x) = \min\{A(x), \underset{y \in U}{\mathcal{T}_\beta} \langle \mathcal{I}(R(y,x), A(y)) \rangle\},$$

$$(R\uparrow^a_{\mathcal{C},\mathcal{S}_\beta} A)(x) = \max\{A(x), \underset{y \in U}{\mathcal{S}_\beta} \langle \mathcal{C}(R(y,x), A(y)) \rangle\}.$$

We refer to this model as the a-β-PREC-model.

In the a-β-PREC-model, (D), (SM) and (RM) still hold and moreover, the properties (INC) and (CS) for $\alpha = 0$ and $\alpha = 1$ hold.

Proposition 6. *Let* $\mathcal{T}$ *be a t-norm and* $\mathcal{S}$ *its* $\mathcal{N}_S$-*dual t-conorm. Let* $\beta \in [0,1]$. *If the pair* $(\mathcal{I}, \mathcal{C})$ *consists of an implicator* $\mathcal{I}$ *and the conjunctor induced by* $\mathcal{I}$ *and* $\mathcal{N}_S$, *then* (D) *w.r.t.* $\mathcal{N}_S$ *holds for the a-β-PREC-model.*

Proposition 7. *The a-β-PREC-model satisfies (INC), (SM), (RM) and (CS) for $\alpha = 0$ and $\alpha = 1$.*

The a-β-PREC-model does not satisfy (A), (IU), (ID) and (CS) for $\alpha \in]0, 1[$.
In a similar way, we adjust the OWA based model:

Definition 7. *Given an implicator $\mathcal{I}$, a conjunctor $\mathcal{C}$ and OWA weight vectors W_1 and W_2 of length n, with $n = |U|$, and such that andness$(W_1) > 0.5$ and orness$(W_2) > 0.5$. The adapted (W_1, W_2)-fuzzy rough approximation of A by R is the pair of fuzzy sets $(R{\downarrow}^a_{\mathcal{I},W_1} A, R{\uparrow}^a_{\mathcal{C},W_2} A)$ defined by, for $x \in U$:*

$$(R{\downarrow}^a_{\mathcal{I},W_1} A)(x) = \min\{A(x), \underset{y \in U}{\text{OWA}_{W_1}} \langle \mathcal{I}(R(y, x), A(y)) \rangle \},$$

$$(R{\uparrow}^a_{\mathcal{C},W_2} A)(x) = \max\{A(x), \underset{y \in U}{\text{OWA}_{W_2}} \langle \mathcal{C}(R(y, x), A(y)) \rangle \}.$$

We refer to this model as the a-OWA-model.
    The a-OWA-model still satisfies (D), (SM) and (RM) and additionally, it satisfies (INC) and (CS) for all α in $[0, 1]$.

Proposition 8. *Let W_1 be a weight vector such that andness$(W_1) > 0.5$ and let W_2 be of the same length n such that $(W_2)_i = (W_1)_{n-i+1}$ for $i \in \{1, \dots, n\}$. Let $(\mathcal{I}, \mathcal{C})$ be a pair consisting of an implicator $\mathcal{I}$ and the conjunctor induced by $\mathcal{I}$ and $\mathcal{N}_S$, then (D) w.r.t. $\mathcal{N}_S$ holds for the a-OWA-model.*

Proposition 9. *The a-OWA-model satisfies (INC), (SM), (RM) and (CS) for $\alpha = 0$ and $\alpha = 1$. If the implicator $\mathcal{I}$ and the conjunctor $\mathcal{C}$ satisfy $\mathcal{I}(1, x) = x$ and $\mathcal{C}(1, x) = x$ for all x in $[0, 1]$, then the a-OWA-model satisfies (CS) for all α in $[0, 1]$.*

The a-OWA-model does not satisfy (A), (IU) and (ID).
    To end, we compare the two adapted models to the IC-model in Table 2. If a property holds, we denote this with ✓; if a property does not hold, we indicate this by ✗ and if a property holds under certain conditions, we write ☆. Note that the property (UE) stands for the (CS)-property with $\alpha \in \{0, 1\}$. In both cases we see that the adapted models satisfy more properties than the original ones, but the IC-model remains the best model from theoretical point of view.

4   Analysis of Robustness

Besides a theoretical comparison of the five models, we evaluate their robustness to noisy data. For this, we use five real-valued data sets from the KEEL data set repository[6]: 'Diabetes' ($|U| = 43$, $|\mathcal{A}| = 2$), 'Ele-1' ($|U| = 495$, $|\mathcal{A}| = 2$), 'AutoMPG6' ($|U| = 392$, $|\mathcal{A}| = 5$), 'MachineCPU' ($|U| = 209$, $|\mathcal{A}| = 6$) and 'Baseball' ($|U| = 337$, $|\mathcal{A}| = 16$). Each data set can be considered as a decision system $(U, \mathcal{A} \cup \{d\})$, where U is the finite set of instances, $\mathcal{A}$ is the set

---

[6] www.keel.es

Table 2. Overview of properties for the different fuzzy rough set models

Property	IC	β-PREC	a-β-PREC	OWA	a-OWA
(D)	☆	☆	☆	☆	☆
(A)	☆	✗	✗	✗	✗
(INC)	☆	✗	✓	✗	✓
(SM)	✓	✓	✓	✓	✓
(RM)	✓	✓	✓	✓	✓
(IU)	✓	✗	✗	✗	✗
(ID)	☆	✗	✗	✗	✗
(CS)	☆	✗	✗	✗	☆
(UE)	☆	✗	✓	✗	✓

of features (conditional attributes) and $d \notin \mathcal{A}$ the decision attribute. We only consider regression problems, so the decision attributes in the five data sets are continuous.

In many data mining tasks based on fuzzy rough set theory [17,18], the *positive region* is used, defined by, for x in U, $\text{POS}(x) = \sup_{y \in U}(R \downarrow R_d y)(x)$. In this paper, $\downarrow$ is as in one of the definitions in Section 3, R is the indiscernibilitiy relation defined by

$$\forall x, y \in U : R(x, y) = \frac{1}{|\mathcal{A}|} \cdot \left(\sum_{a \in \mathcal{A}} 1 - \frac{|a(x) - a(y)|}{\text{range}(a)} \right)$$

and $R_d y$ is the indiscernibility class of y by R_d defined by

$$\forall x \in U : R_d y(x) = 1 - \frac{|d(x) - d(y)|}{\text{range}(d)}.$$

Here, the range of an attribute $a \in \mathcal{A} \cup \{d\}$ is given by the difference between the maximum and the minimum value of a.

If the positive region based on a certain fuzzy rough model does not change drastically when small errors in the data occur, we call the model robust. These errors can occur both in the features values (attribute noise) and in the decision attribute values (class noise). To evaluate the robustness of the fuzzy rough set models discussed in this paper, we compare the values of the fuzzy rough positive region calculated based on the original dataset and based on the same dataset with artificial noise added.

Given a decision system $(U, \mathcal{A} \cup \{d\})$ and a certain noise level n, we define the altered decision system $(U, \mathcal{A}^n \cup \{d\})$ with artificial attribute noise as the decision system where each attribute value $a(x)$ has an $n\%$ chance of being altered to an attribute value in the range of a. For instance, if a takes values in the interval $[0, 10]$ and if the noise level is 10%, for each instance $x \in U$ there is a 10 percent chance that the value of $a(x)$ in the altered decision system is not equal to $a(x)$ but is a random value in the interval $[0, 10]$. We add artificial class noise in a similar manner. The altered decision system $(U, \mathcal{A} \cup \{d^n\})$ is the decision system where each decision value $d(x)$ has an $n\%$ chance of being altered to a random value in the range of d.

We denote by $\text{POS}_n^a(x)$ the fuzzy rough positive region of $x \in U$ based on the decision system $(U, \mathcal{A}^n \cup \{d\})$, where n percent artificial attribute noise is added. The value $\text{POS}_n^d(x)$ refers to the fuzzy rough positive region of $x \in U$ based on the decision system $(U, \mathcal{A} \cup \{d^n\})$, where n percent artificial class noise is added.

As we are interested in how much the value of the fuzzy rough positive region based on the altered decision systems with artificial noise deviates from the original values, we define the following error measures:

$$\text{error}_n^a = \frac{\sum\limits_{x \in U} |\,\text{POS}(x) - \text{POS}_n^a(x)|}{|U|},$$

$$\text{error}_n^d = \frac{\sum\limits_{x \in U} |\,\text{POS}(x) - \text{POS}_n^d(x)|}{|U|}.$$

These measures reflect the average deviation of the fuzzy rough positive region when the decision system has $n\%$ attribute or class noise.

In Algorithm 1 we outline the experiment that we carry out for each dataset, represented by a decision system, and each fuzzy rough set model. We consider 30 noise levels, and calculate the average errors over 10 runs associated with this noise level and dataset in lines 3 to 10. Note that the processes in line 6 and 7 are stochastic, and therefore this procedure is repeated 10 times. As a result, for each dataset and fuzzy rough model, we have 60 error values, namely the average error related to attribute noise and the average error related to class noise for each noise level n in $1, \ldots, 30$.

Algorithm 1. Procedure carried out in our experimental evaluation to assess the robustness of a given fuzzy rough set model on a dataset

---

1: **Input:** Dataset represented by a decision system $(U, \mathcal{A} \cup \{d\})$,
   fuzzy rough set model
2: **for** $n = 1, \ldots, 30$ **do**
3:   av-error$_n^a \leftarrow 0$
4:   av-error$_n^d \leftarrow 0$
5:   **for** $i = 1, \ldots, 10$ **do**
6:     av-error$_n^a \leftarrow$ av-error$_n^a +$ error$_n^a$
7:     av-error$_n^d \leftarrow$ av-error$_n^d +$ error$_n^d$
8:   **end for**
9:   av-error$_n^a \leftarrow$ av-error$_n^a / 10$
10:   av-error$_n^d \leftarrow$ av-error$_n^d / 10$
11:   **Output:** av-error$_n^a$ and av-error$_n^d$
12: **end for**

The parameters used in the different fuzzy rough set models are as follows: in all the models we use the Łukasiewicz implicator $\mathcal{I}(a, b) = \min(1, 1 - a + b)$ for

$a, b \in [0, 1]$. In the β-PREC- and a-β-PREC-model, $\mathcal{T} = \min$ and $\beta = 0.96$ are used. In the OWA- and a-OWA-model we use the weight vector W defined by

$$w_i = \frac{1}{\sum\limits_{i=1}^{n} w_i} \cdot \frac{1}{n - i + 1}.$$

The results are shown in Figures 1 and 2. For both attribute and class noise, we see that the IC-model is the most sensitive model. We observe that the adapted models are more or less equally robust as their respective original models. The practical benefits of the β-PREC- and OWA-model are not lost due to the adaptations: both the a-β-PREC- and a-OWA-model perform well in the robustness analysis. However, it is not possible to decide which robust model performs best. We note that the robustness of the β-PREC- and OWA-models come with an extra computational cost. Assuming that the values of the similarity relation are known, the complexity of the IC-model is $\mathcal{O}(|U|)$, whereas the complexity of the β-PREC- and OWA-models is $\mathcal{O}(|U| \log(|U|))$ due to the sorting operations required by these models. The adapted β-PREC- and OWA-models have the same complexity as their respective original models, which means that the extra theoretical properties do not come with a higher complexity.

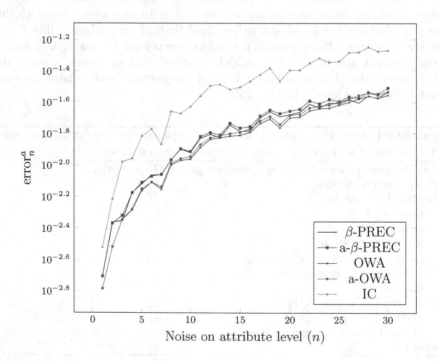

Fig. 1. Average error over the five data sets for attribute noise

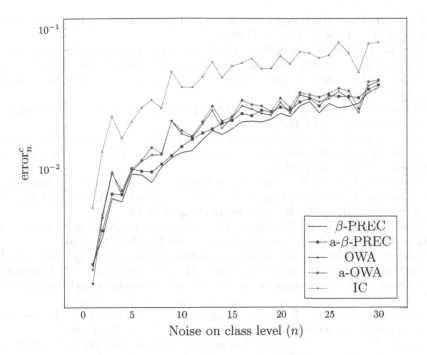

Fig. 2. Average error over the five data sets for class noise

5 Conclusion and Future Work

In this work, we adapted two state-of-the-art robust fuzzy rough set models such that the inclusion property is guaranteed, in order to obtain the required theoretical properties for using the models in feature selection. Furthermore, we compared the robust models to a general implicator-conjunctor-based fuzzy rough set model from a practical point of view. The benefits of the original models are not lost due to the proposed adaptation.

Future work consists of studying a formal framework for data reduction techniques based on fuzzy rough set models, and in particular, for the implicator-conjunctor-based model and the two adapted robust fuzzy rough set models.

Acknowledgment. Lynn D'eer has been supported by the Ghent University Special Research Fund.

References

1. Pawlak, Z.: Rough sets. International Journal of Computer and Information Sciences 11(5), 341–356 (1982)
2. Zadeh, L.: Fuzzy sets. Information and Control 8, 338–353 (1965)

3. Dubois, D., Prade, H.: Rough fuzzy sets and fuzzy rough sets. International Journal of General Systems 17, 191–209 (1990)
4. D'eer, L., Verbiest, N., Cornelis, C., Godo, L.: Implicator-conjunctor based models of fuzzy rough sets: definitions and properties. In: Ciucci, D., Inuiguchi, M., Yao, Y., Ślęzak, D., Wang, G. (eds.) RSFDGrC 2013. LNCS, vol. 8170, pp. 169–179. Springer, Heidelberg (2013)
5. Fernández Salido, J., Murakami, S.: On β-precision aggregation. Fuzzy Sets and Systems 139, 547–558 (2003)
6. Fernández Salido, J., Murakami, S.: Rough set analysis of a general type of fuzzy data using transitive aggregations of fuzzy similarity relations. Fuzzy Sets and Systems 139, 635–660 (2003)
7. Cornelis, C., Verbiest, N., Jensen, R.: Ordered weighted average based fuzzy rough sets. In: Yu, J., Greco, S., Lingras, P., Wang, G., Skowron, A. (eds.) RSKT 2010. LNCS, vol. 6401, pp. 78–85. Springer, Heidelberg (2010)
8. Cornelis, C., Jensen, R., Hurtado Martín, G., Slezak, D.: Attribute selection with fuzzy decision reducts. Information Sciences 180(2), 209–224 (2010)
9. Inuiguchi, M.: Classification- versus approximation-oriented fuzzy rough sets. In: Bouchon-Meunier, B., Coletti, G., Yager, R. (eds.) Proceedings of Information Processing and Management of Uncertainty in Knowledge-Based Systems. CD-ROM (2004)
10. Mieszkowicz-Rolka, A., Rolka, L.: Fuzzy rough approximations of process data. International Journal of Approximate Reasoning 49, 301–315 (2008)
11. Hu, Q., An, S., Yu, X., Yu, D.: Robust fuzzy rough classifiers. Fuzzy Sets and Systems 183, 26–43 (2011)
12. Yao, Y., Mi, J., Li, Z.: A novel variable precision (θ, σ)-fuzzy rough set model based on fuzzy granules. Fuzzy Sets and Systems 236, 58–72 (2014)
13. Cornelis, C., De Cock, M., Radzikowska, A.: Vaguely quantified rough sets. In: An, A., Stefanowski, J., Ramanna, S., Butz, C.J., Pedrycz, W., Wang, G. (eds.) RSFDGrC 2007. LNCS (LNAI), vol. 4482, pp. 87–94. Springer, Heidelberg (2007)
14. Zhao, S., Tsang, E., Chen, D.: The model of fuzzy variable precision rough sets. IEEE 17(2), 451–467 (2009)
15. D'eer, L., Verbiest, N., Cornelis, C., Godo, L.: Modelos de conjuntos rugosos difusos tolerantes al ruido: definiciones y propiedades. In: Proceedings of XVII Congreso Español sobre Tecnologías y Lógica Fuzzy (ESTYLF 2014), pp. 27–32 (2014)
16. Yager, R.: On ordered weighted averaging aggregation operators in multicriteria decision making. IEEE Transactions on Systems, Man and Cybernetics 18, 183–190 (1988)
17. Verbiest, N., Cornelis, C., Herrera, F.: FRPS: a fuzzy rough prototype selection method. Pattern Recognition 46(10), 2770–2782 (2013)
18. Verbiest, N., Cornelis, C., Jensen, R.: Quality, frequency and similarity based fuzzy nearest neighbor classification. In: Proceedings of the IEEE International Conference on Fuzzy Systems (2013)

Nearest Neighbour-Based Fuzzy-Rough Feature Selection

Richard Jensen and Neil Mac Parthaláin

Dept. of Computer Science
Aberystwyth University
Aberystwyth, Ceredigion, Wales, UK
{rkj,ncm}@aber.ac.uk

Abstract. Research in the area of fuzzy-rough set theory and its application to various areas of learning have generated great interest in recent years. In particular, there has been much work in the area of feature or attribute selection. Indeed, as the number of dimensions increases, the number of data objects required in order to generate accurate models increases exponentially. Thus, feature selection (FS) has become an increasingly necessary step in model learning. The use of fuzzy-rough sets as dataset pre-processors offers much in the way of flexibility, however the underlying complexity of the subset evaluation metric often presents a problem and can result in a great deal of potentially unnecessary computational effort. This paper proposes two different novel ways to address this problem using a neighbourhood approximation step in order to alleviate the processing overhead and reduce the complexity of the evaluation metric. The experimental evaluation demonstrates that much computational effort can be avoided, and as a result the efficiency of the FS process can be improved considerably.

Keywords: fuzzy-rough sets, feature selection, nearest neighbours.

1 Introduction

In rough sets, the concepts of the lower and upper approximations are central to the theory [11] and many of its applications. They are constructed using the indiscernibility of data objects and determine those objects that certainly and possibly belong to any given concept. A data object is said to belong to the lower approximation of a concept if all of the data objects indiscernible from it also belong to the lower approximation. Furthermore, it also is said to belong to its upper approximation if at least one data object that is indiscernible from it belongs to the concept.

Fuzzy-rough set theory extends the rough set approximation operators by fuzzifying the indiscernibility relation as well as the concept itself. This generalisation provides much greater flexibility, however, the most commonly utilised definitions of fuzzy-rough approximations ignore some important aspects. In particular, the process which determines the membership to each of the approximations still depends upon the contribution of a single data object, as governed

C. Cornelis et al. (eds.): RSCTC 2014, LNAI 8536, pp. 35–46, 2014.

by the sup and inf operators. In traditional fuzzy-rough sets, all data objects in the dataset must be considered in generating the approximations. This means that considerable computational effort is expended each time the lower approximation memberships are calculated. For feature selection, this occurs with the consideration of each candidate subset meaning that a large number of membership calculations are made needlessly. In addition, when considering all objects in the dataset even small changes in the data distribution can often mean that the generated approximations can vary greatly. This can also have a negative impact on the stability of any technique based upon such definitions.

In this paper, an alternative nearest-neighbour fuzzy-rough set approach is proposed, in which membership degrees to the approximations are computed by including only those data objects which are k-nearest neighbours and are also *not* of the same decision class as the data object under consideration. As such, the technique offers an important benefit: it reduces computational complexity by allowing only those close neighbours which affect the outcome of the fuzzy-rough approximations to be considered. In addition, a further extension to the nearest neighbours approach which employs fuzzy discernibility matrices allows it to avoid the situation of existing definitions where only 1-NN is used in determining lower approximation membership.

The remainder of the paper is structured as follows: the preliminaries for fuzzy-rough set theory and feature selection are covered in Section 2, while in Section 3 nearest neighbour-based fuzzy-rough sets are presented and two different approaches to implementing the ideas are introduced. In Section 4, an experimental evaluation is carried out. Finally, in Section 5, the paper is concluded and topics for future work are discussed.

2   Theoretical Background

In the original work of [11], the lower approximation of a set X using a subset of the conditional attributes $P \subseteq \mathbb{C}$ w.r.t. a crisp equivalence relation is defined as $\underline{P}X = \{x : [x]_P \subseteq X\}$. The positive region can then be constructed, which contains those data objects in the universe $\mathbb{U}$ for which the values of P allow to predict the decision classes in $\mathbb{D}$ unequivocally: $POS_P(\mathbb{D}) = \bigcup_{X \in \mathbb{U}/\mathbb{D}} \underline{P}X$. Based on the positive region, the rough set degree of dependency of the decision attribute(s) $\mathbb{D}$ on a set of attributes P can be calculated: $\gamma_P(\mathbb{D}) = \frac{|POS_P(\mathbb{D})|}{|\mathbb{U}|}$. This measure can then be used to gauge subset quality for rough set-based FS.

Fuzzy-Rough Sets. A fuzzy-rough set [4] is defined by two fuzzy sets, fuzzy lower and upper approximations, obtained by extending the corresponding crisp rough set notions. In the crisp case, elements that belong to the lower approximation (i.e. have a membership of 1.0) are said to belong to the approximated set with absolute certainty. In the fuzzy-rough case, elements may have a membership in the range [0,1], thus allowing greater flexibility in modelling uncertainty. Definitions for the fuzzy lower and upper approximations can be found in [13]. Here, only the fuzzy lower approximation is used, where a fuzzy indiscernibility relation is used to approximate a fuzzy concept X:

$$\mu_{\underline{R_P}X}(x) = \inf_{y\in U} \mathcal{I}(\mu_{R_P}(x,y), \mu_X(y)) \qquad (1)$$

where $\mathcal{I}$ is a fuzzy implicator. A fuzzy implicator is any $[0,1]^2 \to [0,1]$ mapping which satisfies $\mathcal{I}(0,0) = 1, \mathcal{I}(1,x) = x$ (for border implicators) for all x in $[0,1]$. R_P is the fuzzy similarity relation induced by the subset of features P:

$$\mu_{R_P}(x,y) = \mathcal{T}_{a\in P}\{\mu_{R_a}(x,y)\} \qquad (2)$$

where $\mu_{R_a}(x,y)$ is the degree to which objects x and y are similar for feature a, and may be defined in many ways [13], and $\mathcal{T}$ is a t-norm, a function $\mathcal{T}: [0,1] \times [0,1] \to [0,1]$ which satisfies the commutativity, associativity and monotonicity properties [10]. In a similar way to the original crisp rough set approach, the fuzzy positive region [8] can be defined as:

$$\mu_{POS_P(\mathbb{D})}(x) = \sup_{X\in U/\mathbb{D}} \mu_{\underline{R_P}X}(x) \qquad (3)$$

An important issue in data analysis is discovering dependencies between features. The fuzzy-rough degree of dependency of $\mathbb{D}$ on the attribute subset P can be defined in the following way:

$$\gamma'_P(\mathbb{D}) = \frac{\sum\limits_{x\subset U} \mu_{POS_P(\mathbb{D})}(x)}{|U|} \qquad (4)$$

A fuzzy-rough reduct Red is a minimal subset of features (i.e. there is no redundancy) that preserves the dependency degree of the entire dataset, i.e. $\gamma'_{Red}(\mathbb{D}) = \gamma'_C(\mathbb{D})$. Based on this, a fuzzy-rough greedy hill-climbing algorithm can be constructed that uses equation (4) to gauge subset quality.

Fuzzy Discernibility. Crisp discernibility matrices, often used in rough set feature selection, may also be extended to fuzzy-rough feature selection (FRFS) [8]. Entries in a fuzzy discernibility matrix (FDM) are a fuzzy set, to which every feature belongs to a certain degree. The extent to which a feature a belongs to the fuzzy clause C_{ij} is determined by the following:

$$\mu_{C_{ij}}(a) = N(\mu_{R_a}(i,j)) \qquad (5)$$

where N denotes fuzzy negation and $\mu_{R_a}(i,j)$ is the fuzzy similarity of objects i and j, and hence $\mu_{C_{ij}}(a)$ is a measure of the fuzzy discernibility. For the crisp case, if $\mu_{C_{ij}}(a) = 1$ then the two objects are distinct for this feature; if $\mu_{C_{ij}}(a) = 0$, the two objects are identical. For fuzzy cases where $\mu_{C_{ij}}(a) \in (0,1)$, the objects are partly discernible. Each entry (or clause) in the fuzzy indiscernibility matrix is a set of attributes and their memberships:

$$C_{ij} = \{a_x | a \in \mathbb{C}, x = N(\mu_{R_a}(i,j))\}\, i,j = 1, ..., |U| \qquad (6)$$

For example, an entry C_{ij} in the fuzzy discernibility matrix may be: $\{a_{0.4}, b_{0.8}, c_{0.2}, d_{0.0}\}$. This denotes that $\mu_{C_{ij}}(a) = 0.4$, $\mu_{C_{ij}}(b) = 0.8$, etc. In crisp discernibility matrices, these values are either 0 or 1 as the underlying relation is an equivalence relation. The example clause can be viewed as indicating the significance value of each feature - the extent to which the feature discriminates between the two instances i and j.

Fuzzy Discernibility Function. As with the crisp approach, the entries in the matrix can be used to construct the fuzzy discernibility function:

$$f_D(a_1^*, ..., a_m^*) = \wedge\{\vee C_{ij}^* | 1 \leq j < i \leq |\mathbb{U}|\} \tag{7}$$

where $C_{ij}^* = \{a_x^* | a_x \in C_{ij}\}$. The function returns values in $[0, 1]$, which can be seen to be a measure of the extent to which the function is satisfied for a given assignment of truth values to variables. To discover reducts from the fuzzy discernibility function, the task is to find the minimal assignment of the value true to the variables such that the formula is maximally satisfied. By setting all variables to true, the maximal value for the function can be obtained as this provides the greatest discernibility between objects.

Decision-Relative Fuzzy Discernibility Matrix. For a decision system, the decision feature must be taken into account for achieving reductions; only those clauses with different decision values are included. For the fuzzy version, this is encoded as:

$$f_D(a_1^*, ..., a_m^*) = \{\wedge\{\{\vee C_{ij}^*\} \leftarrow q_{N(\mu_{R_q}(i,j))}\} | 1 \leq j < i \leq |\mathbb{U}|\} \tag{8}$$

where $C_{ij}^* = \{a_x^* | a_x \in C_{ij}\}$, for decision feature q, where $\leftarrow$ denotes fuzzy implication. If $\mu_{C_{ij}}(q) = 1$ then this clause provides maximum discernibility (i.e., the two objects are maximally different according to the fuzzy similarity measure). When the decision is crisp and crisp equivalence is used, $\mu_{C_{ij}}(q)$ becomes either 0 or 1. The degree of satisfaction for a clause C_{ij} for a given subset of features P is defined as:

$$SAT_P(C_{ij}) = \mathcal{S}_{a \in P}\{\mu_{C_{ij}}(a)\} \tag{9}$$

for a t-conorm $\mathcal{S}$. In traditional (crisp) propositional satisfiability, a clause is fully satisfied if at least one variable in the clause has been set to true. For the fuzzy case, clauses may be satisfied to a certain degree depending on which variables have been assigned the value true. By setting $P = \mathbb{C}$, the maximum satisfiability degree of a clause can be obtained:

$$maxSAT_{ij} = SAT_{\mathbb{C}}(C_{ij}) = \mathcal{S}_{a \in \mathbb{C}}\{\mu_{C_{ij}}(a)\} \tag{10}$$

In this setting, a fuzzy-rough reduct corresponds to a (minimal) truth assignment to variables such that each clause has been satisfied to its maximum extent.

3   Nearest Neighbour-Based Fuzzy-Rough Sets

It was noted in [6] and [12] that the standard approach to fuzzy-rough sets uses only the nearest data object of a different class when considering the membership of a data object to the lower approximation. This is due to a natural property of fuzzy implicators and their use for calculating membership degrees; when the second component is 1.0 (i.e. true) then the implication result will evaluate to 1.0. The second component here corresponds to the degree to which an object belongs to a given decision class; a value of 1.0 indicates that the object is of the same decision class. Therefore, the only data objects to have an impact on the result of the implications are those of classes other than that of the object under consideration. Of these, the nearest object of a different class will produce the smallest value for the implication operation, and therefore, it is this value only that is used, due to the fact that eqn. (1) results in the minimum of all implications. This process (as mentioned previously), is quite time-consuming, as it requires the calculation of the nearest neighbours for each feature subset candidate that is considered. Hence, there is very little saving in time when employing such a nearest neighbour approach. The approach presented here, however, seeks to approximate these calculations by computing the nearest neighbour(s) for each object beforehand, and only using these in the lower approximation calculations. Although the final subsets produced may not be true reducts (in the fuzzy-rough sense), their computation will be much less intensive and thus methods based on this framework should be applicable to larger data.

3.1   nnFRFS

Using the approach described above, the original FRFS method can be altered to only consider the nearest neighbours, termed nnFRFS hereafter. The lower approximation is thus defined, for fuzzy concept X, feature subset P and fuzzy implicator $\mathcal{I}$:

$$\mu_{\underline{R_P^k}X}(x) = \inf_{y \in NN_x^k} \mathcal{I}(\mu_{R_P}(x,y), 0) \tag{11}$$

Each neighbour in NN_x^k has been determined beforehand using $R_\mathbb{C}$ to measure similarity and only considering those k nearest objects that belong to a different class than x. Those features present in the subset P are used for determining the similarity R_P. For standard nnFRFS, only the closest neighbour is required, so $|NN_x^1| = 1$ for all x, reducing the number of calculations drastically. This framework can be used for other extensions (such as VQRS and OWA-based fuzzy-rough feature selectors); for these, all neighbours will have some impact on the final calculation and so parameter k needs to be set appropriately.

In oredr to demonstrate that the parameter k has no impact on nnFRFS: assume that an object x has k neighbours. The fuzzy lower approximation using these is $\inf_{y \in NN_x^k} \mathcal{I}(\mu_{R_P}(x,y), 0)$, and hence the smallest implication evaluation will be the resultant membership of x to the lower approximation. This will always be the result of using the largest value for $\mu_{R_P}(x,y)$ due to the property of implicators, which is generated by considering the closest neighbour to x. Using the nearest neighbour-based fuzzy lower approximation, the fuzzy positive region can be redefined as:

$$\mu_{POS_P^k(\mathbb{D})}(x) = \sup_{X \in \mathbb{U}/\mathbb{D}} \mu_{\underline{R_P^k}X}(x) \tag{12}$$

The fuzzy-rough degree of dependency of $\mathbb{D}$ on the attribute subset P can then be redefined:

$$\gamma_P^k(\mathbb{D}) = \frac{\sum\limits_{x \in \mathbb{U}} \mu_{POS_P^k(\mathbb{D})}(x)}{|\mathbb{U}|} \tag{13}$$

or, the normalised version (as the data may be inconsistent):

$$\gamma_P^k(\mathbb{D}) = \frac{1}{\mathbb{U}} \sum_{x \in \mathbb{U}} \frac{\mu_{POS_P^k(\mathbb{D})}(x)}{\mu_{POS_\mathbb{C}^k(\mathbb{D})}(x)} \tag{14}$$

This measure of dependency can be used in the same way as the original definition as a basis for guiding search toward optimal subsets. In this paper, a greedy hill-climbing search method is used.

NNFRFS($\mathbb{C}$,$\mathbb{D}$,k).
$\mathbb{C}$, the set of all conditional attributes;
$\mathbb{D}$, the set of decision attributes;
k, the number of nearest neighbours to consider.

(1) $R \leftarrow \{\}$; $\gamma_{best}^k = 0$;
(2)   **foreach** $x \in \mathbb{U}$, calculate NN_x^k
(3)   **do**
(4) $T \leftarrow R$
(5)   **foreach** $x \in (\mathbb{C} - R)$
(6)   **if** $\gamma_{R \cup \{x\}}^k(\mathbb{D}) > \gamma_T^k(\mathbb{D})$
(7) $T \leftarrow R \cup \{x\}$
(8) $\gamma_{best}^k = \gamma_T^k(\mathbb{D})$
(9) $R \leftarrow T$
(10) **until** $\gamma_{best}^k == \gamma_\mathbb{C}^k(\mathbb{D})$
(11) **return** R

Fig. 1. The nnFRFS algorithm

3.2   nnFDM

The fuzzy discernibility matrix-based approach described earlier can also be altered to form a more computationally-efficient process. Recall that the discernibility matrix is constructed by the pairwise comparison of all objects in a dataset, and for the decision-relative discernibility matrix, clauses are only generated when pairs of objects belong to different decision classes. Conditional features that differ in value between object pairs are recorded in the clauses; a subset of features then is required such that all clauses are satisfied, meaning that all objects can be discerned. For the fuzzy-rough approach, the importance of features for a pair of objects is determined by the negation of the fuzzy similarity. Pairs of objects which are very similar but belong to different decision classes are therefore problematic, and the features that differ the most in value between them are very important.

The most important clauses for an object are those that are generated by the nearest neighbours of a different class. As more dissimilar objects are considered, the more features will appear in the clauses (or will belong to a higher degree), meaning that the clause is more easily satisfiable. Hence, the most useful information is contained in the nearest few neighbours for each object, as these are the most difficult to discern. The modified FDM approach presented here attempts to approximate the full set of clauses by only considering the most important clauses, generated by nearest neighbours of objects of different classes. The parameter k determines how many of the nearest objects are used to generate such clauses. Setting k to $|\mathbb{U}| - 1$ will produce all possible clauses, and the algorithm will collapse to the original FDM approach.

Each entry in the fuzzy discernibility matrix is generated by comparing pairs of objects. Here, only the k nearest objects of a different class are considered. Clauses are generated in the same way as for the fuzzy discernibility matrix approach described previously. Based on this, the full set of clauses can be generated as follows:

$$Clauses^k = \{C_{ij}| \ j \in NN_i^k \vee i \in NN_j^k\} \qquad (15)$$

where NN_i^k is the set of k nearest neighbours for object i, generated in the same way as for nnFRFS previously. Therefore, a clause is generated from object pair i, j if at least one of the objects appears in the other's nearest neighbour list.

The degree of satisfaction of a clause C for a subset of features P is defined as:

$$SAT_P(C) = \mathcal{S}_{a \in P}\{\mu_C(a)\} \qquad (16)$$

for t-conorm $\mathcal{S}$. By setting $P = \mathbb{C}$, the maximum satisfiability degree of a clause C can be obtained:

$$maxSAT_C = SAT_{\mathbb{C}}(C) = \mathcal{S}_{a \in \mathbb{C}}\{\mu_C(a)\} \qquad (17)$$

Finally, the following subset evaluation measure can be used to gauge the worth of a subset of features P:

$$\tau^k(P) = \frac{1}{|Clauses^k|} \sum_{C \in Clauses^k} \frac{SAT_P(C)}{maxSAT_C} \qquad (18)$$

This measure checks the extent to which each clause is satisfied by P compared to the total satisfiability for all generated clauses. When this reaches 1, all clauses have been satisfied maximally, and the underlying search can stop; the set of features in P discern all considered object pairs.

Using this framework, a search amongst feature subsets can be conducted that aims to maximise the satisfiability of all generated clauses. In this work, a hill-climbing approach is adopted (see Figure 2). Initially, the k nearest neighbours are computed for each object x and stored in the list NN_x^k. The clauses are generated from these lists via generateClauses(NN_x^k,k). The process then follows the typical hill-climbing algorithm, where the addition of individual features to the current subset candidate is evaluated using the measure τ^k.

NNFDM ($\mathbb{C}$,$\mathbb{D}$,k).
$\mathbb{C}$, the set of all conditional attributes;
$\mathbb{D}$, the set of decision attributes;
k, the number of nearest neighbours to consider.

(1) $R \leftarrow \{\}$; $\tau_{best}^k = 0$;
(2)   **foreach** $x \in \mathbb{U}$, calculate NN_x^k
(3)   generateClauses(NN_x^k,k);
(4)   **do**
(5) $T \leftarrow R$
(6)   **foreach** $x \in (\mathbb{C} - R)$
(7)   **if** $\tau^k(R \cup \{x\}) > \tau^k(T)$
(8) $T \leftarrow R \cup \{x\}$
(9) $\tau_{best}^k = \tau^k(T)$
(10) $R \leftarrow T$
(11)   **until** $\tau_{best}^k == 1$
(12)   **return** R

Fig. 2. The nnFDM algorithm

The nnFRFS and nnFDM algorithms are just two of the possible ways in which nearest neighbour approaches to fuzzy-rough set feature selection can be implemented, employing the two main concepts of dependency degree and the discernibility matrices of rough set theory. However, there are many other potential extensions and applications for the proposed work and these are briefly outlined in the conclusion.

4   Experimental Evaluation

The experiments detailed here employed a total of 10 different datasets, described in Table 1. Eight of these datasets are drawn from [5], whilst the remaining two are real-world mammographic risk-assessment tasks which are related to data derived from [7].

For comparison, three other fuzzy-rough approaches for feature selection [8] are included along with three different reduct search methods: greedy hill-climbing (GHC), genetic algorithm-based search (GA), and particle swarm optimisation-based search (PSO). For the fuzzy-rough subset evaluation metric, the Łukasiewicz t-norm ($\max(x + y - 1, 0)$) and the Łukasiewicz fuzzy implicator ($\min(1 - x + y, 1)$) are adopted to implement the fuzzy connectives and the similarity relation. The similarity measure of eqn. (26) in [8] is also used

Table 1. Benchmark data

Dataset	Features	Objects
MIAS	281	322
DDSM	281	832
web	2557	149
cleveland	13	297
glass	9	214
heart	13	270
olitos	25	120
water2	39	390
water3	39	390
wine	13	178

Table 2. Classification results (%) using the JRip classifier learner

Dataset	Unred.	GHC	nnFRFS	nnFDM ($k =$)			GA	PSO
				1	3	5		
MIAS	63.74	60.94	58.02	58.02	61.96	60.66	64.41	53.34
DDSM	52.78	49.22	50.71	50.71	52.30	52.40	51.79	50.69
web	54.74	49.68	46.71	46.71	45.35	46.46	61.45	50.70
cleveland	54.23	54.48	54.55	54.55	54.28	54.34	54.02	54.09
glass	67.17	67.17	66.06	66.06	67.17	67.17	65.25	65.25
heart	72.96	74.15	74.22	74.44	74.67	75.41	72.30	73.85
olitos	68.50	62.83	63.33	64.00	65.67	66.83	59.33	61.17
water2	82.15	83.28	82.87	82.87	82.97	82.36	82.00	81.90
water3	82.72	81.23	81.23	81.23	81.28	82.15	78.82	78.00
wine	93.54	91.46	89.56	89.69	91.35	91.69	86.60	90.41

here. Also, for the novel nnFDM approaches, values of 1, 3, and 5 are used for k respectively. Note that nnFRFS is not affected by the choice of value for k, as it always relies upon the closest neighbour. For the generation of classification results, two different classifier learners have been employed: JRip, a rule-based classifier [2]; and IBk [1], a nearest-neighbour classifier (with $k = 3$). Five stratified randomisations of 10-fold cross-validation were employed in generating the classification results. It is important to note that feature selection is performed as part of the cross-validation and each fold results in a new selection of features.

The GA search has an initial population size of 200, a maximum number of generations/iterations of 40, crossover probability: 0.6 and mutation probability: 0.033. The number of generations/iterations for PSO search was set to 40, whilst the number of particles was set to 200, with acceleration constants $c1 = 1$ and $c2 = 2$. These parameters may not be ideal for all of the datasets employed here and an optimisation phase may well result in an improvement in performance. However, such an optimisation step would need to be performed on a

Table 3. Classification results (%) using the IBk (kNN) classifier learner (k=3)

Dataset	Unred.	GHC	nnFRFS	nnFDM (k =)			GA	PSO
				1	3	5		
MIAS	69.57	63.29	64.10	64.10	62.54	63.83	65.40	53.48
DDSM	51.55	45.85	51.07	51.07	51.56	50.69	52.13	46.71
web	37.98	44.11	42.66	42.66	37.10	36.98	46.72	36.65
cleveland	56.98	52.96	68.77	68.77	55.97	56.10	53.89	53.83
glass	69.24	69.24	68.77	68.77	69.24	69.24	68.51	68.51
heart	80.96	78.15	80.89	80.96	80.96	80.96	78.15	76.96
olitos	81.00	65.67	71.00	70.67	72.17	71.83	66.50	72.33
water2	85.33	84.56	83.28	83.28	82.26	82.26	78.26	80.10
water3	82.97	81.23	82.00	82.00	81.08	82.05	77.44	77.23
wine	95.97	96.42	95.92	95.97	95.61	95.41	91.82	94.71

Table 4. Average subset sizes

Dataset	GHC	nnFRFS	nnFDM (k =)			GA	PSO
			1	3	5		
MIAS	6.08	13.70	13.70	17.18	19.86	9.0	7.70
DDSM	7.12	33.48	33.48	41.40	44.60	10.96	9.56
web	19.02	4.08	4.08	8.22	10.52	186.00	141.20
cleveland	7.64	11.08	11.10	11.80	11.82	9.0	7.70
glass	9.00	9.00	8.78	8.78	9.00	8.36	8.36
heart	7.06	10.44	10.48	10.32	10.60	7.00	7.38
olitos	5.00	7.52	7.64	8.78	9.34	5.24	5.00
water2	6.00	12.82	12.82	15.04	16.54	6.96	6.44
water3	6.08	11.42	11.42	13.40	14.70	7.00	6.50
wine	5.00	7.26	7.26	8.40	9.40	4.70	4.92

dataset-by-dataset basis which would involve a significant investment of effort and time and would form part of a more comprehensive future investigation.

Tables 2 and 3 detail the classification results for the JRip and IBk classifier learners respectively. Examining the classification results, it is clear that nnFRFS and nnFDM return very similar results to GHC. Indeed, when a paired t-test is employed to examine the statistical significance of the results generated for the proposed approaches, even though the absolute figures are slightly lower in some cases, statistically there are no inferior results. It is worth noting from Table 4, however, that the average subset sizes for nnFRFS and nnFDM are greater than GHC and the GA and PSO methods. One notable exception to this are the results for the *web* dataset, where the novel methods all return average subset sizes which are much smaller than those of all of the standard approaches.

It is in terms of execution times that both nnFRFS and nnFDM have the most to offer in terms of improvement in performance. The speed-up in performance is considerable and demonstrates that the NN methods show potential

Table 5. Average execution times per fold (sec.)

Dataset	GHC	nnFRFS	nnFDM ($k =$)			GA	PSO
			1	3	5		
MIAS	12.04	1.05	1.07	1.67	2.99	3.11	22.60
DDSM	110.44	7.12	6.97	15.58	26.13	23.94	173.93
web	98.42	1.635	2.33	3.70	5.72	3.51	24.07
cleveland	0.39	0.037	0.048	0.059	0.070	16.20	3.83
glass	0.14	0.0202	0.030	0.038	0.045	1.55	1.08
heart	0.30	0.0257	0.0359	0.046	0.0552	14.48	3.46
olitos	0.11	0.018	0.030	0.037	0.049	2.36	1.26
water2	2.16	0.125	0.134	0.195	0.269	20.14	19.71
water3	2.17	0.141	0.153	0.206	0.276	19.57	17.25
wine	0.11	0.017	0.0325	0.043	0.055	7.55	1.29

for application to very large data. Again, the *web* dataset seems to be the exception for nnFDM at least for $k=3$ and $k=5$. (It should be noted however that the corresponding subsets discovered by GA and PSO are at least 14 times the size of those discovered by nnFDM.) This behaviour may arise as a result of the characteristics of the data itself, however, which has a large number of features and a very small number of data objects. Such datasets always present a challenge to learning algorithms regardless of the approach applied.

One of the primary motivations behind the development of the nearest neighbour fuzzy-rough approaches detailed here was that of a reduction in computational overhead. Many of the fuzzy-rough metrics suffer in this regard when applied to larger datasets. It is clear from Table 5, that the proposed methods offer much potential in addressing this problem.

5   Conclusion

Two new approaches for reducing computational effort for feature selection based on nearest neighbour fuzzy-rough sets have been presented. These approaches are based upon the idea of calculating the nearest neighbours prior to the search and then only using these neighbours for the calculations. The time complexity therefore is essentially an order of magnitude smaller for the number of data objects. The results detailed in the previous section show that whilst the subset sizes are larger than those of existing approaches, the speed-up in terms of performance offers much potential for further development. Indeed there are a number of possible avenues of exploration which may be able to offer improvements in the performance in terms of subset size, whilst retaining the saving in computational effort that is clear from the results shown previously. For example, using SAT techniques to find the smallest reducts in the clauses generated from the nnFDM approach, applying the approaches to unsupervised FS, and improving the efficiency of recent fuzzy-rough object/instance selection methods, etc. In addition, an approach which considers not only the nearest neighbours but the neighbourhood *structure/distribution* of the objects which form the fuzzy-rough lower and upper approximation membership values offer much potential for a series of topics for future work. Although the experimental evaluation in this paper features at least three large datasets, it would be interesting to apply nn-FRFS/nnFDM to larger data, in the order of thousands of features and objects. This would also form the basis for a more comprehensive investigation.

References

1. Aha, D., Kibler, D.: Instance-based learning algorithms. Machine Learning 6, 37–66 (1991)
2. Cohen, W.W.: Fast effective rule induction. In: Proceedings of the 12th International Conference on Machine Learning, pp. 115–123 (1995)
3. Cornelis, C., Jensen, R., Hurtado Martín, G., Ślęzak, D.: Attribute Selection with Fuzzy Decision Reducts. Information Sciences 180(2), 209–224 (2010)
4. Dubois, D., Prade, H.: Putting Rough Sets and Fuzzy Sets Together. In: Intelligent Decision Support, pp. 203–232 (1992)
5. Frank, A., Asuncion, A.: UCI Machine Learning Repository. University of California, School of Information and Computer Science, Irvine, CA (2010), http://archive.ics.uci.edu/ml
6. Hu, Q., Zhang, L., An, S., Zhang, D., Yu, D.: On Robust Fuzzy Rough Set Models. IEEE Transactions on Fuzzy Systems 20(4), 636–651 (2012)
7. Oliver, A., Freixenet, J., Marti, R., Pont, J., Perez, E., Denton, E.R.E., Zwiggelaar, R.: A Novel Breast Tissue Density Classification Methodology. IEEE Transactions on Information Technology in Biomedicine 12(1), 55–65 (2008)
8. Jensen, R., Shen, Q.: New Approaches to Fuzzy-Rough Feature Selection. IEEE Transactions on Fuzzy Systems 17(4), 824–838 (2009)
9. Jensen, R., Tuson, A., Shen, Q.: Finding rough and fuzzy-rough set reducts with SAT. Information Sciences 255, 100–120 (2014)
10. Klement, E.P., Radko, M., Endre, P.: Triangular norms. Position paper I: basic analytical and algebraic properties. Fuzzy Sets and Systems 143(1), 5–26 (2004)
11. Pawlak, Z.: Rough Sets: Theoretical Aspects of Reasoning About Data. Kluwer Academic Publishing (1991)
12. Qu, Y., Shen, Q., Mac Parthaláin, N., Shang, C., Wu, W.: Fuzzy similarity-based nearest-neighbour classification as alternatives to their fuzzy-rough parallels. International Journal of Approximate Reasoning 54(1), 184–195 (2012)
13. Radzikowska, A.M., Kerre, E.E.: A comparative study of fuzzy rough sets. Fuzzy Sets and Systems 126(2), 137–155 (2002)

Some Fundamental Laws of Partial First-Order Logic Based on Set Approximations

Tamás Kádek and Tamás Mihálydeák

Department of Computer Science,
Faculty of Informatics, University of Debrecen
Egyetem tér 1, H-4010 Debrecen, Hungary
{kadek.tamas,mihalydeak.tamas}@inf.unideb.hu

Abstract. The authors show that a very general framework of set approximation can be the set-theoretical base of semantics of a partial first-order logic. The most general problem is what happens if in the semantics of first-order logic one uses the approximations of sets as semantic values of predicate parameters instead of sets given by their total interpretation in order to determine the truth values of formulas? The authors show some unexpected properties connected with logical constants directly. The goal of the investigation is to show the possible connections between the result of different approximative and exact evaluation of formulas – or the lack of them. At the end, the authors present the practical example, in which we can see the discussed behavior of approximation.

Keywords: Approximation of sets, rough set, partial logic, partial semantics.

1 Introduction

Rough set theory (proposed by Pawlak (see in [1], [2]))and its different generalizations (see, e.g. in [3]) provide a powerful foundation to reveal and discover important structures and patterns in data, and to classify complex objects.[1]

In most cases, we have a family of base sets — as subsets of a universe of discourse. In philosophy these sets represent our available knowledge, we consider them as the extensions of our available concepts/properties.

Rough set theory can be considered as the foundation of various kinds of deductive reasoning (see in [5], [6]). In this paper the authors go on their logical investigations about the possibility of using different systems of set approximation in first-order logical semantics (see for example in [8]).

The main question of the present paper is the following: What can be said about the consequence relations connected with logical functors directly in a partial first-order logic relying on approximations of sets.

Initially, the authors introduce a syntax of a partial first-order logic, including approximative sentence functors, which give us a possibility to declare the

---

[1] An overview of some research trends on rough set foundations and rough set–based methods can be found in [5].

C. Cornelis et al. (eds.): RSCTC 2014, LNAI 8536, pp. 47–58, 2014.
© Springer International Publishing Switzerland 2014

need of approximative evaluation. Based on the extended first-order language, the semantical definitions show a way how rough sets can be used for first-order formula evaluation. After presenting the semantics, including the consequence relations, the goal of the investigation is to show the possible connections between the evaluation results of formulas including different sentence functors, or in other words, the relation between the lower and upper approximation and the traditional evaluation. The theorems discussed later focus mostly on the weaknesses of the approximation, and the proofs try to discover the causes. Finally, the authors describe a practical example of how we can implement the semantics and use the approximative evaluation on an existing database. The working sample is made to show the possible benefits (like faster evaluation based on the approximation) and the lack of connections between the real and approximative evaluation.

2   General Systems of Tool-Based Partial Approximation of Sets

In the following definition a most fundamental (and very general) notion of an approximation space is given. This core notion serves as the set-theoretical background of semantics of partial first-order logic with approximative operators.

Definition 1. *The ordered 5–tuple* $\langle U, \mathfrak{B}, \mathfrak{D}_{\mathfrak{B}}, \mathsf{l}, \mathsf{u} \rangle$ *is a general partial approximation space if*

1. U *is a nonempty set;*
2. $\mathfrak{B} \subseteq 2^U \setminus \emptyset, \mathfrak{B} \neq \emptyset;$
3. $\mathfrak{D}_{\mathfrak{B}}$ *is an extension of* $\mathfrak{B}$, *i.e.* $\mathfrak{B} \subseteq \mathfrak{D}_{\mathfrak{B}}$, *such that* $\emptyset \in \mathfrak{D}_{\mathfrak{B}};$
4. *the functions* l, u *forms an approximation pair* $\langle \mathsf{l}, \mathsf{u} \rangle$, *i.e.*
   (a) $\mathsf{l}, \mathsf{u} : 2^U \to 2^U;$
   (b) $\mathsf{l}(2^U), \mathsf{u}(2^U) \subseteq \mathfrak{D}_{\mathfrak{B}}$ *(definability of* l, u*);*
   (c) *the functions* l *and* u *are monotone, i.e. for all* $S_1, S_2 \in 2^U$ *if* $S_1 \subseteq S_2$ *then* $\mathsf{l}(S_1) \subseteq \mathsf{l}(S_2)$ *and* $\mathsf{u}(S_1) \subseteq \mathsf{u}(S_2);$
   (d) $\mathsf{u}(\emptyset) = \emptyset$ *(normality of* u*)*
   (e) *if* $S \in \mathfrak{D}_{\mathfrak{B}}$, *then* $\mathsf{l}(S) = S$ *(granularity of* $\mathfrak{D}_{\mathfrak{B}}$, *i.e.* l *is standard);*
   (f) *if* $S \in 2^U$, *then* $\mathsf{l}(S) \subseteq \mathsf{u}(S)$ *(weak approximation property).*

Informally, the set U is the universe of approximation; $\mathfrak{B}$ is a nonempty set of base sets, it represents our knowledge used in the whole approximation process; $\mathfrak{D}_{\mathfrak{B}}$ (i.e. the set of definable sets) contains not only the base sets, but those which we want to use to approximate any subset of U; the functions l, u determine the lower and upper approximation of any set with the help of representations of our primitive or available concepts/properties. The nature of an approximation pair[2] depends on how to relate the lower and upper approximations of a set to

---

[2] One of the most general notion of weak and strong approximation pairs can be found in Düntsch and Gediga [4].

the set itself. The condition 4. (e) in the definition is not typical: if we look at the sets belonging to $\mathfrak{D}_{\mathfrak{B}}$ as our tools to approximate any set, then it can be a requirement that a tool should be approximated by itself from the lower side. If we give it up, we decrease the roles of our tools/base sets (members of $\mathfrak{D}_{\mathfrak{B}}$).

3   Tool-Based Partial First-Order Logic (TbPFoL) with Approximative Functors

3.1   Language of TbPFoL with Approximative Functors

At first we need a given language of first-order logic, and a finite nonempty set $\mathcal{T}$ of predicate parameters. Its members are called tools.

Definition 2. $L = \langle LC, Var, Con, \mathfrak{T}, Pred, Form \rangle$ *is a language of TbPFoL with approximative functors, if it is an extension of standard first-order language:*

1. $LC = \{\neg, \supset, +, \forall, \exists, \downarrow, \uparrow, (,)\}$, *LC is the set of logical constants.*
2. *Con (for simplicity) the set of 1–argument predicate parameters.*
3. $\mathfrak{T} \subseteq Con$, $\mathfrak{T} \neq \emptyset$ *and the set $\mathfrak{T}$ (the set of tools) is finite.*
4. *Pred is the set of predicates, and it is given by the following inductive definition: $Con \subseteq Pred$, and if $P \in Pred$, then $P^{\downarrow}, P^{\uparrow}, \in Pred$.*
5. *The set Form (the set of formulae) is given by usual inductive definition.*

Definition 3. *Approximative sentence functors $\downarrow$, $\uparrow$ can be introduced in the following inductive contextual way:*

1. *It $P \in Pred$, and $x \in Var$, then $\downarrow P(x) =_{def} P^{\downarrow}(x)$, $\uparrow P(x) =_{def} P^{\uparrow}(x)$.*
2. *If $A \in Form$, then*
   (a) $\uparrow \neg A =_{def} \neg \uparrow A$, and $\downarrow \neg A =_{def} \neg \downarrow A$
   (b) $\uparrow +A =_{def} + \uparrow A$, and $\downarrow +A =_{def} + \downarrow A$.
3. *If $A, B \in Form$, then $\uparrow (A \supset B) =_{def} (\uparrow A \supset \uparrow B)$ and $\downarrow (A \supset B) =_{def} (\downarrow A \supset \downarrow B)$.*
4. *If $A \in Form$ and $x \in Var$ then $\uparrow \forall x A =_{def} \forall x \uparrow A$ and $\uparrow \exists x A =_{def} \exists x \uparrow A$.*

3.2   Interpretations of TbPFol with Approximative Functors

Definition 4. *Let $L = \langle LC, Var, Con, \mathfrak{T}, Pred, Form \rangle$ be a language of TbPFoL with approximative functors. The ordered pair $\langle U, \varrho \rangle$ is an interpretation of L, if U is a nonempty set, ϱ is a function such that $\varrho : Con \to \{0, 1\}^{U}$.*

In order to give semantic rules we only need the notions of assignment and modified assignment:

Definition 5. *Function v is an assignment relying on the interpretation $\langle U, \varrho \rangle$ if $v : Var \to U$.*

Definition 6. *Let v be an assignment relying on the interpretation $\langle U, \varrho \rangle$, $x \in Var$ and $u \in U$. $v[x : u]$ is a modified assignment of v, if*

$$v[x:u](y) = \begin{cases} u & \text{if } x = y \\ v(y) & \text{otherwise} \end{cases}$$

Tools (the members of set $\mathcal{T}$) determine a logically relevant general partial approximation space with respect to a given interpretation:

Definition 7. *Let $Ip = \langle U, \varrho \rangle$ be an interpretation of L such that if $T \in \mathcal{T}$, then $\varrho(T) \neq \emptyset$. The 5–tuple $\mathcal{PAS}(\mathcal{T}) = \langle U, \mathfrak{B}, \mathfrak{D}_{\mathfrak{B}}, \mathsf{l}, \mathsf{u} \rangle$ is a logically relevant general partial approximation space generated by set $\mathcal{T}$ of tools with respect to the interpretation $\langle U, \varrho \rangle$ if $\mathfrak{B} = \{\varrho(P) : P \in \mathcal{T}\}$, $\mathfrak{D}_{\mathfrak{B}} \subseteq 2^{U}$ and if $S \subseteq U$, then $\mathsf{l}(S), \mathsf{u}(S) \in \mathfrak{D}_{\mathfrak{B}}$.*

3.3 Semantic Rules of TbPFoL with Approximative Functors

In the semantics of TbPFol the semantic value of an expression depends on a given interpretation $Ip = \langle U, \varrho \rangle$, a given logically relevant general partial approximation space $\mathcal{PAS}(\mathfrak{T}) = \langle U, \mathfrak{B}, \mathfrak{D}_{\mathfrak{B}}, \mathsf{l}, \mathsf{u} \rangle$ generated by set $\mathfrak{T}$ of tools with respect to the interpretation $\langle U, \varrho \rangle$ and a given assignment v. Relying on these tree components one can give the following partial semantic rules. For the sake of simplicity we use a null entity to represent partiality of semantic rules. We use number 0 for falsity, number 1 for truth and number 2 for null entity. The semantic value of an expression A with respect to $Ip = \langle U, \varrho \rangle$, $\mathcal{PAS}(\mathfrak{T})$ and the assignment v is denoted by $[\![A]\!]_{v}^{Ip, \mathcal{PAS}(\mathfrak{T})}$. For the sake of simplicity the superscripts are omitted.

Definition 8. *The semantic rules of TbPFoL with approximative functors are the following:*

1. *If $x \in Var$, then $[\![x]\!]_{v} = v(x)$.*
2. *If $T \in \mathcal{T}$ is a predicate parameter, then $[\![T]\!]_{v} = s$, where $s : U \rightarrow \{0, 1, 2\}$ is a function such that*
$$s(u) = \begin{cases} 1 & \text{if } u \in \varrho(T) \\ 0 & \text{if } u \in \mathsf{l}(U \setminus \varrho(T)) \\ 2 & \text{otherwise} \end{cases}$$
3. *If $P \in Con \setminus \mathcal{T}$, then $[\![P]\!]_{v} = \varrho(P)$.*
4. *If $P \in Pred$, then $[\![P^{\downarrow}]\!]_{v} = s$, where $s : U \rightarrow \{0, 1, 2\}$ is a function such that*
$$s(u) = \begin{cases} 1 & \text{if } u \in \mathsf{l}([\![P]\!]_{v}) \\ 0 & \text{if } u \in \mathsf{l}(U \setminus \mathsf{u}([\![P]\!]_{v})) \\ 2 & \text{otherwise} \end{cases}$$
5. *If $P \in Pred$, then $[\![P^{\uparrow}]\!]_{v} = s$, where $s : U \rightarrow \{0, 1, 2\}$ is a function such that*
$$s(u) = \begin{cases} 1 & \text{if } u \in \mathsf{u}([\![P]\!]_{v}) \\ 0 & \text{if } u \in \mathsf{l}(U \setminus \mathsf{u}([\![P]\!]_{v})) \\ 2 & \text{otherwise} \end{cases}$$
6. *If $P \in Con$, $x \in Var$, then $[\![P(x)]\!]_{v} = [\![P]\!]_{v}([\![x]\!]_{v})$*

7. *If $A \in Form$, then* $[\![+A]\!]_v = \begin{cases} 1 & if \ [\![A]\!]_v = 1 \\ 0 & otherwise \end{cases}$

8. *If $A \in Form$, then* $[\![\neg A]\!]_v = \begin{cases} 2 & if \ [\![A]\!]_v = 2 \\ 1 - [\![A]\!]_v & otherwise \end{cases}$

9. *If $A, B \in Form$, then*

$$[\![(A \supset B)]\!]_v = \begin{cases} 0 & if \ [\![A]\!]_v = 1, \ and \ [\![B]\!]_v = 0 \\ 2 & if \ [\![A]\!]_v = 2, \ or \ [\![B]\!]_v = 2 \\ 1 & otherwise \end{cases}$$

10. *If $A \in Form, x \in Var$, then*

$$[\![\forall x A]\!]_v = \begin{cases} 0 & if \ there \ is \ an \ u \in U \colon \ [\![A]\!]_{v[x:u]} = 0 \\ 2 & if \ for \ all \ u \in U \colon \ [\![A]\!]_{v[x:u]} = 2 \\ 1 & otherwise \end{cases}$$

$$[\![\exists x A]\!]_v = \begin{cases} 1 & if \ there \ is \ an \ u \in U \colon \ [\![A]\!]_{v[x:u]} = 1 \\ 2 & if \ for \ all \ u \in U \colon \ [\![A]\!]_{v[x:u]} = 2 \\ 0 & otherwise \end{cases}$$

The semantic rules of classical logical constants as negation and implication are the conservative extensions of classical two-value ones. Conjunction and disjunction can be introduced by contextual definition: If $A, B \in Form$, then $(A \wedge B) =_{def} \neg(A \supset \neg B)$, $(A \vee B) =_{def} (\neg A \supset B)$.

For example if P is a one-argument predicate parameter which is not a tool and $u \in U$, then

- the functor $\downarrow$ gives the following results:
    1. $P^{\downarrow}$ is true at u if u belongs to the lower approximation of semantic value of P (i.e. our tools evaluate P as certainly true at u);
    2. $P^{\downarrow}$ is false at u if u belongs to the lower approximation of the complement of upper approximation of semantic value of P (i.e. our tools evaluate P as certainly false at u);
    3. otherwise $P^{\downarrow}$ is undefined u (i.e. our tools are not enough to decide whether P is certainly true or certainly false at u);
- the functors $\uparrow$ gives the following results:
    1. $P^{\uparrow}$ is true at u if u belongs to the upper approximation of semantic value of P (i.e. our tools evaluate P as maybe true at u);
    2. $P^{\uparrow}$ is false at u if u belongs to the lower approximation of the complement of upper approximation of semantic value of P (i.e. our tools evaluate P as certainly false at u);
    3. otherwise $P^{\uparrow}$ is undefined at u (i.e. our tools are not enough to decide whether P is maybe true or certainly false at u).

3.4   Central Semantic Notions

The notion of models plays a fundamental role in the semantic definition of consequence relation. $\langle U, \varrho, \mathcal{PAS}(\mathfrak{T}), v \rangle$ is a possible representation of a set $\Gamma (\subseteq$

Form) if $\langle U, \varrho \rangle$ is an interpretation, $\mathcal{PAS}(\mathfrak{T})$ is a logically relevant general partial approximation space generated by the set of tools $\mathfrak{T}$ with respect to the interpretation $\langle U, \varrho \rangle$ and v is an assignment. $\langle U, \varrho, \mathcal{PAS}(\mathfrak{T}), v \rangle$ is a representation of Γ if it is a possible representation such that for all $A \in \Gamma$, $[\![A]\!]_v \neq 2$. $\langle U, \varrho, \mathcal{PAS}(\mathfrak{T}), v \rangle$ is a model of Γ if it is a representation of Γ such that for all $A \in \Gamma$, $[\![A]\!]_v = 1$.

Definition 9
Let $\mathcal{PR}$ be a set of possible representations of language L.

1. *Γ is satisfiable with respect to the set $\mathcal{PR}$ if it has a model in $\mathcal{PR}$.*
2. *A is a strong semantic consequence of Γ with respect to the set $\mathcal{PR}$ (in notation $\Gamma \vDash_s^{\mathcal{PR}} A$) if*
   (a) Γ has a representation in $\mathcal{PR}$;
   (b) every $\mathcal{PR}$–representation of Γ is a representation of $\{A\}$;
   (c) every $\mathcal{PR}$–model of Γ is a model of $\{A\}$.
3. *A is valid with respect to the set $\mathcal{PR}$ (in notation $\vDash_s^{\mathcal{PR}} A$) if $\emptyset \vDash_s^{\mathcal{PR}} A$.*
4. *A is a weak semantic consequence of Γ with respect to the set $\mathcal{PR}$ (in notation $\Gamma \vDash_w^{\mathcal{PR}} A$) if $\langle U, \varrho, \mathcal{PAS}(\mathfrak{T}), v \rangle$ is a $\mathcal{PR}$–modell of Γ, then $[\![A]\!]_v \neq 0$ (i.e. A is not false in any $\mathcal{PR}$–model of Γ).*
5. *A is irrefutable with respect to the set $\mathcal{PR}$ (in notation $\vDash_w^{\mathcal{PR}} A$) if $\emptyset \vDash_w^{\mathcal{PR}} A$ (i.e. A is never false in $\mathcal{PR}$).*

More details and further definitions about central semantic notions can be found in [8].

4   Logical Properties Connected Logical Constants Directly

Suppose, that we have a $P \in Con$ one-argument predicate parameter. We are able to construct an interpretation $\langle U, \varrho \rangle$ and an assignment v where we suppose, there exists an $u \in U$ such that $u \notin \mathsf{l}([\![P]\!])$, $u \in \mathsf{u}([\![P]\!])$. The evaluation results of the formulas $\uparrow P(x)$ and $\downarrow P(x)$ (with respect to $\langle U, \varrho \rangle$ and v) are different:

$$[\![\uparrow P(x)]\!]_{v[x:u]} = [\![P^{\uparrow}(x)]\!]_{v[x:u]} = 1, \quad [\![\downarrow P(x)]\!]_{v[x:u]} = [\![P^{\downarrow}(x)]\!]_{v[x:u]} = 2.$$

This could change the usual behavior of logical constants. For example, the well known low of modus-ponens still exists,

If we establish that A and $A \supset B$ hold, then B holds.

but, when different sentence functors apperas, the above sentence may sound like

If we establish that the lower approximation of A and $A \supset B$ hold, is it true, that the upper approximation B holds?

The theorems below show some unexpected behavior in the case when we use different sentence functors.

Theorem 1. *The set* $\{\downarrow (A \supset B), \ \downarrow A, \ \uparrow \neg B\}$ *is satisfiable and so*

$$\{\downarrow (A \supset B), \ \downarrow A\} \nvDash_w \uparrow B$$

It means that in the general case, the use of the lower approximation has no consequence on the result of upper approximation. Even if the lower approximation of B is a logical consequence of the set $\{\downarrow (A \supset B), \ \downarrow A\}$, we can't say the same about the upper approximation of B.

Proof. By the definition of approximative sentence functors, $[\![\downarrow (A \supset B)]\!] = 1 \Rightarrow [\![(\downarrow A \supset\downarrow B)]\!] = 1$. Because $[\![(\downarrow A \supset\downarrow B)]\!] = 1$ and $[\![\downarrow A]\!] = 1$ following the semantic rules $[\![\downarrow B]\!] = 1$ and because of $[\![\uparrow \neg B]\!] = 1$ then $[\![\uparrow B]\!] = 0$. To fulfill the requirements above, now we construct an A and a B formula, based on the previously introduced $P(x)$:

$$A =_{def} +P(x) \supset +P(x) \qquad B =_{def} \neg + P(x)$$

Therefore (with respect to the interpretation $\langle U, \varrho \rangle$ and assignment v):

$$[\![\downarrow A]\!]_{v[x:u]} = [\![\downarrow (+P(x) \supset +P(x))]\!]_{v[x:u]} = [\![+ \downarrow P(x) \supset + \downarrow P(x)]\!]_{v[x:u]} = 1.$$

For simplicity we can say, the formula A is defined to be valid in any case.

$$[\![\downarrow B]\!]_{v[x:u]} = [\![\downarrow \neg + P(x)]\!]_{v[x:u]} = [\![\neg + \downarrow P(x)]\!]_{v[x:u]} = 1,$$
$$[\![\downarrow (A \supset B)]\!]_{v[x:u]} = [\![(\downarrow A \supset\downarrow B)]\!]_v = 1,$$
$$[\![\uparrow \neg B]\!]_{v[x:u]} = [\![\uparrow \neg\neg + P(x)]\!]_{v[x:u]} = [\![\neg\neg + \uparrow P(x)]\!]_{v[x:u]} = 1.$$

Remark 1. Note, that the law of non-contradiction is still valid, therefore the same formula B cannot be both true and not true at the same time, and so $\{B, \neg B\}$ is unsatisfiable. Even so, the set

$$\{\downarrow B, \uparrow \neg B\}$$

is satisfiable. It is caused by the approximative sentence functors.

Using the same idea as above, in the next two theorem, we want to prove the lack of connection between the lower and upper approximation.

Theorem 2. *The set* $\{\downarrow (A \supset C), \ \downarrow \neg C, \ \uparrow A\}$ *is satisfiable and so*

$$\{\downarrow (A \supset C), \ \downarrow \neg C\} \nvDash_w \uparrow \neg A$$

Proof. As a consequence of the above fact, that the set $\{\downarrow B, \uparrow \neg B\}$ is satisfiable, we are able to define $A =_{def} \neg B$ such that $\{\downarrow \neg A, \uparrow A\}$ satisfiable. Using the $C =_{def} \neg(+P(x) \supset +P(x))$ formula, and the interpretation which satisfies $\{\downarrow \neg A, \uparrow A\}$: $[\![\downarrow (A \supset C)]\!]_{v[x:u]} = 1$, $[\![\downarrow \neg C]\!]_{v[x:u]} = 1$, $[\![\uparrow A]\!]_{v[x:u]} = 1$.

Theorem 3. *The set* $\{\uparrow (A \supset B), \ \uparrow A, \ \downarrow \neg B\}$ *is satisfiable and so*

$$\{\uparrow (A \supset B), \ \uparrow A\} \nvDash_w \downarrow B$$

Proof. Now we also construct an A and a B formula, based on the previously introduced $P(x)$:

$$A =_{def} +P(x) \supset +P(x) \qquad B =_{def} +P(x)$$

Therefore (with respect to the interpretation $\langle U, \varrho \rangle$ and assignment v):

$$[\![\uparrow A]\!]_{v[x:u]} = [\![\uparrow (+P(x) \supset +P(x))]\!]_{v[x:u]} = [\![+ \uparrow P(x) \supset + \uparrow P(x)]\!]_{v[x:u]} = 1,$$
$$[\![\uparrow B]\!]_{v[x:u]} = [\![\uparrow +P(x)]\!]_{v[x:u]} = [\![+ \uparrow P(x)]\!]_{v[x:u]} = 1,$$
$$[\![\uparrow (A \supset B)]\!]_{v[x:u]} = [\![\uparrow A \supset \uparrow B]\!]_v = 1,$$
$$[\![\downarrow \neg B]\!]_{v[x:u]} = [\![\downarrow \neg +P(x)]\!]_{v[x:u]} = [\![\neg + \downarrow P(x)]\!]_{v[x:u]} = 1.$$

Theorem 4. *The set* $\{\uparrow (A \supset C), \ \uparrow \neg C, \ \downarrow A\}$ *is satisfiable and so*

$$\{\uparrow (A \supset C), \ \uparrow \neg C\} \nvDash_w \downarrow \neg A$$

Proof. As a consequence of the $\{\downarrow B, \uparrow \neg B\}$ satisfiability, we are able to define $A =_{def} B$ such that $\{\downarrow A, \uparrow \neg A\}$ satisfiable. Using the $C =_{def} \neg(+P(x) \supset +P(x))$ formula, and the interpretation which satifies $\{\downarrow A, \uparrow \neg A\}$:
$[\![\uparrow (A \supset C)]\!]_{v[x:u]} = 1$, $[\![\uparrow \neg C]\!]_{v[x:u]} = 1$, $[\![\downarrow A]\!]_{v[x:u]} = 1$.

Remark 2. Note, the set $\{\uparrow B, \downarrow \neg B\}$ is also satisfiable, even if the language does not contain the $+$ logical constant. The background of the proofs before was the fact, that the $+$ logical connective doesn't keep the truth value gap. It is also true for the quantifiers. As a result, the same behavior appers in first-order, even if the language doesn't contain the $+$ logical connective.

Proof. Let us suppose, that we have a $Q \in \mathcal{P}(1)$ one-argument predicate parameter and $u_1 \in U$ and $u_2 \in U$ objects, such that

$$[\![Q^{\uparrow}(x)]\!]_{v[x:u]} = 1 \text{ and } [\![Q^{\downarrow}(x)]\!]_{v[x:u]} = 2 \text{ if } v(x) = u_1,$$
$$[\![Q^{\uparrow}(x)]\!]_{v[x:u]} = 0 \text{ and } [\![Q^{\downarrow}(x)]\!]_{v[x:u]} = 0 \text{ if } v(x) = u_2 \text{ and}$$
$$\text{in any other cases } [\![Q^{\uparrow}(x)]\!]_{v[x:u]} = 2 \text{ and } [\![Q^{\downarrow}(x)]\!]_{v[x:u]} = 2.$$

Therefore:

$$[\![\uparrow \exists x Q(x)]\!]_{v[x:u]} = [\![\exists x \uparrow Q(x)]\!]_{v[x:u]} = [\![\exists x Q^{\uparrow}(x)]\!]_{v[x:u]} = 1,$$
$$[\![\downarrow \neg \exists x Q(x)]\!]_{v[x:u]} = [\![\neg \exists x \downarrow Q(x)]\!]_{v[x:u]} = [\![\neg \exists x Q^{\downarrow}(x)]\!]_{v[x:u]} = 1.$$

So, if we choose $B =_{def} \exists x Q(x)$ then $\{\uparrow B, \downarrow \neg B\}$ satisfiable.

Theorem 5. $\{\uparrow \forall x (A(x) \supset B(x)), \uparrow A(y), \downarrow \neg B(y)\}$ *satisfiable and so*

$$\{\uparrow \forall x (A(x) \supset B(x)), \uparrow A(y)\} \nvDash_w \downarrow B(y)$$

Proof. Now we construct again an A and a B formula, based on the previously introduced $P(x)$, supposing, that U has only just one member. In that case, $[\![x]\!]_{v[x:u]} = [\![x]\!]_v = [\![y]\!]_v$.

$$A =_{def} +P(x) \supset +P(x) \qquad B =_{def} +P(x)$$

Therefore:

$[\uparrow B(x)]_v = 1$ and $[\downarrow B(x)]_v = 0$,
$[\uparrow (A(x) \supset B(x))]_v = 1$, so $[\uparrow \forall x(A(x) \supset B(x))]_v = 1$,
$[\downarrow B(y)]_v = [\downarrow +P(y)]_v = [\downarrow +P(x)]_v = 0$, so $[\downarrow \neg B(y)]_v = 1$.

Theorem 6. *The set* $\{\downarrow \forall x(A(x) \supset B(x)), \downarrow A(y), \uparrow \neg B(y)\}$ *is satisfiable and so*

$$\{\downarrow \forall x(A(x) \supset B(x)), \downarrow A(y)\} \not\models_w \uparrow B(y)$$

Proof. It is enough to change the definition of B during the previous proof:

$$B(x) =_{def} \neg +P(x)$$

In this case:

$[\uparrow B(x)]_v = 0$ and $[\downarrow B(x)]_v = 1$,
$[\downarrow (A(x) \supset B(x))]_v = 1$, so $[\downarrow \forall x(A(x) \supset B(x))]_v = 1$,
$[\uparrow B(y)]_v = [\uparrow \neg +P(y)]_v = [\uparrow \neg +P(x)]_v = 0$, so $[\downarrow \neg B(y)]_v = 1$.

Theorem 7

$$\models_s +\neg A \supset \neg +A \quad \text{but} \quad \not\models_w \neg +A \supset +\neg A.$$

Proof. First note that $[+\neg A \supset \neg +A]_{v[x:u]} \in \{0,1\}$.
Indirectly, we assume that $[+\neg A \supset \neg +A]_{v[x:u]} = 0$, therefore

$[+\neg A]_{v[x:u]} = 1 \Rightarrow [\neg A]_{v[x:u]} = 1 \Rightarrow [A]_{v[x:u]} = 0$, and
$[\neg +A]_{v[x:u]} = 0 \Rightarrow [+A]_{v[x:u]} = 1 \Rightarrow [A]_{v[x:u]} = 1$.

This contradiction yields that our assumption was wrong, hence the first half of the theorem is proved. To prove the second half of statement, we suppose, that $[A]_{v[x:u]} = 2$, therefore:

$[A]_{v[x:u]} = 2 \Rightarrow [+A]_{v[x:u]} = 0 \Rightarrow [\neg +A]_{v[x:u]} = 1$, and
$[A]_{v[x:u]} = 2 \Rightarrow [\neg A]_{v[x:u]} = 2 \Rightarrow [+\neg A]_{v[x:u]} = 0$.

Finally, $[\neg +A \supset +\neg A]_{v[x:u]} = 0$.

Theorem 8. *The set* $\{\forall x(P(x) \supset Q(x)), \exists x P(x), \neg \exists x Q(x)\}$ *is satisfiable and so*

$$\forall x(P(x) \supset Q(x)), \exists x P(x) \not\models_w \exists x Q(x).$$

Proof. Let us use an interpretation where $U = \{u_1, u_2\}$, and which is defined so that

$[P(x)]_{v[x:u_1]} = 1$ and $[Q(x)]_{v[x:u_1]} = 2$, and
$[P(x)]_{v[x:u_2]} = 0$ and $[Q(x)]_{v[x:u_2]} = 0$.

By this, $[\forall x(P(x) \supset Q(x))]_v = 1$, $[\exists x P(x)]_v = 1$ furthermore $[\exists x Q(x)]_v = 0$.

Theorem 9

$$\forall x(P(x) \supset Q(x)), \forall x P(x) \models_w \exists x Q(x).$$

Proof. Indirectly, we assume that $[\![\forall x(P(x) \supset Q(x))]\!]_v = 1$ and $[\![\forall x P(x)]\!]_v = 1$, but $[\![\exists x Q(x)]\!]_v = 0$. Therefore:

$[\![\forall x(P(x) \supset Q(x))]\!]_v = 1 \Rightarrow [\![P(x) \supset Q(x)]\!]_{v[x:u]} = 1$ for at least one $u \in U$,
$[\![P(x)]\!]_{v[x:u]} \neq 2$, because $[\![P(x) \supset Q(x)]\!]_{v[x:u]} \neq 2$,
$[\![P(x)]\!]_{v[x:u]} \neq 0$, because $[\![\forall x P(x)]\!]_v \neq 0$.

As the consequence of the previous results, $[\![P(x)]\!]_{v[x:u]} = 1$. Furthermore

$[\![P(x) \supset Q(x)]\!]_{v[x:u]} = 1$ and $[\![P(x)]\!]_{v[x:u]} = 1 \Rightarrow [\![Q(x)]\!]_{v[x:u]} = 1$,
$[\![Q(x)]\!]_{v[x:u]} = 1 \Rightarrow [\![\exists x Q(x)]\!]_v = 1$.

Because $[\![\exists x Q(x)]\!]_v = 0$ was our assumption, hence the theorem is proved.

Theorem 10
$$\exists x(P(x) \vee Q(x)) \vDash_w \exists x P(x) \vee \exists x Q(x).$$

Proof. If we assume, that $[\![\exists x(P(x) \vee Q(x))]\!]_v = 1$, it causes $[\![P(x) \vee Q(x)]\!]_{v[x:u]} = 1$ for at least one $u \in U$. So both $[\![P(x)]\!]_{v[x:u]} \neq 2$ and $[\![Q(x)]\!]_{v[x:u]} \neq 2$, and at least one of them evaluated to 1.

- if $[\![P(x)]\!]_{v[x:u]} = 1$ then $[\![\exists x P(x)]\!]_v = 1$ and $[\![\exists x Q(x)]\!]_v \neq 2$, or
- if $[\![Q(x)]\!]_{v[x:u]} = 1$ then $[\![\exists x Q(x)]\!]_v = 1$ and $[\![\exists x P(x)]\!]_v \neq 2$.

In both of the previous cases, $[\![\exists x P(x) \vee \exists x Q(x)]\!]_v = 1$.

Theorem 11. *The set $\{\forall x(P(x) \wedge Q(x)), \neg\forall x P(x), \neg\forall x Q(x)\}$ is satisfiable and so*
$$\forall x(P(x) \wedge Q(x)) \nvDash_w \forall x P(x) \vee \forall x Q(x).$$

Proof. Let us use an interpretation where $U = \{u_1, u_2, u_3\}$, and which is defined so that $[\![P(x)]\!]_{v[x:u_1]} = 1$, $[\![Q(x)]\!]_{v[x:u_1]} = 1$, $[\![P(x)]\!]_{v[x:u_2]} = 2$, $[\![Q(x)]\!]_{v[x:u_2]} = 0$, $[\![P(x)]\!]_{v[x:u_3]} = 0$ and $[\![Q(x)]\!]_{v[x:u_3]} = 2$.

By this, $[\![\forall x P(x)]\!]_v = 0$ and $[\![\forall x Q(x)]\!]_v = 0$, so $[\![\forall x P(x) \vee \forall x Q(x)]\!]_v = 0$. Furthermore $[\![\forall x(P(x) \wedge Q(x))]\!]_v = 1$, because $[\![P(x) \wedge Q(x)]\!]_{v[x:u_1]} = 1$, $[\![P(x) \wedge Q(x)]\!]_{v[x:u_2]} = 2$ and $[\![P(x) \wedge Q(x)]\!]_{v[x:u_3]} = 2$.

The conclusion we have reached may not be the one that may we had hoped for. The theorems shows the lack of connection between the lower or upper approximative evaluation.

5 Practical Use of TbPFoL Semantics

As a demonstration to show some practical benefits and weaknesses of the TbP-FoL semantics, now we introduce an interpretation based on an existing relational database.

In the following example, we suppose that the data we are interested in is aggregated into one data table. We also suppose that the aggregated table is relatively huge in both dimensions, so we have not only numerous records but also several attributes. Based on the attribute values, we are able to define interpetations for the one-argument predicates[3]. That kind of interpretation can combine several attribute values being connected to one database record to form semantic values of predicates in the form:

$$f : U \to \{0, 1\} \quad \text{or} \quad f : U \to \{0, 1, 2\}.$$

The predicates are assigned to Boolean functions defined over the table record type. (This approach use the table records as members of the set U.) Partiality may appear when we allow `null` result for the Boolean functions, which behavior is usually acceptable in relational database systems but not necessary to take the advantages of the TbPFoL semantics.

Tools (the members of set $\mathcal{T}$) are collected from the *Pred* set. They determine a logically relevant general partial approximation space. We want to use only the tools during the evaluation of expressions with approximate sentence functors. For this purpose, database records pertaining to the truth set of a tool should be collected into a temporary table, and the connection between the tools and the common predicates must also be discovered. We store the number of members in $P_i \cap T_j$ and the number of members in $T_j \setminus P_i$, where P_i is the truth set of the ith predicate, and T_j is the truth set of the jth tool.

When $P_i \cap T_j \neq \emptyset$, the T_j tool

- belongs to the lower approximation of P_i, if $T_j \setminus P_i = \emptyset$,
- belongs to the upper approximation of P_i, if $T_j \setminus P_i \neq \emptyset$.

The pre-evaluation of tools gives us the ability to omit the evaluation of the f_i functions during the evaluation of the formulae starting with approximate sentence functors.

The following example shows a matrix, which describes the connection between the truth set of P and the truth set of tools T_{11} and T_{12}. The matrix generated from the working example.

$$\begin{pmatrix} |T_{11} \cap P|, & |T_{11} \setminus P|, & |T_{12} \cap P|, & |T_{12} \setminus P| \\ |T_{11} \cap T_{11}|, & |T_{11} \setminus T_{11}|, & |T_{12} \cap T_{11}|, & |T_{12} \setminus T_{11}| \\ |T_{11} \cap T_{12}|, & |T_{11} \setminus T_{12}|, & |T_{12} \cap T_{12}|, & |T_{12} \setminus T_{12}| \end{pmatrix} = \begin{pmatrix} 101, & 231, & 126, & 308 \\ 332, & 0, & 0, & 434 \\ 0, & 332, & 434, & 0 \end{pmatrix}$$

As the first line shows, both tools belong to the upper approximation of P. If there is no another tool selected, the following evaluation results occure:

$$[\![\uparrow \exists x P(x)]\!] = 1, \quad [\![\downarrow \exists x P(x)]\!] = 2, \quad [\![\uparrow \forall x P(x)]\!] = 1, \quad [\![\downarrow \forall x P(x)]\!] = 2$$

Obviously the pre-evaluation process is required, but it is enough to do it once for each different interpretation. Here the partial semantics is necessary in a case of object which doesn't belongs to any tool. Note, that 2 as a null entity which represents truth value gap can appear during the evaluation process. The

---

[3] A working example can be found on the website
http://www.inf.unideb.hu/progcont/tbpfol/

truth value gap e.g. as a result of the lower approximation, like above, can also indicate that on object belongs to the upper, but does not belogs to the lower approximation. It gives the idea most of the proofs earlier.

Acknowledgements. The publication was supported by the TÁMOP–4.2.2.C–11/1/KONV–2012–0001 project. The project has been supported by the European Union, co–financed by the European Social Fund.

Many thanks to the anonymous referees for their valuable suggestions.

The authors are thankful to Davide Ciucci for his insightful comments and suggestions.

References

1. Pawlak, Z.: Rough sets. International Journal of Information and Computer Science 11(5), 341–356 (1982)
2. Pawlak, Z.: Rough Sets: Theoretical Aspects of Reasoning about Data. Kluwer Academic Publishers, Dordrecht (1991)
3. Yao, Y.Y.: On generalizing rough set theory. In: Wang, G., Liu, Q., Yao, Y., Skowron, A. (eds.) RSFDGrC 2003. LNCS (LNAI), vol. 2639, pp. 44–51. Springer, Heidelberg (2003)
4. Düntsch, I., Gediga, G.: Approximation operators in qualitative data analysis. In: de Swart, H., Orłowska, E., Schmidt, G., Roubens, M. (eds.) Theory and Applications of Relational Structures as Knowledge Instruments. LNCS, vol. 2929, pp. 214–230. Springer, Heidelberg (2003)
5. Pawlak, Z., Skowron, A.: Rudiments of rough sets. Information Sciences 177(1), 3–27 (2007)
6. Polkowski, L.: Rough Sets: Mathematical Foundations. Advances in Soft Computing. Physica-Verlag, Heidelberg (2002)
7. Csajbók, Z., Mihálydeák, T.: Partial approximative set theory: A generalization of the rough set theory. International Journal of Computer Information System and Industrial Management Applications 4, 437–444 (2012)
8. Mihálydeák, T.: Partial first-order logic with approximative functors based on properties. In: Li, T., Nguyen, H.S., Wang, G., Grzymala-Busse, J., Janicki, R., Hassanien, A.E., Yu, H. (eds.) RSKT 2012. LNCS, vol. 7414, pp. 514–523. Springer, Heidelberg (2012)

Aristotle's Syllogisms in Logical Semantics Relying on Optimistic, Average and Pessimistic Membership Functions

Tamás Mihálydeák

Department of Computer Science,
Faculty of Informatics, University of Debrecen
Egyetem tér 1, H-4010 Debrecen, Hungary
mihalydeak.tamas@inf.unideb.hu

Abstract. After giving the precise partial first-order logical semantics relying on different membership functions, the notion of decision driven consequence relations with parameters is introduced. Aristotle's valid syllogisms of the first figure are investigated. The author shows what kind of decisions is necessary and how parameters have to be chosen in order that a consequence relation remains valid.

Keywords: Approximation of sets, rough set, partial logic, Aristotle's syllogisms.

1 Introduction

Decision-theoretic rough set models can be considered as the probabilistic extensions of algebraic rough set models. Many papers deal with DTRS based on different systems of theory of rough sets (see, for example, [14]). Three different types of membership functions are introduced: optimistic, pessimistic and average. The main objective of this paper is to show how Aristotle's syllogisms of the first figure can be used when one wants to rely on different membership functions. In order to achieve the goals one has

1. to show that a decision-theoretic rough set model can be based on a very general version of partial approximation spaces by introducing optimistic, average and pessimistic partial membership functions relying on partial approximations of sets;
2. to present a partial first-order logic with precise semantics relying on different partial membership functions;
3. to introduce different notions of logical consequence relations which can be used in order to make clear the consequences of our decisions.

After introducing the general partial approximation space as a generalization of different systems appeared in the theory of rough sets, three different types of partial membership functions are defined. They can be embedded in the semantics of partial first-order logic. At the end different notions of logical consequence are produced in order to investigate Aristotle's syllogisms.

C. Cornelis et al. (eds.): RSCTC 2014, LNAI 8536, pp. 59–70, 2014.

2 Theoretical Background

In the paper [8] a partial first-order logic relying on optimistic, pessimistic and average membership functions was introduced. Some characteristic features of the logical system are the following:

- General approximation spaces (see in [1], [2]) are in the center of its logical semantics. The system of base sets plays a crucial role in the whole process of approximation, because it represents background (or available) knowledge. In many practical cases background knowledge is not total: maybe we have no any (useful) information or knowledge about some objects of universe. In general the system of base sets is not a partition of a given universe and it does not fulfill covering property.
- Relying on standard Pawlakian approximation pair three different partial fuzzy membership function are introduced: optimistic, pessimistic and average ones. The possibilities show different ways of fuzzyfication and the usage of background knowledge can be controlled by membership functions.
- In the definition of semantic rules optimistic, pessimistic and average membership functions are used in order to evaluate formulas. The type of applied membership function is determined on meta level and so its role appears in central semantic notions (as consequence relation, irrefutability, validity).

2.1 General Systems of Tool-Based Approximation of Sets

In the following definition a most fundamental (and very general) notion of an approximation space is given. This core notion serves as the set-theoretical background of semantics of partial first-order logic relying on different partial membership functions.

Definition 1. *The ordered 5-tuple $\langle U, \mathfrak{B}, \mathfrak{D}_{\mathfrak{B}}, \mathsf{l}, \mathsf{u} \rangle$ is a general partial approximation space with a Pawlakian approximation pair if*

1. U *is a nonempty set;*
2. $\mathfrak{B} \subseteq 2^U$, $\mathfrak{B} \neq \emptyset$ *and if $B \in \mathfrak{B}$, then $B \neq \emptyset$;*
3. $\mathfrak{D}_{\mathfrak{B}}$ *is an extension of $\mathfrak{B}$, and it is given by the following inductive definition:*
    (a) $\mathfrak{B} \subseteq \mathfrak{D}_{\mathfrak{B}}$;
    (b) $\emptyset \in \mathfrak{D}_{\mathfrak{B}}$;
    (c) if $D_1, D_2 \in \mathfrak{D}_{\mathfrak{B}}$, then $D_1 \cup D_2 \in \mathfrak{D}_{\mathfrak{B}}$.
4. *the functions l, u form a Pawlakian approximation pair $\langle \mathsf{l}, \mathsf{u} \rangle$, i.e.*
    (a) $\mathsf{l}(S) = \bigcup \mathcal{C}^{\mathsf{l}}(S)$, where $\mathcal{C}^{\mathsf{l}}(S) = \{ B \mid B \in \mathfrak{B} \text{ and } B \subseteq S \}$;
    (b) $\mathsf{u}(S) = \bigcup \mathcal{C}^{\mathsf{u}}(S)$, where $\mathcal{C}^{\mathsf{u}}(S) = \{ B \mid B \in \mathfrak{B} \text{ and } B \cap S \neq \emptyset \}$.

Definition 2. *If $\langle U, \mathfrak{B}, \mathfrak{D}_{\mathfrak{B}}, \mathsf{l}, \mathsf{u} \rangle$ is a general partial approximation space with a Pawlakian approximation pair and $S \subseteq U$, then $\mathsf{b}(S) = \bigcup (\mathcal{C}^{\mathsf{u}}(S) \setminus \mathcal{C}^{\mathsf{l}}(S))$ is the border set of S.*

Informally, the set U is the universe of approximation; $\mathfrak{B}$ is a nonempty set of base sets, it represents our knowledge used in the whole approximation process; $\mathfrak{D}_{\mathfrak{B}}$ (i.e. the set of definable sets) contains not only the base sets, but those which we want to use to approximate any subset of U; the functions l, u (and b) determine the lower and upper approximation (and the border) of any set with the help of representations of our primitive or available concepts/properties. The nature of an approximation pair[1] depends on how to relate the lower and upper approximations of a set to the set itself. A general partial approximation space can be specified by giving some requirements for the base sets.

2.2 Optimistic, Average and Pessimistic Partial Membership Functions

Relying on a given general approximation space $\mathsf{GAS} = \langle U, \mathfrak{B}, \mathfrak{D}_{\mathfrak{B}}, \mathsf{l}, \mathsf{u} \rangle$ three different partial membership functions (μ_s^o for optimistic, μ_s^a for average and μ_s^p for pessimistic) can be introduced. For the sake of simplicity we use a null entity (the number 2) to show that a function is undefined for an object u, i.e. to represent partiality of membership functions. Three different functions (of finite sets of numbers) are used in the definition:

– the function min gives the minimum value of a finite set of numbers;
– the function avg gives the average value of a finite set of numbers, i.e.
  $\mathsf{avg}(\{n_1, \ldots, n_k\}) = \frac{\Sigma_{i=1}^{k} n_i}{k}$;
– the function max gives the maximum value of a finite set of numbers.

Definition 3. *Let U be a finite nonempty set, $S \subseteq U$, $u \in U$, $\mathcal{C}(u) = \{B \mid u \in B, B \in \mathfrak{B}\}$ and $\mathsf{V}(u) = \left\{ \frac{|B \cap S|}{|B|} \mid B \in \mathcal{C}(u) \right\}$.*

$$\mu_S^o(u) = \begin{cases} 1 & \text{if } u \in \mathsf{l}(S) \\ \max(\mathsf{V}(u)) & \text{if } u \in \mathsf{b}(S) \setminus \mathsf{l}(S) \\ 0 & \text{if } u \in \bigcup \mathfrak{B} \setminus \mathsf{u}(S) \\ 2 & \text{otherwise} \end{cases}$$

$$\mu_S^a(u) = \begin{cases} 1 & \text{if } u \in \mathsf{l}(S) \\ \mathsf{avg}(\mathsf{V}(u)) & \text{if } u \in \mathsf{b}(S) \setminus \mathsf{l}(S) \\ 0 & \text{if } u \in \bigcup \mathfrak{B} \setminus \mathsf{u}(S) \\ 2 & \text{otherwise} \end{cases}$$

$$\mu_S^p(u) = \begin{cases} 1 & \text{if } u \in \mathsf{l}(S) \\ \min(\mathsf{V}(u)) & \text{if } u \in \mathsf{b}(S) \setminus \mathsf{l}(S) \\ 0 & \text{if } u \in \bigcup \mathfrak{B} \setminus \mathsf{u}(S) \\ 2 & \text{otherwise} \end{cases}$$

For the sake of simplicity it is useful to introduce the crisp membership function (μ_S^c) for any set and any object.

---

[1] One of the most general notion of weak and strong approximation pairs can be found in Düntsch and Gediga [3].

Definition 4. *If S ($S \subseteq U$) and $u \in U$, then*

$$\mu_S^c(u) = \begin{cases} 1 \ if \ u \in S \\ 0 \ if \ u \in \bigcup \mathfrak{B} \setminus S \\ 2 \ otherwise \end{cases}$$

2.3   Partial First-Order Logic (PFoL) Relying on Different Membership Functions

At first we need a given language of first-order logic, and a finite nonempty set $\mathcal{T}$ of one-argument predicate parameters. Its members are called tools.

Definition 5. *L is a language of PFoL with the set $\mathcal{T}$ of tools, if*

1. $L = \langle LC, Var, Con, Term, \mathcal{T}, Form \rangle$
2. $L^{(1)} = \langle LC, Var, Con, Term, Form \rangle$ *is a language of classical first-order logic;*
3. $\mathcal{T} \subseteq \mathcal{P}(1)$, *where $\mathcal{P}(1)$ is the set of one-argument predicate parameters;*
4. $\mathcal{T}$ *is finite and $\mathcal{T} \neq \emptyset$.*

The members of set $\mathcal{T}$ are called tools, and their semantic values play a crucial role in giving different types of rough membership functions because they serve as the base of generated approximation space.

Definition 6. *Let L be a language of PFoL with the set $\mathcal{T}$ of tools. The ordered pair $\langle U, \varrho \rangle$ is a tool-based interpretation of L, if*

1. *U is a finite nonempty set;*
2. ϱ *is a function such that $Dom(\varrho) = Con$ and*
   (a) *if $a \in \mathcal{N}$ ($\mathcal{N}$ is the set of name parameters), then $\varrho(a) \in U$;*
   (b) *if $p \in \mathcal{P}(0)$ ($\mathcal{P}(0)$ is the set of proposition parameters), then $\varrho(p) \in \{0,1\}$;*
   (c) *if $P \in \mathcal{P}(n)$ ($n = 1, 2, \dots$) ($\mathcal{P}(n)$ is the set of n-argument predicate parameters), then $\varrho(P) \subseteq U^{(n)}$;*
   (d) *if $T \in \mathcal{T}$, then $\varrho(T) \neq \emptyset$.*

In order to give semantic rules we only need the notions of assignment and modified assignment:

Definition 7. *Function v is an assignment relying on the interpretation $\langle U, \varrho \rangle$ if $v : Var \to U$.*

Definition 8. *Let v be an assignment relying on the interpretation $\langle U, \varrho \rangle$, $x \in Var$ and $u \in U$. $v[x : u]$ is a modified assignment of v, if $v[x : u]$ is an assignment, $v[x : u](y) = v(y)$ if $x \neq y$, and $v[x : u](x) = u$.*

The semantic values of tools (the members of set $\mathcal{T}$) determine a general (maybe partial) approximation space with respect to the given interpretation. The generated approximation space is logically relevant in the sense, that it gives the lower and upper approximations (what is more, the different partial membership functions) of any predicate P to be taken into consideration in the definition of semantic rules.

Definition 9. *Let L be a language of PFoL with the set $\mathcal{T}$ of tools and $\langle U, \varrho \rangle$ be a tool-based interpretation of L.*

The ordered 5-tuple

$$\mathsf{GAS}(\mathcal{T}) = \langle \mathcal{PR}(U), \mathcal{B}(\mathcal{T}), \mathfrak{D}_{\mathcal{B}(\mathcal{T})}, \mathsf{l}, \mathsf{u} \rangle$$

is a logically relevant general partial approximation space generated by set $\mathcal{T}$ of tools with respect to the interpretation $\langle U, \varrho \rangle$ if

1. *$\mathcal{PR}(U) = \bigcup_{n=1}^{\infty} U^{(n)}$, where $U^{(1)} = U$, $U^{(n)} = U \times U \times \cdots \times U$;*
2. *$\mathcal{B}(\mathcal{T}) = \bigcup_{n=1}^{\infty} \mathcal{B}_n(\mathcal{T})$ where $\mathcal{B}_n(\mathcal{T}) = \{\varrho(T_1) \times \cdots \times \varrho(T_n) \mid T_i \in \mathcal{T}\}$;*

The semantic values of tools (given by the interpretation) generate the set $\mathcal{B}(\mathcal{T})$. It contains those sets by which the semantic value of any predicate parameter is approximated.

In the semantics of PFol the semantic value of an expression depends on a given interpretation, and a given logically relevant general partial approximation space generated by set of tools with respect to the interpretation. For the sake of simplicity we use a null entity to represent partiality of semantic rules. We use number 0 for falsity, number 1 for truth, numbers greater than 0 and less than 1 for true degree and number 2 for null entity. In many cases, four possibly different semantic values can be given: optimistic, average, pessimistic and crisp ones. The forms of semantic rules are similar in different cases and so the superscript $\star$ can be used to denote one of them ($\star \in \{^o, ^a, ^p, ^c\}$). The semantic value of an expression A is denoted by $[\![A]\!]_v^\star$.

The most important semantic rules are the following:

1. If $P \in P(n)$ $(n \neq 0)$, i.e. P is an n-argument predicate parameter and $t_1, t_2, \ldots, t_n \in Term$, then
   $[\![P(t_1, \ldots, t_n)]\!]_v^\star = \mu_{\varrho(P)}^\star(\langle [\![t_1]\!]_v^\star, \ldots, [\![t_n]\!]_v^\star \rangle)$.
2. If $A \in Form$, then
   $$[\![\neg A]\!]_v^\star = \begin{cases} 2 & \text{if } [\![A]\!]_v^\star = 2 \\ 1 - [\![A]\!]_v^\star & \text{otherwise} \end{cases}$$
3. If $A, B \in Form$, then
   $$[\![(A \wedge B)]\!]_v^\star = \begin{cases} 2 & \text{if } [\![A]\!]_v^\star = 2, \text{ or } [\![B]\!]_v^\star = 2; \\ \min\{[\![A]\!]_v^\star, [\![B]\!]_v^\star\} & \text{otherwise} \end{cases}$$
   $$[\![(A \vee B)]\!]_v^\star = \begin{cases} 2 & \text{if } [\![A]\!]_v^\star = 2, \text{ or } [\![B]\!]_v^\star = 2; \\ \max\{[\![A]\!]_v^\star, [\![B]\!]_v^\star\} & \text{otherwise} \end{cases}$$
   $$[\![(A \supset B)]\!]_v^\star = \begin{cases} 2 & \text{if } [\![A]\!]_v^\star = 2, \text{ or } [\![B]\!]_v^\star = 2; \\ \max\{[\![\neg A]\!]_v^\star, [\![B]\!]_v^\star\} & \text{otherwise} \end{cases}$$
4. If $A \in Form$, $x \in Var$ and $\mathcal{V}(A) = \{[\![A]\!]_{v[x:u]}^\star \mid u \in U, [\![A]\!]_{v[x:u]}^\star \neq 2\}$, then
   $$[\![\forall x A]\!]_v^\star = \begin{cases} 2 & \text{if } \mathcal{V}(A) = \emptyset, \\ \min\{\mathcal{V}(A)\} & \text{otherwise} \end{cases}$$
   $$[\![\exists x A]\!]_v^\star = \begin{cases} 2 & \text{if } \mathcal{V}(A) = \emptyset, \\ \max\{\mathcal{V}(A)\} & \text{otherwise} \end{cases}$$

From the logical point of view, flexibility is the main advantage of defined logical framework: the different membership functions can be used for different formulae. So the consequence relation can rely on possibly different semantic values determined by total, optimistic, average and pessimistic membership functions and so the investigations of different decisions (represented by on different membership functions) are possible.

The notion of models plays a fundamental role in the semantic definition of consequence relation:

Definition 10. *Let L be a language of PFoL with the set $\mathcal{T}$ of tools, and $\Gamma = \langle A_1, A_2, \ldots, A_n \rangle$ be an ordered n-tuple of formulae ($A_1, A_2, \ldots, A_n \in Form$).*

1. *The ordered n-tuple $\Delta = \langle \delta_1, \ldots, \delta_n \rangle$ is a decision type of Γ if $\delta_1, \ldots, \delta_n \in \{o, a, p, c\}$.*
2. *Let $\Delta = \langle \delta_1, \ldots, \delta_n \rangle$ be a decision type Γ. Then*
    (a) *$\langle U, \varrho, v \rangle$ is a Δ-type model of Γ with parameter α ($0 < \alpha \leq 1$), if*
        i. *$\langle U, \varrho \rangle$ is an interpretation of L; v is an assignment relying on $\langle U, \varrho \rangle$;*
        ii. *$[\![A_i]\!]_v^{\delta_i} \neq 2$ for all i ($i = 1, 2, \ldots, n$)*
        iii. *$[\![A_i]\!]_v^{\delta_i} \geq \alpha$ for all i ($i = 1, 2, \ldots, n$).*
    (b) *$\langle U, \varrho, v \rangle$ is a Δ-type partial model of Γ with parameter α ($0 < \alpha \leq 1$) if*
        i. *$\langle U, \varrho \rangle$ is an interpretation of L; v is an assignment relying on $\langle U, \varrho \rangle$;*
        ii. *$[\![A_i]\!]_v^{\delta_i} \geq \alpha$ for all i ($i = 1, 2, \ldots, n$).*

Definition 11. *Let L be a language of PFoL with the set $\mathcal{T}$ of tools, $\Gamma = \langle A_1, A_2, \ldots, A_n \rangle$ be an ordered n-tuple of formulae ($A_1, A_2, \ldots, A_n \in Form$) and $B \in Form$ be a formula.*

1. *$\Delta \to \delta$ is a decision driven consequence type from Γ to B if*
    (a) *Δ is a decision type of Γ;*
    (b) *δ is a decision type of $\{B\}$.*
2. *Let $\Delta \to \delta$ is a decision driven consequence type from Γ to B.*
    (a) *B is a parametrized consequence of Γ driven by $\Delta \to \delta$ with the parameter pair $\langle \alpha, \beta \rangle$ if all Δ-type models of Γ with the parameter α are δ-type models of B with the parameter β ($\Gamma \vDash_{\Delta \to \delta}^{\langle \alpha, \beta \rangle} B$)*
    (b) *B is a partial parametrized consequence of Γ driven by $\Delta \to \delta$ with the parameter pair $\langle \alpha, \beta \rangle$ if all Δ-type partial models of Γ with the parameter α are δ-type partial models of B with the parameter β ($\Gamma \vDash_{\Delta \to \delta}^{p, \langle \alpha, \beta \rangle} B$).*

The introduced notion of decision driven consequence relations may be useful in many practical cases: we can say with respect to available knowledge that we want

- to use the 'core' of a premiss, i.e. the corresponding formula may be evaluated by pessimistic membership function;
- to take a premiss by and large, i.e. the corresponding formula may be evaluated by average membership function;

− to consider a premiss as a possibility, i.e. the corresponding formula may be
evaluated by optimistic membership function.

What is more, the expected level of truth of premisses can be determined. In
these flexible situations the main question is the following: how premisses and
conclusions may be evaluated in order that consequence relations hold. In the
next section Aristotle's syllogisms of first figure are investigated.

3   Aristotle's Syllogisms of the First Figure

It is obvious that $[\![P(x)]\!]^o \geq [\![P(x)]\!]^a \geq [\![P(x)]\!]^p$ with respect to any interpretation and assignment, and so

1. $[\![\forall x P(x)]\!]^o \geq [\![\forall x P(x)]\!]^a \geq [\![\forall x P(x)]\!]^p$,
2. $[\![\exists x P(x)]\!]^o \geq [\![\exists x P(x)]\!]^a \geq [\![\exists x P(x)]\!]^p$.

There is no a determined place of the crisp value of predicate P (i.e. $[\![P(x)]\!]^c$)
in the chain. The pessimistic membership function is the strongest, and the optimistic membership function is the weakest. The average membership function
is in middle. The following notation is used: $p \geq a \geq o$. The first theorem shows
what happens with two standard rules in the logical system relying on different
membership functions.

Theorem 1. *Let P be a one-argument predicate parameter, x be a variable, b
be a name-parameter, $\delta_1, \delta_2 \in \{o, a, p\}$ and $0 < \beta \leq \alpha \leq 1$. Then*

1. $\forall x P(x) \vDash_{\delta_1 \to \delta_2}^{\langle \alpha, \beta \rangle} \exists x P(x)$
2. $\forall x P(x) \vDash_{\delta_1 \to \delta_2}^{par, \langle \alpha, \beta \rangle} P(a)$

Proof. Let $\langle U, \varrho, v \rangle$ be δ_1-type model of $\forall x P(x)$ with parameter α. In this case
$\mathcal{V}(P(x)) \neq \emptyset$. The task is to prove that $\langle U, \varrho, v \rangle$ is a δ_2-type model of $\exists x P(x)$
with parameter β.
  If $[\![\forall x P(x)]\!]^{\delta_1} \neq 2$ and $[\![\forall x P(x)]\!]^{\delta_1} \geq \alpha$, then

$$\min_{u \in \mathcal{V}(P(x))} \{[\![P(x)]\!]^{\delta_1}_{v[x:u]}\} \geq \alpha$$

It means that if $\delta_1 \geq \delta_2$ (and so $[\![P(x)]\!]^{\delta_2}_{v[x:u]} \geq [\![P(x)]\!]^{\delta_1}_{v[x:u]}$)

$$\max_{u \in \mathcal{V}(P(x))} \{[\![P(x)]\!]^{\delta_2}_{v[x:u]}\} \geq \alpha \geq \beta$$

The second proposition of the theorem is obvious from the first one (maybe
that $[\![P(b)]\!]^{\delta_2}_v = 2$).   □

Aristotle's syllogisms represent many typical cases of consequence relations containing classical existential and universal quantifiers and monadic predicates.
There is no enough place to investigate all valid syllogisms, Barbara, Darii,
Celarent and Ferio syllogisms of the first figure are analyzed in order to show
possible answers for the question appeared at the end of previous section.

Theorem 2. *Barbara syllogism*
Let $\delta \in \{o, a, p\}$, $\alpha > 1/2$, $\alpha \geq \beta > 0$ and P, Q, R be predicate parameters. Then

$$\langle \forall x(P(x) \supset Q(x)), \forall x(R(x) \supset P(x)) \rangle \models^{\langle \alpha, \beta \rangle}_{\langle \delta, \delta \rangle \to \delta} \forall x(R(x) \supset Q(x))$$

Proof. Let $\langle U, \varrho, v \rangle$ be $\langle \delta, \delta \rangle$-type model of $\langle \forall x(P(x) \supset Q(x)), \forall x(R(x) \supset P(x)) \rangle$ with parameter α and $V(U) = \mathcal{V}(P(x)) \cap \mathcal{V}(Q(x)) \cap \mathcal{V}(R(x))$. In this case $V(U) \neq \emptyset$. The task is to prove that $\langle U, \varrho, v \rangle$ is a δ-type model of $\forall x(R(x) \supset Q(x))$ with parameter β.

If $[\![\forall x(P(x) \supset Q(x))]\!]^{\delta} \neq 2$ and $[\![\forall x(P(x) \supset Q(x))]\!]^{\delta} \geq \alpha$, then

$$\min_{u \in V(U)} \left\{ \max\{[\![\neg P(x)]\!]^{\delta}_{v[x:u]}, [\![Q(x)]\!]^{\delta}_{v[x:u]}\} \right\} \geq \alpha$$

If $[\![\forall x(R(x) \supset P(x))]\!]^{\delta} \neq 2$ and $[\![\forall x(R(x) \supset P(x))]\!]^{\delta} \geq \alpha$, then

$$\min_{u \in V(U)} \left\{ \max\{[\![\neg R(x)]\!]^{\delta}_{v[x:u]}, [\![P(x)]\!]^{\delta}_{v[x:u]}\} \right\} \geq \alpha$$

1. For all $u \in V(U)$: $1 - [\![P(x)]\!]^{\delta}_{v[x:u]} \geq \alpha$ or $[\![Q(x)]\!]^{\delta}_{v[x:u]} \geq \alpha$.
2. For all $u \in V(U)$: $1 - [\![R(x)]\!]^{\delta}_{v[x:u]} \geq \alpha$ or $[\![P(x)]\!]^{\delta}_{v[x:u]} \geq \alpha$.

Four cases appear for all $u \in V(U)$:

1. $1 - [\![P(x)]\!]^{\delta}_{v[x:u]} \geq \alpha$ and $1 - [\![R(x)]\!]^{\delta}_{v[x:u]} \geq \alpha$ and so
   $\max\{[\![\neg R(x)]\!]^{\delta}_{v[x:u]}, [\![Q(x)]\!]^{\delta}_{v[x:u]}\} \geq \alpha$ i.e. $[\![\forall x(R(x) \supset Q(x))]\!]^{\delta} \geq \alpha \geq \beta$;
2. $1 - [\![P(x)]\!]^{\delta}_{v[x:u]} \geq \alpha$ and $[\![P(x)]\!]^{\delta}_{v[x:u]} \geq \alpha$: if $\alpha > 1/2$, then it is impossible;
3. $[\![Q(x)]\!]^{\delta}_{v[x:u]} \geq \alpha$ and $1 - [\![R(x)]\!]^{\delta}_{v[x:u]} \geq \alpha$ and so
   $\max\{[\![\neg R(x)]\!]^{\delta}_{v[x:u]}, [\![Q(x)]\!]^{\delta}_{v[x:u]}\} \geq \alpha$ i.e. $[\![\forall x(R(x) \supset Q(x))]\!]^{\delta} \geq \alpha \geq \beta$;
4. $[\![Q(x)]\!]^{\delta}_{v[x:u]} \geq \alpha$ and $[\![P(x)]\!]^{\delta}_{v[x:u]} \geq \alpha$ and so
   $\max\{[\![\neg R(x)]\!]^{\delta}_{v[x:u]}, [\![Q(x)]\!]^{\delta}_{v[x:u]}\} \geq \alpha$ i.e. $[\![\forall x(R(x) \supset Q(x))]\!]^{\delta} \geq \alpha \geq \beta$.   □

Remark 1. It is obvious from the proof that in the case of Barbara syllogism the same membership function has to be used for premisses and conclusion.

Theorem 3. *Celarent syllogism*
Let $\delta \in \{o, a, p\}$, $\alpha > 1/2$, $\alpha \geq \beta > 0$ and P, Q, R be predicate parameters. Then

$$\langle \forall x(P(x) \supset \neg Q(x)), \forall x(R(x) \supset P(x)) \rangle \models^{\langle \alpha, \beta \rangle}_{\langle \delta, \delta \rangle \to \delta} \forall x(R(x) \supset \neg Q(x))$$

Proof. Let $\langle U, \varrho, v \rangle$ be $\langle \delta, \delta \rangle$-type model of
$$\langle \forall x(P(x) \supset \neg Q(x)), \forall x(R(x) \supset P(x)) \rangle$$
with parameter α and $V(U) = \mathcal{V}(P(x)) \cap \mathcal{V}(Q(x)) \cap \mathcal{V}(R(x))$. In this case $V(U) \neq \emptyset$. The task is to prove that $\langle U, \varrho, v \rangle$ is a δ-type model of $\forall x(R(x) \supset \neg Q(x))$ with parameter β.

If $[\![\forall x(P(x) \supset \neg Q(x))]\!]^{\delta} \neq 2$ and $[\![\forall x(P(x) \supset \neg Q(x))]\!]^{\delta} \geq \alpha$, then

$$\min_{u \in V(U)} \left\{ \max\{[\![\neg P(x)]\!]^{\delta}_{v[x:u]}, [\![\neg Q(x)]\!]^{\delta}_{v[x:u]}\} \right\} \geq \alpha$$

If $[\![\forall x(R(x) \supset P(x))]\!]^\delta \neq 2$ and $[\![\forall x(R(x) \supset P(x))]\!]^\delta \geq \alpha$, then

$$\min_{u \in V(U)} \left\{ \max\{[\![\neg R(x)]\!]^\delta_{v[x:u]}, [\![P(x)]\!]^\delta_{v[x:u]}\} \right\} \geq \alpha$$

1. For all $u \in V(U)$: $1 - [\![P(x)]\!]^\delta_{v[x:u]} \geq \alpha$ or $1 - [\![Q(x)]\!]^\delta_{v[x:u]} \geq \alpha$.
2. For all $u \in V(U)$: $1 - [\![R(x)]\!]^\delta_{v[x:u]} \geq \alpha$ or $[\![P(x)]\!]^\delta_{v[x:u]} \geq \alpha$.

Four cases appear for all $u \in V(U)$:

1. $1 - [\![P(x)]\!]^\delta_{v[x:u]} \geq \alpha$ and $1 - [\![R(x)]\!]^\delta_{v[x:u]} \geq \alpha$ and so
   $\max\{[\![\neg R(x)]\!]^\delta_{v[x:u]}, [\![\neg Q(x)]\!]^\delta_{v[x:u]}\} \geq \alpha$ i.e. $[\![\forall x(R(x) \supset Q(x))]\!]^\delta \geq \alpha \geq \beta$;
2. $1 - [\![P(x)]\!]^\delta_{v[x:u]} \geq \alpha$ and $[\![P(x)]\!]^\delta_{v[x:u]} \geq \alpha$: if $\alpha > 1/2$, then it is impossible;
3. $1 - [\![Q(x)]\!]^\delta_{v[x:u]} \geq \alpha$ and $1 - [\![R(x)]\!]^\delta_{v[x:u]} \geq \alpha$ and so
   $\max\{[\![\neg R(x)]\!]^\delta_{v[x:u]}, [\![\neg Q(x)]\!]^\delta_{v[x:u]}\} \geq \alpha$ i.e. $[\![\forall x(R(x) \supset \neg Q(x))]\!]^\delta \geq \alpha \geq \beta$;
4. $1 - [\![Q(x)]\!]^\delta_{v[x:u]} \geq \alpha$ and $[\![P(x)]\!]^\delta_{v[x:u]} \geq \alpha$ and so
   $\max\{[\![\neg R(x)]\!]^\delta_{v[x:u]}, [\![\neg Q(x)]\!]^\delta_{v[x:u]}\} \geq \alpha$ i.e. $[\![\forall x(R(x) \supset \neg Q(x))]\!]^\delta \geq \alpha \geq \beta$.

$\square$

Theorems concerning Barbara and Celarent syllogisms and their proofs show that in order to get valid consequence relations one has to

1. evaluate premises and conclusion by using the same (optimistic, average or pessimistic) membership function and so there is no real freedom for making different decisions concerning two premisses and the conclusion;
2. suppose that the level of truth of premises is greater than $1/2$, i.e. the premises have to be closer to truth then falsity.

Theorem 4. *Darii syllogism*
Let $\delta_1, \delta_2 \in \{o, a, p\}$, $\delta_1 \geq \delta_2$ $\alpha > 1/2$, $\alpha \geq \beta > 0$ and P, Q, R be predicate parameters. Then

$$\langle \forall x(P(x) \supset Q(x)), \exists x(R(x) \wedge P(x)) \rangle \vDash^{\langle \alpha, \beta \rangle}_{\langle \delta_2, \delta_1 \rangle \to \delta_2} \exists x(R(x) \wedge Q(x))$$

Proof. Let $\langle U, \varrho, v \rangle$ be $\langle \delta_2, \delta_1 \rangle$-type model of $\langle \forall x(P(x) \supset Q(x)), \exists x(R(x) \wedge P(x)) \rangle$ with parameter α and $V(U) = \mathcal{V}(P(x)) \cap \mathcal{V}(Q(x)) \cap \mathcal{V}(R(x))$. In this case $V(U) \neq \emptyset$. The task is to prove that $\langle U, \varrho, v \rangle$ is a δ_2-type model of $\exists x(R(x) \wedge Q(x))$ with parameter β.

If $[\![\forall x(P(x) \supset Q(x))]\!]^{\delta_2} \neq 2$ and $[\![\forall x(P(x) \supset Q(x))]\!]^{\delta_2} \geq \alpha$, then

$$\min_{u \in V(U)} \left\{ \max\{[\![\neg P(x)]\!]^{\delta_2}_{v[x:u]}, [\![Q(x)]\!]^{\delta_2}_{v[x:u]}\} \right\} \geq \alpha$$

If $[\![\exists x(R(x) \wedge P(x))]\!]^{\delta_1} \neq 2$ and $[\![\exists x(R(x) \wedge P(x))]\!]^{\delta_1} \geq \alpha$, then

$$\max_{u \in V(U)} \left\{ \min\{[\![R(x)]\!]^{\delta_1}_{v[x:u]}, [\![P(x)]\!]^{\delta_1}_{v[x:u]}\} \right\} \geq \alpha$$

1. For all $u \in V(U)$: $1 - [\![P(x)]\!]^{\delta_2}_{v[x:u]} \geq \alpha$ or $[\![Q(x)]\!]^{\delta_2}_{v[x:u]} \geq \alpha$.
2. There is a $u \in V(U)$: $[\![R(x)]\!]^{\delta_1}_{v[x:u]} \geq \alpha$ and $[\![P(x)]\!]^{\delta_1}_{v[x:u]} \geq \alpha$.

Two cases appear for the fixed $u \in V(U)$ given by the second premiss:

1. $1 - [\![P(x)]\!]^{\delta_2}_{v[x:u]} \geq \alpha$, $[\![R(x)]\!]^{\delta_1}_{v[x:u]} \geq \alpha$ and $[\![P(x)]\!]^{\delta_1}_{v[x:u]} \geq \alpha$. If $\alpha > 1/2$, then it is impossible because $[\![P(x)]\!]^{\delta_2}_{v[x:u]} \geq [\![P(x)]\!]^{\delta_1}_{v[x:u]}$.
2. $[\![Q(x)]\!]^{\delta_2}_{v[x:u]} \geq \alpha$, $[\![R(x)]\!]^{\delta_1}_{v[x:u]} \geq \alpha$ and $[\![P(x)]\!]^{\delta_1}_{v[x:u]} \geq \alpha$. $\delta_1 \geq \delta_2$, and so $[\![R(x)]\!]^{\delta_2}_{v[x:u]} \geq \alpha$.
   $\min\{[\![R(x)]\!]^{\delta_2}_{v[x:u]}, [\![Q(x)]\!]^{\delta_2}_{v[x:u]}\} \geq \alpha$ i.e. $[\![\exists x(R(x) \wedge Q(x))]\!]^{\delta_2} \geq \alpha \geq \beta$.   □

The theorem concerning Darii syllogisms and its proof show that the validity of the consequence relation requires that

1. the decision for the first premiss can not be stronger than the decision for the second one;
2. the decisions for the first premiss and the conclusion have to be the same;
3. the first premiss plays a more important role than the second one.

Theorem 5. *Ferio syllogism*
Let $\delta_1, \delta_2 \in \{o, a, p\}$, $\delta_2 \geq \delta_1$ $\alpha > 1/2$, $\alpha \geq \beta > 0$ and P, Q, R be predicate parameters. Then

$$\langle \forall x(P(x) \supset \neg Q(x)), \exists x(R(x) \wedge P(x))\rangle \vDash^{\langle \alpha, \beta \rangle}_{\langle \delta_1, \delta_2 \rangle \rightarrow \delta_2} \exists x(R(x) \wedge \neg Q(x))$$

Proof. Let $\langle U, \varrho, v \rangle$ be $\langle \delta_1, \delta_2 \rangle$-type model of $\langle \forall x(P(x) \supset \neg Q(x)), \exists x(R(x) \wedge P(x))\rangle$ with parameter α and $V(U) = \mathcal{V}(P(x)) \cap \mathcal{V}(Q(x)) \cap \mathcal{V}(R(x))$. In this case $V(U) \neq \emptyset$. The task is to prove that $\langle U, \varrho, v \rangle$ is a δ_2-type model of $\exists x(R(x) \wedge \neg Q(x))$ with parameter β.
   If $[\![\forall x(P(x) \supset \neg Q(x))]\!]^{\delta_1} \neq 2$ and $[\![\forall x(P(x) \supset \neg Q(x))]\!]^{\delta_1} \geq \alpha$, then

$$\min_{u \in V(U)} \left\{ \max\{[\![\neg P(x)]\!]^{\delta_1}_{v[x:u]}, [\![\neg Q(x)]\!]^{\delta_1}_{v[x:u]}\} \right\} \geq \alpha$$

If $[\![\exists x(R(x) \wedge P(x))]\!]^{\delta_2} \neq 2$ and $[\![\exists x(R(x) \wedge P(x))]\!]^{\delta_2} \geq \alpha$, then

$$\max_{u \in V(U)} \left\{ \min\{[\![R(x)]\!]^{\delta_2}_{v[x:u]}, [\![P(x)]\!]^{\delta_2}_{v[x:u]}\} \right\} \geq \alpha$$

1. For all $u \in V(U)$: $1 - [\![P(x)]\!]^{\delta_1}_{v[x:u]} \geq \alpha$ or $1 - [\![Q(x)]\!]^{\delta_1}_{v[x:u]} \geq \alpha$.
2. There is a $u \in V(U)$: $[\![R(x)]\!]^{\delta_2}_{v[x:u]} \geq \alpha$ and $[\![P(x)]\!]^{\delta_2}_{v[x:u]} \geq \alpha$.

Two cases appear at the fixed $u \in V(U)$ given by the second premiss:

1. $1 - [\![P(x)]\!]^{\delta_1}_{v[x:u]} \geq \alpha$, $[\![R(x)]\!]^{\delta_2}_{v[x:u]} \geq \alpha$ and $[\![P(x)]\!]^{\delta_2}_{v[x:u]} \geq \alpha$. If $\alpha > 1/2$, then it is impossible because $[\![P(x)]\!]^{\delta_1}_{v[x:u]} \geq [\![P(x)]\!]^{\delta_2}_{v[x:u]}$.

2. $1 - [\![Q(x)]\!]^{\delta_1}_{v[x:u]} \geq \alpha$, $[\![R(x)]\!]^{\delta_2}_{v[x:u]} \geq \alpha$ and $[\![P(x)]\!]^{\delta_2}_{v[x:u]} \geq \alpha$.
   $\delta_2 \geq \delta_1$, and so $[\![Q(x)]\!]^{\delta_1}_{v[x:u]} \geq [\![Q(x)]\!]^{\delta_2}_{v[x:u]}$. Therefore $1 - [\![Q(x)]\!]^{\delta_2}_{v[x:u]} \geq \alpha$.
   $\min\{[\![R(x)]\!]^{\delta_2}_{v[x:u]}, [\![\neg Q(x)]\!]^{\delta_2}_{v[x:u]}\} \geq \alpha$ i.e. $[\![\exists x(R(x) \wedge \neg Q(x))]\!]^{\delta_2} \geq \alpha \geq \beta$. $\square$

The theorem concerning Ferio syllogisms and its proof show that the validity of the consequence relation requires that

1. the decision for the first premiss can not be stronger than the decision for the second one;
2. the decisions for the second premiss and the conclusion have to be the same;
3. the second premiss plays a more important role than the first one.

4   Conclusion

After giving the precise first-order logical semantics relying on different membership function, the notion of decision driven consequence relations was introduced. Aristotle's valid syllogisms of the first figure were investigated. The author showed what kind of decisions is necessary and how parameters can be chosen in order that a consequence relation remains valid. A general observation: in any decision driven consequence of Aristotle's syllogisms of the first figure the parameters which give the level of truth of premisses have to be greater than $1/2$, therefore there is no valid consequence relation with parameters less than or equal to $1/2$. It means that in order to say something about the conclusion the premisses have to be closer to truth than to falsity. Some results are unexpected, and indicate the necessity of further research dealing with other syllogisms. The other direction is to investigate intermediate quantifiers (most, many, etc.) examined in fuzzy logical systems.

Acknowledgements. The publication was supported by the TÁMOP–4.2.2.C–11/1/KONV–2012–0001 project. The project has been supported by the European Union, co-financed by the European Social Fund.

Many thanks to the anonymous referees for their valuable suggestions.

The author is thankful to Davide Ciucci for his insightful comments and suggestions.

References

1. Csajbók, Z., Mihálydeák, T.: Partial approximative set theory: A generalization of the rough set theory. International Journal of Computer Information System and Industrial Management Applications 4, 437–444 (2012)
2. Csajbók, Z., Mihálydeák, T.: A General Set Theoretic Approximation Framework. In: Greco, S., Bouchon-Meunier, B., Coletti, G., Fedrizzi, M., Matarazzo, B., Yager, R.R. (eds.) IPMU 2012, Part I. CCIS, vol. 297, pp. 604–612. Springer, Heidelberg (2012)

3. Düntsch, I., Gediga, G.: Approximation operators in qualitative data analysis. In: de Swart, H., Orłowska, E., Schmidt, G., Roubens, M. (eds.) Theory and Applications of Relational Structures as Knowledge Instruments. LNCS, vol. 2929, pp. 214–230. Springer, Heidelberg (2003)

4. Lin, T.Y., Liu, Q.: First-order rough logic I: approximate reasoning via rough sets. Fundamenta Informaticae 27(23), 7–154 (1996)

5. Lin, T.Y., Liu, Q.: First order rough logic-revisited. In: Zhong, N., Skowron, A., Ohsuga, S. (eds.) RSFDGrC 1999. LNCS (LNAI), vol. 1711, pp. 276–284. Springer, Heidelberg (1999)

6. Mihálydeák, T.: Partial first–order logical semantics based on approximations of sets. In: Cintula, P., Ju, S., Vita, M. (eds.) Non-Classical Modal and Perdicate Logics 2011, Guangzhou (Canton), China, F solutions, Prague, pp. 85–90 (2011)

7. Mihálydeák, T.: Partial first-order logic with approximative functors based on properties. In: Li, T., Nguyen, H.S., Wang, G., Grzymala-Busse, J., Janicki, R., Hassanien, A.E., Yu, H. (eds.) RSKT 2012. LNCS, vol. 7414, pp. 514–523. Springer, Heidelberg (2012)

8. Mihálydeák, T.: Partial first-order logic relying on optimistic pessimistic and average partial membership functions. In: Pasi, G., Montero, J., Ciucci, D. (eds.) Proceedings of the 8th Conference of the European Society for Fuzzy Logic and Technology (EUSFLAT 2013) (2013) ISBN (on-line): 978-90786-77-78-9, doi:10.2991/eusflat.2013.53

9. Pawlak, Z.: Rough sets. International Journal of Information and Computer Science 11(5), 341–356 (1982)

10. Pawlak, Z.: Rough Sets: Theoretical Aspects of Reasoning about Data. Kluwer Academic Publishers, Dordrecht (1991)

11. Pawlak, Z., Skowron, A.: Rudiments of rough sets. Information Sciences 177(1), 3–27 (2007)

12. Polkowski, L.: Rough Sets: Mathematical Foundations. Advances in Soft Computing. Physica-Verlag, Heidelberg (2002)

13. Yao, Y.Y.: On generalizing rough set theory. In: Wang, G., Liu, Q., Yao, Y., Skowron, A. (eds.) RSFDGrC 2003. LNCS (LNAI), vol. 2639, pp. 44–51. Springer, Heidelberg (2003)

14. Yao, Y.Y.: Decision-theoretic rough set models. In: Yao, J., Lingras, P., Wu, W.-Z., Szczuka, M.S., Cercone, N.J., Ślęzak, D. (eds.) RSKT 2007. LNCS (LNAI), vol. 4481, pp. 1–12. Springer, Heidelberg (2007)

Fuzzy Sets for a Declarative Description of Multi-adjoint Logic Programming

Ginés Moreno, Jaime Penabad, and Carlos Vázquez

University of Castilla-La Mancha
Faculty of Computer Science Engineering
02071, Albacete, Spain
{Gines.Moreno,Jaime.Penabad,Carlos.Vazquez}@uclm.es

Abstract. A powerful research line in the design of declarative languages consists in the introduction of expressive resources with a fuzzy taste on their cores, in order to provide comfortable computational constructs for easily solving real-world scientific/engineering problems. Into the fuzzy logic programming arena, the so-called *multi-adjoint approach* (MALP in brief) has emerged as an interesting paradigm for which our research group has developed during the last years the $\mathcal{FLOPER}$ programming environment and the *FuzzyXPath* application in the field of the semantic web. Since the practicality of declarative languages is strongly dependent of their theoretical foundations, here we focus on topics related with the declarative semantics of the MALP framework. So, under an innovative point of view relying on fuzzy sets theory, in this paper we re-formulate in a very simple and elegant way our original model theory-based notions of *least fuzzy Herbrand model* and *(fuzzy) correct answer*. Apart for simplifying the proofs relating these concepts, our results are nicely strengthened with homologous ones in the field of pure logic programming, but largely surpassing them thanks to the fuzzy dimension of the MALP language.

Keywords: Fuzzy sets and fuzzy logic, Fuzzy logic programming, Fuzzy Herbrand model, Fuzzy correct answers, Soundness, Fuzzy information systems.

1 Introduction

There exist a lot of contributions in the specialized literature related to fuzzy logic programming which pay attention to declarative (fix-point, model-theoretic, etc.) semantics which surprisingly make not explicit use of fuzzy sets. In this paper we provide a declarative description, based on fuzzy sets, of the least Herbrand model and correct answer for MALP programs.

In what follows, we present a short summary of the main features of our language (we refer the reader to [8,9,10] for a complete formulation, including completeness and other correctness properties). We work with a first order language, $\mathcal{L}$, containing variables, function symbols, predicate symbols, constants,

C. Cornelis et al. (eds.): RSCTC 2014, LNAI 8536, pp. 71–82, 2014.

quantifiers and several (arbitrary) connectives to increase language expressiveness. In our fuzzy setting, we use *implication connectives* ($\leftarrow_1, \leftarrow_2, \ldots, \leftarrow_m$) and also other connectives: *conjunctions* (denoted by $\wedge_1, \wedge_2, \ldots, \wedge_k$), *disjunctions* ($\vee_1, \vee_2, \ldots, \vee_l$)[1] and *aggregators* (usually denoted by $@_1, @_2, \ldots, @_n$) which are used to combine/propagate truth values through the rules. The general definition of an n-ary aggregator connective $@$ (that extends conjunctions, disjunctions) states for its truth function $\dot{@}$, that $\dot{@} : L^n \to L$ is required to be monotone and fulfills $\dot{@}(\top, \ldots, \top) = \top$, $\dot{@}(\bot, \ldots, \bot) = \bot$[2]. Although the connectives $\wedge_i$, $\vee_i$ and $@_i$ are binary operators, we usually generalize them as functions with an arbitrary number of arguments.

Additionally, our language $\mathcal{L}$ contains the elements of a *multi-adjoint lattice*, $(L, \leq, \leftarrow_1, \&_1, \ldots, \leftarrow_n, \&_n)$ (see Definition 3), equipped with a collection of adjoint pairs ($\leftarrow_i, \&_i$), where each $\&_i$ is a conjunctor intended to the evaluation of *modus ponens*.

A *rule* is a logic formula $H \leftarrow_i \mathcal{B}$, where H is an atomic formula (called the *head*) and $\mathcal{B}$ (which is called the *body*) is a formula built from atomic formulas $B_1, \ldots, B_n$ ($n \geq 0$), truth values of L and conjunctions, disjunctions and aggregators. Rules with an empty body are called *facts*. A *goal* is a body submitted as a query to the system. Variables in a rule are assumed to be governed by universal quantifier and in a goal by existential quantifier. A *multi-adjoint formula* is a rule or a goal. A *multi-adjoint logic program* $\mathcal{P}$ is a set of pairs $\mathcal{R} : \langle R; v \rangle$, where R is a (logic) rule and v is a *truth degree* (a value of L) expressing the confidence which the user of the system has in the truth of the R.

In order to describe the procedural semantics of the multi-adjoint logic language, in the following we denote by $\mathcal{C}[A]$ a formula where A is a sub-expression (usually an atom) which occurs in the –possibly empty– context $\mathcal{C}[]$ whereas $\mathcal{C}[A/A']$ means the replacement of A by A' in context $\mathcal{C}[]$. Moreover, $Var(s)$ denotes the set of distinct variables occurring in the syntactic object s, $\theta[Var(s)]$ refers to the substitution obtained from θ by restricting its domain to $Var(s)$ and $mgu(E)$ denotes the *most general unifier* of an equation set E. In the next definition, we always consider that A is the selected atom in goal $\mathcal{Q}$ and L is the multi-adjoint lattice associated to $\mathcal{P}$.

Definition 1 (Admissible Steps). *Let $\mathcal{Q}$ be a goal and let σ be a substitution. The pair $\langle \mathcal{Q}; \sigma \rangle$ is a state. Given a program $\mathcal{P}$, an* admissible computation *is formalized as a state transition system, whose transition relation $\to_{AS}$ is the smallest relation satisfying the following* admissible rules:

1) $\langle \mathcal{Q}[A]; \sigma \rangle \to_{AS} \langle (\mathcal{Q}[A/v \& _i \mathcal{B}]) \theta; \sigma \theta \rangle$ *if $\theta = mgu(\{H = A\})$, $\langle H \leftarrow_i \mathcal{B}; v \rangle$ in $\mathcal{P}$ and $\mathcal{B}$ is not empty.*

2) $\langle \mathcal{Q}[A]; \sigma \rangle \to_{AS} \langle (\mathcal{Q}[A/v]) \theta; \sigma \theta \rangle$ *if $\theta = mgu(\{H = A\})$, $\langle H \leftarrow_i; v \rangle$ in $\mathcal{P}$.*

3) $\langle \mathcal{Q}[A]; \sigma \rangle \to_{AS} \langle (\mathcal{Q}[A/\bot]); \sigma \rangle$ *if there is no rule in $\mathcal{P}$ whose head unifies with A (this case copes with possible unsuccessful branches).*

---

[1] We assume that $\wedge_i$ is a *t*-norm, $\vee_i$ is a *t*-conorm, $\leftarrow_i$ is a implication, as conceived in [15].

[2] L is a lattice according to the later Definition 3 and $\top = sup(L), \bot = inf(L)$.

Definition 2 (Admissible Derivation). *Let $\mathcal{P}$ be a program with an associated multi-adjoint lattice $(L, \leq)$ and let $\mathcal{Q}$ be a goal. An admissible derivation $\langle \mathcal{Q}; id \rangle \rightarrow^*_{AS} \langle \mathcal{Q}'; \theta \rangle$ is an arbitrary sequence of admissible steps.*

When $\mathcal{Q}'$ is a formula not containing atoms and $r \in L$ is the result of interpreting $\mathcal{Q}'$ in $(L, \leq)$, the pairs $\langle \mathcal{Q}'; \sigma \rangle$ and $\langle r; \sigma \rangle$, where $\sigma = \theta[Var(\mathcal{Q})]$, are called admissible computed answer (a.c.a.) and fuzzy computed answer (f.c.a.), respectively (see [4] for details).

Moreover, in the MALP framework [10,8,9,7], each program has its own associated multi-adjoint lattice, that we define in the following, and each program rule is "weighted" with an element of this one.

Definition 3. *A multi-adjoint lattice is a tuple $(L, \leq, \leftarrow_1, \&_1, \ldots, \leftarrow_n, \&_n)$ such that:*

   i) *$(L, \leq)$ is a complete lattice, i.e., for all $S \subset L$, exist $\inf(S)$ and $\sup(S)$[3].*
   ii) *$(\leftarrow_i, \&_i)$ is an adjoint pair in $(L, \leq)$, namely:*
      1) *$\&_i$ is increasing in both arguments, for all i, $i \in \{1, \ldots, n\}$.*
      2) *$\leftarrow_i$ is increasing in the first argument and decreasing in the second, for all i.*
      3) *$x \leq (y \leftarrow_i z)$ iff $(x \&_i z) \leq y$, for any $x, y, z \in L$ (adjoint property).*
   iii) *$\top \&_i v = v \&_i \top = v$, for all $v \in L, i \in \{1, \ldots, n\}$, where $\top = sup(L)$.*

We refer the reader to [13] where we focus on two relevant mathematical concepts for this kind of domains useful for evaluating multi-adjoint logic programs, and, on the one side, we adapt the classical notion of *Dedekind-MacNeille completion* in order to relax some usual hypothesis on such kind of ordered sets.

The structure of this paper is as follows. The notion of least fuzzy Herbrand model by using fuzzy sets is presented in Section 2. Next, in Section 3 we focus on fuzzy correct answers expressed again in terms of fuzzy sets and moreover, we then prove the soundness property of the framework. Section 4 summarizes some preliminary results of logical consequences after being reformulated by means of fuzzy sets. Finally, Section 5 concludes with our on-going work.

2   Fuzzy Sets and Least Fuzzy Herbrand Model

The concept of fuzzy set, due to [21], frequently occurs when we tend to organize, summarize and generalize knowledge about objects [16]. On this concept is based the theory of uncertainty with classic references on fuzzy logic programming [14,17,18,19,20].

In this section, we use the theory of fuzzy sets in order to define, for the first time in literature, the notion of least fuzzy Herbrand model as a certain fuzzy subset of the Herbrand base. We start the development of contents with two basic notions, namely, the concept of fuzzy set and the one of L-fuzzy set.

---

[3] Then, it is a bounded lattice, that is, it has bottom and top elements, denoted by $\perp$ and $\top$, respectively.

Definition 4. *[16] A fuzzy set A of a (crisp or ordinary) set $\mathcal{U}$, may be represented as a set of ordered pairs with first component $x \in \mathcal{U}$ and second component its degree of membership $\mu_A(x)$*[4]*, that is, $A = \{x|\mu_A(x) : \mu_A(x) \neq 0, x \in \mathcal{U}\}$, where the map $\mu_A : \mathcal{U} \to [0,1]$ is called the membership function of A.*

Thus, the fuzzy set A is characterized by function μ_A. For every $x \in \mathcal{U}$, $\mu_A(x) \in [0,1]$ is a real number that describes the *degree of membership of x in A*. Also, if we observe that a ordinary set $A \subset \mathcal{U}$ is determined by the indicator function or *characteristic function* χ_A,

$$\chi_A : \mathcal{U} \to \{0,1\}, \quad \chi_A(x) = \begin{cases} 1, \text{ if } x \in A \\ 0, \text{ if } x \notin A \end{cases}$$

and, since the function μ_A is a generalization of the function χ_A, a fuzzy set is a generalization of the concept of an ordinary set or the notion of crisp set is extended by the corresponding notion of fuzzy set.

Given A, B fuzzy sets of an universe $\mathcal{U}$, A is said included in B (A is a subset of B) if, and only if, the membership function of A is less than that of B, that is, $A \subset B \Leftrightarrow \mu_A(x) \leq \mu_B(x), \forall x \in \mathcal{U}$.

If, in the above definition, we use a complete lattice L instead of interval $[0,1]$, then it arises the following concept of L-fuzzy set.

Definition 5. *[15] Let $(L, \leq)$ be a complete lattice. An L-fuzzy set A of an universe $\mathcal{U}$, is defined by the membership function $\mu_A : \mathcal{U} \to L$.*

In particular, we are interested in expressing the Herbrand base also as a L-fuzzy set, that is, if $\mathcal{B_P} = \{A_1, \ldots, A_n, \ldots\}$ is the (crisp) Herbrand base of $\mathcal{P}$, we denote by $\mathcal{B}_\mathcal{P}^L = \{A_1|\top, \ldots, A_n|\top, \ldots\}$ the fuzzy Herbrand base and we have $\mu_{\mathcal{B}_\mathcal{P}^L} : \mathcal{F} \to L$ is such that $\mu_{\mathcal{B}_\mathcal{P}^L}(A) = \top = sup(L)$, if $A = A_i$, for any i, and $\mu_{\mathcal{B}_\mathcal{P}^L}(A) = \bot$, otherwise. Here, and thereafter, $\mathcal{F}$ denotes the set of all the formulae of the multi-adjoint language, namely, the set of all formulae generated by the set of symbols of a given multi-adjoint logic program $\mathcal{P}$.

In what follows we formulate, in an original way, the notion of fuzzy Herbrand model conceived as L-fuzzy set of the Herbrand base of the multi-adjoint program.

Definition 6. *A fuzzy Herbrand interpretation*[5] *$\mathcal{I}$ is a L-fuzzy set of the universe $\mathcal{B_P}$ or, equivalently, a map $\mu_\mathcal{I} : \mathcal{B_P} \to L$ (in fact, $\mu_\mathcal{I}$ is the membership function of L-fuzzy set), where $\mathcal{B_P}$ is the Herbrand base of $\mathcal{P}$ and $(L, \leq)$ is the multi-adjoint lattice associated to $\mathcal{P}$.*

Indeed, the above function $\mu_\mathcal{I}$ can be extended in a natural way to the set of all formulas $\mathcal{F}$. In particular, for every (closed) formula $A \in \mathcal{F}$, $\mu_\mathcal{I}(A) = inf_\xi\{\mu_\mathcal{I}(A\xi) : A\xi$ is a ground instance of A$\}$.

---

[4] We follow the notation due to [21] expressing this pair by $x|\mu_A(x)$. It is customary to confuse the predicate $A(x)$ with the degree of membership $\mu_A(x)$, we prefer to explicitly distinguish these two concepts.

[5] We will also say Herbrand interpretation.

Definition 7. *A fuzzy Herbrand interpretation* $\mathcal{I}$ *satisfies (or is* Herbrand model *of) a rule* $\mathcal{R}_i : \langle R_i; \alpha_i \rangle$ *if, and only if,* $\alpha_i \leq \mu_{\mathcal{I}}(\mathcal{R}_i)$. *An Herbrand interpretation* $\mathcal{I}$ *is* Herbrand model of $\mathcal{P}$ *iff all rules in* $\mathcal{P}$ *are satisfied by* $\mathcal{I}$.

Obviously, if $\mathcal{I}$ is a Herbrand model of $\mathcal{P}$, we have $\mu_{\mathcal{I}}(A) \leq \top$, for all $A \in \mathcal{B}_{\mathcal{P}}^L$. Then, using Definition 6, $\mathcal{I}$ is a fuzzy subset of the Herbrand base $\mathcal{B}_{\mathcal{P}}^L$.

Let $\mathcal{H}^L$ be the set of Herbrand interpretations whose order, induced from the order of L, is given by $\mathcal{I}_j \subset \mathcal{I}_k \iff \mu_{\mathcal{I}_j}(F) \leq \mu_{\mathcal{I}_k}(F)$, for all formula $F \in \mathcal{F}$. It is easy to check that $(\mathcal{H}^L, \subset)$ inherits the structure of complete lattice from the multi-adjoint lattice $(L, \leq)$. Also, note that $\mathcal{H}^L$ is a set of L-fuzzy sets of universe $\mathcal{F}$.

It is important to observe that, using L-fuzzy sets, the least fuzzy Herbrand model for multi-adjoint logic programing, can be characterized by the following definition, that is, exactly the same terms as expressed in [6] for pure logic programming.

Definition 8. *Let* $\mathcal{P}$ *be a multi-adjoint logic program with associated lattice* $(L, \leq)$. *The* L-fuzzy set $\mathcal{I}_{\mathcal{P}}^L = \bigcap \mathcal{I}_j$, *where* $\mathcal{I}_j$ *is a Herbrand model of* $\mathcal{P}$, *is called* least fuzzy Herbrand model *of* $\mathcal{P}$.

The previous interpretation $\mathcal{I}_{\mathcal{P}}^L$ can be thought indeed as the least fuzzy Herbrand model, by virtue of the following result.

Theorem 1. *Let* $\mathcal{P}$ *be a multi-adjoint program with associated lattice* $(L, \leq)$. *Then,* $\mathcal{I}_{\mathcal{P}}^L = \bigcap \mathcal{I}_j$, *where* $\mathcal{I}_j$ *is a Herbrand model of* $\mathcal{P}$, *is the least Herbrand model of* $\mathcal{P}$.

Proof. Let $\mathcal{K}$ be the set of Herbrand model of $\mathcal{P}$, that is, the set $\mathcal{K} = \{\mathcal{I}_j : \mathcal{I}_j$ is a Herbrand model of $\mathcal{P}\}$. $\mathcal{I}_{\mathcal{P}}^L$ is a Herbrand interpretation by construction. Since $(\mathcal{H}^L, \subset)$ is a complete lattice, there exist the infimum of the set $\mathcal{K}$, it is a member of $\mathcal{H}^L$ and is given by the intersection of all Herbrand models I_j.

Moreover, $\mathcal{I}_{\mathcal{P}}^L$ is also a Herbrand model of $\mathcal{P}$. By definition of intersection, $\mathcal{I}_{\mathcal{P}}^L \subset \mathcal{I}_j$ for each Herbrand model $\mathcal{I}_j$ of $\mathcal{P}$. Therefore, $\mu_{\mathcal{I}_{\mathcal{P}}^L}(A) \leq \mu_{\mathcal{I}_j}(A)$ for each atom A. On the other hand, since each $\mathcal{I}_j$ is a model of $\mathcal{P}$, by definition of Herbrand model, each rule $\mathcal{R} : \langle A \leftarrow_i B; v \rangle$ in $\mathcal{P}$ is satisfied by $\mathcal{I}_j$, that is, $v \leq \mu_{\mathcal{I}_j}(A \leftarrow_i B)$. Now, by definition of Herbrand interpretation, the monotonic properties of adjoint pairs in a multi-adjoint lattice, and because $\mu_{\mathcal{I}_{\mathcal{P}}^L}(A) \leq \mu_{\mathcal{I}_j}(A)$:

$$ v \leq \mu_{\mathcal{I}_j}(A \leftarrow_i B) = \mu_{\mathcal{I}_j}(A) \dot{\leftarrow}_i \mu_{\mathcal{I}_j}(B) \leq \mu_{\mathcal{I}_{\mathcal{P}}^L}(A) \dot{\leftarrow}_i \mu_{\mathcal{I}_j}(B), $$

where $\dot{\leftarrow}_i$ denote the truth function of the connective $\leftarrow_i$. By the adjoint property, $v \leq \mu_{\mathcal{I}_{\mathcal{P}}^L}(A) \dot{\leftarrow}_i \mu_{\mathcal{I}_j}(B)$ iff $v \dot{\&}_i \mu_{\mathcal{I}_j}(B) \leq \mu_{\mathcal{I}_{\mathcal{P}}^L}(A)$. Also, since the operation $\dot{\&}_i$ is increasing in both arguments and $\mu_{\mathcal{I}_{\mathcal{P}}^L}(B) \leq \mu_{\mathcal{I}_j}(B)$, $v \dot{\&}_i \mu_{\mathcal{I}_{\mathcal{P}}^L}(B) \leq \mu_{\mathcal{I}_{\mathcal{P}}^L}(A)$. Also, applying the adjoint property once again, $v \dot{\&}_i \mu_{\mathcal{I}_{\mathcal{P}}^L}(B) \leq \mu_{\mathcal{I}_{\mathcal{P}}^L}(A)$ iff $v \leq \mu_{\mathcal{I}_{\mathcal{P}}^L}(A) \dot{\leftarrow}_i \mu_{\mathcal{I}_{\mathcal{P}}^L}(B) = \mu_{\mathcal{I}_{\mathcal{P}}^L}(A \leftarrow_i B)$. Therefore, $\mathcal{I}_{\mathcal{P}}^L$ satisfies each rule $\mathcal{R}$ in $\mathcal{P}$, being a Herbrand model of $\mathcal{P}$.

Trivially, since $\mathcal{I}_\mathcal{P}^L$ is the infimum of complete lattice $(\mathcal{K}, \sqsubset)$ (indeed, $inf\{\mathcal{I}_j : \mathcal{I}_j$ is a Herbrand model$\} = \bigcap I_j$), it is the least Herbrand model of $\mathcal{P}$, which concludes the proof.   □

Example 1. Consider the following multi-adjoint logic program $\mathcal{P}$ composed by facts (rules whose bodies are implicitly assumed to be $\top$) and an associate lattice $(L, \leq)$ described by the Hasse's diagram of the figure:

$$\mathcal{R}_1 : \langle p(a) \leftarrow ; \alpha \rangle$$

$$\mathcal{R}_2 : \langle p(a) \leftarrow ; \beta \rangle$$

$$\mathcal{R}_3 : \langle q(a) \leftarrow ; \beta \rangle$$

	$\mathcal{I}_1$	$\mathcal{I}_2$	$\mathcal{I}_3$	$\mathcal{I}_4$	$\mathcal{I}_5$	$\mathcal{I}_6$
$p(a)$	γ	γ	γ	$\top$	$\top$	$\top$
$q(a)$	β	γ	$\top$	β	γ	$\top$

Here, $(\&_\mathsf{G}, \leftarrow_\mathsf{G})$ is the pair of connectives following the *Gödel's intuitionistic* logic, whose truth functions are defined as:

$$\&_\mathsf{G}(x, y) = \inf\{x, y\} \quad \text{and} \quad \leftarrow_\mathsf{G}(y, x) = \begin{cases} \top, \text{if } x \leq y \\ y, \text{ otherwise} \end{cases}$$

It is important to note that with these definitions, the pair $(\leftarrow_\mathsf{G}, \&_\mathsf{G})$ verifies the condition for conforming an adjoint pair regarding lattice $(L, \leq)$ of the figure above.

There exist six different Herbrand models (see $\mathcal{I}_1, \ldots, \mathcal{I}_6$ in the previous table) being $\mathcal{I}_\mathcal{P}^L = \mathcal{I}_1$ the least fuzzy Herbrand model. It is easy to see that $\mathcal{I}_\mathcal{P}^L$ is the L-fuzzy

$$\mathcal{I}_\mathcal{P}^L = \{p(a)|\gamma, q(a)|\beta\} \subset \mathcal{B}_\mathcal{P}^L = \{p(a)|\top, q(a)|\top\}$$

3   Correct Answers by Using Fuzzy Sets

In this section we study the characterization of the notion of correct answer based on L-fuzzy sets. Moreover, we see also for the new formulation of least Herbrand model, that this L-fuzzy set is (like in pure logic programming, see [6]) the set of formulas in the Herbrand base which follow logically from the formulas of the MALP program[6]. The following theorem shows this characterization for correct answer $\langle \lambda; \theta \rangle$.

---

[6] It is not difficult to prove that the fuzzy least Herbrand model coincides with the set of logical consequences, similarly to pure logic programming.

Theorem 2. *Let $\mathcal{P}$ be a multi-adjoint logic program and G a goal. The pair $\langle \lambda; \theta \rangle$ is a correct answer for $\mathcal{P}$ and G if, and only if, $\lambda \leq \mu_{\mathcal{I}_{\mathcal{P}}^L}(G\theta)$, where $\mathcal{I}_{\mathcal{P}}^L$ is the least fuzzy Herbrand model of $\mathcal{P}$, $\lambda \in L$ and θ is a substitution.*

Proof. Since $\mu_{\mathcal{I}_{\mathcal{P}}^L}$ is the membership function of $\mathcal{I}_{\mathcal{P}}^L$, it is enough to use the definitions of least fuzzy Herbrand model and correct answer.   □

The following example adequately suggests how correct answers can be obtained from least fuzzy Herbrand model, the L-fuzzy set $\mathcal{I}_{\mathcal{P}}^L$.

Example 2. Consider the following program $\mathcal{P} = \{\mathcal{R}_1, \mathcal{R}_2, \mathcal{R}_3\}$, whose associated lattice $(L, \leq)$ is given by its depicted Hasse diagram:

$$\mathcal{R}_1 : \langle p(a) \leftarrow \ ; \qquad \alpha \rangle$$

$$\mathcal{R}_2 : \langle p(b) \leftarrow \ ; \qquad \beta \rangle$$

$$\mathcal{R}_3 : \langle q(a) \leftarrow_G p(a); \gamma \rangle$$

Following *Gödel*'s logic, the truth functions of connectives ($\leftarrow_G, \&_G$) are defined in Example 1, thus verifying the conditions for conforming an adjoint pair regarding lattice $(L, \leq)$ of the above figure.

The Herbrand base of the program $\mathcal{P}$ is $\mathcal{B}_{\mathcal{P}} = \{p(a), p(b), q(a), q(b)\}$, hence $\mathcal{B}_{\mathcal{P}}^L = \{p(a)|\top, p(b)|\top, q(a)|\top, q(b)|\top\}$. All Herbrand model of $\mathcal{P}$ is a fuzzy subset of $\mathcal{B}_{\mathcal{P}}^L$, in particular $\mathcal{I}_{\mathcal{P}}^L \subset \mathcal{B}_{\mathcal{P}}^L$. It is easy to check that the least Herbrand model $\mathcal{I}_{\mathcal{P}}^L$ can be given by the L-fuzzy set $\mathcal{I}_{\mathcal{P}}^L = \{p(a)|\alpha, p(b)|\beta, q(a)|\alpha, q(b)|\bot\}$. Then:

i) For goal $p(a)$ the set of correct answers is $\{\langle \lambda; id \rangle : \lambda \in L, \lambda \leq \alpha\}$.

ii) For goal $p(b)$ the set of correct answers is $\{\langle \lambda; id \rangle : \lambda \in L, \lambda \leq \beta\}$.

iii) For goal $q(a)$ the set of correct answers is $\{\langle \lambda; id \rangle : \lambda \in L, \lambda \leq \alpha\}$.

iv) For goal $p(x)$, the set of correct answers is $\{\langle \lambda; \theta \rangle : \lambda \in L, \lambda \leq \mu_{\mathcal{I}_{\mathcal{P}}^L}(p(x)\theta)\} = \{\langle \bot; \{x/a\} \rangle, \langle \alpha; \{x/a\} \rangle, \langle \bot; \{x/b\} \rangle, \langle \beta; \{x/b\} \rangle\}$.
   Note that the membership of $p(x)$ to fuzzy set $\mathcal{I}_{\mathcal{P}}^L$ is
   $\mu_{\mathcal{I}_{\mathcal{P}}^L}(p(x)) = \inf\{\mu_{\mathcal{I}_{\mathcal{P}}^L}(p(x)\sigma) : p(x)\sigma \text{ is ground}\} =^7 \inf\{\mu_{\mathcal{I}_{\mathcal{P}}^L}(p(a)), \mu_{\mathcal{I}_{\mathcal{P}}^L}(p(b))\}$
   $= \inf\{\alpha, \beta\} = \bot$.

v) For goal $q(x)$, the set of correct answers is $\{\langle \lambda; \theta \rangle : \lambda \in L, \lambda \leq \mu_{\mathcal{I}_{\mathcal{P}}^L}(q(x)\theta)\} = \{\langle \alpha; \{x/a\} \rangle, \langle \bot; \{x/b\} \rangle\}$.
   We have now that $\mu_{\mathcal{I}_{\mathcal{P}}^L}(q(x)) = \inf\{\mu_{\mathcal{I}_{\mathcal{P}}^L}(q(a)), \mu_{\mathcal{I}_{\mathcal{P}}^L}(q(b))\} = \inf\{\alpha, \bot\} = \bot$
   (it is easy justify that the least Herbrand model has to be defined this way from $q(a), q(b)$ formulae).

---

[7] Substitutions will only consider terms from the Herbrand universe of the program instead of variables.

In the following theorem we provide an original demonstration for the soundness of the procedural semantics of multi-adjoint programming. Therein we observe a certain analogy with the one included in [6] for the pure logic programming case, despite that the non refutational feature of our language and its fuzzy nature determine very significant differences between both ones. Before tackling the mentioned result, we state the following lemma, that has an instrumental character.

Lemma 1. *Let $(L, \leq)$ be a complete lattice. For all A, B subsets of L, $A \subset B$, implies $inf(B) \leq inf(A)$.*

Proof. It suffices to consider the definition of the infimum and the complete character of lattice $(L, \leq)$. □

Observe that, thanks to the previous lemma, we have $\mu_{\mathcal{I}}(A) \leq \mu_{\mathcal{I}}(A\theta)$[8], for all substitution θ and for all Herbrand interpretation $\mathcal{I}$, whenever the set of ground instances of formula $A\theta$ is a subset of the set of ground instances of A.

Theorem 3 (Soundness). *Let $\mathcal{P}$ be a multi-adjoint logic program, A an atomic goal and $\langle \lambda; \theta \rangle$ a fuzzy computed answer for A in $\mathcal{P}$. Then, $\langle \lambda; \theta \rangle$ is a correct answer for $\mathcal{P}$ and A.*

Proof. Let $D : [G_1, \ldots, G_n]$ be a derivation where $G_1 = \langle A; id \rangle \rightarrow^n_{AS/IS} \langle \lambda; \theta \rangle = G_n$. We prove the claim by induction on n, being n length of D.

We see that, in first place, the result holds for $n = 1$. Indeed, if for goal A exists the derivation $\langle A; id \rangle \rightarrow_{AS} \langle \lambda; \theta \rangle$, then rule $\mathcal{R} : \langle H \leftarrow_i; \lambda \rangle \in \mathcal{P}$ and $A\theta = H\theta$. In that case, every Herbrand model $\mathcal{I}$ of $\mathcal{P}$ satisfies rule $\mathcal{R}$ and, then, $\lambda \leq \mu_{\mathcal{I}}(H \leftarrow_i)$, namely, $\lambda \leq \mu_{\mathcal{I}}(H)$. Furthermore, from the equality $A\theta = H\theta$ it follows that $\mu_{\mathcal{I}}(A\theta) = \mu_{\mathcal{I}}(H\theta)$ and by Lemma 1, we obtain $\mu_{\mathcal{I}}(H) \leq \mu_{\mathcal{I}}(H\theta)$. Consequently, we have $\lambda \leq \mu_{\mathcal{I}}(A\theta)$ and $\langle \lambda; \theta \rangle$ is a correct answer for $\mathcal{P}$ and A, as wanted.

Next suppose that the result is true for all derivation with length k and let us see that it is verified for an arbitrary derivation of length $k + 1$, $D : [G_1, \ldots, G_{k+1}]$. Noting the first step of derivation D, we have $G_1 = \langle A; id \rangle \rightarrow_{AS} \langle v \&_i \mathcal{B}\sigma; \sigma \rangle = G_2$. That is, the admissible step has been executed using the program rule $\mathcal{R} : \langle H \leftarrow_i \mathcal{B}; v \rangle$, where atom A unifies with the head of rule $\mathcal{R}$ through the mgu σ. For each atom $B_i\sigma$[9], $i = 1, \ldots, n$, of $\mathcal{B}\sigma$ exists a derivation whose length is less or equal to k, which gives the computed answer $\langle b_i; \tau_i \rangle$.

More precisely, taking into account that $Dom(\sigma) \cap Ran(B_i) = \emptyset$, D includes the following admisible/interpretive derivation steps[10]:

---

[8] See, for instance, paragraphs $iv), v)$ of Example 2.

[9] Without lost of generality, we can suppose that in the considered derivation all admissible steps are executed before applying interpretive steps.

[10] If $\mathcal{Q}$ is a goal and σ is a substitution, an *interpretive computation* is a state transition system, whose transition relation $\rightarrow_{IS} \subseteq (\mathcal{E} \times \mathcal{E})$ is defined as $\langle Q[@(r_1, r_2)]; \sigma \rangle \rightarrow_{IS} \langle Q[@(r_1, r_2)/\dot{@}(r_1, r_2)]; \sigma \rangle$ where $\dot{@}$ is the truth function of connective $@$. If $\mathcal{Q}$ is a goal not containing atoms, an *interpretive derivation* is a sequence $\langle Q; \sigma \rangle \rightarrow^*_{IS} \langle Q'; \sigma \rangle$ of arbitrary interpretive steps.

$$D : \langle A; id \rangle \qquad\qquad\qquad \rightarrow_{AS}$$

$$\langle v \&_i \mathcal{B}\sigma; \sigma \rangle \qquad\qquad\qquad =$$

$$\langle v \&_i @(B_1\sigma, \ldots, B_n\sigma); \sigma \rangle \qquad \rightarrow_{AS/IS}^{l_1}$$

$$\langle v \&_i @(b_1, \ldots, B_n\sigma); \sigma \circ \tau_1 \rangle \qquad \rightarrow_{AS/IS}^{l_n}$$

$$\langle v \dot{\&}_i @(b_1, \ldots, b_n); \sigma \circ \tau_1 \circ \cdots \circ \tau_n \rangle \rightarrow_{IS}$$

$$\langle v \dot{\&}_i b; \sigma \circ \tau \rangle \qquad\qquad\qquad \rightarrow_{IS}$$

$$\langle \lambda; \theta \rangle]$$

where $\tau = \tau_1 \circ \tau_2 \circ \cdots \circ \tau_n$, $\theta = \sigma \circ \tau$, $\lambda = v \dot{\&}_i b$, $l_1 + l_2 + \ldots + l_n = k - 2$ y $b = @(b_1, \ldots, b_n)$, being @ the combination of all conjunctions, disjunctions and aggregators that links the elements $b_i \in L$ in order to obtain the correct answer $\langle b; \tau \rangle$ for program $\mathcal{P}$ and goal $\mathcal{B}\sigma$.

By the induction hypothesis, for each $B_i\sigma$, $\langle b_i; \tau_i \rangle$ is a correct answer and, then, $b_i \leq \mu_\mathcal{I}(B_i\sigma\tau_i)$, for all Herbrand interpretation $\mathcal{I}$ that is model of $\mathcal{P}$. In that case, from $b_i \leq \mu_\mathcal{I}(B_i\sigma\tau_i)$ it follows that $b \leq \mu_\mathcal{I}(\mathcal{B}\sigma)$ since $\mu_\mathcal{I}(\mathcal{B}\sigma)$ is obtained from $\mu_\mathcal{I}(B_i\tau_i)$ as a result of applying the truth functions of conjunctions, disjunctions or aggregators, being all them monotone in each component.

Then, the equality $A\sigma = H\sigma$ entails $A\theta = H\theta$ and, therefore, $\mu_\mathcal{I}(A\theta) = \mu_\mathcal{I}(H\theta)$. Besides, by firstly using Lemma 1 having into account later that $(\leftarrow_i, \&_i)$ is an adjoint pair, it results $\lambda = v \dot{\&}_i b \leq v \dot{\&}_i \mu_\mathcal{I}(\mathcal{B}\sigma) \leq \mu_\mathcal{I}(H) \leq \mu_\mathcal{I}(H\theta)$.

Consequently, $\lambda \leq \mu_\mathcal{I}(A\theta)$ and $\langle \lambda; \theta \rangle$ is a correct answer for program $\mathcal{P}$ and atom A, as claimed. $\qquad\qquad\qquad\qquad\qquad\qquad\qquad\qquad\qquad\qquad\qquad$ □

4   Logical Consequences by Using Fuzzy Sets

Now, we present a concept strongly related with the developments seen in the core of the paper. We include our approach of fuzzy logical consequences via fuzzy sets in this appendix due to lack of space in the body of the work.

We formalize the concept of logical consequence in terms of the fuzzy set $\mathcal{I}_\mathcal{P}^L$ and we relate it with the notion of correct answer. Moreover, we prove that the least Herbrand model $\mathcal{I}_\mathcal{P}^L$ coincides with the set of formulae from the Herbrand base $\mathcal{B}_\mathcal{P}^L$ that are a logical consequence of the set of rules of a multi-adjoint program. This result allows to extend, for multi-adjoint framework, the classical and well-known formulation of least Herbrand model to logic programming.

In what follows, we propose to state a characterization of the concept of logical consequence through the least Herbrand model $\mathcal{I}_\mathcal{P}^L$. Moreover, from a conceptual standpoint, this characterization will be formulated in a completely similar to the classical case.

Given the multi-adjoint logic program $\mathcal{P} = \{\mathcal{R}_1, \ldots, \mathcal{R}_n\}$ with $\mathcal{R}_i : \langle R_i; \alpha_i \rangle$, $i = 1, \ldots, n$, we have that $\mathcal{A} = \langle A; \alpha \rangle$ is a logical consequence of $\mathcal{P}$ if, and only if, $\alpha_i \leq \mathcal{I}_j(R_i) \Rightarrow \alpha \leq \mathcal{I}_j(A), \forall i, j$. Now, the following theorem gives a characterization of this concept in terms of fuzzy set $\mathcal{I}_\mathcal{P}^L$.

Theorem 4. *Let $\mathcal{P}$ be a multi-adjoint logic program and $\mathcal{A} = \langle A; \alpha \rangle$ a multi-adjoint formula. $\mathcal{A}$ is a logical consequence of $\mathcal{P}$ if, and only if, $\mathcal{I}_\mathcal{P}^L$ is a Herbrand model of $\mathcal{A}$.*

Proof. It is enough to consider the definition of (least fuzzy Herbrand model) $\mathcal{I}_\mathcal{P}^L$ in order to obtain the equivalence: $\mathcal{A}$ is a logical consequence of $\mathcal{P}$ if, and only if, $\alpha \leq \mu_{\mathcal{I}_\mathcal{P}^L}(A)$. □

In the following results we relate the concepts of logical consequence and correct answer.

Theorem 5. *Let $\mathcal{P}$ be a multi-adjoint logic program and G a goal. If $\langle \lambda; \theta \rangle$ is a correct answer for $\mathcal{P}$ and G then $\langle G\theta; \lambda \rangle$ is a logical consequence of $\mathcal{P}$.*

Proof. Let $\mathcal{I}_\mathcal{P}^L$ be the least Herbrand model of $\mathcal{P}$ and see that $\mathcal{I}_\mathcal{P}^L$ is Herbrand model of $\langle G\theta; \lambda \rangle$. However, by definition of correct answer is verified that $\lambda \leq \mu_{\mathcal{I}_\mathcal{P}^L}(G\theta)$, as wanted. □

Theorem 6. *Let $\mathcal{P}$ be a multi-adjoint logic program and $\mathcal{A} = \langle A; \alpha \rangle$ a multi-adjoint formula such that A is a goal. If $\mathcal{A}$ is a logical consequence of $\mathcal{P}$, then the pair $\langle \alpha; id \rangle$ is a correct answer for $\mathcal{P}$ and A.*

Proof. By the Theorem 4, $\mathcal{I}_\mathcal{P}^L$ is Herbrand model of $\mathcal{A}$, so that $\alpha \leq \mu_{\mathcal{I}_\mathcal{P}^L}(A)$ and therefore $\langle \alpha; id \rangle$ is a correct answer for $\mathcal{P}$ and A as claimed. □

Theorem 7. *Let $\mathcal{P}$ be a multi-adjoint logic program and $\mathcal{A} = \langle A\theta; \alpha \rangle$ a multi-adjoint formula such that $A\theta$ is a goal. If $\mathcal{A}$ is a logical consequence of $\mathcal{P}$, then the pair $\langle \alpha; \theta \rangle$ is a correct answer for $\mathcal{P}$ and A.*

Proof. Analogous to the above theorem. □

The next result is a natural adaptation, to multi-adjoint logic programing, of the corresponding theorem of pure logic programming, (see [6]), which characterizes the least Herbrand model as the set of formulae from the Herbrand base that are logical consequences of the multi-adjoint program. In this theorem we express a formula multi-adjoint $\mathcal{A} = \langle A; \alpha \rangle$ as the pair $A|\alpha$ (α is degree of membership of A in fuzzy set $\{A|\alpha\}$). Observe that this syntax is also allowed for rules in multi-adjoint program.

Theorem 8. *Let $\mathcal{I}_\mathcal{P}^L$ be the least fuzzy Herbrand model of a multi-adjoint program $\mathcal{P}$ with associated lattice L. If we choose formulae $\mathcal{A} = A|\alpha$ with $\alpha = \mu_{\mathcal{I}_j}(A)$, for some Herbrand model $\mathcal{I}_j$, then $\mathcal{I}_\mathcal{P}^L = \{\mathcal{A} \in \mathcal{B}_\mathcal{P}^L : \mathcal{A}$ is a logical consequence of $\mathcal{P}\}$.*

Proof. If $\mathcal{A} \in \mathcal{I}_\mathcal{P}^L \subset \mathcal{B}_\mathcal{P}^L$, then $\alpha \leq \mu_{\mathcal{I}_\mathcal{P}^L}(A)$ so $\mathcal{A}$ is logical consequence of $\mathcal{P}$ and this shows that $\mathcal{I}_\mathcal{P}^L \subset \{\mathcal{A} \in \mathcal{B}_\mathcal{P}^L : \mathcal{A}$ is a logical consequence of $\mathcal{P}\}$. For the reverse inclusion, let $\mathcal{A} = A|\alpha$ be a formulae, with $\alpha = \mu_{\mathcal{I}_j}(A)$, for some Herbrand model $\mathcal{I}_j$; then, $\mu_{\mathcal{I}_\mathcal{P}^L}(A) \leq \mu_{\mathcal{I}_j}(A) = \alpha$, because $\mathcal{I}_\mathcal{P}^L \subset \mathcal{I}_j$. Moreover, since $\mathcal{A}$ is a logical consequence of $\mathcal{P}$, $\alpha \leq \mu_{\mathcal{I}_\mathcal{P}^L}(A)$ and, consequently, $\alpha = \mu_{\mathcal{I}_\mathcal{P}^L}(A)$, as required.   □

5   Conclusions and Future Work

This paper has focused on the MALP framework, for which during the last years we have produced a wide range of results regarding both theoretical [2,3,4,13], and practical [11,12,1] developments. After recalling from [5] our concept of least Herbrand model for MALP, we have characterized, through the concept of fuzzy set, notions of Herbrand model, least Herbrand model and correct answer, thus extending the classic concepts of pure logic programming to this kind of fuzzy logic programs. The main goals of this work have been both the re-formulation of all these concepts as well as their strong relationships (by also including an original proof of the soundness for the procedural semantics of MALP) by means of the well-known fuzzy sets theory, thus providing more natural and clearer results which directly resemble the properties of pure logic programming described in [6], but lifted now to the modern case of fuzzy logic programming. We are nowadays implementing most notions defined in this paper inside our *"Fuzzy LOgic Programming Environment for Research"* $\mathcal{FLOPER}$ (visit `http://dectau.uclm.es/floper/` where some real-world examples are available too).

Acknowledgements. We are grateful to anonymous reviewers for providing us valuable suggestions which have been used for improving the material compiled so far. This work was supported by the EU (FEDER), and the Spanish MINECO Ministry (*Ministerio de Economía y Competitividad*) under grant TIN2013-45732-C4-2-P.

References

1. Almendros-Jiménez, J.M., Luna Tedesqui, A., Moreno, G.: A Flexible XPath-based Query Language Implemented with Fuzzy Logic Programming. In: Bassiliades, N., Governatori, G., Paschke, A. (eds.) RuleML 2011 - Europe. LNCS, vol. 6826, pp. 186–193. Springer, Heidelberg (2011)
2. Julián, P., Moreno, G., Penabad, J.: On Fuzzy Unfolding. A Multi-adjoint Approach. Fuzzy Sets and Systems 154, 16–33 (2005)
3. Julián, P., Moreno, G., Penabad, J.: Operational/Interpretive Unfolding of Multi-adjoint Logic Programs. Journal of Universal Computer Science 12(11), 1679–1699 (2006)
4. Julián, P., Moreno, G., Penabad, J.: An Improved Reductant Calculus using Fuzzy Partial Evaluation Techniques. Fuzzy Sets and Systems 160, 162–181 (2009)
5. Julián, P., Moreno, G., Penabad, J.: On the declarative semantics of multi-adjoint logic programs. In: Cabestany, J., Sandoval, F., Prieto, A., Corchado, J.M. (eds.) IWANN 2009, Part I. LNCS, vol. 5517, pp. 253–260. Springer, Heidelberg (2009)

6. Lloyd, J.W.: Foundations of Logic Programming. Springer, Berlin (1987)
7. Medina, J., Ojeda-Aciego, M.: Multi-adjoint logic programming. In: Proc. of the 10th International Conference on Information Processing and Managment of Uncertainty in Knowledge-Based Systems, IPMU 2004, Perugia, pp. 823–830 (2004)
8. Medina, J., Ojeda-Aciego, M., Vojtáš, P.: Multi-adjoint logic programming with continuous semantics. In: Eiter, T., Faber, W., Truszczyński, M. (eds.) LPNMR 2001. LNCS (LNAI), vol. 2173, pp. 351–364. Springer, Heidelberg (2001)
9. Medina, J., Ojeda-Aciego, M., Vojtáš, P.: A procedural semantics for multi-adjoint logic programming. In: Brazdil, P.B., Jorge, A.M. (eds.) EPIA 2001. LNCS (LNAI), vol. 2258, pp. 290–297. Springer, Heidelberg (2001)
10. Medina, J., Ojeda-Aciego, M., Vojtáš, P.: Similarity-based Unification: a multi-adjoint approach. Fuzzy Sets and Systems 146, 43–62 (2004)
11. Morcillo, P.J., Moreno, G.: Programming with fuzzy logic rules by using the FLOPER tool. In: Bassiliades, N., Governatori, G., Paschke, A. (eds.) RuleML 2008. LNCS, vol. 5321, pp. 119–126. Springer, Heidelberg (2008)
12. Morcillo, P.J., Moreno, G., Penabad, J., Vázquez, C.: Fuzzy Computed Answers Collecting Proof Information. In: Cabestany, J., Rojas, I., Joya, G. (eds.) IWANN 2011, Part II. LNCS, vol. 6692, pp. 445–452. Springer, Heidelberg (2011)
13. Morcillo, P.J., Moreno, G., Penabad, J., Vázquez, C.: Dedekind-MacNeille completion and cartesian product of multi-adjoint lattices. International Journal of Computer Mathematics, 1–11 (2012)
14. Mukaidono, M., Shen, Z., Ding, L.: Fundamentals of fuzzy prolog. International Journal Approximate Reasoning 3(2), 179–193 (1989)
15. Nguyen, H.T., Walker, E.A.: A First Course in Fuzzy Logic, 3rd edn. Chapman & Hall/CRC, Boca Ratón, Florida (2006)
16. Pedrycz, W., Gomide, F.: Introduction to fuzzy sets. MIT Press, Cambridge (1998)
17. Shapiro, E.Y.: Logic programs with uncertainties: A tool for implementing rule-based systems. In: Proc. of the 8th International Joint Conference on Artificial Intelligence, IJCAI 1983, Karlsruhe, pp. 529–532. University of Trier (1983)
18. van Emden, M.H.: Quantitative deduction and its fixpoint theory. Journal of Logic Programming 3(1), 37–53 (1986)
19. Vojtáš, P.: Fuzzy Logic Programming. Fuzzy Sets and Systems 124(1), 361–370 (2001)
20. Vojtáš, P., Paulík, L.: Soundness and completeness of non-classical extended SLD-resolution. In: Herre, H., Dyckhoff, R., Schroeder-Heister, P. (eds.) ELP 1996. LNCS, vol. 1050, pp. 289–301. Springer, Heidelberg (1996)
21. Zadeh, L.A.: Fuzzy Sets. Information and Control 8(3), 338–353 (1965)

Multi Threshold FRPS: A New Approach to Fuzzy Rough Set Prototype Selection

Nele Verbiest

Department of Applied Mathematics, Computer Science and Statistics,
Ghent University, Krijgslaan 281 (S9), B-9000 Gent, Belgium
Nele.Verbiest@UGent.be

Abstract. Prototype Selection (PS) is the preprocessing technique for
K nearest neighbor classification that selects a subset of instances before
classification takes place. The most accurate state-of-the-art PS method
is Fuzzy Rough Prototype Selection (FRPS), which assesses the quality
of the instances by means of the fuzzy rough positive region and automatically selects a good threshold to decide if instances should be retained
in the prototype subset. In this paper we introduce a new PS method
based on FRPS, called Multi Threshold FRPS (MT-FRPS) . Instead of
determining one threshold against which the quality of every instance is
compared, we consider one threshold for each class.

We evaluate MT-FRPS on 40 standard classification datasets and
compare it against MT-FRPS and the state-of-the-art PS methods and
show that MT-FRPS improves the accuracy of the state-of-the-art PS
methods.

Keywords: fuzzy rough set theory, classification, prototype selection.

1 Introduction

Classification, the process where unlabeled data described by conditional attributes is classified using labeled training data, is an important task in data
mining. One of the most intuitive and most widely used classifiers is the K Nearest Neighbor (KNN, [1]) classifier, which classifies a target instance by looking up
its K nearest neighbors in the training data and labeling it as the most frequently
occurring class among these nearest neighbors.

There are two main drawbacks associated with KNN classification. The first
is that noisy or mislabeled data directly influences the classification of new instances. The second is that KNN is time consuming as the distances between
the target instances and all instances in the training data need to be calculated.
Prototype Selection (PS, [2]) is an answer to both problems. It removes noisy
and/or superfluous instances from the data and provides the KNN classifier with
an improved training dataset, called the prototype subset.

Many PS methods have been proposed in the literature, a good overview
is given in [2]. The most accurate PS techniques are evolutionary algorithms
[3,4,5,6] and Fuzzy Rough Prototype Selection (FRPS, [7,8]). The advantage of

C. Cornelis et al. (eds.): RSCTC 2014, LNAI 8536, pp. 83–91, 2014.

evolutionary algorithms over FRPS is that they reduce the data more, most of these evolutionary algorithms remove up to 90 percent of the data. On the other hand, FRPS is more accurate and faster than the evolutionary algorithms.

As our interest lies in improving the accuracy of the classification process, we focus on improving FRPS here. The FRPS algorithm assesses the quality of the instances by means of fuzzy rough set theory [9,10,11] and automatically determines a good threshold to decide if instances should be removed or not.

Instead of using one threshold to decide if instances should be included in the prototype subset, our proposal, called Multi Threshold Fuzzy Rough Prototype Selection (MT-FRPS), considers one threshold for each class, motivated by the idea that not every class in the dataset looks the same and that the threshold should be relaxed or rigidified for particular classes.

The remainder of this paper is organized as follows: in Section 2 we first recall FRPS and then explain in detail how MT-FRPS improves FRPS. In Section 3 we experimentally evaluate MT-FRPS and we conclude in Section 4.

2   Multi Threshold Fuzzy Rough Prototype Selection

In this section we introduce our proposal MT-FRPS. We first recall the FRPS algorithm in Section 2.1 and explain how we improve FRPS in Section 2.2.

2.1   Fuzzy Rough Prototype Selection

From now on we assume that we are given a decision system $(U, \mathcal{A} \cup \{d\})$ where U is the universe of instances described by the conditional attributes $\mathcal{A}$ and the decision attribute d. The value of an instance x for an attribute $b \in \mathcal{A} \cup \{d\}$ is denoted by $b(x)$.

The FRPS algorithm consists of two main components. First, for each instance $x \in U$ its quality is assessed using fuzzy rough set theory, and secondly, a good threshold is determined to decide if instances should be removed from the data or not.

Assessing The Quality Of Instances Using Fuzzy Rough Set Theory. Recall that, for classification problems, the membership degree of an instance $x \in U$ to the fuzzy rough positive region is given as follows:

$$\forall x \in U : POS(x) = \min_{y \in U} \mathcal{I}(R(x, y), [x]_d(y)), \qquad (1)$$

with $\mathcal{I}$ a fuzzy implicator, R an indiscernibility relation and $[x]_d$ the decision class of x. As for all x and y in U $[x]_d(y)$ only takes values in $\{0, 1\}$, we can rewrite this membership degree as follows[1] [12]:

$$\forall x \in U : POS(x) = \min_{y \in U \setminus [x]_d} (1 - R(x, y)), \qquad (2)$$

---

[1] We assume that $\forall a \in [0, 1] \mathcal{I}(a, 0) = 1 - a$ which is the case for the customary fuzzy implicators like the Kleene-Dienes and Łukasiewicz implicator

that is, only the instances in a class different from the class of x need to be considered. The fuzzy rough positive region expresses for each instance $x \in U$ to what extent instances belonging to a different class from x are discernible from it, and can be used to assess the quality of instances.

Unfortunately the traditional fuzzy rough positive region is based on the strict minimum operator. As a result, small changes in the decision system can result in drastic changes in the fuzzy rough positive region membership values, which is not desirable when working with real-world datasets.

To overcome this problem, the traditional fuzzy rough sets on which the fuzzy rough positive region is based can be replaced by Ordered Weighted Average (OWA, [13]) fuzzy rough sets [14]. They replace the strict minimum and maximum operators in the fuzzy rough lower and upper approximation by their soft OWA analogues. Given a weight vector $W = \langle w_1, \dots, w_p \rangle$ such that each weight is in $[0, 1]$ and such that all weights sum to 1, the OWA_W aggregation of p values $v_1, \dots, v_p$ is given by

$$\text{OWA} \sum_{i=1}^{p} w_i t_i \tag{3}$$

where t_i is the ith largest value in $v_1, \dots, v_p$. That is, the OWA_W aggregation orders the values decreasingly and then assigns the weights in W to these values in that order.

The OWA aggregation can be used to soften the minimum and maximum operator. Note that, if $\langle 0, \dots, 0, 1 \rangle$ is used as weight vector, the minimum is retrieved. This strict operator can be softened by using a vector with increasing weights. As a result, high values are associated with a low weight, while low values are associated with a high weight. The resulting operator behaves like the minimum but takes into account more values. The same strategy can be applied to obtain an OWA operator that softens the maximum operator.

After replacing the minimum and maximum operator in the traditional fuzzy rough sets by their OWA analogues, the fuzzy rough positive region is given by:

$$\forall x \in U : POS(x) = \text{OWA}_W \mathcal{I}(R(x, y), [x]_d(y)), \tag{4}$$
$$\text{\small } y \in U$$

where OWA_W is a soft analogue of the minimum operator. As for all $x, y \in U$ the value of $\mathcal{I}(R(x, y), [x]_d(y))$ is one if y and x are in the same class, the instances y in the same class as x are omitted:

$$\forall x \in U : POS(x) = \text{OWA}_W (1 - R(x, y)). \tag{5}$$
$$\text{\small } y \in U \setminus [x]_d$$

This OWA fuzzy rough positive region is used to assess the quality of instances. An instance $x \in U$ for which the instances y of other classes are highly discernible from it (i. e. the indiscernibility values $R(x, y)$ are low) will have a high quality value. These are instances that are *typical* for their class. Instances on the borders between classes and mislabeled instances will get a low quality value as there exist many instances from a different class highly indiscernible from them.

Determining a Good Threshold. Once the quality of the instances is determined, the question raises which threshold to use to decide if instances should be retained or not. The FRPS algorithm proceeds as follows:

1. First all candidate thresholds are determined, these are all the quality values of the instances without duplicates:

$$T = \{POS(x)|x \in U\} \tag{6}$$

2. Next, for each threshold $\tau \in T$ the corresponding prototype subset containing all instances for which the quality is at least τ is calculated:

$$\forall \tau \in T : S_\tau = \{x|x \in U \text{ and } POS(x) \geq \tau\} \tag{7}$$

3. The training accuracy corresponding to each subset S is calculated. In order to classify an instance x in U the 1NN classification rule is applied using S as pool of candidate nearest neighbors if $x \notin S$ and using $S \setminus \{x\}$ as pool of candidate nearest neighbors otherwise.
4. The prototype subset S_τ with the highest training accuracy is returned. If multiple subsets have the same training accuracy, the largest one is returned.

Summarized, all possible thresholds are considered and the threshold corresponding to the highest training accuracy is used to decide if instances should be removed or not.

2.2   Improving FRPS Using Multiple Thresholds

FRPS determines one threshold and compares the quality of each instance in U against it to decide if it should be included in the final prototype subset or not. As a single threshold is used for all instances, some information and properties of the dataset might get lost. Classes within a dataset can be different, for instance, some classes consist of one solid block of instances close to each other, while other classes consist of smaller groups of instances spread out over the feature space. The quality of the first group of instances will in general be higher, while the last class contains low quality instances. If the same threshold is used for both classes, this information gets lost.

MT-FRPS deals with this problem by determining a separate threshold for each class. Assume that there are C classes and denote these classes by $c_1, \ldots, c_C$. Instead of using one threshold τ like FRPS does, C thresholds $(\tau_1, \ldots, \tau_C)$ are used. An instance x in class c_i is included in the prototype subset if the quality of x is at least τ_i.

One option to determine the C thresholds could be to consider all possible combinations of thresholds, but the computational cost related to this strategy is too high. For instance, if the universe consists of 1000 instances divided in two classes of each size 500, the number of threshold combinations that needs to be evaluated is 250 000, which is much more than the 1000 evaluations that FRPS needs. Moreover, this strategy would very likely lead to over-fitting.

Instead, MT-FRPS uses a search strategy that aims to optimize the threshold combination $(\tau_1, \ldots, \tau_C)$. The outline of MT-FRPS is given below:

– The starting point of the search algorithm is based on the threshold τ that is returned by FRPS. That is, FRPS is applied to U and the optimal threshold τ is retrieved. The threshold combination $\tau_{best} = (\tau_1, \ldots, \tau_C)$ is initialized as $(\tau, \ldots, \tau)$ and the training accuracy corresponding to this threshold combination is denoted by acc_{best}.
– This threshold combination is now optimized iteratively. The following steps are carried out N times, where N is a parameter:
  1. Determine a random class c_i in $c_1, \ldots, c_C$.
  2. Determine a random value between -0.01 and 0.01, call it r
  3. Consider the new threshold combination $\tau_{new} = (\tau_1, \ldots, \tau_i + r, \ldots, \tau_C)$, where one threshold in the combination is slightly changed.
  4. Consider the subset S corresponding to this threshold combination:

$$S - \{x \in U | POS(x) \geq \tau_{d(x)}\} \tag{8}$$

  5. Determine the training accuracy acc of S by classifying each instance $x \in U$ using S as pool of candidate nearest neighbors if $x \notin S$ and using $S \setminus \{x\}$ as pool of candidate nearest neighbors otherwise.
  6. If $acc \geq acc_{best}$, the new threshold combination is stored in the old one $\tau_{comb} = \tau_{new}$ and the best accuracy is updated: $acc_{best} = acc$.
– The prototype subset $S = \{x \in U | POS(x) \geq \tau_{d(x)}\}$ corresponding to τ_{best} is returned.

Summarized, the threshold that is returned by FRPS is fine-tuned for every class by introducing small random changes and evaluating the resulting subsets.

3   Experimental Evaluation

In this section we verify if MT-FRPS improves FRPS. We discuss the experimental set-up of our evaluation in Section 3.1 and discuss the results in Section 3.2.

3.1   Experimental Set-Up

We consider 40 datasets from the UCI [15] and KEEL [16] repository. Their properties are listed in Table 1. We follow a 10 fold cross validation scheme, that is, the data is divided into 10 folds of equal size, and the instances in each (test) fold are classified using the 1NN rule on the remaining 9 folds preprocessed by the PS algorithm. We report the average accuracy, Cohen's kappa[2] [17] and the running time covering the PS algorithm, not the 1NN classification applied afterwards.

The quality measure used in the FRPS and MT-FRPS algorithm depends on the indiscernibility relation R. A preliminary study has shown that the following

---

[2] Cohen's kappa is an evaluation measure that compensates for correct classifications due to chance

measure performs well for FRPS (we assume that the values of the continuous attributes are normalized such that they are in $[0,1]$):

$$\forall x, y \in U : R(x,y) = \frac{\sum\limits_{a \in \mathcal{A}} R_a(x,y)}{|\mathcal{A}|}, \tag{9}$$

where $R_a(x,y) = 1 - |a(x) - a(y)|$ for continuous attributes, and $R_a(x,y)$ is 1 if x and y have the same values for a and 0 otherwise for a discrete attribute a. We use this indiscernibility relation for both FRPS and MT-FRPS.

The quality measure also depends on the weights used for the OWA aggregation, we use:

$$W = \langle \frac{1}{p \sum\limits_{i=1}^{p} \frac{1}{i}}, \frac{1}{(p-1) \sum\limits_{i=1}^{p} \frac{1}{i}}, \dots, \frac{1}{2 \sum\limits_{i=1}^{p} \frac{1}{i}}, \frac{1}{1 \sum\limits_{i=1}^{p} \frac{1}{i}} \rangle, \tag{10}$$

as FRPS performed well with these weights in a preliminary study.

In order to limit the extra computional time of MT-FRPS over FRPS we set the parameter N that determines how many evaluations are carried out to 100.

Table 1. Data used in the experimental evaluation

	# inst.	# feat.	# class.		# inst.	# feat.	# class.
appendicitis	106	7	2	housevotes	232	16	2
australian	690	14	2	iris	150	4	3
automobile	150	25	6	led7digit	500	7	10
balance	625	4	3	lymphography	148	18	4
bands	365	19	2	mammographic	830	5	2
breast	277	9	2	monk-2	432	6	2
bupa	345	6	2	movement_libras	360	90	15
car	1728	6	4	newthyroid	215	5	3
cleveland	297	13	5	pima	768	8	2
contraceptive	1473	9	3	saheart	462	9	2
crx	653	15	2	sonar	208	60	2
dermatology	358	34	6	spectfheart	267	44	2
ecoli	336	7	8	tae	151	5	3
flare	1066	11	6	tic-tac-toe	958	9	2
german	1000	20	2	vehicle	846	18	4
glass	214	9	7	vowel	990	13	11
haberman	306	3	2	wine	178	13	3
hayes-roth	160	4	3	wisconsin	683	9	2
heart	2270	13	2	yeast	1484	8	10
hepatitis	80	19	2	zoo	101	16	7

Table 2. Results obtained by FRPS and MT-FRPS

	FRPS			MT-FRPS		
	Acc.	Cohen's κ	Time (s)	Acc.	Cohen's κ	Time (s)
appendicitis	0.8509	0.5283	0.44	0.8609	0.5523	4.56
australian	0.8565	0.7098	9.85	0.8522	0.7015	58.56
automobile	0.7755	0.7066	1.50	0.7755	0.7066	15.00
balance	0.8913	0.7992	2.00	0.8897	0.7965	16.78
bands	0.7029	0.3348	9.59	0.7147	0.3670	27.11
breast	0.7684	0.3389	0.73	0.7650	0.3289	11.67
bupa	0.6017	0.1763	2.26	0.5988	0.1710	18.67
car	0.8686	0.6985	10.71	0.8640	0.6939	120.89
cleveland	0.5897	0.2881	0.33	0.5897	0.2881	3.00
contraceptive	0.4366	0.1337	7.32	0.4284	0.1041	75.89
crx	0.8554	0.7092	6.03	0.8538	0.7069	36.00
dermatology	0.9634	0.9538	4.84	0.9607	0.9504	36.56
ecoli	0.8248	0.7558	2.93	0.8217	0.7534	18.00
flare	0.6436	0.5442	7.91	0.6548	0.5575	100.93
german	0.7280	0.2459	6.71	0.7290	0.2236	30.56
glass	0.7176	0.6128	0.92	0.7176	0.6128	17.78
haberman	0.6959	0.0929	0.81	0.7091	0.1085	6.56
hayes-roth	0.5188	0.2580	0.36	0.5250	0.2633	3.11
heart	0.8074	0.6083	2.41	0.8148	0.6214	8.44
hepatitis	0.8160	0.2451	0.83	0.8160	0.2451	6.44
housevotes	0.9218	0.8424	1.87	0.9218	0.8424	13.22
iris	0.9400	0.9100	1.63	0.9400	0.9100	14.56
led7digit	0.5820	0.5343	2.54	0.5820	0.5344	19.33
lymphography	0.8415	0.6966	0.80	0.8554	0.7228	0.33
mammographic	0.8047	0.6108	4.65	0.8023	0.6056	42.22
monk-2	0.8228	0.6419	2.81	0.8135	0.6237	0.78
movement_libras	0.8028	0.7882	3.94	0.8111	0.7972	34.00
newthyroid	0.9675	0.9278	0.71	0.9675	0.9278	5.78
pima	0.7371	0.4026	5.46	0.7396	0.4102	35.33
saheart	0.7165	0.3649	2.57	0.7252	0.3560	25.11
sonar	0.8362	0.6686	3.12	0.8457	0.6880	19.44
spectfheart	0.7906	0.1155	7.73	0.7832	0.1233	9.33
tae	0.5313	0.2974	0.41	0.5313	0.2974	3.22
tic-tac-toe	0.7985	0.4885	7.12	0.8497	0.6657	41.89
vehicle	0.6962	0.5948	8.31	0.6891	0.5855	69.44
vowel	0.9949	0.9944	7.66	0.9949	0.9944	88.89
wine	0.9552	0.9326	1.98	0.9608	0.9409	11.22
wisconsin	0.9680	0.9297	9.23	0.9680	0.9297	55.33
yeast	0.5277	0.3896	6.01	0.5439	0.4056	92.11
zoo	0.9347	0.9122	0.48	0.9347	0.9122	8.00
AVERAGE:	**0.7771**	**0.5696**	**3.93**	**0.7800**	**0.5756**	**30.15**

3.2 Results

The results are listed in Table 2. On average, MT-FRPS improves FRPS both with respect to accuracy and Cohen's kappa. The difference is larger for Cohen's kappa, which may be due to the fact that Cohen's kappa takes into account the classification accuracy per class, which is in favor of MT-FRPS as it considers the classes separately to determine good thresholds. The improved performance comes with a higher computational cost, due to the fact that extra evaluations need to be carried out.

4 Conclusion

In this paper we proposed a new PS method called MT-FRPS that improves FRPS. MT-FRPS assesses the quality of the instances using fuzzy rough set theory and then determines for each class a threshold to decide if the instances in that class should be removed or not. An experimental evaluation shows that MT-FRPS improves FRPS.

References

1. Cover, T., Hart, P.: Nearest neighbor pattern classification. IEEE Transactions on Information Theory 13, 21–27 (1967)
2. García, S., Derrac, J., Cano, J., Herrera, F.: Prototype selection for nearest neighbor classification: Taxonomy and empirical study. IEEE Transactions on Pattern Analysis and Machine Intelligence 34, 417–435 (2012)
3. Kuncheva, L.I., Jain, L.C.: Nearest neighbor classifier: Simultaneous editing and feature selection. Pattern Recognition Letters 20, 1149–1156 (1999)
4. Kuncheva, L.I.: Editing for the k-nearest neighbors rule by a genetic algorithm. Pattern Recognition Letters 16, 809–814 (1995)
5. García, S., Cano, J.R., Herrera, F.: A memetic algorithm for evolutionary prototype selection: A scaling up approach. Pattern Recognition 41, 2693–2709 (2008)
6. Casillas, J., Cordón, O., del Jesus, M., Herrera, F.: Genetic feature selection in a fuzzy rule-based classification system learning process. Information Sciences 136, 135–157 (2001)
7. Verbiest, N., Cornelis, C., Herrera, F.: Frps: a fuzzy rough prototype selection method. Pattern Recognition 46, 2770–2782 (2013)
8. Verbiest, N., Cornelis, C., Herrera, F.: A prototype selection method based on ordered weighted average fuzzy rough set theory. In: Proceedings of the 14th International Conference on Rough Sets, Fuzzy Sets, Data Mining and Granular Computing, pp. 180–190 (2013)
9. Cornelis, C., De Cock, M., Radzikowska, A.M.: Fuzzy rough sets: from theory into practice. In: Pedrycz, W., Skowron, A., Kreinovich, V. (eds.) Handbook of Granular Computing, pp. 533–552 (2008)
10. Dubois, D., Prade, H.: Rough fuzzy sets and fuzzy rough sets. International Journal of General Systems 17, 191–209 (1990)
11. Dubois, D., Prade, H.: Putting fuzzy sets and rough sets together. In: Slowinskii, R. (ed.) Intelligent Decision Support- Handbook of Applications and Advances of the Rough Sets Theory, pp. 203–232 (1992)

12. Cornelis, C., Jensen, R., Hurtado Martın, G., Slezak, D.: Attribute selection with fuzzy decision reducts. Information Sciences 180, 209–224 (2010)
13. Yager, R.: On ordered weighted averaging aggregation operators in multicriteria decisionmaking. IEEE Transactions on Systems, Man and Cybernetics 18, 183–190 (1988)
14. Cornelis, C., Verbiest, N., Jensen, R.: Ordered weighted average based fuzzy rough sets. In: Yu, J., Greco, S., Lingras, P., Wang, G., Skowron, A. (eds.) RSKT 2010. LNCS, vol. 6401, pp. 78–85. Springer, Heidelberg (2010)
15. Bache, K., Lichman, M.: UCI machine learning repository (2013)
16. Alcalá, J., Fernandez, A., Luengo, J., Derrac, J., García, S., Sánchez, L., Herrera, F.: Keel data-mining software tool: Data set repository, integration of algorithms and experimental analysis framework. Journal of Multiple-Valued Logic and Soft Computing 17, 255–287 (2011)
17. Ben-David, A.: Comparison of classification accuracy using cohen's weighted kappa. Expert Systems with Applications 34, 825–832 (2008)

Measurable Structures of $\mathcal{I}$-Fuzzy Rough Sets

Wei-Zhi Wu[1,2], Xiao-Feng Mu[1], You-Hong Xu[1,2], and Xia Wang[1]

[1] School of Mathematics, Physics and Information Science,
Zhejiang Ocean University, Zhoushan, Zhejiang, 316022, China
[2] Chongqing Key Laboratory of Computational Intelligence,
Chongqing University of Posts and Telecommunications, Chongqing, 400065, China
{wuwz,xyh}@zjou.edu.cn, 295586218@qq.com, bblylm@126.com

Abstract. In this paper, dual fuzzy rough approximation operators determined by a fuzzy implication operator $\mathcal{I}$ in infinite universes of discourse are first introduced. Measurable structures of $\mathcal{I}$-fuzzy rough sets are then discussed. It is shown that the family of all definable sets in an $\mathcal{I}$-fuzzy rough set algebra derived from a reflexive fuzzy space forms a σ-fuzzy algebra. In a finite universe of discourse, the family of all definable sets in a serial $\mathcal{I}$-fuzzy rough set algebra is a fuzzy algebra, and conversely if a σ-fuzzy algebra is generated by a crisp algebra, then there must exist an $\mathcal{I}$-fuzzy rough set algebra such that the family of all definable sets is exactly the given σ-fuzzy algebra.

Keywords: Approximation spaces, Fuzzy rough sets, Measurable spaces, Rough sets, σ-algebras.

1 Introduction

The concept of σ-algebra is of importance in mathematical analysis as the foundation for Lebesgue integration, and in probability or measure theory, where it is interpreted as the collection of events which can be assigned probabilities. The study of relationship between approximation spaces in rough set theory and measurable spaces in probability theory is an interesting issue. In [9], Pawlak proved that the family of all definable sets in a Pawlak approximation space forms a σ-algebra. In [12,15], it was shown that the family of all definable sets in a serial crisp approximation space forms a σ-algebra, and, conversely, for any crisp measurable space in a finite universe of discourse there must exist a crisp approximation space such that the family of all definable sets is the given crisp algebra. As for the fuzzy environment, Wu examined that the family of all definable sets in a $\mathcal{T}$-fuzzy rough algebra derived from a reflexive fuzzy approximation space is a σ-fuzzy algebra [11].

In [13], Wu et al. presented a general framework for the study of dual pair of lower and upper fuzzy rough approximation operators determined by a general fuzzy implicator $\mathcal{I}$ in infinite universes of discourse. In this paper, we will investigate the measurable structures of $\mathcal{I}$-fuzzy rough sets. We will establish the relationship between fuzzy approximation spaces and fuzzy measurable spaces.

C. Cornelis et al. (eds.): RSCTC 2014, LNAI 8536, pp. 92–99, 2014.

2   Fuzzy Logical Operators

Throughout this paper, U will be a nonempty set called the universe of discourse. The class of all subsets (respectively, fuzzy subsets) of U will be denoted by $\mathcal{P}(U)$ (respectively, by $\mathcal{F}(U)$). For $\alpha \in I$ (where $I = [0,1]$ is the unit interval), $\widehat{\alpha}$ will denote the constant fuzzy set: $\widehat{\alpha}(x) = \alpha$, for all $x \in U$. Zadeh's fuzzy union and fuzzy intersection will be denoted by $\cup$ and $\cap$ respectively.

A *triangular norm* [4], or t-norm in short, is an increasing, associative and commutative mapping $\mathcal{T} : I^2 \to I$ that satisfies the boundary condition: for all $\alpha \in I, \mathcal{T}(\alpha, 1) = \alpha$.

A *triangular conorm* (t-conorm in short) is an increasing, associative and commutative mapping $\mathcal{S} : I^2 \to I$ that satisfies the boundary condition: for all $\alpha \in I, \mathcal{S}(\alpha, 0) = \alpha$.

A *negator* $\mathcal{N}$ is a decreasing $I \to I$ mapping satisfying $\mathcal{N}(0) = 1$ and $\mathcal{N}(1) = 0$. The negator $\mathcal{N}_S(\alpha) = 1 - \alpha$ is usually referred to as the *standard negator*. A negator $\mathcal{N}$ is called *involutive* iff $\mathcal{N}(\mathcal{N}(\alpha)) = \alpha$ for all $\alpha \in [0, 1]$. It is well-known that every involutive negator is continuous [5].

Given a negator $\mathcal{N}$, a t-norm $\mathcal{T}$ and a t-conorm $\mathcal{S}$ are called dual w.r.t. $\mathcal{N}$ iff the De Morgan's laws are satisfied, i.e.

$$\begin{aligned} \mathcal{S}(\mathcal{N}(\alpha), \mathcal{N}(\beta)) &= \mathcal{N}(\mathcal{T}(\alpha, \beta)), \ \forall \alpha, \beta \in I, \\ \mathcal{T}(\mathcal{N}(\alpha), \mathcal{N}(\beta)) &= \mathcal{N}(\mathcal{S}(\alpha, \beta)), \ \forall \alpha, \beta \in I. \end{aligned} \tag{1}$$

It is well known [5] that for an involutive negator $\mathcal{N}$ and a t-conorm $\mathcal{S}$, the function $\mathcal{T}_S(\alpha, \beta) = \mathcal{N}(\mathcal{S}(\mathcal{N}(\alpha), \mathcal{N}(\beta))), \alpha, \beta \in I$, is a t-norm such that $\mathcal{T}$ and $\mathcal{S}$ are dual w.r.t. $\mathcal{N}$. It will be referred to as a t-norm dual to $\mathcal{S}$ w.r.t. $\mathcal{N}$.

In what follows, $\sim_{\mathcal{N}}$ will be used to denote fuzzy complement determined by a negator $\mathcal{N}$, i.e. for every $A \in \mathcal{F}(U)$ and every $x \in U$, $(\sim_{\mathcal{N}} A)(x) = \mathcal{N}(A(x))$. If $\mathcal{N} = \mathcal{N}_s$, then we will write $\sim A$ instead of $\sim_{\mathcal{N}} A$.

By *an implicator* (fuzzy implication operator) we mean a function $\mathcal{I} : I^2 \to I$ satisfying $\mathcal{I}(1, 0) = 0$ and $\mathcal{I}(1, 1) = \mathcal{I}(0, 1) = \mathcal{I}(0, 0) = 1$. An implicator $\mathcal{I}$ is called *left monotonic* (respectively, *right monotonic*) iff for every $\alpha \in I, \mathcal{I}(\cdot, \alpha)$ is decreasing (respectively, $\mathcal{I}(\alpha, \cdot)$ is increasing). If $\mathcal{I}$ is both left monotonic and right monotonic, then it is called *hybrid monotonic*.

Remark 1. It is easy to verify that $\mathcal{I}(\alpha, 1) = 1$ for all $\alpha \in I$ when $\mathcal{I}$ is a left monotonic implicator, and if $\mathcal{I}$ is right monotonic then $\mathcal{I}(0, \alpha) = 1$ for all $\alpha \in \mathcal{I}$.

An implicator $\mathcal{I}$ is said to be a *border implicator* (or it satisfies the neutrality principle [1]) if $\mathcal{I}(1, x) = x$ for all $x \in I$.

An implicator $\mathcal{I}$ is said to be *an EP implicator* (EP stands for exchange principle [10]) if it satisfies for all $\alpha, \beta, \gamma \in I$

$$\mathcal{I}(\alpha, \mathcal{I}(\beta, \gamma)) = \mathcal{I}(\beta, \mathcal{I}(\alpha, \gamma)). \tag{2}$$

An implicator $\mathcal{I}$ is said to be *a CP implicator* (CP stands for confinement principle [1]) if it satisfies for all $\alpha, \beta \in I$

$$\alpha \leq \beta \Longleftrightarrow \mathcal{I}(\alpha, \beta) = 1. \tag{3}$$

Several classes of implicators have been studied in the literature. We recall here the definitions of two main classes of operators [5].

Let $\mathcal{T}, \mathcal{S}$ and $\mathcal{N}$ be a t-norm, a t-conorm and a negator respectively. An implicator $\mathcal{I}$ is called

- an *S-implicator* based on $\mathcal{S}$ and $\mathcal{N}$ iff

$$\mathcal{I}(x, y) = \mathcal{S}(\mathcal{N}(x), y) \text{ for all } x, y \in I. \tag{4}$$

- an *R-implicator* (residual implicator) based on a left-continuous t-norm $\mathcal{T}$ iff for every $x, y \in [0, 1]$,

$$\mathcal{I}(x, y) = \sup\{\lambda \in [0, 1] : \mathcal{T}(x, \lambda) \leq y\}. \tag{5}$$

Given a negator $\mathcal{N}$ and a border implicator $\mathcal{I}$, one can define an $\mathcal{N}$-dual operator of $\mathcal{I}$, $\theta_{\mathcal{I}, \mathcal{N}} : I^2 \to I$ as follows [8]:

$$\theta_{\mathcal{I}, \mathcal{N}}(x, y) = \mathcal{N}(\mathcal{I}(\mathcal{N}(x), \mathcal{N}(y))), \quad x, y \in I. \tag{6}$$

Proposition 1. [13] *For a border implicator $\mathcal{I}$ and a negator $\mathcal{N}$, the following properties hold:*

(1) $\theta_{\mathcal{I}, \mathcal{N}}(1, 0) = \theta_{\mathcal{I}, \mathcal{N}}(1, 1) = \theta_{\mathcal{I}, \mathcal{N}}(0, 0) = 0.$

(2) $\theta_{\mathcal{I}, \mathcal{N}}(0, 1) = 1.$

(3) *If $\mathcal{N}$ is involutive, then $\theta_{\mathcal{I}, \mathcal{N}}(0, x) = x$ for all $x \in I$.*

(4) $\theta_{\mathcal{I}, \mathcal{N}}$ *is left monotonic (resp. right monotonic) whenever $\mathcal{I}$ is left monotonic (resp. right monotonic).*

(5) *If $\mathcal{I}$ is left monotonic, then $\theta_{\mathcal{I}, \mathcal{N}}(x, 0) = 0$ for all $x \in I$; and if $\mathcal{I}$ is right monotonic, then $\theta_{\mathcal{I}, \mathcal{N}}(1, x) = 0$ for all $x \in I$.*

(6) *If $\mathcal{I}$ is an EP implicator, then $\theta_{\mathcal{I}, \mathcal{N}}$ satisfies the exchange principle, i.e.*

$$\theta_{\mathcal{I}, \mathcal{N}}(x, \theta_{\mathcal{I}, \mathcal{N}}(y, z)) = \theta_{\mathcal{I}, \mathcal{N}}(y, \theta_{\mathcal{I}, \mathcal{N}}(x, z)), \quad \forall x, y, z \in I. \tag{7}$$

(7) *If $\mathcal{I}$ is a CP implicator, then $y \leq x$ iff $\theta_{\mathcal{I}, \mathcal{N}}(x, y) = 0$.*

3 $\mathcal{I}$-Fuzzy Rough Sets

Definition 1. *Let U be a nonempty universe of discourse and $\mathcal{T}$ a t-norm on I. A fuzzy subset $R \in \mathcal{F}(U \times U)$ is referred to as a fuzzy binary relation on U, $R(x, y)$ is the degree of relation between x and y, where $(x, y) \in U \times U$. If for each $x \in U$, $\bigvee_{y \in U} R(x, y) = 1$, then R is a serial fuzzy relation on U. R is a reflexive fuzzy relation if $R(x, x) = 1$ for all $x \in U$.*

Definition 2. *Let $\mathcal{I}$ be an implicator and $\mathcal{N}$ an involutive negator on I. For a fuzzy approximation space (U, R), i.e. R is a fuzzy binary relation on U, and any fuzzy set $A \in \mathcal{F}(U)$, the lower and upper $\mathcal{I}$-fuzzy rough approximations of A w.r.t. the approximation space (U, R), denoted as $\underline{R_{\mathcal{I}}}(A)$ and $\overline{R_{\mathcal{I}}}(A)$ respectively, are fuzzy sets of U whose membership functions are defined respectively by*

$$\underline{R_{\mathcal{I}}}(A)(x) = \bigwedge_{y \in U} \mathcal{I}(R(x, y), A(y)), \quad x \in U. \tag{8}$$

$$\overline{R}_{\mathcal{I}}(A)(x) = \bigvee_{y \in U} \theta_{\mathcal{I},\mathcal{N}}(\mathcal{N}(R(x,y)), A(y)), \quad x \in U. \tag{9}$$

The operators $\underline{R}_{\mathcal{I}}$ and $\overline{R}_{\mathcal{I}}$ from $\mathcal{F}(U)$ to $\mathcal{F}(U)$ are referred to as lower and upper $\mathcal{I}$-fuzzy rough approximation operators of (U, R) respectively, and the pair $(\underline{R}_{\mathcal{I}}(A), \overline{R}_{\mathcal{I}}(A))$ is called the $\mathcal{I}$-fuzzy rough set of A w.r.t. (U, R). We will call the system $(\mathcal{F}(U), \cup, \cap, \sim_{\mathcal{N}}, \underline{R}_{\mathcal{I}}, \overline{R}_{\mathcal{I}})$ an $\mathcal{I}$-fuzzy rough set algebra. When $\underline{R}_{\mathcal{I}}(A) = A = \overline{R}_{\mathcal{I}}(A)$, A is said to be definable w.r.t. (U, R).

The following theorem shows that the lower and upper $\mathcal{I}$-fuzzy rough approximation operators determined by a border implicator $\mathcal{I}$ and an involutive negator $\mathcal{N}$ are dual with each other.

Theorem 1. [13] Let (U, R) be a fuzzy approximation space, $\mathcal{I}$ a border implicator and $\mathcal{N}$ an involutive negator, then

$$\begin{aligned} &\text{(DFL)} \quad \underline{R}_{\mathcal{I}}(A) = \sim_{\mathcal{N}} \overline{R}_{\mathcal{I}}(\sim_{\mathcal{N}} A), \forall A \in \mathcal{F}(U), \\ &\text{(DFU)} \quad \overline{R}_{\mathcal{I}}(A) = \sim_{\mathcal{N}} \underline{R}_{\mathcal{I}}(\sim_{\mathcal{N}} A), \forall A \in \mathcal{F}(U). \end{aligned} \tag{10}$$

Remark 2. (1) When $\mathcal{N} = \mathcal{N}_s$, $\mathcal{I}$ is an R-implicator determined by a t-norm $\mathcal{T}$, and $\mathcal{S}$ is the t-conorm dual to $\mathcal{T}$, then it can be verified that the lower and upper approximation operators in Definition 2 degenerate to the dual fuzzy rough approximation operators defined by Mi and Zhang in [7], i.e.

$$\begin{aligned} \underline{R}_{\mathcal{I}}(A)(x) &= \bigwedge_{y \in U} \theta(R(x,y), A(y)), A \in \mathcal{F}(U), x \in U, \\ \overline{R}_{\mathcal{I}}(A)(x) &= \bigvee_{y \in U} \sigma(1 - R(x,y), A(y)), A \in \mathcal{F}(U), x \in U, \end{aligned} \tag{11}$$

where

$$\begin{aligned} \theta(a,b) &= \sup\{c \in I : \mathcal{T}(a,c) \le b\}, \ a, b \in I. \\ \sigma(a,b) &= \inf\{c \in I : \mathcal{S}(a,c) \ge b\}, \ a, b \in I. \end{aligned} \tag{12}$$

In [17], the lower and upper fuzzy rough approximation operators defined in Eq. (11) are called θ-lower and σ-upper approximation operators respectively.

(2) When $\mathcal{N} = \mathcal{N}_s$, $\mathcal{T}$ is a t-norm, $\mathcal{S}$ is the t-conorm dual to $\mathcal{T}$, and $\mathcal{I}$ is the $\mathcal{S}$-implicator determined by the t-norm $\mathcal{T}$, i.e. $\mathcal{I}(a,b) = \mathcal{S}(1 - a, b)$, then it can be verified that the lower and upper approximation operators in Definition 2 degenerate to the dual fuzzy rough approximation operators defined by Mi et al. [6] and Wu [11], i.e.

$$\begin{aligned} \underline{R}_{\mathcal{I}}(A)(x) &= \bigwedge_{y \in U} \mathcal{S}(1 - R(x,y), A(y)), A \in \mathcal{F}(U), x \in U, \\ \overline{R}_{\mathcal{I}}(A)(x) &= \bigvee_{y \in U} \mathcal{T}(R(x,y), A(y)), A \in \mathcal{F}(U), x \in U, \end{aligned} \tag{13}$$

In literature, the lower and upper approximation operators defined by Eq. (13) are called the $\mathcal{S}$-lower and $\mathcal{T}$-upper approximation operators respectively [17].

More specifically, when U is a finite set, if $\mathcal{T} = \min$ and $\mathcal{S} = \max$, then the lower and upper approximation operators in Definition 2 are no other than the dual fuzzy rough approximation operators defined by Wu and Zhang [14], i.e.

$$\underline{R}_{\mathcal{I}}(A)(x) = \bigwedge_{y \in U} ((1 - R(x,y)) \vee A(y)), A \in \mathcal{F}(U), x \in U,$$
$$\overline{R}_{\mathcal{I}}(A)(x) = \bigvee_{y \in U} (R(x,y) \wedge A(y)), A \in \mathcal{F}(U), x \in U. \tag{14}$$

By Definition 2, the following Proposition 2 can be derived.

Proposition 2. *Let $\mathcal{I}$ be a continuous border and CP implicator, and $\mathcal{N}$ an involutive negator. If R is a crisp binary relation on U, then*

$$\underline{R}_{\mathcal{I}}(A)(x) = \bigwedge_{y \in R_s(x)} A(y), A \in \mathcal{F}(U), x \in U,$$
$$\overline{R}_{\mathcal{I}}(A)(x) = \bigvee_{y \in R_s(x)} A(y), A \in \mathcal{F}(U), x \in U. \tag{15}$$

where $R_s(x) = \{y \in U : (x,y) \in R\}$ is the successor neighborhood of x w.r.t. R [16].

The $\mathcal{I}$-fuzzy rough approximation operators satisfy following properties [13]:
(FL1) $\underline{R}_{\mathcal{I}}(\bigcap_{j \in J} A_j) = \bigcap_{j \in J} \underline{R}_{\mathcal{I}}(A_j), A_j \in \mathcal{F}(U)(\forall j \in J, J$ is an index set).
(FU1) $\overline{R}_{\mathcal{I}}(\bigcup_{j \in J} A_j) = \bigcup_{j \in J} \overline{R}_{\mathcal{I}}(A_j), A_j \in \mathcal{F}(U)(\forall j \in J, J$ is an index set).
(FL2) $A \subseteq B \Longrightarrow \underline{R}_{\mathcal{I}}(A) \subseteq \underline{R}_{\mathcal{I}}(B), A, B \in \mathcal{F}(U).$
(FU2) $A \subseteq B \Longrightarrow \overline{R}_{\mathcal{I}}(A) \subseteq \overline{R}_{\mathcal{I}}(B), A, B \in \mathcal{F}(U).$
(FL3) $\underline{R}_{\mathcal{I}}(\bigcup_{j \in J} A_j) \supseteq \bigcup_{j \in J} \underline{R}_{\mathcal{I}}(A_j), A_j \in \mathcal{F}(U)(\forall j \in J, J$ is an index set).
(FU3) $\overline{R}_{\mathcal{I}}(\bigcap_{j \in J} A_j) \subseteq \bigcap_{j \in J} \overline{R}_{\mathcal{I}}(A_j), A_j \in \mathcal{F}(U)(\forall j \in J, J$ is an index set).

Theorem 2. [13] *Let (U, R) be a fuzzy approximation space, $\mathcal{I}$ a continuous border and CP implicator, and $\mathcal{N}$ an involutive negator. Then*

$$R \text{ is serial} \iff \text{(FL0) } \underline{R}_{\mathcal{I}}(\widehat{\alpha}) = \widehat{\alpha}, \; \forall \alpha \in I.$$
$$\iff \text{(FU0) } \overline{R}_{\mathcal{I}}(\widehat{\alpha}) = \widehat{\alpha}, \; \forall \alpha \in I.$$

According to Definition 2, Theorem 3 below can be induced.

Theorem 3. *Let $\mathcal{I}$ be a continuous border and CP implicator, and $\mathcal{N}$ an involutive negator. If U is a finite universe of discourse, and R a serial fuzzy relation on U, Then*

$$\text{(FLU0) } \underline{R}_{\mathcal{I}}(A) \subseteq \overline{R}_{\mathcal{I}}(A), \; \forall A \in \mathcal{F}(U).$$

Theorem 4. [13] *Let (U, R) be a fuzzy approximation space (i.e. R is a fuzzy relation on U), $\mathcal{I}$ a border and CP implicator, and $\mathcal{N}$ an involutive negator. Then*

$$R \text{ is reflexive} \iff \text{(FLR) } \underline{R}_{\mathcal{I}}(A) \subseteq A, \quad \forall A \in \mathcal{F}(U).$$
$$\iff \text{(FUR) } A \subseteq \overline{R}_{\mathcal{I}}(A), \quad \forall A \in \mathcal{F}(U).$$

4   Measurable Structures of $\mathcal{I}$-Fuzzy Rough Sets

In this section, we investigate the measurable structures of $\mathcal{I}$-fuzzy rough sets. We will establish the relationship between fuzzy approximation spaces and fuzzy measurable spaces.

Definition 3. [2] *Let U be a nonempty set. A subset $\mathcal{A}$ of $\mathcal{P}(U)$ is called a σ-algebra on U iff*
(A1) $U \in \mathcal{A}$,
(A2) $\{X_n : n \in \mathbf{N}\} \subset \mathcal{A}$ *(where $\mathbf{N}$ is the set of positive integer numbers)*
$\Longrightarrow \bigcup_{n\in\mathbf{N}} X_n \in \mathcal{A}$,
(A3) $X \in \mathcal{A} \Longrightarrow \sim X \in \mathcal{A}$.
The sets in $\mathcal{A}$ are called measurable sets and the pair $(U, \mathcal{A})$ is referred to as a measurable space.

With the definition we can see that $\emptyset \in \mathcal{A}$ and
(A2)$'$ $\{X_n : n \in \mathbf{N}\} \subset \mathcal{A} \Longrightarrow \bigcap_{n\in\mathbf{N}} X_n \in \mathcal{A}$.
If U is a finite universe of discourse, then condition (A2) in Definition 3 can be replaced by
(A2)$''$ $X, Y \in \mathcal{A} \Longrightarrow X \cup Y \in \mathcal{A}$.
In such a case, $\mathcal{A}$ is also called an *algebra*. If we denote

$$\mathcal{A}(x) = \cap\{X \in \mathcal{A} : x \in X\}, \quad x \in U, \tag{16}$$

then $\mathcal{A}(x) \in \mathcal{A}$, it can be checked that $\{\mathcal{A}(x) : x \in U\} \subseteq \mathcal{A}$ forms a partition of U and $\mathcal{A}(x)$ is called the *atom* of $\mathcal{A}$ containing x.

Definition 4. [3] *Let U be a nonempty set. A subset $\mathcal{F}$ of $\mathcal{F}(U)$ is called a fuzzy σ-algebra on U iff*
(FA1) $\widehat{\alpha} \in \mathcal{F}$ *for all $\alpha \in I$,*
(FA2) $\{A_n : n \in \mathbf{N}\} \subset \mathcal{F} \Longrightarrow \bigcup_{n\in\mathbf{N}} A_n \in \mathcal{F}$,
(FA3) $A \in \mathcal{F} \Longrightarrow \sim A \in \mathcal{F}$.
The sets in $\mathcal{F}$ are called fuzzy measurable sets and the pair $(U, \mathcal{F})$ is referred to as a fuzzy measurable space.

Clearly, condition (FA1) in Definition 4 implies that $\emptyset, U \in \mathcal{F}$. And similar to Definition 3, in terms of (FA3), condition (FA2) can be replaced by the following condition (FA2)$'$:
(FA2)$'$ $\{A_n : n \in \mathbf{N}\} \subset \mathcal{F} \Longrightarrow \bigcap_{n\in\mathbf{N}} A_n \in \mathcal{F}$.
If U is a finite universe of discourse, then condition (FA2) in Definition 4 can be replaced by
(FA2)$''$ $A, B \in \mathcal{F} \Longrightarrow A \cup B \in \mathcal{F}$.
In such a case, $\mathcal{F}$ is also called a *fuzzy algebra*.

Definition 5. *Let U be a finite universe of discourse. A fuzzy algebra $\mathcal{F}$ on U is said to be generated by a crisp algebra $\mathcal{A}$ iff for each $A \in \mathcal{F}$ there exist $a_i \in [0, 1], i = 1, 2, \ldots, k$, such that*

$$A(x) = \sum_{i=1}^{k} a_i 1_{C_i}(x), \quad x \in U, \tag{17}$$

where $\{C_1, C_2, \ldots, C_k\} = \{\mathcal{A}(x) : x \in U\}$ is the atoms of $\mathcal{A}$ and 1_{C_i} is the characteristic function of the set C_i, i.e., $1_{C_i}(x) = 1$ for $x \in C_i$ and 0 otherwise.

Remark 3. If a fuzzy algebra $\mathcal{F}$ on U is generated by a crisp algebra $\mathcal{A}$, we can see that, for each $A \in \mathcal{F}$, $A : U \to [0,1]$ is measurable w.r.t. $\mathcal{A}$-$\mathcal{B}([0,1])$, where $\mathcal{B}([0,1])$ is the Borel subsets of $[0,1]$, alternatively, A is measurable in sense of Zadeh [18].

By employing Theorems 3 and 4, we can conclude following Theorems 5 and 6 respectively.

Theorem 5. *Assume that $\mathcal{I}$ is a continuous border and CP implicator, and $\mathcal{N}$ an involutive negator. If U is a finite universe of discourse, and R is a serial fuzzy relation on U, denote*

$$\mathcal{F} = \{A \in \mathcal{F}(U) : \underline{R}_{\mathcal{I}}(A) = A = \overline{R}_{\mathcal{I}}(A)\}. \tag{18}$$

Then $\mathcal{F}$ is a fuzzy algebra on U.

Theorem 6. *Assume that $\mathcal{I}$ is a continuous border and CP implicator, and $\mathcal{N}$ an involutive negator. If (U, R) is a reflexive fuzzy approximation space, denote*

$$\mathcal{F} = \{A \in \mathcal{F}(U) : \underline{R}_{\mathcal{I}}(A) = A = \underline{R}_{\mathcal{I}}(A)\}. \tag{19}$$

Then $\mathcal{F}$ is a σ-fuzzy algebra on U.

By employing Proposition 2 we can obtain following Theorem 7 which presents the conditions that a fuzzy measurable space can be associated with a fuzzy approximation space such that the family of all definable sets induced from the fuzzy approximation space is exactly the class of all measurable sets in the given fuzzy measurable space.

Theorem 7. *Assume that $\mathcal{I}$ is a continuous border and CP implicator, and $\mathcal{N}$ an involutive negator. If U is a finite universe of discourse and $(U, \mathcal{F})$ a fuzzy measurable space which is generated by a crisp algebra $\mathcal{A}$, then there exists a reflexive fuzzy binary relation R on U such that*

$$\mathcal{F} = \{A \in \mathcal{F}(U) : \underline{R}_{\mathcal{I}}(A) = A = \overline{R}_{\mathcal{I}}(A)\}. \tag{20}$$

5   Conclusion

We have introduced a general type of relation-based $\mathcal{I}$-fuzzy rough sets by using constructive approach. By employing fuzzy logical operators on $[0,1]$, we have introduced $\mathcal{I}$-lower and $\mathcal{I}$-upper approximations of fuzzy sets with respect to a generalized fuzzy approximation space. We have further discussed the relationships between $\mathcal{I}$-fuzzy rough set algebras derived from fuzzy approximation spaces and fuzzy σ-algebras in fuzzy measurable spaces. We have shown that the family of all definable sets in a reflexive fuzzy approximation space forms a fuzzy σ-algebra. On the other hand, for a fuzzy algebra generated by a crisp algebra in a finite universe of discourse there must exist a fuzzy approximation space such that the family of all $\mathcal{I}$-definable sets is exactly the class of all measurable sets in the given fuzzy measurable space.

Acknowledgement. This work was supported by grants from the National Natural Science Foundation of China (Nos. 61272021, 61202206, and 61173181), the Zhejiang Provincial Natural Science Foundation of China (No. LZ12F03002), and Chongqing Key Laboratory of Computational Intelligence (No. CQ-LCI-2013-01).

References

1. Cornelis, C., Deschrijver, G., Kerre, E.E.: Implication in intuitionistic fuzzy and interval-valued fuzzy set theory: construction, classification, application. International Journal of Approximate Reasoning 35, 55–95 (2004)
2. Halmos, P.R.: Measure Theory. Van Nostrand-Reinhold, New York (1950)
3. Klement, E.P.: Fuzzy σ-algebras and fuzzy measurable functions. Fuzzy Sets and Systems 4, 83–93 (1980)
4. Klement, E.P., Mesiar, R., Pap, F.: Triangular Norms. In: Trends in Logic, vol. 8, Kluwer Academic Publishers, Dordrecht (2000)
5. Klir, G.J., Yuan, B.: Fuzzy Logic: Theory and Applications. Prentice-Hall, Englewood Cliffs (1995)
6. Mi, J.-S., Leung, Y., Zhao, H.-Y., Feng, T.: Generalized fuzzy rough sets determined by a triangular norm. Information Sciences 178, 3203–3213 (2008)
7. Mi, J.-S., Zhang, W.-X.: An axiomatic characterization of a fuzzy generalization of rough sets. Information Sciences 160, 235–249 (2004)
8. Ouyang, Y., Wang, Z.D., Zhang, H.-P.: On fuzzy rough sets based on tolerance relations. Information Sciences 180, 532–542 (2010)
9. Pawlak, Z.: Rough Sets: Theoretical Aspects of Reasoning about Data. Kluwer Academic Publishers, Boston (1991)
10. Ruan, D., Kerre, E.E.: Fuzzy implication operators and generalized fuzzy method of cases. Fuzzy Sets and Systems 54, 23–37 (1993)
11. Wu, W.-Z.: On some mathematical structures of T-fuzzy rough set algebras in infinite universes of discourse. Fundamenta Informaticae 108, 337–369 (2011)
12. Wu, W.-Z., Mi, J.-S.: Some mathematical structures of generalized rough sets in infinite universes of discourse. In: Peters, J.F., Skowron, A., Chan, C.-C., Grzymala-Busse, J.W., Ziarko, W.P. (eds.) Transactions on Rough Sets XIII. LNCS, vol. 6499, pp. 175–206. Springer, Heidelberg (2011)
13. Wu, W.-Z., Leung, Y., Shao, M.-W.: Generalized fuzzy rough approximation operators determined by fuzzy implicators. International Journal of Approximate Reasoning 54, 1388–1409 (2013)
14. Wu, W.-Z., Zhang, W.-X.: Constructive and axiomatic approaches of fuzzy approximation operators. Information Sciences 159, 233–254 (2004)
15. Wu, W.-Z., Zhang, W.-X.: Rough set approximations vs. measurable spaces. In: IEEE International Conference on Granular Computing, pp. 329–332. IEEE Press, New York (2006)
16. Yao, Y.Y.: Generalized rough set model. In: Polkowski, L., Skowron, A. (eds.) Rough Sets in Knowledge Discovery 1. Methodology and Applications, pp. 286–318. Physica-Verlag, Heidelberg (1998)
17. Yeung, D.S., Chen, D.G., Tsang, E.C.C., Lee, J.W.T., Wang, X.Z.: On the generalization of fuzzy rough sets. IEEE Transactions on Fuzzy Systems 13, 343–361 (2005)
18. Zadeh, L.A.: Probability measures of fuzzy events. Journal of Mathematical Analysis and Applications 23, 421–427 (1968)

Multi-label Attribute Evaluation
Based on Fuzzy Rough Sets

Lingjun Zhang, Qinghua Hu, Yucan Zhou, and Xiaoxue Wang

School of Computer Science and Technology, Tianjin University, Tianjin 150001, China
huqinghua@tju.edu.cn

Abstract. In multi-label learning task, each sample may be assigned with one or more labels. Moreover multi-label classification tasks are often characterized by high-dimensional and inconsistent attributes. Fuzzy rough sets are an effective mathematic tool for dealing with inconsistency and attribute reduction. In this work, we discuss multi-label attribute reduction within the frame of fuzzy rough sets. We analyze the definitions of fuzzy lower approximation in multi-label classification and give several improvements of the traditional algorithms. Furthermore, the attribute dependency function is defined to evaluate condition attributes. A multi-label attribute reduction algorithm is constructed based on the dependency function. Numerical experiments show the effectiveness of the proposed technique.

Keywords: Multi-label learning, attribute evaluation, fuzzy rough set, attribute dependency.

1 Introduction

There are many multi-label learning tasks in practice, such as web page classification, text categorization and image annotation. In these tasks, each sample may have one or multiple labels. For instance, a web page about technology may belong to computer and internet in web page classification [1-2, 14]; an image may be described not only the desert, but also the sunset in natural scene image annotation [3,16]. Moreover, multi-label learning tasks are usually described with high-dimensional, heterogeneous, and inconsistent features, which make the task more difficult.

Attribute reduction is of great significance for improving the performance of a multi-label learning task. The curse of dimensionality, caused by high-dimensionality, may not only require expensive cost to acquire and store, but also deteriorate the classification performance. The rough set theory, proposed by Pawlak in 1982 [4], is a powerful mathematical tool for modeling the imprecise and inconsistent information in knowledge acquisition, and has been successfully used in attribute reduction, dependency analysis and rule extraction. However, this model cannot directly deal with numerical and fuzzy attributes.

Fuzzy rough set model, introduced by Dubois and Prade [5], is developed to deal with numerical and fuzzy data, where fuzzy equivalence relations satisfying

C. Cornelis et al. (eds.): RSCTC 2014, LNAI 8536, pp. 100–108, 2014.

reflexivity, symmetry, and max-min transitivity, were considered. In addition, t-norms and s-conorms were used in defining fuzzy lower and upper approximation operators [6]. Fuzzy rough sets have been widely discussed in fuzzy attribute reduction [7-10].

In this work, we extend the fuzzy rough sets to learn multi-label tasks. We discuss the algorithm for computing fuzzy lower approximations in the context of multi-label classification. We study how to find a sample coming from a different class in multi-label datasets. Furthermore, the novel attribute dependency function is defined and an algorithm for multi-label attribute reduction is constructed. The performance our method is tested on the Web page classification and scene image annotation task.

2　Multi-label Attribute Reduction Based on Fuzzy Rough Sets

In this section, we discuss the definition of fuzzy lower approximation of multi-labels.

2.1　Fuzzy Rough Sets

In traditional two-class or multi-class classification, a decision table is formulated as $DT=<U, C, D>$, where U is a finite nonempty set of objects, C is a set of condition attributes, and D is a decision attribute. U is divided D into several subsets A_i. $R(x,y)$ is the fuzzy similar relation between sample x and y. The s-norm fuzzy lower approximation operator was defined as [11]:

$$\underline{R}_S A_i(x) = \inf_{y \in U} S(N(R(x,y)), A_i(y)) . \tag{1}$$

There are some commonly encountered operators used in fuzzy reasoning, where $S_M(a,b) = \max(a,b)$ is called the standard max operator; $S_P(a,b) = a+b-a*b$ is probabilistic sum operator; $S_L(a,b) = \min(a+b,1)$ is Lukasiewicz norm; while $S_{\cos}(a,b) = \min(a+b-a*b+\sqrt{2a-a^2}\sqrt{2b-b^2},0)$ is cosine norm.

2.2　Attribute Dependency in Multi-label Classification

Multi-label classification is a special task, which sample is associated with one or more class labels. An intuitive approach to solving multi-label learning problem is to convert it into multiple independent binary class [13] or into a multi-class single-label task [14].

$MDT=<U, C, D>$ is a multi-label decision table. $D=\{d_1,d_2,...,d_L\}$ contains L labels. The proper labels associated with sample x constitute an L-dimensional decision vector. If x has label d_k, the k-th dimension of x's decision vector is 1; and 0 otherwise. A decision label $d_k \in D$ divides U into two subsets P_k and N_k, If the k-th dimension of x_i's decision vector is 1, then x_i belongs to P_k ; otherwise, x_i belongs to N_k .

Take sample x_i as an example. The Near Miss of x_i, a subset of samples with different class label with x_i, denoted by $NM(x_i)$, is defined as:

$$NM_1(x_i) = \bigcap_k N_k, \text{ where } \{k \mid x_i \in p_k\} \quad \text{or}$$

$$NM_2(x_i) = \{x \mid \exists k, x_i \in P_k \wedge x \in N_k\}, \tag{2}$$

where $NM_1(x_i)$ *is* the subset of samples without any label in x_i; $NM_2(x_i)$ is the subset of samples which have different class labels with x_i.

Table 1. An artificial multi-label table, and the Near Miss of x_i

U	C	D			$NM_1(x_i)$	$NM_2(x_i)$
		Desert	Sunset	Tree		
x_1	...	0	1	0	$\{x_3, x_4, x_5\}$	$\{x_2, x_3, x_4, x_5, x_6\}$
x_2	...	1	1	0	$\{x_4\}$	$\{x_1, x_3, x_4, x_5, x_6, x_7\}$
x_3	...	1	0	1	$\{x_1, x_7\}$	$\{x_1, x_2, x_4, x_5, x_6, x_7\}$
x_4	...	0	0	1	$\{x_1, x_2, x_5, x_7\}$	$\{x_1, x_2, x_3, x_5, x_6, x_7\}$
x_5	...	1	0	0	$\{x_1, x_4, x_7\}$	$\{x_1, x_2, x_3, x_4, x_6, x_7\}$
x_6	...	1	1	1	$\varnothing$	$\{x_1, x_2, x_3, x_4, x_5, x_7\}$
x_7	...	0	1	0	$\{x_3, x_4, x_5\}$	$\{x_2, x_3, x_4, x_5, x_6\}$

An example is showed in Table 1 and Table 2. There are desert and sunset in an image x, and there is not a tree. If we want to pick out the images with different labels of x, you will first find the images which have not desert and sunset. However, if there is only sunset, you may first find the image without sunset, just like $NM_1(x)$. In extreme case, every label combination can be regarded as one class. The subset of samples which has different class labels with x is $NM_2(x)$.

In formula (1), U is divided D into two subsets A_1 and A_2. If $x \in A_1$, $y \in A_1$, i.e. $A_1(y) = 1$, we get $\underline{R_s}A_1(x) = 1$; If $x \in A_1$, $y \notin A_1$, i.e. $A_1(y)=0$, we get $\underline{R_s}A_1(x) = \inf_{y \notin A_1}(1 - R(x, y))$. As $x \notin A_2$, $\underline{R_s}A_2(x) = 0$. We get $\underline{R_s}D(x) = \sum_{A_i} \inf_{y \notin A_i}(1 - R(x, y)) = \inf_{y \notin A_i}(1 - R(x, y))$

Definition 2. Given a *MDT*=<U, C, D >, R is fuzzy similar relation on U with the conditional attribute set. The fuzzy lower approximation operators are defined as:

$$\underline{R_s}D(x) = \inf_{y \in NM(x)}(1 - R(x, y)) . \tag{3}$$

Formula (3) shows that the membership of x to the lower approximation of x's decision is determined by the closest sample with different class labels, while the membership of x to the lower approximation of the other decision is zero.

Definition 3. Given *MDT*=<U, C, D >, R is T-equivalence relation on U with the conditional attribute set $B \subseteq C$. The attribute dependency of B is defined as:

$$\gamma_B^S(D) = \frac{\sum_{i=1}^{n} \underline{R_S} D(x_i)}{|U|}.$$ (4)

The dependency between the decision and condition attributes reflects the approximation ability of the condition attributes.

2.3 Multi-label Attribution Reduction Algorithm Based on KFRS

There are two key problems in constructing an algorithm for attribute reduction: attribute evaluation and search strategies. The first one is used to evaluate the quality of condition attributes, and the second one is to search the optimal features with respect to the given evaluation function. We develop a ranking based algorithm for feature selection according to attribute dependency. It is easy to extend it to forward or backward greedy search.

```
Algorithm: Feature selection for Multi-label task
Input: U, C, D
Output: Attribute Reduct R
Begin
  for each condition attribute c∈C do
    for each sample x∈C  do
      choose different class sample set H of x
      Computing fuzzy low approximation R_S D(x)

      if H is empty set
        R_S D(x) = 1
      end
    end
    Computing attribute dependency degree γ_c^S(D)
  end
  attribute c∈C is ranked by γ_c^S(D)
end
```

3   Experiments

In this section, we evaluate the performance of the proposed method by comparing with other multi-label attribute reduction methods on web page categorization and scene image annotation.

3.1 Experimental Dataset and Setting

Two multi-label datasets are used in this work. The web page data set consist of 14 top-level categories and each category is classified into a number of second-level subcategories. We use 11 out of the 14 web page datasets. Each set is divided into a training set with 2000 documents and a test set with 3000 documents. The scene image annotation dataset contains 2407 natural scene images with 294 attributes and 6 labels.

The multi-label k-nearest neighbor method (ML-kNN) [15] is used for classification after attribute reduction. The parameter of ML-kNN controlling the strength of uniform prior is set 1, which yields the Laplace smoothing, and k=10. Five criteria are computed to evaluate the performance of the compared methods, including hamming loss (HL), ranking loss (RL), one-error (OE), coverage (CV) and average precision (AP).

3.2   Experimental Results

Table 2. The details of 11 web page classification datasets, their classification results(%) in the original attribute spaces and attribute reduction spaces, and the dimension present of attribute when average precision reached the maximum. " ↓ " indicates the smaller the better. " ↑ " indicates the larger the better.

Label	Attribute	HL(*10) ↓	RL ↓	OE ↓	CV(*0.01) ↓	AP ↑
1 26	462	60.69±0.94	15.02±0.45	62.13±3.15	54.04±0.95	51.64±1.66
	43%	59.83±0.37	14.87±0.20	61.16±0.86	53.84±0.61	51.99±0.36
2 30	438	26.76±0.65	3.74±0.21	11.79±0.62	21.81±0.71	88.23±0.41
	100%	26.71±0.56	3.72±0.19	11.73±0.56	21.65±0.70	88.26±0.41
3 33	681	40.85±0.42	8.99±0.10	44.06±0.54	42.86±0.76	63.37±0.28
	39%	39.32±0.48	8.88±0.22	43.26±0.65	42.59±0.98	64.14±0.40
4 33	550	39.24±0.33	8.16±0.17	52.42±0.49	36.02±0.67	59.73±0.34
	100%	41.43±0.31	8.12±0.19	52.76±0.83	36.13±0.82	59.54±0.49
5 21	640	62.81±1.82	12.49±0.73	59.18±6.46	33.68±1.46	56.50±3.50
	100%	61.89±2.06	12.26±0.72	58.33±6.68	33.09±1.48	57.21±3.80
6 32	612	39.40±1.75	5.29±0.23	33.24±1.40	29.92±0.94	73.24±1.06
	37%	42.76±0.48	6.08±0.25	40.08±1.36	32.73±1.07	68.89±1.03
7 22	606	61.61±0.56	19.18±0.41	69.75±1.01	50.67±0.82	46.12±0.67
	28%	59.70±0.57	18.09±0.26	64.52±0.78	48.40±0.72	49.66±0.49
8 33	793	32.14±0.44	9.31±0.17	48.86±0.77	35.51±0.62	61.13±0.51
	34%	30.32±0.31	8.61±0.18	46.35±0.68	33.10±0.62	63.20±0.46
9 40	743	33.48±0.43	11.82±0.20	58.18±0.73	61.20±1.02	53.00±0.43
	100%	33.62±0.49	11.83±0.29	57.90±0.97	61.67±1.20	53.24±0.69
10 39	1047	21.85±0.33	5.66±0.19	32.78±0.61	30.17±0.99	74.71±0.49
	90%	21.75±0.32	5.66±0.17	32.74±0.82	30.37±0.78	74.69±0.51
11 27	636	54.48±0.51	13.33±0.18	43.98±0.74	53.80±0.63	60.86±0.43
	100%	54.56±0.48	13.44±0.24	43.95±0.81	54.28±0.75	60.63±0.59

The results on the web page categorization are summarized in Table 3, where ML-kNN is used as the classifier. We can see that the classification performance is improved in most of the cases. Therefore, the candidate attributes are reduced. This result shows the effectiveness of our methods.

▨ 100100	▨ 100010	╲ 100001
⋍ 100000	≣ 010000	▨ 001110
▧ 001100	⋎ 001010	▨ 001000
⧄ 000110	‖ 000101	◱ 000100
⧂ 000011	⧇ 000010	⁘ 000001

(x-axis: 0, 100, 200, 300, 400)

Fig. 1. Number of images associated with different label sets

Fig. 1 describes the number of samples associated with different label sets. The candidate labels are desert, mountains, sea, sunset and trees. The x axis is the number of samples. The y axis is the different combinations of labels. The corresponding position is 1 if the sample has this label; 0 otherwise.

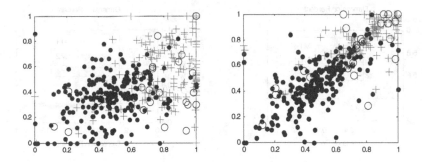

Fig. 2. The scatter chart of samples in two-attribute space

Fig. 2 gives the sample distribution in three kinds of label combination, which are "000001", "000010", and "000011". The sample marked with "." and "+" has one label, and class "O" have two labels including the label of "." and "+". Both of two condition attributes of Fig. 1. (a) and (b) cannot distinguish class "O" with other classes. However, our method showed in Fig. 1 (b) is better than the simple multi-label attribute reduction method (SEI) [15].

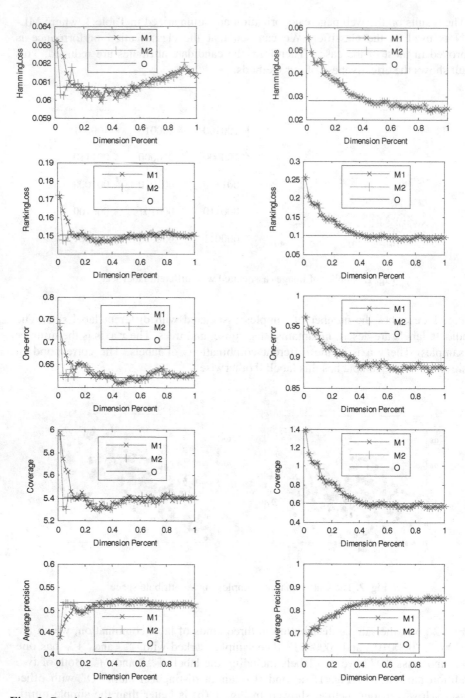

Fig. 3. Results of five evaluation metrics of M1 and M2 on the first web page dataset (left plots) and scene image dataset (right plots) (M1 and M2 are NM_1 and NM_2)

Fig. 3 shows that the five evaluation metrics of the proposed algorithms on the web page dataset and scene image dataset. We compute the classification performance with 1% to 100% of the features after the attributes have been sorted according to their attribute dependency NM_1 and NM_2. We can see the classification performance almost doesn't change when the attributes dimension present reached 20%. We find that the performance curve on image annotation is not as smooth as that on web page categorization and the sharp fluctuation weakened gradually with increase of attribute. In addition, the experimental result shows that the performances of classification used different features selected by NM_1 and NM_2 are almost the same.

4   Conclusion

This paper extends the fuzzy rough sets model to deal with multi-label attribute evaluation based on attribute dependency. The results show that the proposed method produces a higher performance in reduction attribute spaces on five multi-label classification evaluation criteria than original attribute space. In addition, both of the proposed methods are effective. The multi-label attribute reduction based on the fuzzy rough sets model and extended fuzzy rough sets models will be discussed in the future.

References

1. Ueda, N., Saito, K.: Parametric mixture models for multi-labeled text. In: Advances in Neural Information Processing Systems, pp. 737–744 (2003)
2. Yu, K., Yu, S., Tresp, V.: Multi-label informed latent semantic indexing. In: 28th Annual International ACM SIGIR Conference on Research and Development in Information Retrieval, pp. 258–265. ACM (2005)
3. Boutell, M.R., Luo, J., Shen, X., et al.: Learning multi-label scene classification. Pattern Recognition 37, 1757–1771 (2004)
4. Pawlak, Z., Grzymala-Busse, J., Slowinski, R., et al.: Rough sets. Communications of the ACM 38, 88–95 (1995)
5. Dubois, D., Prade, H.: Rough fuzzy sets and fuzzy rough sets. International Journal of General System 17, 191–209 (1990)
6. Morsi, N.N., Yakout, M.M.: Axiomatics for fuzzy rough sets. Fuzzy sets and Systems 100, 327–342 (1998)
7. Hu, Q., Yu, D., Xie, Z.: Information-preserving hybrid data reduction based on fuzzy-rough techniques. Pattern Recognition Letters 27, 414–423 (2006)
8. Hu, Q., Xie, Z., Yu, D.: Hybrid attribute reduction based on a novel fuzzy-rough model and information granulation. Pattern Recognition 40, 3509–3521 (2007)
9. Jensen, R., Shen, Q.: Fuzzy-rough sets assisted attribute selection. IEEE Transactions on Fuzzy Systems 15, 73–89 (2007)
10. Jensen, R., Shen, Q.: Semantics-preserving dimensionality reduction: rough and fuzzy-rough-based approaches. IEEE Transactions on Knowledge and Data Engineering 16, 1457–1471 (2004)
11. Yeung, D.S., Chen, D., Tsang, E.C.C., et al.: On the generalization of fuzzy rough sets. IEEE Transactions on Fuzzy Systems 13, 343–361 (2005)

12. Boutell, M.R., Luo, J., Shen, X., et al.: Learning multi-label scene classification. Pattern Recognition 37, 1757–1771 (2004)
13. Tsoumakas, G., Vlahavas, I.P.: Random *k*-labelsets: An ensemble method for multilabel classification. In: Kok, J.N., Koronacki, J., Lopez de Mantaras, R., Matwin, S., Mladenič, D., Skowron, A. (eds.) ECML 2007. LNCS (LNAI), vol. 4701, pp. 406–417. Springer, Heidelberg (2007)
14. Zhang, Y., Zhou, Z.H.: Multilabel dimensionality reduction via dependence maximization. ACM Transactions on Knowledge Discovery from Data 14, 14–26 (2010)
15. Zhang, M.L., Zhou, Z.H.: ML-KNN: A lazy learning approach to multi-label learning. Pattern Recognition 40, 2038–2048 (2007)
16. Yu, Y., Pedrycz, W., Miao, D.: Multi-label classification by exploiting label correlations. Expert Systems with Applications 41, 2989–3004 (2014)

A Comparison of Two Versions of the MLEM2 Rule Induction Algorithm Extended to Probabilistic Approximations

Patrick G. Clark[1] and Jerzy W. Grzymala-Busse[1,2]

[1] Department of Electrical Engineering and Computer Science,
University of Kansas, Lawrence, KS 66045, USA
[2] Department of Expert Systems and Artificial Intelligence,
University of Information Technology and Management,
35-225 Rzeszow, Poland
patrick.g.clark@gmail.com, jerzy@ku.edu

Abstract. A probabilistic approximation is a generalization of the standard idea of lower and upper approximations, defined for equivalence relations. Recently probabilistic approximations were additionally generalized to an arbitrary binary relation so that probabilistic approximations may be applied for incomplete data. We discuss two ways to induce rules from incomplete data using probabilistic approximations, by applying true MLEM2 algorithm and an emulated MLEM2 algorithm. In this paper we report novel research on a comparison of both approaches: new results of experiments on incomplete data with three interpretations of missing attribute values. Our results show that both approaches do not differ much.

Keywords: Probabilistic approximations, generalization of probabilistic approximations, concept probabilistic approximations, true MLEM2 algorithm, emulated MLEM2 algorithm.

1   Introduction

Probabilistic approximations were studied in a number of papers, mostly for completely specified data sets, in areas such as Bayesian rough sets, decision-theoretic rough sets, variable precision rough sets [11,13,14,15,16,17,18,19,20]. The indiscernibility relation, describing complete data, is an equivalence relation. Recently probabilistic approximations were extended to an arbitrary relation so that probabilistic approximations may be applied to incomplete data [8]. The first papers reporting experimental results on probabilistic approximations were [1,2].

In this paper we discuss two ways to induce rules from incomplete data using probabilistic approximations. The obvious way is to develop the rule induction system from scratch. For any concept X and given parameter α we compute its probabilistic approximation $appr_\alpha(X)$ and then induce rules from $appr_\alpha(X)$ using MLEM2 strategy. This approach will be called *true* MLEM2.

C. Cornelis et al. (eds.): RSCTC 2014, LNAI 8536, pp. 109–119, 2014.

On the other hand, we may use existing data mining system LERS (Learning from Examples using Rough Set theory) with the incorporated MLEM2 algorithm. LERS computes standard lower and upper approximations for any concept. Using this approach it is necessary to compute the probabilistic approximation, for given α, of the concept first, like in the former approach, and then pass it to LERS. The approximations are computed twice (the second time it is computed internally by the LERS system), so it is necessary to adjust the strength of all induced rules according to the original probabilistic approximation $appr_\alpha(X)$. This approach will be called *emulated* MLEM2. It is much easier to use the emulated approach than the true MLEM2 approach since while the set $appr_\alpha(X)$ is computed, the entire rule set is induced by the existing system, strength adjustment of rules is simple.

The open problem was how good such an emulated MLEM2 algorithm is. Our objective was to compare the true and emulated versions of the MLEM2 algorithm on the basis of an error rate, computed by ten-fold cross validation for eight incomplete data sets, all with 35% of missing attribute values. We used three interpretations of missing attribute values: a lost value, denoted by ?; attribute-concept value, denoted by $-$, and "do not care" condition, denoted by *. Incomplete data sets were obtained by random replacement of actual, specified attribute values by signs of ? for lost values, $-$ for attribute-concept values, and * for "do not care" conditions. Therefore we used 24 data sets for our experiments.

In the lost value interpretation of missing attribute values we assume that the original attribute value is lost, e.g., was erased, and that we should induce rules form existing, specified attribute values. The second interpretation of missing attribute values, the attribute-concept value, is based on the idea that such missing attribute values may be replaced by any actual attribute value restricted to the concept to which the case belongs. For example, if our concept is a specific disease, an attribute is a diastolic pressure, and all patients affected by the disease have high or very high diastolic pressure, a missing attribute value of the diastolic pressure for a sick patient will be high or very-high. The third interpretation of missing attribute values, the "do not care" condition, is interpreted as a situation in which it does not matter what is the attribute value. Such value may be replaced by any value from the set of all possible attribute values.

In our experiments we compared true and emulated versions of the MLEM2 algorithm using the error rate, a result of ten-fold cross validation, as the quality criterion. Among 24 data sets used for experiments, for six data sets, for all eleven values of the parameter α, results were identical; for other 14 data sets results did not differ significantly (we used the Wilcoxon matched-pairs signed rank test, 5% significance level, two-tailed test). For three other data sets, the true MLEM2 algorithm was better than emulated, for remaining one data set the emulated MLEM2 algorithm was better than the true one.

2   Incomplete Data Sets

An example of incomplete data set is presented in Table 1. In Table 1, the set A of all attributes consists of three variables *Temperature*, *Headache* and *Cough*.

A *concept* is a set of all cases with the same decision value. There are two concepts in Table 1, the first one contains cases 1, 2, 3 and 4 and is characterized by the decision value *no* of decision *Flu*. The other concept contains cases 5, 6, 7 and 8 and is characterized by the decision value *yes*.

Table 1. An incomplete data set

Case	Attributes			Decision
	Temperature	Headache	Cough	Flu
1	normal	no	no	no
2	*	yes	no	no
3	normal	–	yes	no
4	high	*	?	no
5	high	yes	*	yes
6	very-high	*	no	yes
7	*	no	yes	yes
8	very-high	no	yes	yes

The fact that an attribute a has the value v for the case x will be denoted by $a(x) = v$. The set of all cases will be denoted by U. In Table 1, $U = \{1, 2, 3, 4, 5, 6, 7, 8\}$.

For complete data sets, an attribute-value pair $(a, v) = t$, a *block* of t, denoted by $[t]$, is a set of all cases from U such that for attribute a have value v. An *indiscernibility relation* R on U is defined for all $x, y \in U$ by

$$xRy \text{ if and only if } a(x) = a(y) \text{ for all } a \in A.$$

For incomplete decision tables the definition of a block of an attribute-value pair must be modified in the following way [5,7]:

- If for an attribute a there exists a case x such that $a(x) =?$, i.e., the corresponding value is lost, then the case x should not be included in any blocks $[(a, v)]$ for all values v of attribute a,
- If for an attribute a there exists a case x such that the corresponding value is an attribute-concept value, i.e., $a(x) = -$, then the corresponding case x should be included in blocks $[(a, v)]$ for all specified values $v \in V(x, a)$ of attribute a, where

$$V(x, a) = \{a(y) \mid a(y) \text{ is specified}, y \in U, d(y) = d(x)\},$$

and d is the decision.
- If for an attribute a there exists a case x such that the corresponding value is a "do not care" condition, i.e., $a(x) = *$, then the case x should be included in blocks $[(a, v)]$ for all specified values v of attribute a.

For a case $x \in U$ the *characteristic set* $K_B(x)$ is defined as the intersection of the sets $K(x, a)$, for all $a \in B$, where B is a subset of the set A of all attributes and the set $K(x, a)$ is defined in the following way:

- If $a(x)$ is specified, then $K(x, a)$ is the block $[(a, a(x))]$ of attribute a and its value $a(x)$,
- If $a(x) =?$ or $a(x) = *$ then the set $K(x, a) = U$,
- If $a(x) = -$, then the corresponding set $K(x, a)$ is equal to the union of all blocks of attribute-value pairs (a, v), where $v \in V(x, a)$ if $V(x, a)$ is nonempty. If $V(x, a)$ is empty, $K(x, a) = U$.

The characteristic set $K_B(x)$ may be interpreted as the set of cases that are indistinguishable from x using all attributes from B and using a given interpretation of missing attribute values.

For the data set from Table 1, the set of blocks of attribute-value pairs is

$[(Temperature, normal)] = \{1, 2, 3, 7\}$,
$[(Temperature, high)] = \{2, 4, 5, 7\}$,
$[(Temperature, very - high)] = \{2, 6, 7, 8\}$,
$[(Headache, no)] = \{1, 3, 4, 6, 7, 8\}$,
$[(Headache, yes)] = \{2, 3, 4, 5, 6\}$,
$[(Cough, no)] = \{1, 2, 5, 6\}$,
$[(Cough, yes)] = \{3, 5, 7, 8\}$.

The corresponding characteristic sets are
$K_A(1) = \{1\}$,
$K_A(2) = \{2, 5, 6\}$,
$K_A(3) = \{3, 7\}$,
$K_A(4) = \{2, 4, 5, 7\}$,
$K_A(5) = \{2, 4, 5\}$,
$K_A(6) = \{2, 6\}$,
$K_A(7) = \{3, 7, 8\}$,
$K_A(8) = \{7, 8\}$.

3   Probabilistic Approximations

For incomplete data set there exists a number of different definitions of approximations, in this paper we will use only *concept* approximations, we will skip the word *concept*.

The B-*lower approximation* of X, denoted by $\underline{appr}(X)$, is defined as follows

$$\cup \{K_B(x) \mid x \in X, K_B(x) \subseteq X\}.$$

Such lower approximations were introduced in [5,6].

The B-*upper approximation* of X, denoted by $\overline{appr}(X)$, is defined as follows

$$\cup \{K_B(x) \mid x \in X, K_B(x) \cap X \neq \emptyset\} = \cup \{K_B(x) \mid x \in X\}.$$

These approximations were studied in [5,6,12].

For incomplete data sets there exist a few definitions of probabilistic approximations, we will use only *concept* probabilistic approximations, again, we will skip the word *concept*.

A B-probabilistic approximation of the set X with the threshold α, $0 < \alpha \leq 1$, denoted by B-*appr*$_\alpha(X)$, is defined as follows

$$\cup \{K_B(x) \mid x \in X, \ Pr(X|K_B(x)) \geq \alpha\},$$

where $Pr(X|K_B(x)) = \frac{|X \cap K_B(x)|}{|K_B(x)|}$ is the conditional probability of X given $K_B(x)$. A-probabilistic approximations of X with the threshold α will be denoted by $appr_\alpha(X)$.

For Table 1 and the concept $X = [(\textit{Flu, no})] = \{1, 2, 3, 4\}$, for any characteristic set $K_A(x)$, $x \in U$, all conditional probabilities $P(X|K_A(x))$ are presented in Table 2.

Table 2. Conditional probabilities

$K_A(x)$	{1}	{2, 4, 5}	{2, 4, 5, 7}	{2, 6}	{3, 7}	{2, 5, 6}	{3, 7, 8}	{7, 8}
$P(\{1,2,3,4\} \mid K_A(x))$	1.0	0.667	0.5	0.5	0.5	0.333	0.333	0

There are four distinct conditional probabilities $P(\{1, 2, 3, 4\} \mid K_A(x))$, $x \in U$: 1.0, 0.667, 0.5, and 0.333. Therefore, there exist at most four distinct probabilistic approximations of $\{1, 2, 3, 4\}$ (in our example, there are only three distinct probabilistic approximations of $\{1, 2, 3, 4\}$). A probabilistic approximation $appr_\beta(\{1, 2, 3, 4\})$ not listed below is equal to the closest probabilistic approximation $appr_\alpha(\{1, 2, 3, 4\})$ with α larger than or equal to α. For example, $appr_{0.2}(\{1, 2, 3, 4\}) = appr_{0.333}(\{1, 2, 3, 4\})$. For Table 1, all distinct probabilistic approximations are

$appr_{0.333}(\{1, 2, 3, 4\}) = \{1, 2, 3, 4, 5, 6, 7\},$

$appr_{0.5}(\{1, 2, 3, 4\}) = \{1, 2, 3, 4, 5, 7\},$

$appr_1(\{1, 2, 3, 4\}) = appr_{0.667}(\{1, 2, 3, 4\}) = \{1\},$

$appr_{0.333}(\{5, 6, 7, 8\}) = \{2, 3, 4, 5, 6, 7, 8\},$

$appr_{0.5}(\{5, 6, 7, 8\}) = \{2, 3, 6, 7, 8\},$

$appr_{0.667}(\{5,6,7,8\}) = \{3,7,8\},$

$appr_1(\{5,6,7,8\}) = \{7,8\}.$

4   Rule Induction

In this section we will discuss two different ways to induce rule sets using probabilistic approximations: true MLEM2 and emulated MLEM2.

4.1   True MLEM2

In the true MLEM2 approach, for a given concept X and parameter α, first we compute the probabilistic approximation $A_\alpha(X)$. The set $A_\alpha(X)$ is a union of characteristic sets, so it is globally definable [10]. Thus, we may use the MLEM2 strategy to induce rule sets [3,4] by inducing rules directly from the set $A_\alpha(X)$. For example, for Table 1, for the concept $[(Flu, yes)] = \{5, 6, 7, 8\}$ and for the probabilistic approximation $appr_1(\{5,6,7,8\}) = \{7,8\}$, using the true MLEM2 approach, the following single rule is induced

2, 2, 2
(Temperature, very-high) & (Cough, yes) -> (Flu, yes).

Rules are presented in the LERS format, every rule is associated with three numbers: the total number of attribute-value pairs on the left-hand side of the rule, the total number of cases correctly classified by the rule during training, and the total number of training cases matching the left-hand side of the rule, i.e., the rule domain size.

4.2   Emulated MLEM2

We will discuss how the existing rough set based data mining systems, such as LERS (Learning from Examples based on Rough Sets), may be used to induce rules using probabilistic approximations. All what we need to do, for every concept, is to modify the input data set, run LERS, and then edit the induced rule set [9]. We will illustrate this procedure by inducing a rule set for Table 1 and the concept $[(Flu, yes)] = \{5, 6, 7, 8\}$ using the probabilistic approximation $appr_1(\{5,6,7,8\}) = \{7,8\}$. First, a new data set should be created in which for all cases that are members of the set $appr_1(\{5,6,7,8\})$ the decision values are copied from the original data set (Table 1). For all remaining cases, those not being in the set $appr_1(\{5,6,7,8\})$, a new decision value is introduced. In our experiments the new decision value was named SPECIAL. Thus a new data set is created, see Table 3.

This data set is input into the LERS data mining system. The LERS system computes the upper concept approximation of the set $\{7, 8\}$to be $\{3, 7, 8\}$, and using this approximation, computes the corresponding final modified data set.

Table 3. A preliminary modified data set

	Attributes			Decision
Case	Temperature	Headache	Cough	Flu
1	normal	no	no	SPECIAL
2	*	yes	no	SPECIAL
3	normal	–	yes	SPECIAL
4	high	*	?	SPECIAL
5	high	yes	*	SPECIAL
6	very-high	*	no	SPECIAL
7	*	no	yes	yes
8	very-high	no	yes	yes

The MLEM2 algorithm induces the following preliminary rule set from the final modified data sets

1, 5, 5
(Headache, yes) -> (Flu, SPECIAL)
1, 3, 4
(Temperature, normal) -> (Flu, SPECIAL)
2, 2, 3
(Cough, yes) & (Headache, no) -> (Flu, yes)

where the three numbers that precede every rule are computed from Table 3. Obviously, only the last rule

2, 2, 3
(Cough, yes) & (Headache, no) -> (Flu, yes)

should be saved and the remaining two rules should be deleted in computing the final rule set.

In the preliminary rule set the three numbers that precede every rule are adjusted taking into account the preliminary modified data set. Thus during classification of unseen cases by the LERS classification system rules describe the original concept probabilistic approximation of the concept X.

5 Experiments

In our experiments we used eight real-life data sets taken from the University of California at Irvine *Machine Learning Repository*, see Table 4. These data sets were enhanced by replacing 35% of existing attribute values by symbols of lost values, i.e., question marks. All data sets with lost values were edited, symbols of lost values were replaced by symbols of attribute-concept values (hyphens), and then by "do not care" conditions (stars). Thus, for any data sets from Table 4,

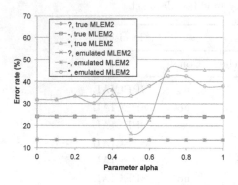

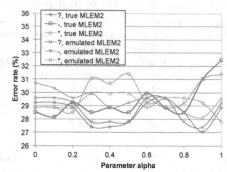

Fig. 1. Error rates for the *Bankruptcy* data set

Fig. 2. Error rates for the *Breast cancer* data set

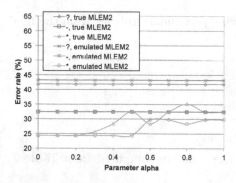

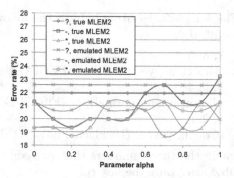

Fig. 3. Error rate for the *Echocardiogram* data set

Fig. 4. Error rate for the *Hepatitis* data set

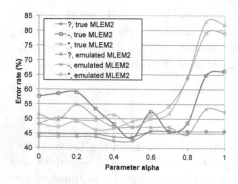

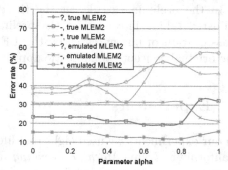

Fig. 5. Rule set size for the *Image segmentation* data set

Fig. 6. Error rate for the *Iris* data set

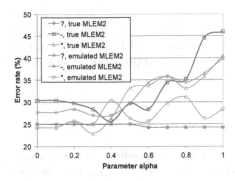 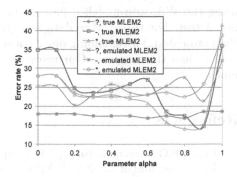

Fig. 7. Error rate for the *Lymphography* data set

Fig. 8. Error rate for the *Wine recognition* data set

three data sets were used for experiments, so that the total number of data sets was 24.

Our main objective was to compare both approaches to rule induction, true MLEM2 and emulated MLEM2, in terms of an error rate. Results of our experiments are presented in Figures 1–8, with lost values denoted by ?, attribute-concept values denoted by −, and "do not care" conditions denoted by *.

Table 4. Data sets used for experiments

Data set	Number of		
	cases	attributes	concepts
Bankruptcy	66	5	2
Breast cancer	277	9	2
Echocardiogram	74	7	2
Hepatitis	155	19	2
Image segmentation	210	19	7
Iris	150	4	3
Lymphography	148	18	4
Wine recognition	178	13	3

6   Conclusions

In our experiments we compared true and emulated versions of the MLEM2 algorithm using the error rate, a result of ten-fold cross validation, as the quality criterion. Among the 24 data sets used for experiments, for six data sets and for all eleven values of the parameter α, error rates were identical between the two methods. For the other 14 data sets results did not differ significantly when using the Wilcoxon matched-pairs signed rank test, 5% significance level, two-tailed test. For three other data sets, the true MLEM2 algorithm was better than emulated, for remaining one data set the emulated MLEM2 algorithm was better than the true one. Our results show that both approaches do not differ much.

References

1. Clark, P.G., Grzymala-Busse, J.W.: Experiments on probabilistic approximations. In: Proceedings of the 2011 IEEE International Conference on Granular Computing, pp. 144–149 (2011)
2. Clark, P.G., Grzymala-Busse, J.W.: Rule induction using probabilistic approximations and data with missing attribute values. In: Proceedings of the 15-th IASTED International Conference on Artificial Intelligence and Soft Computing, ASC 2012, pp. 235–242 (2012)
3. Grzymala-Busse, J.W.: A new version of the rule induction system LERS. Fundamenta Informaticae 31, 27–39 (1997)
4. Grzymala-Busse, J.W.: MLEM2: A new algorithm for rule induction from imperfect data. In: Proceedings of the 9th International Conference on Information Processing and Management of Uncertainty in Knowledge-Based Systems, pp. 243–250 (2002)
5. Grzymala-Busse, J.W.: Rough set strategies to data with missing attribute values. In: Workshop Notes, Foundations and New Directions of Data Mining, in Conjunction with the 3rd International Conference on Data Mining, pp. 56–63 (2003)
6. Grzymala-Busse, J.W.: Data with missing attribute values: Generalization of indiscernibility relation and rule induction. Transactions on Rough Sets 1, 78–95 (2004)
7. Grzymala-Busse, J.W.: Three approaches to missing attribute values—a rough set perspective. In: Proceedings of the Workshop on Foundation of Data Mining, in Conjunction with the Fourth IEEE International Conference on Data Mining, pp. 55–62 (2004)
8. Grzymała-Busse, J.W.: Generalized parameterized approximations. In: Yao, J., Ramanna, S., Wang, G., Suraj, Z. (eds.) RSKT 2011. LNCS, vol. 6954, pp. 136–145. Springer, Heidelberg (2011)
9. Grzymala-Busse, J.W.: Generalized probabilistic approximations. Transactions on Rough Sets 16, 1–16 (2013)
10. Grzymala-Busse, J.W., Rzasa, W.: Local and global approximations for incomplete data. In: Greco, S., Hata, Y., Hirano, S., Inuiguchi, M., Miyamoto, S., Nguyen, H.S., Słowiński, R. (eds.) RSCTC 2006. LNCS (LNAI), vol. 4259, pp. 244–253. Springer, Heidelberg (2006)
11. Grzymala-Busse, J.W., Ziarko, W.: Data mining based on rough sets. In: Wang, J. (ed.) Data Mining: Opportunities and Challenges, pp. 142–173. Idea Group Publ., Hershey (2003)
12. Lin, T.Y.: Topological and fuzzy rough sets. In: Slowinski, R. (ed.) Intelligent Decision Support. Handbook of Applications and Advances of the Rough Sets Theory, pp. 287–304. Kluwer Academic Publishers, Dordrecht (1992)
13. Pawlak, Z., Skowron, A.: Rough sets: Some extensions. Information Sciences 177, 28–40 (2007)

14. Pawlak, Z., Wong, S.K.M., Ziarko, W.: Rough sets: probabilistic versus deterministic approach. International Journal of Man-Machine Studies 29, 81–95 (1988)
15. Ślęzak, D., Ziarko, W.: The investigation of the bayesian rough set model. International Journal of Approximate Reasoning 40, 81–91 (2005)
16. Wong, S.K.M., Ziarko, W.: INFER—an adaptive decision support system based on the probabilistic approximate classification. In: Proceedings of the 6-th International Workshop on Expert Systems and their Applications, pp. 713–726 (1986)
17. Yao, Y.Y.: Probabilistic rough set approximations. International Journal of Approximate Reasoning 49, 255–271 (2008)
18. Yao, Y.Y., Wong, S.K.M.: A decision theoretic framework for approximate concepts. International Journal of Man-Machine Studies 37, 793–809 (1992)
19. Ziarko, W.: Variable precision rough set model. Journal of Computer and System Sciences 46(1), 39–59 (1993)
20. Ziarko, W.: Probabilistic approach to rough sets. International Journal of Approximate Reasoning 49, 272–284 (2008)

Mining Significant Granular Association Rules for Diverse Recommendation

Fan Min[1] and William Zhu[2]

[1] Department of Computer Science, Southwest Petroleum University,
Chengdu 610500, China
[2] Lab of Granular Computing, Minnan Normal University, Zhangzhou 363000, China
minfanphd@163.com

Abstract. Granular association rule is a new technique to build recommender systems. The quality of a rule is often evaluated by the confidence measure, namely the probability that users purchase or rate certain items. Unfortunately, the confidence-based approach tends to suggest popular items to users, and novel patterns are often ignored. In this paper, we propose to mine significant granular association rules for diverse and novel recommendation. Generally, a rule is significant if the recommended items favor respective users more than others; while a recommender is diverse if it recommends different items to different users. We define two sets of measures to evaluate the quality of a rule as well as a recommender. Then we propose a significance-based approach seeking top-k significant rules for each user. Results on the MovieLens dataset show that the new approach provides more significant and diverse recommendations than the confidence-based one.

Keywords: Recommender system, granule association rule, confidence, significance, entropy.

1 Introduction

Granular association rule [1,2] is a new relational data mining technique [3,4] to build recommender systems [5] that suggest item of interest to users. From the MovieLens dataset [6], we obtain a granular association rule "$\langle$Gender = F$\rangle$ $\bigwedge\langle$Occupation = student$\rangle$ $\Rightarrow$ $\langle$Release-decade = 1990s$\rangle$ $\bigwedge\langle$Action = 1$\rangle$ (0.28)." The rule is read as "female students rate action movies released in 1990s with a probability of 0.28." Here "female students," "action movies released in 1990s," and "0.28" are the source granule, the target granule, and the confidence, respectively of the rule. Since this type of rules are user and item independent, they are especially appropriate for cold-start recommendation [7].

The confidence measure [2] is natural to evaluate the strength of the rule. Rules with high confidence are desired to build recommender systems with high accuracy. Unfortunately, a confidence-based recommender often suggests popular items which are rather obvious, and there is a need to define novelty measures for practical purposes [8].

C. Cornelis et al. (eds.): RSCTC 2014, LNAI 8536, pp. 120–127, 2014.
© Springer International Publishing Switzerland 2014

In this paper, we propose to mine significant granular association rules for diverse and novel recommendation. For this purpose, we define two sets of measures, namely significance measures and diversity measures. First, the significance of a rule is its confidence divided by the popularity of respective items. In other words, a rule is significant if the recommended items favor respective users more than others. Therefore significant rules are often interesting and novel. Second, based on the item recommendation times, three diversity measures are defined. These are the fraction of recommended items, the maximal recommendation ratio, and the recommendation entropy. Generally, a recommender is diverse if it recommends different items to different users.

We develop a significance-based approach which employs top-k significant rules for each user. The approach is composed of two stages. In the first stages, the rule set is constructed from the training set given the source and the target coverage thresholds. Similar to the threshold employed in frequent itemset mining [4,9], these two thresholds filter out small granules to avoid overfitting. In the second stage, each user in the testing set is matched to the rule set, and the top-k significant rules are employed for recommendation.

Experiments are undertaken on the MovieLens dataset [6] using our open source software Grale [10]. The technique proposed in [11] is employed to deal with multi-valued data. Compared with the confidence-based approach, the significance-based approach provides more significant and diverse, however less accurate recommendations. Therefore we should choose appropriate approaches for different objectives.

2   Preliminaries

In this section, we review some preliminary knowledge such as many-to-many entity-relationship systems and information granules [1,2]. We also review granular association rules with three measures [2].

Definition 1. *[1] A many-to-many entity-relationship system (MMER) is a 5-tuple $ES = (U, A, V, B, R)$, where (U, A) and (V, B) are two information systems, and $R \subseteq U \times V$ is a binary relation from U to V.*

In this context, U is the set of all users, A is the set of all user attributes, V is the set of all items, and B is the set of all item attributes. Users and items can be described by information granules [2,12]. In an information system, any $A' \subseteq A$ induces an equivalence relation $E_{A'} = \{(x, y) \in U \times U | \forall a \in A', a(x) = a(y)\}$, and partitions U into a number of disjoint subsets called *blocks* or *granules*. The granule containing $x \in U$ is $E_{A'}(x) = \{y \in U | \forall a \in A', a(y) = a(x)\}$.

Definition 2. *[12] A granule is a triple*

$$G = (g, i(g), e(g)), \tag{1}$$

where g is the name assigned to the granule, $i(g)$ is a representation of the granule, and $e(g)$ is a set of objects that are instances of the granule.

(A', x) determines a granule in an information system. Hence $g = g(A', x)$ is a natural name to the granule. $i(g)$ can be formalized as the conjunction of respective attribute-value pairs, i.e., $i(g(A', x)) = \bigwedge_{a \in A'} \langle a : a(x) \rangle$. $e(g(A', x)) = E_{A'}(x)$.

Now we discuss the means for connecting users and items. A *granular association rule* [1,2] is an implication of the form

$$(GR) : \bigwedge_{a \in A'} \langle a : a(x) \rangle \Rightarrow \bigwedge_{b \in B'} \langle b : b(y) \rangle, \tag{2}$$

where $A' \subseteq A$ and $B' \subseteq B$.

The set of objects meeting the left-hand side of the granular association rule is $LH(GR) = E_{A'}(x)$; while the set of objects meeting the right-hand side of the granular association rule is $RH(GR) = E_{B'}(y)$.

Let us look at an example granular association rule "female students rate action movies released in 1990s with a probability of 0.28, 6% users are female students and 24% movies are action ones released in 1990s." Here 6%, 24%, and 0.28 are the source coverage, the target coverage, and the confidence, respectively. Formally, the *source coverage* of GR is $scov(GR) = |LH(GR)|/|U|$; while the *target coverage* of GR is $tcov(GR) = |RH(GR)|/|V|$. The *confidence* of GR is the probability that a user chooses an item, namely

$$conf(GR) = \frac{|(LH(GR) \times RH(GR)) \cap R|}{|LH(GR)| \times |RH(GR)|}. \tag{3}$$

A recommender can be viewed a function $RC : U \rightarrow 2^V$. $RC(x)$ is the set of items recommended to user $x \in U$. We propose the following definition.

Definition 3. *The accuracy of RC on $X \subseteq U$ is*

$$acc(X, RC) = \frac{\sum_{x \in X} |(\{x\} \times RC(x)) \cap R|}{\sum_{x \in X} |RC(x)|}. \tag{4}$$

Although not formally defined in [2], $acc(U, RC)$ is essentially employed to evaluate the quality of the recommender. To build a recommender with high accuracy, a set of granular association rules are first generated. Then for each user, all rules matching her are ranked according to the confidence measure, and the top-k rules are employed for recommendation. This approach is called top-k confident rules recommendation. We will compare our new approach with this one through experiments.

3   Two Sets of New Measures

In this section we define two sets of new measures. The significance measures specify the degree a set of items favor one group of users over others. The diversity measures specify how different the recommender makes for different users.

3.1 Significance Measures

In many applications there are popular items. For example, many people buy bread in a food store, or watch movies that won an Oscar award. However, recommending those items to people is uninteresting due to common sense knowledge. For this reason, we would like to recommend special items to special people. We propose the following definitions.

Definition 4. *Let* $ES = (U, A, V, B, R)$ *be an MMER,* $\emptyset \subset X \subseteq U$ *and* $\emptyset \subset Y \subseteq V$. *The significance of recommending* Y *to* X *is*

$$sig(X, Y) = \begin{cases} 1 & Y = \emptyset; \\ acc(X, Y)/acc(U, Y) & otherwise. \end{cases} \tag{5}$$

Here $acc(U, Y)$ indicate the popularity of item set Y. In other words, there is a penalty on popular items such that they are not recommended to everyone. The consideration of emptyset is for the new item case.

The same term can be employed to evaluate the quality of a rule.

Definition 5. *The significance of a granular association rule* GR *is*

$$sig(GR) = sig(LH(GR), RH(GR)). \tag{6}$$

The quality of the recommender RC can be evaluated as follows.

Definition 6. *The significance of recommender* RC *on* $x \in U$ *is*

$$sig(x, RC) = sig(\{x\}, RC(x)) = acc(\{x\}, RC(x))/acc(U, RC(x)); \tag{7}$$

while the significance of RC *over* U *is*

$$sig(U, RC) = \sum_{x \in U} sig(x, RC)/|U|. \tag{8}$$

We will employ $sig(GR)$ to build the rule set, and $sig(U, RC)$ to evaluate the performance of the recommender.

3.2 Diversity Measures

In this subsection, we propose three measures to evaluate the diversity of the recommender RC on an MMER $ES = (U, A, V, B, R)$. First, we calculate how many items are recommended at least once. The fraction of recommended items is given by

$$fri(ES, RC) = |\cup_{x \in U} RC(x)|/|V|. \tag{9}$$

Naturally, higher $fri(ES, RC)$ indicates diverse recommendation since more items are recommended.

Second, we look at most frequently recommended items. Any $v \in V$ is recommended $|\{x \in U | v \in RC(x)\}|$ times. $|\{x \in U | v \in RC(x)\}| = 1$ indicates that the

---

Algorithm 1. Rule set construction

Input: The training set $ES = (U, A, V, B, R)$, ms, mt.
Output: Source granules, target granules, and the rule set, all stored in the memory.
Method: training

1: $SG(ms) = \{(A', x) \in 2^A \times U | \frac{|E_{A'}(x)|}{|U|} \geq ms\};$
2: $TG(mt) = \{(B', y) \in 2^B \times V | \frac{|E_{B'}(y)|}{|V|} \geq mt\};$
3: **for** each $g \in SG(ms)$ **do**
4:    **for** each $g' \in TG(mt)$ **do**
5:       $GR = (i(g) \Rightarrow i(g'));$
6:       compute $sig(GR);$
7:    **end for**
8: **end for**

---

item is recommended to all users. The maximal recommendation ratio is given by

$$rmax(ES, RC) = \max_{v \in V} |\{x \in U | v \in RC(x)\}|/|U|. \tag{10}$$

Higher $rmax(ES, RC)$ indicates less diverse recommendation since other items have fewer chances to be recommended.

Third, we study the recommendation distribution in terms of information entropy. There are $|V|$ items. Let the ith item be recommended r_i times. The recommendation vector is $r = [r_1, r_2, \ldots, r_{|V|}]$. Let $r'_i = r_i / \sum_{j=1}^{|V|} r_j$ and $r' = [r'_1, r'_2, \ldots, r'_{|V|}]$ for normalization. The recommendation entropy is

$$H(r) = H(r') = -\sum_{i=1}^{|V|} r'_i \log r'_i, \tag{11}$$

where $r'_i \log r'_i = 0$ if $r'_i = 0$.

These measures are based on the distribution of recommended items from different perspectives. $fr(ES, RC)$ and $rmax(ES, RC)$ give us intuitive understanding, while $H(r)$ is convincing from a statistical viewpoint.

4   Top-k Significant Rules Based Recommendation

Our approach is composed of two stages, as listed in Algorithms 1 and 2. Algorithm 1 shows the rule set construction stage. The input include the training set and two user-specified thresholds ms and mt. With the Apriori [13] or the FP-growth algorithm [14] $SG(ms)$ can be obtained quickly.

Algorithm 2 shows the recommendation stage. The input is the testing set, which might be equal to or different from the training set. Another hidden input is the rule set which is already stored in the memory. In this stage, top-k significant rules are mined for each user. Respective item granules are recommended to the user.

Algorithm 2. Top-k rules based recommendation

Input: The testing set $ES' = (U', A, V', B, R')$.
Output: Recommendation for each object in U'.
Method: recommending

```
1: for each x ∈ U' do
2:     for each g ∈ SG(ms) do
3:         if x matches g then
4:             for each g' ∈ TG(mt) do
5:                 if g' is already among the top-k recommended granules then
6:                     reserve the higher confidence value;
7:                     continue;
8:                 end if
9:                 GR = (i(g) ⇒ i(g'));
10:                compare GR with other rules matching x, and reserve top-k recom-
                   mended granules according to the significance;
11:            end for
12:        end if
13:    end for
14:    recommend the top-k granules to x;
15: end for
```

5 Experiments

In this section, we try to answer two questions through experimentation. First, does the significance-based approach outperform the confidence-based one in terms of accuracy, significance, and diversity? Second, how does the performance change for different threshold settings?

We tested both approaches on MovieLens [6]. The genre is a multi-valued attribute. Therefore we deal with it using the approach proposed in [11]. We randomly select 60% data (566 users and 1009 movies) as the training set, and the remaining data (377 users and 673 movies) as the testing set. The value of k is set to 1.

Fig. 1(a) compares the accuracy of these two algorithms. We observe that confidence-based approach always outperforms the significance-based one. Fig. 1(b) compares the significance. We observe that the significance-based approach is always better than the confidence-based one. This indicates that the new approach satisfies our requirement well. Fig. 1(b) also helps answering the second question. We observe that the significance-based approach performs best with $ms = mt = 0.008$. If the granular size decreases, the recommender experiences overfitting. That is, it performs well on the training set, however poor on the testing set.

Fig. 2(a) compares the fraction of recommended items $fri(ES, RC)$. It is shown that with the significance-based approach, more items have a chance to be recommended. Fig. 2(b) compares the maximal recommendation ratio $tmax(ES, RC)$. With the confidence-based approach, some items are recommended to almost all

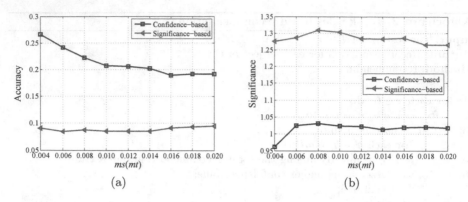

Fig. 1. Accuracy and significance comparison: (a) accuracy, (b) significance

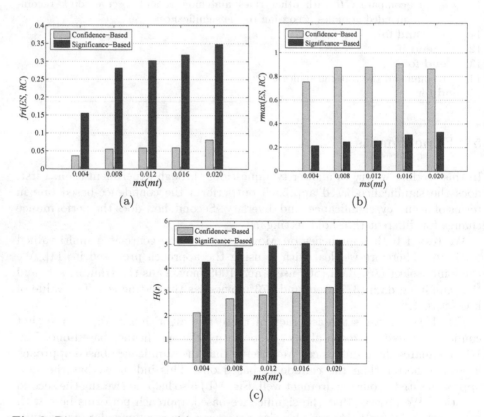

Fig. 2. Diversity comparison: (a) fraction of recommended items, (b) maximal recommendation ratio, (c) recommendation entropy

users. With the significance-based approach, the popularity of items is not overwhelming. Fig. 2(c) compares the recommendation entropy $H(r)$. The entropy of the new recommender system is always much higher than the existing one. To sum up, from Fig. 2 we know that the significance-based approach provides diverse recommendation than the confidence-based one.

6   Conclusions

The paper introduces significance measures and diversity measures for granular association rules and recommender systems. The confidence-based approach provides more accurate recommendations, while the significance-based approach provides more significant and diverse ones. In applications, one may select an appropriate approach, or employ both for better recommendation.

Acknowledgements. This work is in part supported by National Science Foundation of China under Grant Nos. 61379089 and 61379049, the Key Project of Education Department of Fujian Province under Grant No. JA13192.

References

1. Min, F., Hu, Q.H., Zhu, W.: Granular association rules with four subtypes. In: IEEE GrC, pp. 432–437 (2012)
2. Min, F., Zhu, W.: Mining top-k granular association rules for recommendation. In: IEEE IFSA-NAFIPS, pp. 1372–1376 (2013)
3. Džeroski, S., Lavrac, N. (eds.): Relational data mining. Springer (2001)
4. Goethals, B., Page, W.L., Mampaey, M.: Mining interesting sets and rules in relational databases. In: ACM Symposium on Applied Computing, pp. 997–1001 (2010)
5. Linden, G., Smith, B., York, J.: Amazon.com recommendations: Item-to-item collaborative filtering. IEEE Internet Computing 7(1), 76–80 (2003)
6. Internet movie database, http://movielens.umn.edu
7. Schein, A.I., Popescul, A., Ungar, L.H., Pennock, D.M.: Methods and metrics for cold-start recommendations. In: ACM SIGIR, pp. 253–260 (2002)
8. Herlocker, J.L., Konstan, J.A., Terveen, L.G., Riedl, J.T.: Evaluating collaborative filtering recommender systems. ACM Transactions on Information Systems 22(1), 5–53 (2004)
9. Agrawal, R., Imieliński, T., Swami, A.: Mining association rules between sets of items in large databases. ACM SIGMOD Record, 207–216 (1993)
10. Min, F., Zhu, W., He, X.: Grale: Granular association rules (2013), http://grc.fjzs.edu.cn/~fmin/grale/
11. Min, F., Zhu, W.: Granular association rules for multi-valued data. In: IEEE CCECE, pp. 1–5 (2013)
12. Yao, Y.Y., Deng, X.F.: A granular computing paradigm for concept learning. In: Emerging Paradigms in Machine Learning, vol. 13, pp. 307–326. Springer (2013)
13. Agrawal, R., Srikant, R.: Fast algorithms for mining association rules in large databases. In: VLDB, pp. 487–499 (1994)
14. Han, J., Pei, J., Yin, Y.: Mining frequent patterns without candidate generation. In: ACM SIGMOD, pp. 1–12 (2000)

Incremental Three-Way Decisions
with Incomplete Information

Chuan Luo[1,2] and Tianrui Li[1]

[1] School of Information Science and Technology, Southwest Jiaotong University,
Chengdu, 610031, China
luochuan@my.swjtu.edu.cn, trli@swjtu.edu.cn
[2] Department of Computer Science, University of Regina, Regina, Saskatchewan,
S4S 0A2, Canada
luo256@cs.uregina.ca

Abstract. The theory of rough sets proposed a framework for approximating concepts by three pair-wise disjoint regions, namely, the positive, boundary and negative regions. Rules generated by the three regions form three-way decision rules, which are acceptance, deferment and rejection decisions. The periodic updating of decision rules is required due to the dynamic nature of decision systems. Incremental learning technique is an effective way to solve the problem of dynamic data, which is capable of updating the learning results incrementally without recalculation in the total data set from scratch. In this paper, we present a methodology for incremental updating three-way decisions with incomplete information when the object set varies through the time.

Keywords: three-way decisions, incremental updating, incomplete decision system.

1 Introduction

Probabilistic rough set models are generalizations of classical rough sets which are proposed to solve probabilistic decision problems by allowing certain acceptable level errors [10]. By considering a pair of thresholds (α, β) on probabilities for defining probabilistic approximations, Yao et al. proposed a generalized probabilistic model, called a decision-theoretic rough set (DTRS) [11]. Various existing rough set models can be derived through setting different thresholds in DTRS. According α and β, a concept is approximated by three pair-wise disjoint regions that are definable by logic formulas, namely, the positive, the boundary and the negative regions of the concept. Decision rules derived from these three regions are associated with different actions and decisions, which immediately leads to the theory of three-way decisions. Instead of making an immediate acceptance or rejection decision in two-way decision strategy, a third option of making a deferred or non-committed decision is considered in the three-way decisions which is more similar to the human decision strategy in the practical decision problems [12].

C. Cornelis et al. (eds.): RSCTC 2014, LNAI 8536, pp. 128–135, 2014.
© Springer International Publishing Switzerland 2014

In practical applications, decision behaviors need to be reinforced continuously under the dynamic decision environment. Incremental learning technique are designed for mining dynamic databases in which data values are continually being changed, which use previously acquired mining results to facilitate knowledge maintenance in the changed database instead of mining the the whole database from scratch [1]. There exist much research to deal with dynamic information systems based on rough set theory. Li et al. discussed an attribute generalization and its relation to feature selection and feature extraction [9]. They proposed an incremental approach for updating approximations under the characteristic relation-based rough sets. Chen et al. discussed incremental updating properties of information granulation and approximations under the dynamic environment based on variable precision rough set model [2]. Luo et al. proposed incremental algorithms for computing rough approximations in the set-valued information systems [4,5]. Fan et al. proposed an incremental rule-extraction algorithm to deal with the new added data set when the database increased dynamically [3]. Liang et al. developed a group incremental rough feature selection algorithm based on information entropy [8].

In the context of multiple levels of granularity, Yao proposed a framework for sequential three-way decisions based on granular computing [13]. A definite decision of acceptance or rejection for some objects are made with coarse-grained granules and a non-commitment decision for some other objects may be further investigated by using fine-grained granules. Li et al. proposed a sequential strategy for cost-sensitive three-way decisions [7]. Considering the dynamic change of loss functions under dynamic decision environment, Liu et al proposed a dynamic DTRS [6]. An incomplete decision system, which indicates a system with unknown values, is much more common in data and knowledge engineering than the complete decision system. In this paper, we mainly focus on investigating the dynamic change of three-way decisions in the incomplete decision system when the object set varies with time, and present an incremental method for updating three-way decisions.

The remainder of the paper is organized as follows. In Section 2, some basic concepts of the incomplete decision system and three-way decisions based on rough set theory are briefly reviewed. In Section 3, the principles of incremental updating three-way decisions in the incomplete decision system when the object set varies with time are presented. The paper ends with conclusions and further research topics in Section 4.

2   Three-Way Decisions in an Incomplete Information System

In rough set based data analysis, data represented in a tabular form, called an information system. The rows of the system are labelled by objects, whereas columns are labelled by attributes and entries of the system are attribute values. Formally, an information system is a 4-tuple $S = (U, AT, V, f)$, where U is a non-empty finite set of objects, AT is a non-empty finite set of attributes,

$V = \bigcup_{a \in AT} V_a$ and V_a is a domain of attribute a, $f : U \times AT \to V$ is an information function such that $f(x, a) \in V_a$ for every $a \in AT$, $x \in U$.

It may happen that some of attribute values for an object are missing, a so-called null value, denoted by "$*$" is usually assigned to those attributes. An information system is called an incomplete information system if V_a contains null value for at least one attribute $a \in AT$.

Definition 1. Let $S = (U, AT, V, f)$ be an incomplete information system, $\forall A \subseteq AT$, a similarity relation $SIM(A)$ on U is defined as follows:

$$SIM(A) = \{(x, y) \in U \times U | \forall a \in A, f(x, a) = f(y, a) \qquad (1)$$
$$\text{or } f(x, a) = * \text{ or } f(y, a) = *\}$$

$SIM(A)$ is a tolerance relation on U. Let $S_A(x)$ denote the set $\{y \in U | (x, y) \in SIM(A)\}$, called a tolerance class or an information granule. $S_A(x)$ is the maximal set of objects which are possibly indistinguishable by A with respect to x. $U/SIM(A)$ denotes the classification, which is the family set $\{S_A(x) | x \in U\}$. $SIM(A)$ degenerates into an equivalence relation in a complete information system.

Definition 2. Let $S = (U, AT, V, f)$ be an incomplete information system, $\forall x \in U$, $X \subseteq U$ and $A \subseteq AT$, a conditional probability is used to state the degree of overlap between the tolerance class $S_A(x)$ and a set X can be defined as follows

$$Pr(X|S_A(x)) = \frac{|X \cap S_A(x)|}{|S_A(x)|} \qquad (2)$$

where $|\cdot|$ denotes the cardinality of a set. The conditional probability $Pr(X|S_A(x))$ represents the probability that an object belongs to X given that the object is described by $S_A(x)$.

Yao introduced DTRS, a general probabilistic rough set model, in which a pair of thresholds α and β with $\alpha > \beta$ probability is used to define three probabilistic regions [14].

Definition 3. Let $S = (U, AT, V, f)$ be an incomplete information system, $\forall X \subseteq U$, $A \subseteq AT$, The (α, β)-probabilistic positive, boundary and negative regions are defined as follows

$$POS_{\alpha,\beta}(X) = \{x \in U | Pr(X|S_A(x)) \geq \alpha\}$$
$$BND_{\alpha,\beta}(X) = \{x \in U | \beta < Pr(X|S_A(x)) < \alpha\} \qquad (3)$$
$$NEG_{\alpha,\beta}(X) = \{x \in U | Pr(X|S_A(x)) \leq \beta\}$$

With insufficient information as provided by a set of attributes $A \subseteq AT$, DTRS promotes a methodology for three-way decision making. The three regions $POS_{\alpha,\beta}(X_i)$, $BND_{\alpha,\beta}(X_i)$ and $NEG_{\alpha,\beta}(X_i)$ are semantically interpreted as the following three-way decisions. We accept an object x to be a member of X_i if the conditional probability is greater than or equal to α, with an understanding that it comes with an $(1 - \alpha)$-level acceptance error and associated cost. We reject x to be a member of X_i if the conditional probability is less than or equal to β, with an understanding that it comes with an β-level of rejection error and associated cost. We neither accept nor reject x to be a member of X_i if the conditional probability is between of α and β, instead, we make a decision of deferment.

3   Incremental Three-Way Decisions in an Incomplete Decision System

In this section, we introduce the incremental mechanisms for the three-way decisions when the universe varies in the incomplete decision system. To describe a dynamic incomplete decision system, we assume the incremental updating process of the system lasts two periods from time t to time $t + 1$.

3.1   Insertion of an Object

Given an incomplete decision system $S^{(t)} = (U^{(t)}, C \cup \{d\}, V, f)$ at time t, where $U^{(t)} = \{x_1, x_2, \ldots, x_n\}$. Suppose that a new object x_{n+1} is added to $S^{(t)}$ at time $t + 1$, we have $U^{(t+1)} = U^{(t)} \cup \{x_{n+1}\}$.

Proposition 1. *Let $A \subseteq C$, $U^{(t)}/SIM(A) = \{S_A^{(t)}(x_1), S_A^{(t)}(x_2), \ldots, S_A^{(t)}(x_n)\}$. $\forall 1 \leq q \leq n + 1$, we have*

$$S_A^{(t+1)}(x_q) = \begin{cases} S_A^{(t)}(x_q), & 1 \leq q \leq n \text{ and } (x_q, x_{n+1}) \notin SIM(A); \\ S_A^{(t)}(x_q) \cup \{x_{n+1}\}, & 1 \leq q \leq n \text{ and } (x_q, x_{n+1}) \in SIM(A); \\ \{y \in U^{(t+1)} | (x_q, y) \in SIM(A)\}, & q = n + 1. \end{cases}$$

With the addition of a new object x_{n+1}, There are two distinguishing situations about the decision classes.

(1) x_{n+1} belongs to an existing decision class;
(2) x_{n+1} does not belong to any existing decision classes.

For the first situation, let $\pi^{(t)} = \{X_1^{(t)}, X_2^{(t)}, \ldots, X_r^{(t)}\}$, $\forall 1 \leq p \leq r$, we discuss the updating principles of three-way decisions of $X_p^{(t)}$ at time $t + 1$ through the following two cases: (1) $x_{n+1} \in X_p^{(t+1)}$, (2) $x_{n+1} \notin X_p^{(t+1)}$.

Case 1: $x_{n+1} \in X_p^{(t+1)}$, i.e., $X_p^{(t+1)} = X_p^{(t)} \cup \{x_{n+1}\}$.

Lemma 1. *$\forall 1 \leq q \leq n$, we have*

$$Pr(X_p^{(t)} | S_A^{(t)}(x_q)) \leq Pr(X_p^{(t+1)} | S_A^{(t+1)}(x_q))$$

Proposition 2. *Let the positive region of $X_p^{(t)}$ is $POS_{\alpha,\beta}(X_p^{(t)})$. With the addition of x_{n+1} at time $t+1$, $POS_{\alpha,\beta}(X_p^{(t+1)})$ can be updated as follows.*

$$POS_{\alpha,\beta}(X_p^{(t+1)}) = POS_{\alpha,\beta}(X_p^{(t)}) + \triangle'$$

where $\triangle' = \{x \in BND_{\alpha,\beta}(X_p^{(t)}) \cup NEG_{\alpha,\beta}(X_p^{(t)}) | x_{n+1} \in S_A^{(t+1)}(x), Pr(X_p^{(t+1)} | S_A^{(t+1)}(x)) \geq \alpha\}.$

Proposition 3. *Let the boundary region of $X_p^{(t)}$ is $BND_{\alpha,\beta}(X_p^{(t)})$. With the addition of x_{n+1} at time $t+1$, $BND_{\alpha,\beta}(X_p^{(t+1)})$ can be updated as follows.*

$$BND_{\alpha,\beta}(X_p^{(t+1)}) = BND_{\alpha,\beta}(X_p^{(t)}) - \triangle + \triangle'$$

where $\triangle = \{x \in BND_{\alpha,\beta}(X_p^{(t)}) | x_{n+1} \in S_A^{(t+1)}(x), Pr(X_p^{(t+1)} | S_A^{(t+1)}(x)) \geq \alpha\}$, $\triangle' = \{x \in NEG_{\alpha,\beta}(X_p^{(t)}) | x_{n+1} \in S_A^{(t+1)}(x), \beta < Pr(X_p^{(t+1)} | S_A^{(t+1)}(x)) < \alpha\}.$

Case 2: $x_{n+1} \notin X_p^{(t+1)}$, *i.e.*, $X_p^{(t+1)} = X_p^{(t)}$.

Lemma 2. $\forall 1 \leq q \leq n$, *we have*

$$Pr(X_p^{(t)} | S_A^{(t)}(x_q)) \geq Pr(X_p^{(t+1)} | S_A^{(t+1)}(x_q))$$

Proposition 4. *Let the positive region of $X_p^{(t)}$ is $POS_{\alpha,\beta}(X_p^{(t)})$. With the addition of x_{n+1} at time $t+1$, $POS_{\alpha,\beta}(X_p^{(t+1)})$ can be updated as follows.*

$$POS_{\alpha,\beta}(X_p^{(t+1)}) = POS_{\alpha,\beta}(X_p^{(t)}) - \triangle$$

where $\triangle = \{x \in POS_{\alpha,\beta}(X_p^{(t)}) | x_{n+1} \in S_A^{(t+1)}(x), Pr(X_p^{(t+1)} | S_A^{(t+1)}(x)) < \alpha\}.$

Proposition 5. *Let the boundary region of $X_p^{(t)}$ is $BND_{\alpha,\beta}(X_p^{(t)})$. With the addition of x_{n+1} at time $t+1$, $BND_{\alpha,\beta}(X_p^{(t+1)})$ can be updated as follows.*

$$BND_{\alpha,\beta}(X_p^{(t+1)}) = BND_{\alpha,\beta}(X_p^{(t)}) - \triangle + \triangle'$$

where $\triangle = \{x \in BND_{\alpha,\beta}(X_p^{(t)}) | x_{n+1} \in S_A^{(t+1)}(x), Pr(X_p^{(t+1)} | S_A^{(t+1)}(x)) \leq \beta\}$, $\triangle' = \{x \in POS_{\alpha,\beta}(X_p^{(t)}) | x_{n+1} \in S_A^{(t+1)}(x), \beta < Pr(X_p^{(t+1)} | S_A^{(t+1)}(x)) < \alpha\}.$

For the second situation, since the new object x_{n+1} does not belong to any existing decision classes, then x_{n+1} will form a new decision class, *i.e.*, $\pi^{(t+1)} = \pi^{(t)} \cup \{X_{r+1}^{(t+1)}\}$, where $X_{r+1}^{(t+1)} = \{x_{n+1}\}$. At this point, $\forall 1 \leq p \leq r$, since $x_{n+1} \notin X_p^{(t)}$, *i.e.*, $X_p^{(t+1)} = X_p^{(t)}$, then the three-way decisions of $X_p^{(t+1)}$ can be obtained by the case 2 of the first situation. On the other hand, for the new decision class $X_{r+1}^{(t+1)}$, the positive and boundary regions of $X_{r+1}^{(t+1)}$ can be obtained as follows.

Proposition 6. *With the addition of x_{n+1} at time $t + 1$, for the new decision class $X_{r+1}^{(t+1)}$, $POS_{\alpha,\beta}(X_{r+1}^{(t+1)})$ can be obtained as follows.*

$$POS_{\alpha,\beta}(X_{r+1}^{(t+1)}) = \{x_q \in U^{(t+1)} | x_{n+1} \in S_A^{(t+1)}(x_q), |S_A^{(t+1)}(x_q)| < \frac{1}{\alpha}\}$$

Proposition 7. *With the addition of x_{n+1} at time $t + 1$, for the new decision class $X_{r+1}^{(t+1)}$, $BND_{\alpha,\beta}(X_{r+1}^{(t+1)})$ can be obtained as follows.*

$$BND_{\alpha,\beta}(X_{r+1}^{(t+1)}) = \{x_q \in U^{(t+1)} | x_{n+1} \in S_A^{(t+1)}(x_q), \frac{1}{\beta} < |S_A^{(t+1)}(x_q)| < \frac{1}{\alpha}\}$$

Based on the updated (α, β)-probabilistic positive and boundary regions, the negative region can be obtained directly according to Definition 3 when a new object is inserted into the incomplete decision system.

3.2   Deletion of an Object

Given an incomplete decision system $S^{(t)} = (U^{(t)}, C \cup \{d\}, V, f)$ at time t, where $U^{(t)} = \{x_1, x_2, \ldots, x_n\}$. Suppose that the object x_q is deleted from $S^{(t)}$ at time $t + 1$, where $1 \leq q \leq n$, we have $U^{(t+1)} = U^{(t)} - \{x_q\}$.

Proposition 8. *Let $A \subseteq C$, $U^{(t)}/SIM(A) = \{S_A^{(t)}(x_1), S_A^{(t)}(x_2), \ldots, S_A^{(t)}(x_n)\}$. $\forall 1 \leq i \leq n$ and $i \neq q$, we have*

$$S_A^{(t+1)}(x_i) = \begin{cases} S_A^{(t)}(x_i), & (x_i, x_q) \notin SIM(A); \\ S_A^{(t)}(x_i) - \{x_q\}, & (x_i, x_q) \in SIM(A); \end{cases}$$

Let $\pi^{(t)} = \{X_1^{(t)}, X_2^{(t)}, \ldots, X_r^{(t)}\}$. Similar to the addition of object, when an object x_q is removed from the universe, $\forall 1 \leq p \leq r$, we discuss the updating principles of three-way decisions of $X_p^{(t)}$ at time $t + 1$ through the following two cases: (1) $x_q \in X_p^{(t+1)}$, (2) $x_q \notin X_p^{(t+1)}$.

Case 1: $x_q \in X_p^{(t+1)}$, i.e., $X_p^{(t+1)} = X_p^{(t)} - \{x_q\}$.

Lemma 3. $\forall 1 \leq i \leq n$, $i \neq q$, *we have*

$$Pr(X_p^{(t)} | S_A^{(t)}(x_i)) \geq Pr(X_p^{(t+1)} | S_A^{(t+1)}(x_i))$$

Proposition 9. *Let the positive region of $X_p^{(t)}$ is $POS_{\alpha,\beta}(X_p^{(t)})$. With the removal of x_q at time $t + 1$, $POS_{\alpha,\beta}(X_p^{(t+1)})$ can be updated as follows.*

$$POS_{\alpha,\beta}(X_p^{(t+1)}) = POS_{\alpha,\beta}(X_p^{(t)}) - \triangle$$

where $\triangle = \{x \in POS_{\alpha,\beta}(X_p^{(t)}) | x_q \in S_A^{(t+1)}(x), Pr(X_p^{(t+1)} | S_A^{(t+1)}(x)) < \alpha\}$.

Proposition 10. *Let the boundary region of $X_p^{(t)}$ is $BND_{\alpha,\beta}(X_p^{(t)})$. With the removal of x_q at time $t+1$, $BND_{\alpha,\beta}(X_p^{(t+1)})$ can be updated as follows.*

$$BND_{\alpha,\beta}(X_p^{(t+1)}) = BND_{\alpha,\beta}(X_p^{(t)}) - \triangle + \triangle'$$

where $\triangle = \{x \in BND_{\alpha,\beta}(X_p^{(t)})|x_q \in S_A^{(t+1)}(x), Pr(X_p^{(t+1)}|S_A^{(t+1)}(x)) \leq \beta\}$, $\triangle' = \{x \in POS_{\alpha,\beta}(X_p^{(t)})|x_q \in S_A^{(t+1)}(x), \beta < Pr(X_p^{(t+1)}|S_A^{(t+1)}(x)) < \alpha\}$.

Case 2: $x_q \notin X_p^{(t+1)}$, i.e., $X_p^{(t+1)} = X_p^{(t)}$.

Lemma 4. $\forall 1 \leq i \leq n$, $i \neq q$, we have

$$Pr(X_p^{(t)}|S_A^{(t)}(x_i)) \leq Pr(X_p^{(t+1)}|S_A^{(t+1)}(x_i))$$

Proposition 11. *Let the positive region of $X_p^{(t)}$ is $POS_{\alpha,\beta}(X_p^{(t)})$. With the removal of x_q at time $t+1$, $POS_{\alpha,\beta}(X_p^{(t+1)})$ can be updated as follows.*

$$POS_{\alpha,\beta}(X_p^{(t+1)}) = POS_{\alpha,\beta}(X_p^{(t)}) + \triangle$$

where $\triangle = \{x \in BND_{\alpha,\beta}(X_p^{(t)}) \cup NEG_{\alpha,\beta}(X_p^{(t)})|x_q \in S_A^{(t+1)}(x), Pr(X_p^{(t+1)}| S_A^{(t+1)}(x_q)) \geq \alpha\}$.

Proposition 12. *Let the boundary region of $X_p^{(t)}$ is $BND_{\alpha,\beta}(X_p^{(t)})$. With the removal of x_q at time $t+1$, $BND_{\alpha,\beta}(X_p^{(t+1)})$ can be updated as follows.*

$$BND_{\alpha,\beta}(X_p^{(t+1)}) = BND_{\alpha,\beta}(X_p^{(t)}) - \triangle + \triangle'$$

where $\triangle = \{x \in BND_{\alpha,\beta}(X_p^{(t)})|x_q \in S_A^{(t+1)}(x), Pr(X_p^{(t+1)}|S_A^{(t+1)}(x)) \geq \alpha\}$, $\triangle' = \{x \in NEG_{\alpha,\beta}(X_p^{(t)})|x_q \in S_A^{(t+1)}(x), \beta < Pr(X_p^{(t+1)}|S_A^{(t+1)}(x)) < \alpha\}$.

Based on the updated (α, β)-probabilistic positive and boundary regions, the negative region can be obtained directly according to Definition 3 when an object is deleted from the incomplete decision system.

4   Conclusion

Three-way decisions have become an important approach for trading off different types of classification error to support decision making at a minimum cost, which is complementary to binary decision making. A three-way decision of acceptance, rejection and deferment can be derived directly from positive, negative and boundary regions based on rough set theory. In this paper, we proposed an incremental approach for updating three-way decisions in the incomplete decision system. The approach aims at incremental computing the (α, β)-probabilistic positive, boundary and negative regions incrementally when the object set changes. Our future work will focus on algorithm development and experimentation for the validation of the proposed approaches.

Acknowledgements. This work is supported by the National Science Foundation of China (Nos. 61175047, 71201133, 61100117 and 61262058) and NSAF (No. U1230117), the Youth Social Science Foundation of the Chinese Education Commission (Nos. 10YJCZH117, 11YJC630127), the Fundamental Research Funds for the Central Universities (Nos. SWJTU11ZT08, SWJTU12CX091, SWJTU12CX117), and the Doctoral Innovation Foundation of Southwest Jiaotong University (No. 2014LC). The authors would like to thank Professor Yiyu Yao for useful comments on this study.

References

1. Altiparmak, F., Tuncel, E., Ferhatosmanoglu, H.: Incremental maintenance of online summaries over multiple streams. IEEE Transactions on Knowlege and Data Engineering 20(2), 216–229 (2008)
2. Chen, H.M., Li, T.R., Ruan, D., Lin, J.H., Hu, C.X.: A rough-set based incremental approach for updating approximations under dynamic maintenance environments. IEEE Transactions on Knowledge and Data Engineering 25(2), 274–284 (2013)
3. Fan, Y.N., Tseng, T.L., Chern, C.C., Huang, C.C.: Rule induction based on an incremental rough set. Expert Systems with Applications 36(9), 11439–11450 (2009)
4. Luo, C., Li, T.R., Chen, H.M.: Dynamic maintenance of approximations in set-valued ordered decision systems under the attribute generalization. Information Sciences 257, 210–228 (2014)
5. Luo, C., Li, T.R., Chen, H.M., Liu, D.: Incremental approaches for updating approximations in set-valued ordered information systems. Knowledge-Based Systems 50, 218–233 (2013)
6. Liu, D., Li, T., Liang, D.: Three-way decisions in dynamic decision-theoretic rough sets. In: Lingras, P., Wolski, M., Cornelis, C., Mitra, S., Wasilewski, P. (eds.) RSKT 2013. LNCS, vol. 8171, pp. 291–301. Springer, Heidelberg (2013)
7. Li, H., Zhou, X., Huang, B., Liu, D.: Cost-sensitive three-way decision: A sequential strategy. In: Lingras, P., Wolski, M., Cornelis, C., Mitra, S., Wasilewski, P. (eds.) RSKT 2013. LNCS, vol. 8171, pp. 325–337. Springer, Heidelberg (2013)
8. Liang, J., Wang, F., Dang, C., Qian, Y.: A group incremental approach to feature selection applying rough set technique. IEEE Transactions on Knowledge and Data Engineering 26(2), 294–308 (2012)
9. Li, T.R., Ruan, D., Geert, W., Song, J., Xu, Y.: A rough sets based characteristic relation approach for dynamic attribute generalization in data ming. Knowledge-Based Systems 20, 485–494 (2007)
10. Yao, Y.Y.: Probabilistic rough set approximations. International Journal of Approximation Reasoning 49, 255–271 (2008)
11. Yao, Y.Y.: Three-way decisions with probabilistic rough sets. Information Sciences 180, 341–353 (2010)
12. Yao, Y.Y.: The superiority of three-way decision in probabilistic rough set models. Information Sciences 181, 1080–1096 (2011)
13. Yao, Y.: Granular computing and sequential three-way decisions. In: Lingras, P., Wolski, M., Cornelis, C., Mitra, S., Wasilewski, P. (eds.) RSKT 2013. LNCS, vol. 8171, pp. 16–27. Springer, Heidelberg (2013)
14. Yao, Y.Y., Wong, S.K.M.: A decision-theoretic framework for approximating concepts. International Journal of Man-Machine Studies 37(6), 793–809 (1992)

On Determination of Thresholds in Three-Way Approximation of Many-Valued NM-Logic

Yanhong She[1,2]

[1] College of Science, Xi'an Shiyou University Xi'an 710065, China
[2] Department of Computer Science, University of Regina, Regina, Saskatchewan, Canada S4S 0A2
yanhongshe@gmail.com

Abstract. Approximation of a many-valued logic by a logic with less number of truth values is an important topic. Three-way approximation based on a pair of thresholds is an example considered by Yao. However, the determination of thresholds has not been investigated yet. In this paper, we aim to study this issue in the context of many-valued NM-logic with the standard valuation domain $\{0, \frac{1}{n-1}, \frac{2}{n-1}, \cdots, \frac{n-2}{n-1}, 1\}$. The main result is that when n is odd, the thresholds for three-way decision is uniquely determined. When n is even, there is actually no three-way decision, but two-way decision.

Keywords: three-way decision, NM-logic, valuation, thresholds.

1 Introduction

Pawlak's rough set divides the universe into three disjoint regions, i.e., positive region, boundary region and negative region. To interpret these three regions, the concept of three-way decisions was proposed by Yao in [1] and [2]. In the theory of three-way decision, the three regions are viewed, respectively, as the regions of acceptance, rejection, and non-commitment in a ternary classification. The positive and negative regions can be used to induce rules of acceptance and rejection, whenever it is impossible to make an acceptance or a rejection decision, the third non-commitment decision is made.

Many recent studies further investigated extensions and applications of three-way decisions [3,4,5,6,7,8,9]. To extend the concept of three-way decisions of rough sets to a much wider context, [10] outlines a theory of three-way decisions. In this paper, Yao pointed out that three issues regarding evaluations and designated values should be considered, that is, construction and interpretation of a set of values for measuring satisfiability and a set of values for measuring non-satisfiability, construction and interpretation of evaluations and determination and interpretation of designated values for acceptance and designated values for rejection.

By using three-way decision, three-valued approximations of a many-valued logic can be derived based on the two designated sets [10,11]. Precisely, we accept a truth degree as being true if it is in the positively designated set, reject it as

C. Cornelis et al. (eds.): RSCTC 2014, LNAI 8536, pp. 136–143, 2014.

being true if it is the negatively designated set, and neither accept nor reject if it is not in any of the two sets. By so doing, an approximation of the many-valued logic can be obtained. In widely used totally ordered valuation domain, the thresholds can uniquely determine the positively designated set and negatively designated set, and thus play a key role in the three-valued approximation of many-valued logic. However, this issue of thresholds determination has not been investigated yet.

In this paper, we aim to solve this problem in a special kind of many-valued logic, i.e., NM-logic [12] which is equivalently called R_0-logic in [13]. The valuation domain we consider in this paper is $\{0, \frac{1}{n-1}, \cdots, \frac{n-2}{n-1}, 1\}$ instead of its abstract algebra. Moreover, to make the three-valued approximation of many-valued logic consistent with logical structure in NM logic, it is required that the mapping from the set of logic formulae to $\{0, \frac{1}{2}, 1\}$ should be a valuation, that is, it preserves the logical connectives. Such a requirement can uniquely determine the thresholds, as we will show below.

2 Preliminaries

In this section, we briefly review some preliminary knowledge about three-way decision and multiple-valued logic.

2.1 Three-Way Decision

As stated in [10], the essential ideas of three-way decisions are described in terms of a ternary classification according to evaluations of a set of criteria.

Suppose U is a finite nonempty set of objects or decision alternatives and C is a finite set of conditions. Each condition in C may be a criterion, an objective, or a constraint. To better suit our needs, we assume that C is a finite set of criteria in the present paper. Our decision task is to classify objects of U according to whether they satisfy the set of criteria. Note that in widely used two-way decision models, an object either satisfies the criteria or does not satisfy the criteria, accordingly, we can divide the universe into two parts, i.e., the set of objects satisfying the criteria and the set of objects not satisfying the criteria. However, when using this approach, we can encounter some fundamental problems. For instance, it may happen that we cannot decide whether an object satisfy the criteria or not due to the incomplete information. in such a situation, the choice of two-way decision may be costly. Consequently, we have to search for an approximate solution. Instead of making a binary decision, we use thresholds on the degrees of satisfiability to make one of three decisions: (a) accept an object as satisfying the set of criteria if its degree of satisfiability is at or above a certain level; (b) reject the object by treating it as not satisfying the criteria if its degree of satisfiability is at or below another level; and (c) neither accept nor reject the object but opt for a noncommitment. The third option may also be referred to as a deferment decision that requires further information or investigation.

To formally describe the satisfiability of objects, rules for acceptance and rules for rejection, we need to introduce the notion of evaluations of objects and designated values for acceptance and designated values for rejection. Evaluations provide the degrees of satisfiability, designated values for acceptance are acceptable degrees of satisfiability, and designated valued for rejection are acceptable degrees of non-satisfiability. They provide a basis for a theory of three-way decisions.

In Yao's paper [10], three different but closely related types of three-way decision were presented, they are three-way Decisions with a pair of poset-based evaluations, three-way decisions with one poset-based evaluation and three-way decisions with an evaluation using a totally ordered set. In the present paper, we only consider the three-way decision with an evaluation using a totally ordered set, which is also a convenient and widely used approach.

Definition 1 *[10] Suppose $(L, \leq)$ is a totally ordered set, that is, $\leq$ is a total order. For two elements α, β with $\beta < \alpha$ (i.e., $\beta < \alpha \wedge \neg(\alpha < \beta)$), suppose that the set of designated values for acceptance is given by $L^+ = \{t \in L | t \geq \alpha\}$ and the set of designated values for rejection is given by $L^- = \{b \in L | b \leq \beta\}$. For an evaluation function $v : U \to L$, its three regions are defined by:*

$$POS(\alpha, \beta)(v) = \{x \in U | v(x) \geq \alpha\}, \tag{1}$$
$$NEG(\alpha, \beta)(v) = \{x \in U | v(x) \leq \beta\}, \tag{2}$$
$$BND(\alpha, \beta)(v) = \{x \in U | \beta < v(x) < \alpha\}. \tag{3}$$

2.2   Many-Valued NM-Logic

Many-valued NM-logic [12,13] consists of the syntax and semantics of its logical language. In the syntax, atomic formulae $p_1, p_2, \cdots$ (the set of atomic formulae is denoted by AF)are always used to denote simple proposition, the set of all logic formulae $F(S)$ can be defined in the following way:

(i) Each atomic formula belongs to $F(S)$,
(ii) If $B, C \in F(S)$, then $B \vee C, B \wedge C, \neg B, B \to C \in F(S)$,
(iii) Any formula of $F(S)$ can be defined by finitely using (i) and (ii).
In NM-logic, two additional logic connectives $\otimes$ and $\oplus$ can be defined by

$$\forall A, B \in F(S), A \otimes B = \neg(A \to \neg B), A \oplus B = \neg A \to B.$$

The axioms of NM-logic include the following types of formulae:

(i) $B \to (C \to B \wedge C)$,
(ii) $(\neg B \to \neg C) \to (C \to B)$,
(iii) $(B \to (C \to D)) \to (C \to (B \to D))$,
(iv) $(C \to D) \to ((B \to C) \to (B \to D))$,
(v) $B \to \neg\neg B$,
(vi) $B \to B \vee C$,
(vii) $B \vee C \to C \vee B$,
(viii) $(B \to D) \wedge (C \to D) \to (B \vee C \to D)$,
(ix) $(B \wedge C \to D) \to (B \to D) \vee (C \to D)$,
(x) $(B \to C) \vee ((B \to C) \to \neg B \vee C)$,
where $B \wedge C$ is an abbreviation for $\neg(\neg B \vee \neg C)$.

The inference rule is MP rule, i.e., we can infer $\{B\}$ from $\{A, A \to B\}$.

Semantically, we only consider the linearly order truth values containing n elements, i.e., $L_n = \{0, \frac{1}{n-1}, \frac{2}{n-2}, \cdots, \frac{n-2}{n-1}, 1\}$. Some commonly used operations defined on L_n include $\vee, \wedge, \neg$, which are defined by

$$\forall a, b \in L, a \vee b = \max\{a, b\}, a \wedge b = \min\{a, b\}, \neg a = 1 - a.$$

The definition of implication operator $\to$ is defined

$a \to b = 1$ when $a \le b$ and $a \to b = (1 - a) \vee b$, otherwise.

Moreover, a key notion in semantics is valuation, which is a mapping from $F(S)$ to L preserving logical connectives, i.e., it satisfies the following condition: $v(A \vee B) = v(A) \vee v(B), v(A \wedge B) = v(A) \wedge v(B), v(\neg A) = 1 - v(A), v(A \to B) = v(A) \to v(B)$.

Since any valuation preserves the logic connectives in many-valued logic, it can be uniquely determined by its value on the set of atomic formulae $\{v(p_1), \cdots, v(p_n), \cdots\}$.

3   Three-Way Decision in Many-Valued NM-Logic

In this section, we narrow our focus on three-valued approximation of many-valued logic. We choose a special kind of many-valued NM-logic [12,13], that is, the valuation domain L is endowed with the NM implication operator, apart from the other operators.

Let U be the set of logic formulae in NM-logic, i.e., $U = F(S)$, L be the valuation domain, i.e., $L = \{0, \frac{1}{n-1}, \cdots, \frac{n-2}{n-1}, 1\}$ and any valuation $v : F(S) \to L$ be the evaluation function. Then by using Definition 1, we obtain that for any two thresholds α, β ($\alpha > \beta$), the corresponding three regions in the context of many-valued NM-logic are given by

$$POS_{\alpha,\beta}(v) = \{A \in F(S) | v(A) \ge \alpha\}, \tag{4}$$
$$NEG_{\alpha,\beta}(v) = \{A \in F(S) | v(A) \le \beta\}, \tag{5}$$
$$BND_{\alpha,\beta}(v) = \{A \in F(S) | \beta < v(A) < \alpha\}. \tag{6}$$

According to the truth value of A under valuation v, one of three truth values $\{0, \frac{1}{2}, 1\}$ can be assigned to it. Specifically, when A belongs to the acceptance region $POS_{\alpha,\beta}(v)$, the truth value of A is equal to 1, if A appears in the rejection region $NEG_{\alpha,\beta}(v)$, its truth value is equal to 0, and $\frac{1}{2}$ otherwise. The obtained three-valued logic [14] with the truth values $\{0, \frac{1}{2}, 1\}$ is the three-valued approximation of many-valued logic. Such an approximation naturally leads to a new mapping $\varphi : F(S) \to \{0, \frac{1}{2}, 1\}$, as defined by

$$\varphi(A) = \begin{cases} 1, & A \in POS_{\alpha,\beta}(v), \\ \frac{1}{2}, & A \in BND_{\alpha,\beta}(v), \\ 0, & A \in NEG_{\alpha,\beta}(v). \end{cases}$$

To further indicate the close relationship between φ and v, we denote φ by φ_v instead.

It is important to remark here that φ_v is not necessarily a valuation in NM-logic, as shown by the following example.

Example 1. *Let $L = \{0, \frac{1}{4}, \frac{2}{4}, \frac{3}{4}, 1\}, \alpha = \frac{3}{4}, \beta = 0$, then according to the definition of three-way decision, one conclude that*

$$POS_{\alpha,\beta}(v) = \{A \in F(S) | v(A) \geq \frac{3}{4}\}, \tag{7}$$

$$NEG_{\alpha,\beta}(v) = \{A \in F(S) | v(A) \leq 0\}, \tag{8}$$

$$BND_{\alpha,\beta}(v) = \{A \in F(S) | 0 < v(A) < \frac{3}{4}\}. \tag{9}$$

Consequently, φ_v is defined by

$$\varphi_v(A) = \begin{cases} 1, & v(A) \geq \frac{3}{4}, \\ \frac{1}{2}, & 0 < v(A) < \frac{3}{4}, \\ 0, & v(A) \leq 0. \end{cases}$$

Construct a mapping $v : AF \to L$ as follows: $v(p_1) = \frac{3}{4}$ and the valuation value of any other atomic formula under v is arbitrarily chosen. Then v can generate a unique valuation, for simplicity, we still denote it by v. According to the definition of φ_v, we have $\varphi_v(p) = 1$ and $\varphi_v(\neg p) = \frac{1}{2}$. Clearly, it does not satisfy $\varphi_v(\neg p) = 1 - \varphi_v(p)$, that is, φ_v is not a valuation mainly due to the choice of thresholds α, β.

Example 1 shows that under the three-valued approximation of many-valued logic, the obtained new mapping $\varphi_v : F(S) \to \{0, \frac{1}{2}, 1\}$ is not necessarily a valuation, which in turn implies that the obtained approximation may be not consistent with the logical structure of many-valued NM-logic. Such an observation motivates us to discuss the determination of thresholds in three-valued approximation of many-valued logic below.

Theorem 1. *In three-valued approximation of many-valued logic, $\forall v : F(S) \to L_n$, φ_v is a valuation if and only if the thresholds α, β satisfy the following conditions:*
(i) $\alpha > \frac{1}{2} > \beta$,
(ii) $\alpha + \beta = 1$,
(iii) $|(\beta, \alpha) \cap L_n| \leq 1$.

Proof. " $\Rightarrow$ ":

(i) Suppose, on the contrary, that $\alpha \leq \frac{1}{2}$, and let $v_1 : F(S) \to L_n$ be a valuation satisfying $v_1(p_1) = \alpha$. Since $v_1(p_1) = \alpha \geq \alpha$, we then have from the definition of φ_v that $\varphi_{v_1}(p_1) = 1$. Similarly, we have $\varphi_{v_1}(\neg p_1) = 1$ owing to $v_1(\neg p_1) = 1 - \alpha \geq \alpha$, which, however, contradicts with the fact that φ_{v_1} is a valuation.

A similar analysis can show that $\beta < \frac{1}{2}$.

(ii) Let $v_1 : F(S) \to L_n$ be the same valuation defined as above, then by definition, $\varphi_{v_1}(p_1) = 1$. Since φ_{v_1} is a valuation, we immediately have $\varphi_{v_1}(\neg p_1) = 0$, which together with (2)-(5) implies that $v_1(\neg p) \leq \beta$, i.e., $1 - \alpha \leq \beta$.

Similarly, define $v_2 : F(S) \to L_n$ as a valuation satisfying $v_2(p_2) = \beta$, then by definition, $\varphi_{v_2}(p_2) = 0$. Since φ_{v_2} is a valuation, we also have $\varphi_{v_2}(\neg p_2) = 1$, which together with (2)-(5) implies that $v_2(\neg p_2) \geq \alpha$, i.e., $1 - \beta \geq \alpha$.

Putting these two results together, we have that $1 - \beta = \alpha$. i.e., $\alpha + \beta = 1$.

(iii) Suppose, on the contrary, that $(\beta, \alpha) \cap L_n$ contains more than two elements, say as a, b. We assume, without any loss of generality, that $a > b$. Define $v_3 : F(S) \to L_n$ as a valuation satisfying $v(p_3) = a, v(p_4) = b$, then $v(p_3 \to p_4) = v(p_3) \to v(p_4) = (1 - a) \vee b$. We have from (ii) and $\beta < a, b < \alpha$ that $\beta < (1 - a) \vee b < \alpha$, and therefore, $\varphi_v(p_3 \to p_4) = \frac{1}{2}$. However, it follows from $v(p_3) = a, v(p_4) = b$ that $\varphi_v(p_3) = \varphi_v(p_4) = \frac{1}{2}$, consequently, $\varphi_v(p_3) \to \varphi_v(p_4) = 1$, a contradiction with the fact that φ_v is a valuation.

" $\Leftarrow$:" According to the definition of valuation, it suffices to show that $\forall A, B \in F(S), \varphi_v(A \vee B) = \varphi_v(A) \vee \varphi_v(B), \varphi_v(\neg A) = 1 - \varphi_v(A), \varphi_v(A \to B) = \varphi_v(A) \to \varphi_v(B)$. We need only show that $\varphi_v(A \to B) = \varphi_v(A) \to \psi_v(B)$. The other two equalities can be shown similarly.

There are three cases to be considered below.

Case 1: $\varphi_v(A \to B) = 1$, then we have $v(A \to B) \geq \alpha$, i.e., $v(A) \to v(B) \geq \alpha$. According to the definition of R_0 implication, we further have $v(A) \leq v(B)$, or $(1 - v(A)) \vee v(B) \geq \alpha$. Suppose the former holds, then we immediately have $\varphi_v(A) \leq \varphi_v(B)$, consequently, $\varphi_v(A) \to \varphi_v(B) = 1$, showing $\varphi_v(A \to B) = \varphi_v(A) \to \varphi_v(B)$. Suppose the latter holds, there are also three subcases. Subcase 1: $1 - v(A) \geq \alpha, v(B) < \alpha$, then $v(A) \leq 1 - \alpha$. We have from (ii) that $\beta = 1 - \alpha$, and so, $v(A) \leq \beta$, consequently, $\varphi_v(A) = 0$ and $\varphi_v(A) \to \varphi_v(B) = 1$. Subcase 2: $1 - v(A) \geq \alpha, v(B) \geq \alpha$, since $v(A) \leq 1 - \alpha < \alpha \leq v(B)$, we have from the definition of φ_v that $\varphi_v(A) \leq \varphi_v(B)$, consequently, $\varphi_v(A) \to \varphi_v(B) = 1$. Subcase 3: $1 - v(A) \leq \alpha, v(B) \geq \alpha$, we immediately have that $\varphi_v(B) = 1$, and hence, we can conclude that $\varphi_v(A) \to \varphi_v(B) = 1$.

Case 2: $\varphi_v(A \to B) = \frac{1}{2}$, then we have $\beta < v(A \to B) < \alpha$, i.e., $\beta < (1 - v(A)) \vee v(B) < \alpha$, which in turn implies that $1 - v(A) < \alpha, v(B) < \alpha$. There are three subcases to be considered below. Subcase 1: $1 - v(A) > \beta, v(B) \leq \beta$. Combining $1 - v(A) < \alpha, \alpha + \beta = 1$ and (2)-(5), we have $\varphi_v(A) = \frac{1}{2}, \varphi_v(B) = 0$, and hence, $\varphi_v(A) \to \varphi_v(B) = \frac{1}{2} \to 0 = \frac{1}{2} = \varphi_v(A \to B)$. Subcase 2: $1 - v(A) > \beta, v(B) > \beta$. Then we conclude from $1 - v(A) < \alpha, v(B) < \alpha, \alpha + \beta = 1$ and (2)-(5) that both $v(A)$ and $v(B)$ simultaneously belong to the set $(\beta, \alpha) \cap L_n$. The fact that $|(\beta, \alpha) \cap L_n| \leq 1$ implies immediately that $v(A) = v(B)$, consequently, $v(A \to B) = 1$ and $\varphi_v(A \to B) = 1$, which, however, is a contradiction with the precondition $\varphi_v(A \to B) = \frac{1}{2}$. This shows that the subcase does not actually exist. Subcase 3: $1 - v(A) \leq \beta, v(B) > \beta$. Combining $1 - v(A) < \alpha, v(B) < \alpha$ and $\alpha + \beta = 1$, we conclude that both $v(A) \geq \alpha, \beta < v(B) < \alpha$. According to (2)-(5), we have $\varphi_v(A) = 1, \varphi_v(B) = \frac{1}{2}$, and hence, $\varphi_v(A) \to \varphi_v(B) = 1 \to \frac{1}{2} = \frac{1}{2} = \varphi_v(A \to B)$, as desired.

Case 3: $\varphi_v(A \to B) = 0$, then we have $v(A \to B) \leq \beta$, i.e., $(1 - v(A)) \vee v(B) \leq \beta$, which in turn implies that $1 - v(A) \leq \beta, v(B) \leq \beta$, consequently, $\varphi_v(A) = 1, \varphi_v(B) = 0$, and hence, $\varphi_v(A) \to \varphi_v(B) = \varphi_v(A \to B)$, as desired.

This completes the proof of Theorem 1.

Theorem 2. *In three-valued approximation of many-valued logic,* $\forall v : F(S) \to L_n$, φ_v *is a valuation if and only if it satisfies the following condition:*

(i) *When* n *is odd,*

$$\varphi_v(A) = \begin{cases} 1, v(A) \geq \frac{n+1}{2(n-1)}, \\ \frac{1}{2}, v(A) = \frac{1}{2}, \\ 0, v(A) \leq \frac{n-3}{2(n-1)}. \end{cases}$$

(ii) *When* n *is even,*

$$\varphi_v(A) = \begin{cases} 1, v(A) \geq \frac{n}{2(n-1)}, \\ 0, v(A) \leq \frac{n-2}{2(n-1)}. \end{cases}$$

Proof. It follows from Theorem 1 and the structure of L_n.

As shown in [10], the existing models of three-way decision mainly include Pawlak's rough set model, decision-theoretic rough set model, shadowed set, etc. Comparatively, our approach is different from those mentioned above in that it is firstly presented from the viewpoint of formal logic. Moreover, it can improve the classification of logic formulae in that the pair of thresholds (α, β) can be uniquely determined.

The requirement that $\varphi_v : F(S) \to \{0, \frac{1}{2}, 1\}$ should be a valuation mainly comes from the perspective of logic, that is, φ_v should preserve the logical connectives in NM-logic.

4   Conclusion

In this paper, we consider three-valued approximation of many-valued logic from the viewpoint of three-way decision. Our discussion is based on the following assumption: when the set of objects consists of logic formulae, then evaluation function should be the valuation in many-valued logic. Based upon this assumption, some interesting results have been obtained, i.e., when n is odd, the thresholds for three-way decision is uniquely determined. When n is even, there is actually no three-way decision, but two-way decision.

Note that our work in this paper is a tentative approach to combining three-way decision and logic because the considered logic has the linearly ordered truth values. In the future, we will extend the present work to a more general setting, where the truth values may be not linearly ordered, and moreover, NM-logic may be replaced by other types of logic systems.

Acknowledgements. This work is supported by the National Science Foundation of China (No. 61103133) and the Natural Science Program for Basic Research of Shaanxi Province,China (No. 2014JQ1032).

References

1. Yao, Y.Y.: Three-way decisions with probabilistic rough sets. Information Sciences 180, 341–353 (2010)

2. Yao, Y.Y.: The superiority of three-way decisions in probabilistic rough set models. Information Sciences 181, 1080–1096 (2011)
3. Li, H.X., Zhou, X.Z.: Risk decision making based on decision-theoretic rough set: A three-way view decision model. International Journal of Computational Intelligence Systems 4, 1–11 (2011)
4. Liu, D., Yao, Y.Y., Li, T.R.: Three-way investment decisions with decision-theoretic rough sets. International Journal of Computational Intelligence Systems 4, 66–74 (2011)
5. Jia, X., Zheng, K., Li, W., Liu, T., Shang, L.: Three-way decisions solution to filter spam email: An empirical study. In: Yao, J., Yang, Y., Słowiński, R., Greco, S., Li, H., Mitra, S., Polkowski, L. (eds.) RSCTC 2012. LNCS, vol. 7413, pp. 287–296. Springer, Heidelberg (2012)
6. Yang, X.P., Yao, J.T.: Modelling multi-agent three-way decisions with decision-theoretic rough sets. Fundamenta Informaticae 115, 157–171 (2012)
7. Yu, H., Wang, Y.: Three-way decisions method for overlapping clustering. In: Yao, J., Yang, Y., Słowiński, R., Greco, S., Li, H., Mitra, S., Polkowski, L. (eds.) RSCTC 2012. LNCS, vol. 7413, pp. 277–286. Springer, Heidelberg (2012)
8. Yao, Y.Y.: Three-way decision: An interpretation of rules in rough set theory. In: Wen, P., Li, Y., Polkowski, L., Yao, Y., Tsumoto, S., Wang, G. (eds.) RSKT 2009. LNCS, vol. 5589, pp. 642–649. Springer, Heidelberg (2009)
9. Zhou, B., Yao, Y.Y., Luo, J.G.: A three-way decision approach to email spam filtering. In: Farzindar, A., Kešelj, V. (eds.) Canadian AI 2010. LNCS, vol. 6085, pp. 28–39. Springer, Heidelberg (2010)
10. Yao, Y.Y.: An outline of a theory of three-way decisions. In: Yao, J., Yang, Y., Słowiński, R., Greco, S., Li, H., Mitra, S., Polkowski, L. (eds.) RSCTC 2012. LNCS, vol. 7413, pp. 1–17. Springer, Heidelberg (2012)
11. Gottwald, S.: A Treatise on Many-Valued Logics. Research Studies Press, Baldock (2001)
12. Esteva, F., Godo, L.: Monoidal t-norm based logic: towards a logic for left-continuous t-norms. Fuzzy sets and systems 124(3), 271–288 (2001)
13. Wang, G.J., Zhou, H.J.: Introduction to mathematical logic and resolution principle. Science Press, Beijing (2009)
14. Ciucci, D., Dubois, D.: Three-Valued Logics, Uncertainty Management and Rough Sets. In: Peters, J.F., Skowron, A. (eds.) Transactions on Rough Sets XVII. LNCS, vol. 8375, pp. 1–32. Springer, Heidelberg (2014)

Want More? Pay More!

Xibei Yang[1,2,3], Yong Qi[1], Hualong Yu[4], and Jingyu Yang[3]

[1] School of Economics and Management, Nanjing University of Science
and Technology, Nanjing, P.R. China, 210094
yangxibei@hotmail.com
[2] Artificial Intelligence Key Laboratory of Sichuan Province, Zigong,
P.R. China, 643000
[3] Key Laboratory of Intelligent Perception and Systems for High-Dimensional
Information (Nanjing University of Science and Technology), Ministry of Education,
Nanjing, P.R. China, 210094
[4] School of Computer Science and Engineering, Jiangsu University of Science
and Technology, Zhenjiang, P.R. China, 212003

Abstract. Decision-theoretic rough set is a special rough set approach,
which includes both misclassification and delayed decision costs. Though
the property of monotonicity does not always hold in decision-theoretic
rough set, the decision-monotonicity reduct may help us to increase both
lower approximation and upper approximation. The experimental results
in this paper tell us that by comparing with the original decision system,
more cost is required with decision-monotonicity reduct. The implied
philosophy is: **if you want to get more, you should pay more!**

Keywords: cost, decision-monotonicity reduct, decision-theoretic rough
set.

1 Introduction

Presently, Decision-Theoretic Rough Set (DTRS) [1] has attracted much attention due to the following two reasons: 1. it introduced the basic idea of cost
sensitivity into rough set model ; 2. the used thresholds come from cost itself.

However, it must be noticed that the property of monotonicity does not always
hold in DTRS [2,3]. For example, the lower and upper approximations are not
necessarily monotonic increasing or decreasing with the increasing of the conditional attributes. Therefore, the attribute reduction is an valuable problem,
which should to be addressed in DTRS since monotonicity plays a crucial role
in finding reducts of classical rough set. With respect to different requirements,
two main types of the attribute reductions have been considered in DTRS: one
is the reduct related to decision while the other is the reduct related to cost. For
example, Yao and Zhao [2] have proposed decision-monotonicity criterion and
generality criterion based reducts, which can be included into the first category.
Li et al. [3] proposed the concept of non-monotonic attribute reduction, which
aims to keep or increase the number of decision rules supported by positive regions. As far as the cost related reduct is considered, Yao and Zhao [2] proposed

C. Cornelis et al. (eds.): RSCTC 2014, LNAI 8536, pp. 144–151, 2014.
© Springer International Publishing Switzerland 2014

cost criterion based reduct, Jia et al. [4] analyzed three different algorithms to obtain reducts with smaller costs.

In many practical applications, our main objective is to classify an object into a class other than exclude an object from a class. Classification of object is corresponding to the positive rule in DTRS while exclusion of object is corresponding to the negative rule in DTRS. Therefore, we argue that Yao and Zhao's decision-monotonicity reduct is a very important reduct in DTRS because it requires to keep or increase the number of positive rules in a decision system. The decision-monotonicity criterion means the following two conditions:

The purpose of this paper is to further investigate decision-monotonicity reduct. This is mainly because decision-monotonicity reduct only take the decision rules into consideration while cost is also an important aspect in DTRS. Therefore, what we want to explore is the relationship between decision and cost of decision-monotonicity reduct.

2   DTRS

Formally, an information system can be considered as a pair $I = \langle U, AT \rangle$, in which U is a non-empty finite set of the objects called the universe; AT is a non-empty finite set of the attributes (features). $\forall a \in AT$, V_a is the domain of attribute a. $\forall x \in U$, $a(x)$ denotes the value that x holds on a ($\forall a \in AT$). Given an information system I, $\forall A \subseteq AT$, an indiscernibility relation $IND(A)$ may be defined as $IND(A) = \{(x, y) \in U^2 : \forall a \in A, a(x) = a(y)\}$, $[x]_A = \{y \in U : (x, y) \in IND(A)\}$ is the equivalence class of x in terms of $IND(A)$.

For a Bayesian decision procedure, a finite set of the states can be denoted by $\Omega = \{\omega_1, \omega_2, \cdots, \omega_s\}$, a finite set of t possible actions can be denoted by $\mathcal{A} = \{a_1, a_2, \cdots, a_t\}$. $\forall x \in U$, let $Pr(\omega_j | x)$ be the conditional probability of object x being in state ω_j, and $\lambda(a_i | \omega_j)$ be the loss, or cost for taking action a_i when the state is ω_j. Suppose that we take the action a_i for object x, then the expected loss is $\mathcal{R}(a_i | x) = \sum_{j=1}^{s} \lambda(a_i | \omega_j) \cdot Pr(\omega_j | x)$.

For Yao's DTRS, the set of states is composed by two classes such that $\Omega = \{X, \sim X\}$, it can be used to indicate that an object is in class X or out of class X; the set of actions is given by $\mathcal{A} = \{a_P, a_B, a_N\}$, in which a_P, a_B and a_N express three actions: a_P means that x is classified into positive region of X, i.e., $POS^A(X)$; a_B means that x is classified into boundary region of x, i.e., $BND^A(X)$; a_N means that x is classified into negative region of X, i.e., $NEG^A(X)$. The loss function regarding the costs of three actions in two different states is given in Tab. 1.

Table 1. The loss function regarding the costs of three actions in two states

	X	$\sim X$
a_P	λ_{PP}	λ_{PN}
a_B	λ_{BP}	λ_{BN}
a_N	λ_{NP}	λ_{NN}

In Tab. 1, $\lambda_{PP}, \lambda_{BP}$ and λ_{NP} are the losses for taking actions of a_P, a_B and a_N, respectively, when stating x is being in X; $\lambda_{PN}, \lambda_{BN}$ and λ_{NN} are the losses for taking actions of a_P, a_B and a_N, respectively, when stating x is out of X. It should be noticed that both λ_{BP} and λ_{BN} are delayed decision costs. $\forall x \in U$, by using the conditional probability $Pr(X|[x]_A) = \frac{|[x]_A \cap X|}{|[x]_A|}$, the expected losses associated with taking three actions are:

$\mathcal{R}(a_P|[x]_A) = \lambda_{PP} \cdot Pr(X|[x]_A) + \lambda_{PN} \cdot Pr(\sim X|[x]_A);$
$\mathcal{R}(a_B|[x]_A) = \lambda_{BP} \cdot Pr(X|[x]_A) + \lambda_{BN} \cdot Pr(\sim X|[x]_A);$
$\mathcal{R}(a_N|[x]_A) = \lambda_{NP} \cdot Pr(X|[x]_A) + \lambda_{NN} \cdot Pr(\sim X|[x]_A).$

The Bayesian decision procedure leads to the following minimum-risk decision rules:

(P) $\mathcal{R}(a_P|[x]_A) \leq \mathcal{R}(a_B|[x]_A) \wedge \mathcal{R}(a_P|[x]_A) \leq \mathcal{R}(a_N|[x]_A) \to x \in POS^A(X);$
(B) $\mathcal{R}(a_B|[x]_A) \leq \mathcal{R}(a_P|[x]_A) \wedge \mathcal{R}(a_B|[x]_A) \leq \mathcal{R}(a_N|[x]_A) \to x \in BND^A(X);$
(N) $\mathcal{R}(a_N|[x]_A) \leq \mathcal{R}(a_P|[x]_A) \wedge \mathcal{R}(a_N|[x]_A) \leq \mathcal{R}(a_B|[x]_A) \to x \in NEG^A(X).$

Since $Pr(X|[x]_A) + Pr(\sim X|[x]_A) = 1$ and we assume a reasonable loss function with the conditions such that $0 \leq \lambda_{PP} \leq \lambda_{BP} \leq \lambda_{NP}$ and $0 \leq \lambda_{NN} \leq \lambda_{BN} \leq \lambda_{PN}$, then decision rules (P), (B) and (N) can be expressed as:

(P) $Pr(X|[x]_A) \geq \alpha \wedge Pr(X|[x]_A) \geq \gamma \to x \in POS^A(X);$
(B) $Pr(X|[x]_A) < \alpha \wedge Pr(X|[x]_A) < \beta \to x \in BND^A(X);$
(N) $Pr(X|[x]_A) < \gamma \wedge Pr(X|[x]_A) \leq \beta \to x \in NEG^A(X);$

where $\alpha = \frac{(\lambda_{PN} - \lambda_{BN})}{(\lambda_{PN} - \lambda_{BN}) + (\lambda_{BP} - \lambda_{PP})}$, $\beta = \frac{(\lambda_{BN} - \lambda_{NN})}{(\lambda_{BN} - \lambda_{NN}) + (\lambda_{NP} - \lambda_{BP})}$,
$\gamma = \frac{(\lambda_{PN} - \lambda_{NN})}{(\lambda_{PN} - \lambda_{NN}) + (\lambda_{NP} - \lambda_{PP})}$.

Since $0 \leq \beta < \gamma < \alpha \leq 1$, then we have

(P) $Pr(X \mid [x]_A) \geq \alpha \to x \in POS^A(X);$
(B) $\beta < Pr(X \mid [x]_A) < \alpha \to x \in BND^A(X);$
(N) $Pr(X \mid [x]_A) \leq \beta \to x \in NEG^A(X).$

The above three rules are referred to as (P) rule, (B) rule and (N) rule, respectively. Following these rules, the lower approximation, upper approximation of X are $\underline{A}_{DT}(X) = \{x \in U : Pr(X|[x]_A) \geq \alpha\}$; $\overline{A}_{DT}(X) = \{x \in U : Pr(X|[x]_A) > \beta\}$.

For decision-theoretic rough set, we may consider following three costs which are supported by objects in positive region, boundary region and negative region, respectively.

(P) cost: $\sum_{x \in POS^A_{DT}(X)} \left(Pr(X|[x]_A) \cdot \lambda_{PP} + (1 - Pr(X|[x]_A)) \cdot \lambda_{PN}\right).$

(B) cost: $\sum_{x \in BND^A_{DT}(X)} \left(Pr(X|[x]_A) \cdot \lambda_{BP} + (1 - Pr(X|[x]_A)) \cdot \lambda_{BN}\right).$

(N) cost: $\sum_{x \in NEG^A_{DT}(X)} \left(Pr(X|[x]_A) \cdot \lambda_{NP} + (1 - Pr(X|[x]_A)) \cdot \lambda_{NN}\right).$

The overall cost of X is denoted by $\mathbf{COST}_A(X)$ such that $\mathbf{COST}_A(X) = \mathbf{COST}_A^{POS}(X) + \mathbf{COST}_A^{BND}(X) + \mathbf{COST}_A^{NEG}(X)$. Obviously, (P) cost is closely related to (P) rule since the computation of (P) cost is based on the objects in positive region and objects in positive region can induce (P) rules. Similarity, (B) cost is closely related to (B) rule and (N) cost is closely related to (N) rules.

3   Decision-Monotonicity Reduct

Definition 1. *Let $I =< U, AT \cup \{d\} >$ be a decision system in which $A \subseteq AT$, $U/IND(\{d\}) = \{X_1, \cdots, X_n\}$ is the set of the decision classes induced by decision attribute d, A is referred to as a decision-monotonicity reduct if and only if A is the minimal set of the conditional attributes, which preserves $\underline{AT}_{DT}(X_j) \subseteq \underline{A}_{DT}(X_j)$ and $\overline{AT}_{DT}(X_j) \subseteq \overline{A}_{DT}(X_j)$ for each $X_j \in U/IND(\{d\})$.*

It has been known that the (P) rules are supported by objects in positive region. Therefore, for the first condition $\underline{AT}_{DT}(X_j) \subseteq \underline{A}_{DT}(X_j)$, it is required that by reducing attributes, a (P) rule is still a (P) rule with the same decision. Moreover, by the second condition $\overline{AT}_{DT}(X_j) \subseteq \overline{A}_{DT}(X_j)$, $\forall x \in \overline{AT}_{DT}(X_j) - \underline{AT}_{DT}(X_j)$, we have $x \in \overline{A}_{DT}(X_j) - \underline{A}(X_j)$ or $x \in \underline{A}_{DT}(X_j)$, from which we can conclude that by reducing attributes a (B) rule is still a (B) rule, or is upgraded to a (P) rule with the same decision.

In many rough set literatures, two approaches have been widely adopted to find reducts [5]: exhaustive algorithm and heuristic algorithm. Unfortunately, exhaustive algorithm takes exponential time and is not feasible when we deal with large amount of data. This is why heuristic algorithm has attracted much attention. Most heuristic algorithms have the same structure and their differences lie in the different constructions of the heuristic functions. In this section, we will use the heuristic algorithm to compute decision-monotonicity reduct.

Algorithm 1. Heuristic approach to compute decision-monotonicity reduct.

Input: Decision system $I =< U, AT \cup \{d\} >$;
Output: A decision-monotonicity reduct A.
1.   Compute $\underline{AT}_{DT}(X_j)$ and $\overline{AT}_{DT}(X_j)$ for each $X_j \in U/IND(\{d\})$;
2. $A \longleftarrow \emptyset$, $\underline{A}_{DT}(X_j) = \overline{A}_{DT}(X_j) = \emptyset$ for each $X_j \in U/IND(\{d\})$;
       //Addition
3.   **while** $\underline{AT}_{DT}(X_j) \nsubseteq \underline{A}_{DT}(X_j)$ or $\overline{AT}_{DT}(X_j) \nsubseteq \overline{A}_{DT}(X_j)$ **do**
4.   **for** each $a \in AT - A$
5.   Compute the significance of attribute a;
6.   **end for**
7.   Find the maximal significance and the corresponding a;
8. $A = A \cup \{a\}$
9.   **end while**
       //Deletion
10.   **for** each $a \in A$
11.   **if** $\underline{AT}_{DT}(X_j) \subseteq \underline{A - \{a\}}_{DT}(X_j)$ and $\overline{AT}_{DT}(X_j) \subseteq \overline{A - \{a\}}_{DT}(X_j)$
           then
12. $A = A - \{a\}$;
13.   **end if**
14.   **end for**
15.   Return A.

---

4   Experimental Analysis

The experiments in this section is to uncover the relationships between decision and cost of decision-monotonicity reduct in DTRS. All the experiments in this section have been carried out on a personal computer with Windows 7, Intel Core 2 DuoT5800 CPU (2.00 GHz) and 2.00 GB memory. The programming language is Matlab 2010b. The data sets used are outlined in Tab. 2, which were all downloaded from UCI Repository of machine learning databases.

Table 2. A description of data sets

ID	Data sets	Objects	Attributes	Classes
I	Statlog (Australian Credit Approval)	690	14	$X_1(307)/X_2(383)$
II	Congressional Voting Records	435	16	$X_1(267)/X_2(168)$
III	Statlog (German Credit Data)	1000	24	$X_1(300)/X_2(700)$
IV	Ionosphere	351	34	$X_1(126)/X_2(225)$
V	SPECT Heart	267	22	$X_1(55)/X_2(212)$

Tab. 3 shows the experimental results of (P) rules, (P) costs, (B) rules and (B) costs. The number of (P)/(B) rules is equivalent to the numbers of objects in positive/boundary regions, respectively. This is mainly because each object in positive/boundary regions can induce a (P)/(B) decision rules.

Table 3. The comparisons between raw data and reduct data for (P) rules, (P) costs, (B) rules and (B) costs

ID	(P) rules		(P) cost		(B) rules		(B) cost	
	Raw	Reduct	Raw	Reduct	Raw	Reduct	Raw	Reduct
I								
X_1	301	301	0	0	10.8	**11.1**	0.9770	**1.0530**
X_2	377.8	**379.4**	0.5424	**1.2624**	10.2	*9*	0.4456	*0.2673*
II								
X_1	267	**284.6**	0	**5.7533**	0	0	0	0
X_2	168	**195.5**	0	**8.8848**	0	0	0	0
III								
X_1	300	**302.8**	0	**2.0861**	0	0	0	0
X_2	700	**702.8**	0	**2.0861**	0	0	0	0
IV								
X_1	123.6	**146.1**	0.1436	**8.2153**	9.4	**21.3**	12.2353	**15.0917**
X_2	225.6	**238.8**	1.3423	**5.9613**	2.4	*1.8*	0	0
V								
X_1	46.8	**47.5**	4.7331	**5.1563**	16.8	**17.6**	4.9706	**6.0161**
X_2	200.3	203.3	2.7096	**4.0181**	21.6	*20.1*	8.6606	**8.8863**

Based on a careful investigation of Tab. 3, it is not difficult to draw the following conclusions:

1. by comparing with the original data set, decision-monotonicity reduct can generate more (P) rules, this is mainly because the first condition of decision-monotonicity reduct requires that by reducing attributes, a positive rule is still a positive rule with the same decision;
2. with the increasing of number of (P) rules, (P) cost also increases, in other words, **we should pay more to obtain more (P) rules**;
3. by comparing with the original data set, decision-monotonicity reduct can generate more or less (B) rules, this is mainly because the second condition of decision-monotonicity reduct requires that by reducing attributes, a boundary rule is still a boundary rule, or is upgraded to a positive rule with the same decision;
4. by comparing with original (B) cost, the (B) cost obtained by decision-monotonicity reduct may increase or decrease.

Tab. 4 shows the experimental results of (P)+(B) rules, (P)+(B) costs, (N) rules and (N) costs, respectively. The number of (P)+(B) rules is equivalent to the sum of numbers of objects in lower approximation and boundary region, i.e. upper approximation, while number of (N) rules is equivalent to the number of objects in negative region. (P)+(B) cost is the sum of (P) and (B) cost.

Table 4. The comparisons between original data and reduct data for (P)+(B) rules, (P)+(B) costs, (N) rules and (N) costs

ID	(P)+(B) rules		(P)+(B) cost		(N) rules		(N) cost	
	Raw	Reduct	Raw	Reduct	Raw	Reduct	Raw	Reduct
I								
X_1	311.8	**312.1**	0.9970	**1.0530**	378.2	_377.9_	0.7519	0.7519
X_2	388	**388.4**	0.9880	**1.5297**	302	_301.6_	0.4043	_0.2426_
II								
X_1	267	**284.6**	0	**5.7533**	168	_150.4_	0	0
X_2	168	**195.5**	0	**8.8848**	267	_239.5_	0	0
III								
X_1	300	**302.8**	0	**2.0861**	700	_697.2_	0	0
X_2	700	**702.8**	0	**2.0861**	300	_297.2_	0	0
IV								
X_1	133	**167.4**	12.3789	**23.307**	218	_183.6_	1.1296	_0.9349_
X_2	228	**240.6**	1.3423	**5.9613**	123	_110.4_	0	0
V								
X_1	63.6	**65.1**	9.7037	**11.1669**	203.4	_201.9_	3.7049	3.7049
X_2	221.9	**223.4**	11.3702	**12.9044**	45.1	_43.6_	4.2684	_4.1970_

Based on a careful investigation of Tab. 4, it is not difficult to draw the following conclusions:

1. by comparing with the original data set, decision-monotonicity reduct can generate more (P)+(B) rules, this is mainly because the second condition of decision-monotonicity reduct requires that by reducing attributes, a (P)+(B) rule is still a (P)+(B) rule with the same decision;
2. with the increasing of number of (P)+(B) rules, (P)+(B) cost also increases, in other words, **we should pay more to obtain more (P)+(B) rules;**
3. by comparing with the original data set, decision-monotonicity reduct can generate less (N) rules, this is mainly because the second condition of decision-monotonicity reduct requires that by reducing attributes, a negative rule is not necessarily to be a negative rule with the same decision;
4. by comparing with original (N) cost, the (N) cost obtained by decision-monotonicity reduct decrease.

Tab. 5 shows the overall costs for five data sets. We can see that these overall costs are not only computed for each decision classes, but also obtained in whole decision system.

Table 5. The comparison between original data and reduct data for overall costs

ID	Overall costs	
	Raw	Reduct
I	3.1412	**3.6402**
X_1	1.7489	**1.8049**
X_2	1.3923	**1.8353**
II	0	**14.6381**
X_1	0	**5.7533**
X_2	0	**8.8848**
III	0	**4.1722**
X_1	0	**2.0861**
X_2	0	**2.0861**
IV	19.4698	**30.2032**
X_1	13.5085	**24.2419**
X_2	5.9613	5.9613
V	29.0472	**31.9732**
X_1	13.4086	**14.8718**
X_2	15.6386	**17.1014**

Finally, following a careful investigation of Tab. 5, it is not difficult to draw the following conclusions:

1. by comparing with the original data set, the overall cost of decision class obtained by decision-monotonicity reduct increase;
2. by comparing with the original data set, the overall cost of decision system obtained by decision-monotonicity reduct increase.

From discussions above, it is clear that through decision-monotonicity reduct, the numbers of decision rules supported by lower/upper approximation increase, while the corresponding costs also increase.

5   Conclusions

Decision-monotonicity reduct is proposed to keep or increase the lower approximation and upper approximation of decision-theoretic rough set. In this paper, what we want to explore is the relationships between such increasing and the variation of decision cost. Through experimental analysis, we have drawn an important conclusion such that the increasing of decision-theoretic rough approximation leads to the increasing of decision cost.

The present study is the first step to decision-monotonicity reduct in decision-theoretic rough set. The following are challenges for further research.

1. Decision-monotonicity reduct with complex types of cost, e.g. interval-valued costs.
2. Decision-monotonicity reduct by considering both Yao's loss function and test cost.

Acknowledgment. This work is supported by Natural Science Foundation of China (Nos. 61100116, 61272419, 61305058), Natural Science Foundation of Jiangsu Province of China (Nos. BK2011492, BK2012700, BK20130471), Qing Lan Project of Jiangsu Province of China, Key Laboratory of Intelligent Perception and Systems for High-Dimensional Information (Nanjing University of Science and Technology), Ministry of Education (No. 30920130122005), Key Laboratory of Artificial Intelligence of Sichuan Province (No. 2013RYJ03), Natural Science Foundation of Jiangsu Higher Education Institutions of China (Nos. 13KJB520003, 13KJD520008).

References

1. Yao, Y.Y., Wong, S.K.M., Lingras, P.: A decision-theoretic rough set model. In: 5th International Symposium on Methodologies for Intelligent Systems, pp. 17–25 (1990)
2. Yao, Y.Y., Zhao, Y.: Attribute reduction in decision-theoretic rough set models. Information Sciences 178(17), 3356–3373 (2008)
3. Li, H.X., Zhou, X.Z., Zhao, J.B., Liu, D.: Non-nonotonic attribute reduction in decision-theoretic rough sets. Fundamenta Informaticae 126(4), 415–432 (2013)
4. Jia, X.Y., Liao, W.H., Tang, Z.M., Shang, L.: Minimum cost attribute reduction in decision-theoretic rough set models. Information Sciences 219, 151–167 (2013)
5. Min, F., He, H.P., Qian, Y.H., Zhu, W.: Test-cost-sensitive attribute reduction. Information Sciences 181(22), 4928–4942 (2011)

An Incremental Clustering Approach
Based on Three-Way Decisions

Hong Yu, Cong Zhang, and Feng Hu

Chongqing Key Laboratory of Computational Intelligence
Chongqing University of Posts and Telecommunications, Chongqing, 400065, China
yuhong@cqupt.edu.cn

Abstract. Most of the clustering algorithms reported assume a data set does not always change. However, it is often observed that the analyzed data set changes over time. To combat changes, we introduce a new incremental soft clustering approach based on three-way decisions theory in this paper. Firstly, an initial clustering algorithm is proposed by using representative points. Secondly, to eliminate the influence of the processing order on final incremental clustering results, the incremental data is pre-clustered the same way. To quickly search similar areas for incremental data, a searching tree based on the representative points is constructed, and the strategies of searching and updating are presented. Finally, the three-way decisions strategy is used for incremental clustering. The results of the example analysis show the approach is effective to incremental clustering.

Keywords: Incremental clustering, Soft clustering, Searching tree, Three-way decisions.

1 Introduction

Clustering analysis is one of important basic research directions in the field of data mining. Nowadays, with the development of information technology, the amount of data is growing quickly in many industries such as electronic commerce, social networks, medical images, biological information, etc. So the clustering results obtained previously need to be updated with the increasing data. However, the traditional static clustering algorithms are not suitable to deal with these dynamic data sets, because they have to scan all the data again.

Nowadays, more researchers pay attention to the incremental clustering approaches. Pham et al. [1] proposed the clustering objective function based on the center under the framework of K-means clustering algorithm, but the approach suffered from several drawbacks such as it needs predefined number of clusters and only detects spherical clusters. Cheng et al. [2] proposed an incremental grid density-based clustering algorithm, clusters of arbitrary shape could be found according to concept of grid density reachable, but the granularity of grid is difficult to define. Patra et al. [3] proposed an incremental clustering algorithm based on distance and leaders. The incremental data are clustered by calculating distance between it and the surrounding leaders, but the algorithm needs

C. Cornelis et al. (eds.): RSCTC 2014, LNAI 8536, pp. 152–159, 2014.

to search the whole data space to find the surrounding leaders. Ning et al. [4] proposed an incremental spectral clustering approach by efficiently updating the eigen-system, but it could not find the overlapping clusters.

On the other hand, overlapping clustering is more appropriate in a variety of important applications such as network structure analysis, wireless sensor networks and biological information. Therefore, there are a few achievements of overlapping clustering for incremental data. For example, SHC [5] is an incremental clustering algorithm based on concept of Histogram Ratio of a cluster, and SHC analyzes difference of semantic histogram ratio of each cluster when new data are added. DHS [6] algorithm is also a soft incremental clustering algorithm, and it is derived from dynamic hierarchical agglomerative framework, which causes computational complexity is very high. Pérez-Suárea et al. [7] proposed an algorithm based on density and compactness for dynamic overlapping clustering, which is a graph-based clustering algorithm and represents collection of objects as a similarity graph, but it builds a large number of small clusters.

Therefore, the main objective of this paper is to present a novel incremental clustering approach which can also deal with the overlapping region among clusters, where the interval sets are used to represent a cluster [8]. Inspired by the three-way decisions theory [9], we will solve the problem based on the three-way decisions. In order to enhance the performance of computing, some information corresponding to the initial clustering results will be recorded in a tree first, the node of the tree is not an object but a region of some objects. When the incremental data arrives, the tree will be dynamically updated, and some strategies are proposed to reduce the finding space. The experimental results show that the proposed approach is effective to deal with the soft incremental clustering.

2   Basic Theory

2.1   Formulation of Three-Way Decisions Incremental Clustering

Let $U = \{x_1, \cdots, x_i, \cdots, x_N\}$ be a universe of D-dimensions objects, where $x_i = (x_i^1, \cdots, x_i^d, \cdots, x_i^D)$ and $i = 1, 2, \cdots, N$. Let the clustering results of the universe be $C = \{C_1, \ldots, C_k, \ldots, C_K\}$, and clusters are described by an interval set as $C = \{[\underline{C_1}, \overline{C_1}], \cdots, [\underline{C_k}, \overline{C_k}], \cdots, [\underline{C_K}, \overline{C_K}]\}$. For a cluster $C_k = [\underline{C_k}, \overline{C_k}]$, objects in $\underline{C_k}$ belong to the cluster definitely, objects in $\overline{C_k} - \underline{C_k}$ may belong to the cluster, objects in $U - \overline{C_k}$ do not belong to the cluster definitely. In other words, the cluster C_k consists of three regions, i.e. the positive region, the boundary region and the negative region, which are represented as $POS(C_k)$, $BND(C_k)$ and $NEG(C_k)$; and $POS(C_k) = \underline{C_k}$, $BND(C_k) = \overline{C_k} - \underline{C_k}$, $NEG(C_k) = U - \overline{C_k}$. The three domains of the cluster provide strong theoretical basis for three-way decisions, it can be used to construct corresponding three-way decisions rules. That is, an object will belong to $POS(C_k)$, $NEG(C_k)$ or $BND(C_k)$, when we make the decision of positive/acceptance, negative/rejection or deferred/non-promise.

Assuming that the time series are $T_0, \cdots, T_t, \cdots, T_n$. For a moment t, the system can be described by a knowledge expression system $IS^t = (U^t, A^t)$, where U^t means the universe at the moment and A^t means the set of D-dimensions attributes. A dynamic information system (DIS) can be represented by information system series at each moment, namely, $DIS = \{IS_0, \cdots, IS_t, IS_{t+1}, \cdots, IS_n\}$.

The problem of incremental clustering can be formalized as follows. Give an dynamic information system at the time $t + 1$, $IS^{t+1} = (U^{t+1}, A^{t+1})$, $U^{t+1} = U^t + \triangle U$, $A^{t+1} = A^t$, clustering result of time t is $RC^t = \{C_1^t, \ldots, C_i^t, \ldots C_{|RC^t|}^t\}$, and the structure information of each cluster at time t is known, then the question is to find clustering result at time $t+1$, namely, $RC^{t+1} = \{C_1^{t+1}, \ldots, C_i^{t+1}, \ldots , C_{|RC^{t+1}|}^{t+1}\}$. Of course, to combat the soft incremental clustering, the cluster will be represented as an interval set.

2.2 Related Definitions

In a D-dimensions space, when considering a small enough region, the objects are usually well-distributed, thus we can use a fictional point called representative point to represent these objects. Considering the discovery area, the center is o and r is the radius, the number of objects in the area is called the density of o relative to r and denoted by $Density(o, r)$.

Definition 1. Representative points: For point p, distance r and threshold $threshold$, if $Density(p, r) \geq threshold$, then p is a representative point.

Representative points are fictional points, not points/objects in the system.

Definition 2. Representing region: Every representative point p is the representative of a circular area where the point is the center of the area and the radius is r, and the region is representing region of representative point p.

All objects in the representing region of a representative point are seen as an equivalence class, all objects that do not be represented by any representative point are noise. For example, assuming r_k is the kth representative point, $cover_k = \{x_1, \cdots, x_k\}$ is the objects in its representative region. Since the representative point r_k can represent the region, it is reasonable to suppose that the fictional point has D-dimensions attributes. That is $r_k = (r_k^1, \cdots, r_k^d, \cdots, r_k^D)$ and a triple is used to represent the r_k^d, namely, $r_k^d = (r_k^d.left, r_k^d.right, r_k^d.average)$. The following formulas are used to compute them: $r_k^d.left = min(x_1^d, \cdots, x_k^d)$, $r_k^d.right = max(x_1^d, \cdots, x_k^d)$ and $r_k^d.average = \frac{1}{k} \sum_{i=1}^{k} x_i^d$.

Generally speaking, it is possible that there exists overlapping region between the representative regions.

Definition 3. Similarity between representative regions: Let r_i and r_j be two arbitrary representative points, the similarity between their representative regions is defined as follows.

$$Similarity(r_i, r_j) = \sqrt{\sum_{k=1}^{D} (r_i^k.average - r_j^k.average)^2} \tag{1}$$

To speed up the searching similar space for the incremental data, we build the searching tree based on the representative points. The root represents the original space composing of all representative points, then we sort the attributes by significance. According to the most significance attribute, we construct the nodes in the 1st layer. That is, all representative points are split according to these representative points' values in the attribute, and the same way to construct the other layers.

Definition 4. The node: Let $Node_j^i$ be the jth node of the i layer in the searching tree, let $R = \{r_1, \cdots, r_{|Node_j^i|}\}$ be the set of representative points belonging to the node $Node_j^i$. The node is represented by a value range, $Node_j^i = (Node_j^i.left, Node_j^i.right)$, where $Node_j^i.left = min(r_1^i.left, \ldots, r_{|Node_j^i|}^i.left)$, and $Node_j^i.right = max(r_1^i.right, \ldots, r_{|Node_j^i|}^i.right)$.

In addition, we need to measure the similarity between the incremental data representative points and nodes when looking up the searching tree. That is, we need to measure the similarity between 2 mathematical value ranges.

Definition 5. Similarity of value range: For arbitrary value range $Range1$ and $Range2$, wherein $Range1 = (Range1.left, Range1.right)$, so as to $Range2$. Assuming $Range1.left \leq Range2.left$, if we have $Range1.right \geq Range2.left$, then we call that $Range1$ is similar to $Range2$.

3   Incremental Clustering Algorithm Based on Three-Way Decisions

3.1   The Initial Clustering Algorithm

The initial clustering algorithm is a soft clustering algorithm based on the concepts of representative points. Firstly, we search the representative points of the data space according to the Definition 1, and the map between representative points and the objects in its representative region is built. Secondly, the algorithm adds some strong connected edges or weak connected edges among representative points based on three-way decisions. Thirdly, the algorithm finds out the strongly connected subgraph, which is the positive region of the cluster, the union of the regions of representative points, which have the weak connected to strong connected subgraph, constructs the boundary region of the cluster. The algorithm is described as follows.

Algorithm 1: The Initial Clustering Algorithm:
Input: $U = \{x_1, \ldots, x_n\}$; the parameter τ
Output: the initial clustering results RC and the noise set NS.
Step 1. Finding the representative points set RP and noise set NS according to the Def.(1), Def.(2). The step is repeated τ times.
Step 2. Constructing undirected graph G between representative points in RP using the idea of three-way decisions. Here, α, β are thresholds. For all $RP_i, RP_j \in RP$, to compute the $Similarty(RP_i, RP_j)$ according to the Eq.(1). If $Similarty(RP_i, RP_j) \geq \alpha$, there is a strong connected edge between them, if $\beta \leq Similarty(RP_i, RP_j) < \alpha$, there is a weak connected edge between them.
Step 3. Searching strong connected subgraphs in graph G. Every strong connected subgraph represents $POS(RC_i)$. The objects in the union of regions

corresponding to representative points which have weak edges connected to this strong connected subgraph, but not in $POS(RC_i)$ form the $BND(RC_i)$.

3.2   Constructing the Searching Tree

The method of creating searching tree is similar to that of creating the decision tree, which is built top-down. Firstly, we calculate the attribute importance of each attribute and sort them in descending order according to the attribute importance. The entropy index [10] is used to measure the attribute importance in this paper. Assuming that the sorted attributes are denoted as AS. Secondly, we build every layer according to the attribute importance, the more important attribute is prior to construct layer of tree. When there exists two adjacent layer whose numbers of nodes is roughly same, then we stop building the searching tree. Let $|node(i)|$ be the number of nodes of the ith layer.

Algorithm 2: Constructing the Searching Tree
Input: the representative point set RP, the ordered attribute set AS;
Output: the representative points searching tree T.
Step 1. Constructing the root node including all representative points, $i = 0$.
Step 2. Constructing the nodes of the ith($i \geq 1$) layer. Classifying every node $Node_j^{i-1}$ of the $i - 1$th layer into the nodes of the ith layer according to the Definition 5 based on the ith attribute of AS.
Step 3. If $\frac{|node(i)|}{|node(i+1)|} > \lambda (0 < \lambda < 1)$, then stop; otherwise, go to the Step 2.

3.3   Clustering the Incremental Data

Because the objects in new incremental data block are not completely isolated, there exists links among them. Firstly, we cluster the incremental data using the Algorithm 1, and the representative points in the incremental data block will be discussed in the following. Secondly, Algorithm 3 gives the searching and updating processing for every representative point in the incremental data block to find the similar nodes by searching the searching tree. Then, Algorithm 4 proposes an incremental clustering based on three-way decisions. Let r_{wait} be one representative point in the incremental data block, let $Node(j) = \{Node_{k1}^j, \ldots, Node_{kn}^j\}$ be the jth layer similar nodes.

Algorithm 3: Searching and Updating the Searching Tree
Input: the searching tree T, the incremental representative point r_{wait};
Output: the similar representative points $Similarpoint$.
Step 1. Adding r_{wait} to root node. According to the 1th dimension attribute value of r_{wait} and definition 5, determining $Node(1)$ in the 1th layer.
Step 2. According to $Node(j)$ (initial $j = 1$) and the $j + 1$ dimension attribute value of r_{wait}, determining $Node(j + 1)$. There are 3 cases as follows.
(1) There exists only one node, i.e. assuming $Node_{ki}^j$ in $Node(j)$ is similar to r_{wait} according to definition 5, if jth layer is last layer, adding all representative points in $Node_{ki}^j$ to $Similarpoint$, adding r_{wait} to $Node_{ki}^j$, otherwise, adding

r_{wait} to $Node_{ki}^j$, adding child nodes of $Node_{ki}^j$ to $Node(j+1)$, goto Step 3.

(2) There exists more than two nodes, i.e. assuming $Node_{ki}^j, Node_{kj}^j$ in $Node(j)$ are similar to r_{wait}, if jth layer is last, adding all points in $Node_{ki}^j, Node_{kj}^j$ to $Similarpoint$, merging $r_{wait}, Node_{ki}^j, Node_{kj}^j$ into a new node, otherwise, merging $r_{wait}, Node_{ki}^j, Node_{kj}^j$, merging child node of $Node_{ki}^j, Node_{kj}^j$ to form new child node, then adding all child nodes of new node to $Node(j+1)$, goto Step 3.

(3) There exists none, i.e. any $Node_{ki}^j$ in $Node(j)$ is not similar to r_{wait}, if the jth layer is not last, then r_{wait} itself forms a new node of the jth and subsequent several layers.

Step 3. $j = j+1$, goto Step 2. We calculate similarity between r_{wait} and representative points in $Similarpoint$ after finding similar region. Then, r_{wait} is merged into initial clusters or forms new cluster, the detail algorithm is described as follows.

Algorithm 4: Clustering the New Data
Input: $Similarpoint$, the incremental representative point r_{wait}.
Output: the final clustering results.
Step 1. For all $r_i \in Similarpoint$, calculating $Similarity(r_{wait}, r_i)$ according to Eq.(1). Let R_{alpha} be representative points which satisfy $Similarity(r_{wait}, r_i) \geq \alpha$ and R_{beta} be points which satisfy $\beta \leq Similarity(r_{wait}, r_i) < \alpha$.
Step 2. Clustering r_{wait}, there are 2 cases as follows:
(1) if R_{alpha} is null, forming new cluster, the $POS(RC_{new})$ is composed by r_{wait}, the $BND(RC_{new})$ is composed by objects in the union of region corresponding to the representative points in R_{beta}.
(2) if representative points in R_{alpha} belong to same or different clusters, merging r_{wait} and lower bound of these clusters together to form $POS(RC_{merge})$, and objects in the union of region corresponding to the representative points in R_{beta} but not in $POS(RC_{merge})$ into $BND(RC_{merge})$.

4 Example Analysis

In this section, we have carried out a number of experiments on synthetic data sets to validate the performance of the proposed algorithm. The 2-dimension synthetic data set is tested to illustrate the ideas presented in the previous section, which has 1000 objects. 90% of the data set is used as the original data set, and 10% of the data set is used as the incremental data. In this experiment, the parameters $r, threshold, \tau, \alpha, \beta, \lambda$ are set as 0.3,2,4,0.401,0.5,0.9, respectively. In order to show the effectiveness of the proposed algorithm, we design three tests on the synthetic data set.

The first case: the incremental data is selected randomly for every cluster. The clustering results of DS1 are shown in the Fig.1, DS1 is clustered into 5 clusters initially. The incremental clustering results of DS1 are shown in the Fig.2, it is obviously that cluster C_1 has grown and the emergence of new boundary region of cluster C_5 has been observed.

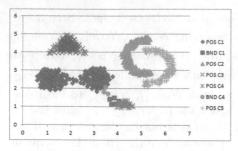

Fig. 1. The original results of DS1   **Fig. 2.** The incremental results of DS1

The second case: we remove the objects that connect with the two parts of the cluster C_5 and C_6 intentionally, and the clustering results are shown in Fig.3 according to Algorithm1, DS2 is clustered into 6 clusters initially. Then, these objects add to the set DS2, the incremental clustering results in the Fig.4 show that the two parts be connected as a cluster, which reveals the inherent structure in the DS2.

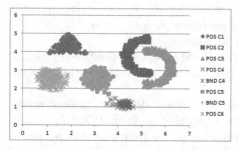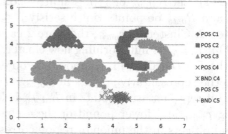

Fig. 3. The original results of DS2   **Fig. 4.** The incremental results of DS2

The third case: we remove a cluster from the test data set as shown in Fig.5 intentionally, the DS3 is clustered into 4 clusters according to Algorithm1. Then, these objects add to set DS3, and the incremental clustering results of DS3 are shown in Fig.6. The approach proposed in this paper can find the new cluster.

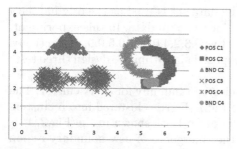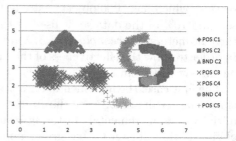

Fig. 5. The original clustering of DS3   **Fig. 6.** The incremental clustering of DS3

5   Conclusions

In this paper, we presented an incremental clustering algorithm based on three-way decisions. We used representative points in initial data set to build the searching tree and defined the conception of the similarity of value range. As a consequence, when incremental data is added up to the initial data set, we can quickly find the similar subspace of new data by looking up the searching tree. We then calculated the similarity between new data and all representative points in the similar area and decided the cluster of the new data. The proposed approach not only can find clusters of arbitrary shapes, but have the ability of processing merging of different clusters and finding new clusters in the incremental data. The results of the example analysis also show the approach is effective for incremental clustering. Future works aim to provide a convenient parameter selection method in the algorithm.

Acknowledgments. This work was supported in part by the National Natural Science Foundation of China under grant No.61379114, No.61309014 and No.61272060.

References

1. Pham, D.T., Dimov, S.S., Nguyen, C.: D.: An Incremental K-means Algorithm. Proceedings of the Institution of Mechanical Engineers, Part C: Journal of Mechanical Engineering Science 218(7), 783–795 (2004)
2. Chen, N., Chen, A., Zhou, L.X.: An Incremental Grid Density-Based Clustering Algorithm. Journal of Software 13(1), 1–7 (2002)
3. Patra, B.K., Ville, O., Launonen, R., Nandi, S., Babu, K.S.: Distance based Incremental Clustering for Mining Clusters of Arbitrary Shapes. In: Maji, P., Ghosh, A., Murty, M.N., Ghosh, K., Pal, S.K. (eds.) PReMI 2013. LNCS, vol. 8251, pp. 229–236. Springer, Heidelberg (2013)
4. Ning, H.Z., Xu, W., Chi, Y., Gong, Y.H., Huang, T.S.: Incremental spectral clustering by efficiently updating the eigen-system. Pattern Recognition 43(1), 113–127 (2010)
5. Gad, W.K., Kamel, M.S.: Incremental clustering algorithm based on phrase-semantic similarity histogram. In: 2010 International Conference on Machine Learning and Cybernetics (ICMLC), pp. 2088–2093. IEEE Pess, Qingdao (2010)
6. Gil-Garcia, R., Pons-porrate, A.: Dynamic Hierarchical Algorithms for Document Clustering. Pattern Recognition Letters 31(6), 469–477 (2010)
7. Pérez-Suárea, A., Martínez-Trinidad, J.F., Carrasco-Ochoa, J.A., Medina-Pagola, J.E.: An Algorithm Based on Density and Compactness for Dynamic Overlapping Clustering. Pattern Recognition 46(11), 3040–3055 (2013)
8. Yao, Y., Lingras, P., Wang, R., Miao, D.: Interval Set Cluster Analysis: A Reformulation. In: Sakai, H., Chakraborty, M.K., Hassanien, A.E., Ślęzak, D., Zhu, W. (eds.) RSFDGrC 2009. LNCS, vol. 5908, pp. 398–405. Springer, Heidelberg (2009)
9. Yao, Y.: An Outline of a Theory of Three-Way Decisions. In: Yao, J., Yang, Y., Słowiński, R., Greco, S., Li, H., Mitra, S., Polkowski, L. (eds.) RSCTC 2012. LNCS, vol. 7413, pp. 1–17. Springer, Heidelberg (2012)
10. Wang, A.P., Wan, G.W., Cheng, Z.Q., Li, S.K.: Incremental Learning Extremely Random Forest Classifier for Online Learning. Journal of Software 22(9), 2059–2074 (2011)

Determining Three-Way Decision Regions with Gini Coefficients

Yan Zhang and JingTao Yao

Department of Computer Science, University of Regina
Regina, Saskatchewan, Canada S4S 0A2
{zhang83y,jtyao}@cs.uregina.ca

Abstract. Three-way decision rules can be constructed from rough set regions, i.e., positive, negative and boundary regions. These rough set regions can be viewed as the acceptance, rejection, and non-commitment decision regions in three-way classification. Interpretation and determination of decision regions are one of the key issues of three-way decision and rough set theories. We investigate the relationship between changes in rough set regions and their impacts on the Gini coefficients of decision regions. Effective decision regions can be obtained by satisfying objective functions of Gini coefficients of decision regions. Three different objective functions are discussed in this paper. The example shows that effective decision regions can be obtained by tuning Gini coefficients of decision regions to satisfy a certain objective function. It is suggested that with the new approach more applicable decision regions and decision rules may be obtained.

1 Introduction

The acceptance and rejection are two options adopted in the commonly used binary decision model. However, making a definite decision of either acceptance or rejection may lead to either a high level of incorrect-acceptance error and a high level incorrect-rejection error [7]. Yao formulated three-way decision theory [21]. It extends binary decision model by adding a non-commitment option. Yao introduced and suggested to interpret three rough set regions, i.e., positive, negative and boundary regions as the regions of acceptance, rejection, and non-commitment in a ternary classification [20]. The rules of acceptance and rejection decisions can be induced from the positive and negative regions, respectively. The non-commitment decisions can be made from the boundary region [20]. In Pawlak model, the decisions of acceptance and rejection are made from positive and negative regions with no errors [13]. The intolerance to any error may lead to smaller positive and negative regions which are only applicable to a small set of objects. Probabilistic rough sets introduce a pair of thresholds (α, β) to weaken the strict condition of the Pawlak model. They provide a trade-off between error and applicability in order to obtain more efficient decision regions [18][19].

Interpretation and determination of decision region thresholds are one of the key issues in three-way or ternary decision theories. Many attempts have been

C. Cornelis et al. (eds.): RSCTC 2014, LNAI 8536, pp. 160–171, 2014.

made to determine region thresholds with various approaches. Decision-theoretic rough set model (DTRS) determines the pair of region thresholds by minimizing overall classification cost [22]. Game-theoretic rough sets (GTRS) obtain region thresholds by formulating competition or cooperation among multiple criteria [1][2][12][17]. In information-theoretic rough sets (ITRS), the effective decision regions are obtained by minimizing Shannon entropy of three probabilistic rough set regions [5]. In this paper, we use Gini coefficient to measure decision regions, which in turn will provide a balance between decision regions. The relationship between changes in rough set regions and their impacts on the Gini coefficients of decision regions is used for this purpose. In particular, three objective functions for Gini coefficients of regions are discussed in this paper. Different effective decision regions can be obtained by satisfying different objective functions. The result in this study may enhance our understanding of rough sets and three-way decision and make them practical in applications.

2   Rough Set Regions and Gini Coefficients

In this section, we briefly introduce the background concepts used in the paper, namely, the relationship between decision regions and thresholds, the concept of Gini coefficients, and Gini coefficients of decision regions.

2.1   Decision Regions and Thresholds

Suppose that the universe U is a finite nonempty set. Let $E \subseteq U \times U$ be an equivalence relation on U, i.e., E is reflexive, symmetric, and transitive [13]. The basic building blocks of rough set theory are the equivalence classes of E. For an element $x \in U$, the equivalence class containing x is given by $[x] = \{y \in U | xEy\}$. The family of all equivalence classes is called the quotient set of U, and is denoted by $U/E = \{[x] | x \in U\}$. It defines a partition of the universe.

For an indescribable concept $C \subseteq U$, lower and upper approximations are used to approximate it. In probabilistic rough sets, a pair of thresholds (α, β) are used to define approximations [19]. We assume that $0 \leq \beta < \alpha \leq 1$, and this condition is hold in this paper. Based on these approximations, positive, negative and boundary regions are defined as [19]:

$$POS_{(\alpha,\beta)}(C) = \bigcup\{[x] \mid [x] \in U/E, Pr(C|[x]) \geq \alpha\},$$
$$NEG_{(\alpha,\beta)}(C) = \bigcup\{[x] \mid [x] \in U/E, Pr(C|[x]) \leq \beta\},$$
$$BND_{(\alpha,\beta)}(C) = \bigcup\{[x] \mid [x] \in U/E, \beta < Pr(C|[x]) < \alpha\}. \tag{1}$$

The acceptance, rejection and non-commitment decisions can be induced from the positive, negative and boundary regions, respectively. This provides an interpretation of the three rough set regions as the acceptance, rejection, and non-commitment decision regions in three-way classification. We denote the three decision regions as $POS_{(\alpha,\beta)}(C)$, $NEG_{(\alpha,\beta)}(C)$ and $BND_{(\alpha,\beta)}(C)$.

Intuitively speaking, given an equivalence class $[x]$, if the probability of the concept C given $[x]$ is greater or equal to α, i.e., $Pr(C|[x]) = \frac{[x] \cap C}{[x]} \geq \alpha$, we accept $[x]$ in the concept C. The incorrect acceptance error is less than or equal to $1 - \alpha$ [5]. If the probability of the concept C given $[x]$ is less or equal to β, i.e., $Pr(C|[x]) = \frac{[x] \cap C}{[x]} \leq \beta$, we reject $[x]$ as the concept C. The incorrect rejection error is less than or equal to β [5].

Pawlak rough set model is a special case of probabilistic rough set model in which $\alpha = 1$ and $\beta = 0$. The decisions of acceptance and rejection are made with no error. The incorrect acceptance error of acceptance region and the incorrect rejection error of rejection region are both 0. But the non-commitment region in the Pawlak model may be too large to apply in practical applications. The thresholds (α, β) in probabilistic rough sets provide a trade-off between errors and applicability of the model.

2.2   Gini Coefficients

A Gini coefficient or Gini index is an inequality measure used in economics [8][10]. Mathematically, Gini coefficient can be used as a measure of divergence of two or more probability distributions [4][9]. Gini coefficients have been applied in many studies. For example, in data mining, it is used to measure the impurity of node when building decision tree and determining the best split attribute [3][16].

Considering objects in the set S having an attribute A with k possible values, $a_1, a_2, ..., a_k$. We use block $\sigma_{A=a_i}(S)$ to denote the set of objects in S with attribute value a_i. The blocks containing objects with different attribute values constitute a partition of S, $\pi_A = \{\sigma_{A=a_1}(S), \sigma_{A=a_2}(S), ..., \sigma_{A=a_k}(S)\}$, i.e., the union of these blocks is the set S and the intersection of any two blocks is an empty set. The probabilistic distribution of a partition π_A can be defined as:

$$P_{\pi_A} = \left(\frac{|\sigma_{A=a_1}(S)|}{|S|}, \frac{|\sigma_{A=a_2}(S)|}{|S|}, ..., \frac{|\sigma_{A=a_k}(S)|}{|S|} \right), \tag{2}$$

where $|\cdot|$ denotes the cardinality of a set, and $|\sigma_{A=a_i}(S)| \setminus |S|$ denotes the probability of the block $\sigma_{A=a_i}(S)$, i.e., $Pr(a_i) = |\sigma_{A=a_i}(S)| \setminus |S|$.

There are two kinds of Gini coefficients, namely, the absolute Gini coefficient and the relative Gini coefficient. The absolute Gini coefficient of S with respect of the probability distribution P_{π_A} measures the inequality of each block of the partition π_A and can be defined by [3]:

$$Gini(S, \pi_A) = \sum_{i=1}^{k} \frac{|\sigma_{A=a_i}(S)|}{|S|} \times \left(1 - \frac{|\sigma_{A=a_i}(S)|}{|S|} \right) = 1 - \sum_{i=1}^{k} \left(\frac{|\sigma_{A=a_i}(S)|}{|S|} \right)^2. \tag{3}$$

The value of absolute Gini coefficient is between 0 to $\frac{k-1}{k}$. When all objects in S have same attribute value a_i, the absolute Gini coefficient obtains the minimum value 0. When P_{π_A} distributes averagely in S, the absolute Gini coefficient obtains the maximum value, $\frac{k-1}{k}$.

Given two attributes A with k possible values, $a_1, a_2, ..., a_k$, and B with m possible values, $b_1, b_2, ..., b_m$, we can get two partitions $\pi_A = \{\sigma_{A=a_1}(S), \sigma_{A=a_2}(S), ..., \sigma_{A=a_k}(S)\}$ and $\pi_B = \{\sigma_{B=b_1}(S), \sigma_{B=b_2}(S), ..., \sigma_{B=b_m}(S)\}$. The two partitions mean dividing the set S into different groups based on the values of attributes A and B. The relative Gini coefficient of each block $\sigma_{B=b_i}(S)$ measures the distribution of P_{π_A} in block $\sigma_{B=b_i}(S)$, and can be calculated as:

$$Gini(\sigma_{B=b_i}(S)) = \frac{|\sigma_{B=b_i}(S)|}{|S|} \times Gini\left(\sigma_{B=b_i}(S), \pi_A\right)$$

$$= \frac{|\sigma_{B=b_i}(S)|}{|S|} \times \left(1 - \sum_{j=1}^{k} \left(\frac{|\sigma_{A=a_j}(S) \cap \sigma_{B=b_i}(S)|}{|\sigma_{B=b_i}(S)|}\right)^2\right). \quad (4)$$

The relative Gini coefficient of each block $\sigma_{B=b_i}(S)$ reaches the maximum value $\frac{k-1}{k}$ when P_{π_A} is averagely distributed in each block $\sigma_{B=b_i}(S)$. It obtains the minimum value 0 when each block $\sigma_{B=b_i}(S)$ only contains the objects with single attribute value a_j.

2.3 Gini Coefficients of Decision Regions

Given a concept C and a threshold pair (α, β), the three decision regions are defined according to Equation (1). As the three decision regions are pair-wise disjoint and their union is the entire universe U, the following partition is formed:

$$\pi_{(\alpha,\beta)} = \{POS_{(\alpha,\beta)}(C), NEG_{(\alpha,\beta)}(C), BND_{(\alpha,\beta)}(C)\}. \quad (5)$$

The probabilistic distribution of the partition $\pi_{(\alpha,\beta)}$ is the probabilities of three decision regions:

$$P_{\pi_{(\alpha,\beta)}} = \left(Pr(POS_{(\alpha,\beta)}(C)), Pr(NEG_{(\alpha,\beta)}(C)), Pr(BND_{(\alpha,\beta)}(C))\right). \quad (6)$$

The probabilities of three decision regions are:

$$Pr(POS_{(\alpha,\beta)}(C)) = \frac{|POS_{(\alpha,\beta)}(C)|}{|U|},$$

$$Pr(NEG_{(\alpha,\beta)}(C)) = \frac{|NEG_{(\alpha,\beta)}(C)|}{|U|},$$

$$Pr(BND_{(\alpha,\beta)}(C)) = \frac{|BND_{(\alpha,\beta)}(C)|}{|U|}. \quad (7)$$

The concept C and its complement set constitute a partition, i.e., $\pi_C = \{C, C^c\}$. The absolute Gini coefficient of each decision region with respect to π_C are determined as [23]:

$$Gini\left(POS_{(\alpha,\beta)}(C), \pi_C\right) = 1 - Pr(C|POS_{(\alpha,\beta)}(C))^2 - Pr(C^c|POS_{(\alpha,\beta)}(C))^2,$$
$$Gini(NEG_{(\alpha,\beta)}(C), \pi_C) = 1 - Pr(C|NEG_{(\alpha,\beta)}(C))^2 - Pr(C^c|NEG_{(\alpha,\beta)}(C))^2,$$
$$Gini(BND_{(\alpha,\beta)}(C), \pi_C) = 1 - Pr(C|BND_{(\alpha,\beta)}(C))^2 - Pr(C^c|BND_{(\alpha,\beta)}(C))^2. \quad (8)$$

The probability $Pr(C|POS_{(\alpha,\beta)}(C))$ in Equation (8) denotes the conditional probability of an object x in C given that the object is in the acceptance decision region $POS_{(\alpha,\beta)}(C)$. The conditional probabilities are computed as:

$$Pr(C|POS_{(\alpha,\beta)}(C)) = \frac{|C \cap POS_{(\alpha,\beta)}(C)|}{|POS_{(\alpha,\beta)}(C)|},$$

$$Pr(C|NEG_{(\alpha,\beta)}(C)) = \frac{|C \cap NEG_{(\alpha,\beta)}(C)|}{|NEG_{(\alpha,\beta)}(C)|},$$

$$Pr(C|BND_{(\alpha,\beta)}(C)) = \frac{|C \cap BND_{(\alpha,\beta)}(C)|}{|BND_{(\alpha,\beta)}(C)|}. \tag{9}$$

The conditional probabilities of C^c are similarly obtained.

The relative Gini coefficients of acceptance, rejection and non-commitment decision regions are [23]:

$$G_P(\alpha, \beta) = Pr(POS_{(\alpha,\beta)}(C)) \times Gini(POS_{(\alpha,\beta)}(C), \pi_C),$$

$$G_N(\alpha, \beta) = Pr(NEG_{(\alpha,\beta)}(C)) \times Gini(NEG_{(\alpha,\beta)}(C), \pi_C),$$

$$G_B(\alpha, \beta) = Pr(BND_{(\alpha,\beta)}(C)) \times Gini(BND_{(\alpha,\beta)}(C), \pi_C). \tag{10}$$

If a half of the objects in a region belong to C and the other half of the objects belong to C^c, the absolute Gini coefficient of a region has the maximum value of $1/2$. The range of the probabilities for a region is between 0 to 1. We can infer that the ranges of these relative Gini coefficients of the three decision regions are between 0 to $1/2$:

$$0 \leq G_P(\alpha, \beta), G_B(\alpha, \beta), G_N(\alpha, \beta) \leq 1/2. \tag{11}$$

In Pawlak model, i.e., $(\alpha, \beta) = (1, 0)$, the acceptance region only contains the objects belonging to the concept C and the rejection region only contains the objects not belonging to the concept C. Therefore, Pawlak model has the minimal relative Gini coefficients of 0 for both acceptance and rejection regions. However, it may have the maximum relative Gini coefficient for non-commitment region, since non-commitment region contains the objects belonging to C and objects belonging to C^c. In addition, the non-commitment region is normally large in size. When α is decreased or β is increased, the relative Gini coefficients of the three regions will change correspondingly. When α equals to β, the size of non-commitment region shrinks to 0, and the sizes of acceptance and rejection regions increase. At this time, three-way decision evolves into binary decision model. The relative Gini coefficient of non-commitment region obtains the minimal value 0, and the sum of the relative Gini coefficients of acceptance and rejection regions is maximal. We can see there exists a restriction mechanism among three decision regions. Gini coefficients of regions may provide a balance among three regions, but the position of balance depends on the requirements of applications.

3   Determining Decision Regions by Setting Objective Functions for Gini Coefficients of Regions

The change of decision regions will have directly affect on the relative Gini coefficients of these decision regions. We can obtain effective decision regions which satisfy various kinds of objective functions for relative Gini coefficients of regions. In this section, we mainly discuss three types of objective functions, minimizing the overall Gini coefficient, minimizing the difference between Gini coefficients of immediate decision regions and non-commitment region, and setting limits for three Gini coefficients. Since we may have to consider the sizes of decision regions during analysis, we will use Gini coefficient to indicate relative Gini coefficient in the rest of the paper when no confusion arises.

3.1   Minimizing the Overall Gini Coefficient

We consider the trade-off among Gini coefficients of the three decision regions and formulate it as an optimization problem. The approach of converting determination of region thresholds to an optimization problem is feasible and effective [21]. Assume that a measure evaluates some desired quality of the induced decision regions and a lower value of the measure represents a more desirable classification. Our goal is to find an optimal pair of thresholds that minimizing the overall quality of three decision regions [7]. Various measures can be employed to evaluate decision regions. Information-theoretic rough sets use Shannon entropy to measure the uncertainty of decision regions and try to obtain region thresholds which can induce decision regions with minimizing the overall uncertainty [5][7]. Here we employ the optimization problem to obtain effective decision regions which can provide a trade-off among Gini coefficients of the decision regions.

Let's use the summation of Gini coefficients of three decision regions to denote the overall Gini coefficients of rough set regions:

$$G_{sum}(\alpha, \beta) = G_P(\alpha, \beta) + G_N(\alpha, \beta) + G_B(\alpha, \beta). \tag{12}$$

The aim is to minimize $G_{sum}(\alpha, \beta)$ to obtain the decision regions that make the distributions $\pi_{(\alpha, \beta)}$ and π_C more convergent. The problem of finding optimal threshold pairs can be formulated as the optimization problem:

$$(\alpha, \beta) = \{(\alpha, \beta) | MIN(G_{sum}(\alpha, \beta))\}. \tag{13}$$

In Pawlak model, the acceptance and rejection regions have the minimal Gini coefficient of 0, i.e., $G_P(1, 0) = G_N(1, 0) = 0$. However, the overall Gini coefficient of decision model $G_{sum}(1, 0)$ may not be the minimum due to the maximal Gini coefficient of non-commitment region. With the decrease of α and the increase of β, the Gini coefficients of acceptance and rejection regions will increase and the Gini coefficient of non-commitment region will decrease. When α equals to β, i.e., $\alpha = \beta$, the non-commitment region has a minimal size of 0 and three-way decision model evolves to a two-way decision model. The Gini coefficient of non-commitment region obtain the minimal value 0, i.e., $G_B(\alpha, \beta) = 0$. However, the overall Gini coefficient of decision model $G_{sum}(1, 0)$ may not be the minimum due to the changes of acceptance and rejection regions. The optimal decision regions that make the overall Gini coefficient minimum depend on the distribution of data. There may be more than one threshold pair corresponding to the minimal overall Gini coefficient. One can search the space of all possible region thresholds to obtain optimal ones that satisfy the objective function. When data set are big and the size of possible thresholds are huge, heuristic strategies can be employed to reduce the search cost and time.

3.2   Minimizing the Difference between Gini Coefficients

We consider acceptance and rejection regions versus non-commitment region as two types of regions. Since we can make decisions and induce rules from acceptance and rejection regions, and we have to defer decisions or make non-commitment decisions from non-commitment region, we call them immediate decision regions and non-commitment decision region, respectively [11]. The Gini coefficients of immediate and non-commitment decision regions influence each other. The increase of one's value may lead to the decrease of the other's value. In Pawlak model, Gini coefficients of immediate decision regions are 0 and Gini coefficient of non-commitment decision region is the maximum. The model is not applicable due to the large size of non-commitment decision region. In binary decision models, Gini coefficients of immediate decision regions are the maximum and Gini coefficient of non-commitment decision region is 0. The model has unacceptable error rates. These two extreme situations lead to unsuitable decision regions. We need to consider a criterion to make a trade-off between Gini coefficients of immediate and non-commitment decision regions.

Dividing three regions into two groups and considering the difference between them previously was employed in shadowed sets [6][14]. Pedrycz proposed that an optimal pair of thresholds should satisfy minimization of the absolute difference of the sum of the elevated and reduced areas and shadow area [14][15]. The sizes of three areas are calculated by the cardinality of the sets. An optimal pair of thresholds can be obtained by taking the arguments that minimize the difference of the sum of the elevated and reduced areas and shadow. In this paper, we use this objective function to consider the difference between Gini coefficients of immediate and non-commitment decision regions.

The difference between Gini coefficients of immediate and non-commitment decision regions is expressed as:

$$G_{diff}(\alpha, \beta) = (G_P(\alpha, \beta) + G_N(\alpha, \beta)) - G_B(\alpha, \beta). \tag{14}$$

We aim to minimize $G_{diff}(\alpha, \beta)$ to obtain the decision regions that make a trade-off between Gini coefficients of immediate and non-commitment decision regions. The problem of finding pairs of optimal region thresholds can be formulated as the optimization problem:

$$(\alpha, \beta) = \{(\alpha, \beta) | MIN\ (|G_{diff}(\alpha, \beta)|)\}. \tag{15}$$

In Pawlak model, the acceptance and rejection regions have the minimal Gini coefficient of 0, i.e., $G_P(1,0) = G_N(1,0){=}0$, the non-commitment region has the maximal Gini coefficient, we can obtain $G_{diff}(1,0) = G_B(1,0)$. With the decrease of α and the increase of β, the difference between two Gini coefficients will decrease. After it reaches the minimal value, the difference between two Gini coefficients will increase again. The shape of the change seems to be concave. When $\alpha = \beta$, the Gini coefficient of non-commitment region has the minimal value 0, i.e., $G_B(\gamma, \gamma) = 0$. We have $G_{diff}(\gamma, \gamma) = G_P(\gamma, \gamma) + G_N(\gamma, \gamma)$. The distribution of data decide the final values of thresholds pairs. More than one pair may be obtained. Similarly, we can search the space of all possible region thresholds to obtain the suitable decision regions. When the search space is large, heuristic search maybe helpful.

3.3   Setting Limits for Three Gini Coefficients

We consider acceptance, rejection and non-commitment decision regions individually. In some applications, the requirements for acceptance decisions and rejection decisions are different. For example, in the medical applications, when diagnosing if a patient suffers from a serious disease, a more conservative assessment should be made for rejection decisions. Rejecting a patient can delay the treatment and may lead to the lose of life. In this kind of applications, we need to treat three decision regions individually and try to keep the Gini coefficients of them less simultaneously. We can set limits for Gini coefficients of acceptance, rejection and non-commitment decision regions, and control them less than specific values simultaneously:

$$(\alpha, \beta) = \{(\alpha, \beta) | G_P(\alpha, \beta) \le c_P \land G_N(\alpha, \beta) \le c_N \land G_B(\alpha, \beta) \le c_B\}. \tag{16}$$

The specific limits c_P, c_B and c_N can be designated by users or experts, or evaluated by statistical results. The suitable limits are important for obtaining effective decision regions. The determination of these specific limits is beyond the scope of this paper, we will explore it in future papers.

Table 1 shows the changes of Gini coefficients of three decision regions when α decreases from 1 to γ and β increases from 0 to γ. In the table, $\nearrow$ denotes the increase, $\searrow$ denotes the decrease, and $\rightarrow$ denotes no change. There are two observations:

Table 1. The relationship between the changes of decision regions and Gini coefficients of regions

	$(\alpha, \beta) = (1,0)$	$(\alpha \downarrow, \beta)$	$(\alpha, \beta \uparrow)$	$(\alpha \downarrow, \beta \uparrow)$	$(\alpha, \beta) = (\gamma, \gamma)$
$G_P(\alpha, \beta)$	0	$\nearrow$	$\rightarrow$	$\nearrow$	$\nearrow$
$G_B(\alpha, \beta)$	max	$\searrow$	$\searrow$	$\searrow$	0
$G_N(\alpha, \beta)$	0	$\rightarrow$	$\nearrow$	$\nearrow$	$\nearrow$

- The threshold α influences the Gini coefficients of acceptance and non-commitment decision regions. The decrease of α causes the increase in Gini coefficient of acceptance region while the decrease in Gini coefficient of non-commitment region;
- The threshold β influences the Gini coefficients of rejection and non-commitment regions. The increase of β causes the increase in Gini coefficient of rejection region while the decrease in Gini coefficient of non-commitment region.

There are three specific limits for three decision regions. If one or two decision regions are not important for applications, the objective function can ignore insignificant decision regions. For example, if an application only cares acceptance decision region, the objective function would be $(\alpha, \beta) = \{(\alpha, \beta)|G_P(\alpha, \beta) \leq c_P\}$. There could be many thresholds pairs satisfying the objective functions. The result thresholds depend on what strategy is used to search.

4   A Demonstrative Example

In this section, we present a demonstrative example to illustrate the relationship between the changes of decision regions and their impacts on Gini coefficients of decision regions. Table 2 summarizes probabilistic data about a concept C. There are 16 equivalence classes denoted by $X_i (i = 1, 2, ..., 16)$, which are listed in a decreasing order of the conditional probabilities $Pr(C|X_i)$ for convenient computations.

Table 2. Summary of the experimental data

	X_1	X_2	X_3	X_4	X_5	X_6	X_7	X_8
$Pr(X_i)$	0.093	0.088	0.093	0.089	0.069	0.046	0.019	0.015
$Pr(C|X_i)$	1	0.978	0.95	0.91	0.89	0.81	0.72	0.61

	X_9	X_{10}	X_{11}	X_{12}	X_{13}	X_{14}	X_{15}	X_{16}
$Pr(X_i)$	0.016	0.02	0.059	0.04	0.087	0.075	0.098	0.093
$Pr(C|X_i)$	0.42	0.38	0.32	0.29	0.2	0.176	0.1	0

When $(\alpha, \beta) = (1, 0)$, the three decision regions are $POS_{(1,0)}(C) = X_1$, $BND_{(1,0)}(C) = X_2 \cup X_3 \cup ... \cup X_{15}$ and $NEG_{(1,0)}(C) = X_{16}$.

The absolute Gini coefficients of acceptance and rejection regions are both 0:

$$Gini(POS_{(1,0)}(C), \pi_C) = Gini(NEG_{(1,0)}(C), \pi_C) = 0.$$

The relative Gini coefficients of acceptance and rejection regions are both 0, i.e., $G_P(1, 0) = 0$ and $G_N(1, 0) = 0$.

For the non-commitment region, the probability is

$$Pr(BND_{(1,0)}(C)) = \sum_{i=2}^{15} Pr(X_i) = 0.814.$$

The conditional probability of C is

$$Pr(C|BND_{(1,0)}(C)) = \frac{\sum_{i=2}^{15} Pr(C|X_i)Pr(X_i)}{\sum_{i=2}^{15} Pr(X_i)} = \frac{0.4621}{0.814} = 0.5677.$$

The absolute Gini coefficient of the non-commitment region is

$$Gini(BND_{(1,0)}(C), \pi_C) = 1 - (0.5677)^2 - (1 - 0.5677)^2 = 0.4908.$$

The relative Gini coefficient of the non-commitment region is

$$G_B(1,0) = Pr(BND_{(1,0)}(C)) \times Gini(BND_{(1,0)}(C), \pi_C) = 0.3995.$$

Table 3 shows the Gini coefficients of decision regions corresponding to different thresholds pairs. Instead of listing all possible thresholds pairs, we only choose $\alpha = (1, 0.9, 0.8, 0.7, 0.6)$ and $\beta = (0, 0.1, 0.2, 0.3, 0.4, 0.5)$. Each cell represents the values of Gini coefficients of three regions, G_P, G_B and G_N. The cell on the top left corner corresponds to the thresholds $(\alpha, \beta) = (1, 0)$ and the Gini coefficients of three regions are $G_P(1,0) = 0$, $G_B(1,0) = 0.3995$ and $G_N(1,0) = 0$.

Table 3. Gini coefficients of regions for different thresholds pairs

β / α	0.0	0.1	0.2
	G_P, G_B, G_N	G_P, G_B, G_N	G_P, G_B, G_N
1.0	0.0000, 0.3995, 0.0000	0.0000, 0.3332, 0.0186	0.0000, 0.2014, 0.0716
0.9	0.0280, 0.2563, 0.0000	0.0280, 0.2199, 0.0186	0.0280, 0.1378, 0.0716
0.8	0.0579, 0.1617, 0.0000	0.0579, 0.1382, 0.0186	0.0579, 0.0811, 0.0716
0.7	0.0672, 0.1453, 0.0000	0.0672, 0.1233, 0.0186	0.0672, 0.0691, 0.0716
0.6	0.0773, 0.1336, 0.0000	0.0773, 0.1125, 0.0186	0.0773, 0.0599, 0.0716

β / α	0.3	0.4	0.5
1.0	0.0000, 0.1658, 0.0902	0.0000, 0.0906, 0.1309	0.0000, 0.0757, 0.1407
0.9	0.0280, 0.1132, 0.0902	0.0280, 0.0572, 0.1309	0.0280, 0.0448, 0.1407
0.8	0.0579, 0.0634, 0.0902	0.0579, 0.0242, 0.1309	0.0579, 0.0150, 0.1407
0.7	0.0672, 0.0521, 0.0902	0.0672, 0.0155, 0.1309	0.0672, 0.0071 0.1407
0.6	0.0773, 0.0432, 0.0902	0.0773, 0.0078, 0.1309	0.0773, 0.0000 0.1407

Based on Table 3, we can obtain effective decision regions which make Gini coefficients of regions satisfy different objective functions.

– When determining decision regions by minimizing the overall Gini coefficient, we can see $G_{sum}(0.7, 0.2) = 0.2079$ is the minimal value in the table, $(\alpha, \beta) = (0.7, 0.2)$ is the thresholds close to optimal. Please be noted that since we did not compute all possible thresholds, we can not say they are optimal.

- When determining decision regions by minimizing the difference between Gini coefficients of immediate and non-commitment decision regions, we can see $G_{diff}(0.7, 0.3) = 0.0291$ is the minimal value in the table, $(\alpha, \beta) = (0.7, 0.3)$ is the thresholds close to optimal.
- When determining decision regions by setting specific limits for $G_P(\alpha, \beta) \leq 0.06$, $G_B(\alpha, \beta) \leq 0.1$, and $G_N(\alpha, \beta) \leq 0.08$, we can obtain two pairs of thresholds, $(\alpha, \beta) = \{(0.8, 0.2), (0.8, 0.3)\}$.

Based on the same data set, different objective functions can produce different decision regions. It is hard to decide which approach is better, since the objective functions depend on the applications. No matter which objective function selected, the decision regions induced from the result thresholds are more applicable and effective than the regions without error when measured by Gini coefficients.

5 Conclusion

Gini coefficient is a kind of entropy calculation. In this paper, we use Gini coefficient to measure the distribution of three decision regions defined by rough set model. We analyze the impacts that the changes of decision regions bring on the Gini coefficients of decision regions. Effective decision regions can be obtained by adjusting Gini coefficients of decision regions to satisfy defined objective functions. In particular, we discuss three objective functions, i.e., minimizing the overall Gini coefficients of three regions, minimizing the difference between Gini coefficients of immediate and non-commitment decision regions, and setting limits for Gini coefficients of acceptance, rejection and non-commitment decision regions. In the first two situations, we formulate the thresholds search as the optimization problems. The third one deals with the relatively flexible requirement, in which each Gini coefficient of decision region can be considered independently. The requirements of applications determine the suitable objective functions.

The future work will focus on the search strategies and learning mechanisms for obtaining effective decision regions, as well as determination of suitable specific limits for each decision region.

Acknowledgements. This work is partially supported by a Discovery Grant from NSERC Canada.

References

1. Azam, N.: Formulating three-way decision making with game-theoretic rough sets. In: Proceedings of the 26th Canadian Conference on Electrical and Computer Engineering (CCECE 2013), pp. 695–698. IEEE (2013)
2. Azam, N., Yao, J.T.: Analyzing uncertainties of probabilistic rough set regions with game-theoretic rough sets. International Journal of Approximate Reasoning 55(1), 142–155 (2014)

3. Breiman, L., Friedman, J., Stone, C.J., Olshen, R.A.: Classification and Regression Trees. Chapman and Hall (1984)
4. Damgaard, C., Weiner, J.: Describing inequality in plant size or fecundity. Ecology 81(4), 1139–1142 (2000)
5. Deng, X.F., Yao, Y.Y.: An information-theoretic interpretation of thresholds in probabilistic rough sets. In: Li, T., Nguyen, H.S., Wang, G., Grzymala-Busse, J., Janicki, R., Hassanien, A.E., Yu, H. (eds.) RSKT 2012. LNCS, vol. 7414, pp. 369–378. Springer, Heidelberg (2012)
6. Deng, X.F., Yao, Y.Y.: Decision-theoretic three-way approximations of fuzzy sets. Information Science (2014), doi:10.1016/j.ins.2014.04.022i
7. Deng, X.F., Yao, Y.Y.: A multifaceted analysis of probabilistic three-way decisions. Fundamenta Informaticae (forthcoming)
8. Gastwirth, J.L.: The estimation of the lorenz curve and gini index. The Review of Economics and Statistics 54(3), 306–316 (1972)
9. Gelfand, S.B., Ravishankar, C.S., Delp, E.J.: An iterative growing and pruning algorithm for classification tree design. IEEE Transactions on Pattern Analysis and Machine Intelligence 13(2), 163–174 (1991)
10. Gini, C.: Variabilia e Mutabilita. Studi Economico-Giuridici dell' Universita di Cagliari 3, 1–158 (1912)
11. Herbert, J.P., Yao, J.T.: Criteria for choosing a rough set model. Computers and Mathematics with Applications 57(6), 908–918 (2009)
12. Herbert, J.P., Yao, J.T.: Game-theoretic rough sets. Fundamenta Informaticae 108(3-4), 267–286 (2011)
13. Pawlak, Z.: Rough Sets: Theoretical Aspects of Reasoning About Data. Kluwer Academic Publishers, Boston (1991)
14. Pedrycz, W.: Shadowed sets: representing and processing fuzzy sets. IEEE Transactions on System, Man and Cybernetics 28(1), 103–109 (1998)
15. Pedrycz, W.: From fuzzy sets to shadowed sets: interpretation and computing. International Journal of Intelligent Systems 24(1), 48–61 (2009)
16. Tan, P.N., Steinbach, M., Kumar, V.: Introduction to Data Mining. Addison-Wesley (2006)
17. Yao, J.T., Herbert, J.P.: A game-theoretic perspective on rough set analysis. Journal of Chongqing University of Posts & Telecommunications 20(3), 291–298 (2008)
18. Yao, J.T., Herbert, J.P.: Probabilistic rough sets: approximations, decision-makings and applications. International Journal of Approximation Reasoning 49(2), 253–254 (2008)
19. Yao, Y.Y.: Probabilistic rough set approximations. International Journal of Approximate Reasoning 49(2), 255–271 (2008)
20. Yao, Y.Y.: The superiority of three-way decisions in probabilistic rough set models. Information Sciences 181(6), 1080–1096 (2011)
21. Yao, Y.Y.: An outline of a theory of three-way decisions. In: Yao, J.T., Yang, Y., Słowiński, R., Greco, S., Li, H., Mitra, S., Polkowski, L. (eds.) RSCTC 2012. LNCS, vol. 7413, pp. 1–17. Springer, Heidelberg (2012)
22. Yao, Y.Y., Wong, S.K.M.: A decision theoretic framework for approximating concepts. International Journal of Man-Machine Studies 37(6), 793–809 (1992)
23. Zhang, Y.: Optimizing gini coefficient of probabilistic rough set regions using game-theoretic rough sets. In: Proceedings of the 26th Canadian Conference on Electrical and Computer Engineering (CCECE 2013), pp. 699–702. IEEE (2013)

Cost-Sensitive Three-Way Decisions Model Based on CCA

Yanping Zhang[1,2,*], Huijin Zou[1,2], Xi Chen[1,2], Xiangyang Wang[3],
Xuqing Tang[4], and Shu Zhao[1,2,**]

[1] Key Laboratory of Intelligent Computing and Signal Processing of Ministry
of Education, Hefei, Anhui Province, 230601, P.R. China
zhaoshuzs2002@hotmail.com
[2] School of Computer Science and Technology, Anhui University, Hefei,
Anhui Province 230601, P.R. China
[3] Anhui Electrical Engineering Professional Technique College, Hefei,
Anhui Province, 230051, P.R. China
[4] Jiangnan University, Wuxi, P.R. China

Abstract. The concept of three-way decisions was proposed by Yao
and most of three-way decisions models use two thresholds α and β to
partition the universe into three parts: the positive region POS(X), the
negative region NEG(X) and the boundary region BND(X). But acquirement of a pair of thresholds is a challenge. In this paper, we propose a
new method to get the thresholds of three-way decisions model. We introduce loss function into three-way decisions model based on constructive
covering algorithm (CCA). A loss function is interpreted as the costs of
making classification decisions. More specifically, for reducing losses of
classification, we change the radius of the cover according to the loss
function, then we compute the thresholds based on the modification.
This paper propose an effective method to compute thresholds α and β,
which is according to cost-sensitive three-way decisions model based on
CCA.

Keywords: Constructive Covering Algorithm, three-way decisions, loss
function, DTRSM.

1 Introduction

The concept of three-way decisions was proposed by Yao and used to interpret
rough set three regions [1][2][3]. Most of the three-way decisions models use two
parameters α and β to divide the universe into three parts. More specially, the
positive region POS(X), the negative region NEG(X) and the boundary region
BND(X). The rules generated by these three regions correspond to the results
of a three-way decisions that the situation is verified positively, negatively, or

---

* Yanping Zhang, Professor, Anhui University, main research in machine learning,
artificial neural network and data mining.
** Corresponding author.

C. Cornelis et al. (eds.): RSCTC 2014, LNAI 8536, pp. 172–180, 2014.

undecidedly based on the evidence. The positive rules express that an object or object sets belong to one decision class when the threshold is more than α; the probabilistic boundary rules express that an object or object sets belong to one decision class when the thresholds are between α and β; the probabilistic negative rules express that an object or object sets not belong to one decision class when the threshold is less than β. The ideas of three-way decisions have in fact been considered and may be further applied to many fields, including medical clinic [4], environmental management [5], e-learning [6], email spam filtering [7], model selection [8], information filtering [9], market timing decisions [10].

A great challenge for the three-way decision models is acquirement of a pair of thresholds. Different thresholds may lead to different decision results. How to choose the proper thresholds thus becomes an important task. Unfortunately, the thresholds are usually given by experts experience in most of the probabilistic rough sets. Instead of this, Yao introduced the risk to probabilistic rough sets and proposed decision theoretic of rough set (DTRS) with bayesian theory [11]. Two thresholds α and β can be directly and systematically calculated by minimizing the decision loss [12]. Herbert and Yao introduced a game-theoretic approach to DTRS for learning optimal parameter values [13][14].

Zhou introduced a cost-sensitive three-way decision approach to email spam filtering. A loss function is interpreted as the costs of making classification decisions. A decision is made for which the overall cost is minimum[15]. In this paper, we propose a new method to get the thresholds of three-way decision models. The new model is cost-sensitive. More specifically, the new method combines loss function with three-way decisions model based on constructive covering algorithm. Radius of the cover is depended on the loss function. We can compute α and β according to the change of the radius. The rest of the paper is organized as follows. In Section 2, we briefly review the related work about decision theoretic of rough set and constructive covering algorithm. In Section 3, a new method that based on constructive covering algorithm to get thresholds of three-way decision models is proposed. Experimental comparison results are shown in Section 4. We conclude the paper and explain future work in Section 5.

2   Related Work

2.1   Probabilistic Rough Set Models

In probabilistic rough set models, we can obtain the (α,β)-probabilistic positive, boundary and negative regions based on the (α,β)-probabilistic lower and upper approximations:

$$
\begin{aligned}
POS_{(\alpha,\beta)}(X) &= \{x \in U | Pr(X|[x]) \geq \alpha\} \\
BND_{(\alpha,\beta)}(X) &= \{x \in U | \beta < Pr(X|[x]) < \alpha\} \\
NEG_{(\alpha,\beta)}(X) &= \{x \in U | Pr(X|[x]) \leq \beta\}
\end{aligned}
\tag{1}
$$

Where the equivalence class $[x]$ of x is viewed as description of x and $Pr(X|[x])$ denotes the conditional probability of the classification, U is the universe.

In most of probabilistic rough sets, the thresholds are given by experts with intuitive understanding or experiences. However, Yao introduces Bayesian decision procedure into rough set theory (RST) and proposes a decision-theoretic rough set model (DTRSM), in which the acceptable level of errors α and β can be automatically computed from losses function, and the optimal decisions with the minimum conditional risk can be directly calculated by using Bayes theory [17].

We briefly review the concepts of DTRS. Let $=\{X, \neg X\}$ be a set of 2 states, indicating that an element in X and not in X respectively. Let $A=\{a_P, a_B, a_N\}$ be a finite set of 3 possible actions. The λ_{PP}, λ_{BP} and λ_{NP} denote the losses incurred for taking action a_P, a_B and a_N when an object belongs to X; the λ_{PN}, λ_{BN} and λ_{NN} denote the losses incurred for taking action a_P, a_B and a_N when an object does not belong to X. After deducing from the Bayesian decision procedure, fundamental result of DTRSM is that the positive, boundary and negative regions are defined by a pair of thresholds (α, β) which is shown as formula (1). We accept an object x to be a member of C if $\Pr(X|[x])$ is greater than α; we reject x if $\Pr(X|[x])$ is less than β; we neither accept nor reject x if $\Pr(X|[x])$ is between α and β.

2.2 Three-Way Decisions Model Based on CCA

Three-Way Decisions Model Based on CCA was proposed by Zhang and Xing [16]. Given a training samples set $X=\{(x_1, y_1), (x_2, y_2), ...(x_p, y_p)\}$, $(i=1,2,...p)$, which is the set in n-dimensional Euclidean space. $x_i=(x_i^1, x_i^2, ...x_i^n)$ is n-dimensional characteristic attribute of the ith sample. y_i is the decision attribute, i.e., category. The specific formation process of the covers has been introduced in [16]. CCA finally obtained a set of covers $C=\{C_1^1, C_1^2, ... ,C_1^{n_1}, C_2^1, C_2^2, ... ,C_2^{n_2}, ... ,C_m^1, ..., C_m^{n_m}\}$, where C_i^j represents the jth cover of the ith category. We assume $C_i = \bigcup C_i^j$, $j=1,2,...n_i$. C_i represents all covers of the ith category samples.

For convenience in discussion, we assume only two categories C_1 and C_2. The covers of C_1 and C_2 are $(C_1^1, C_1^2, . . . , C_1^m)$ and $(C_2^1, C_2^2, . . ., C_2^n)$, respectively, i.e., $C_1=(C_1^1, C_1^2, . . . , C_1^m)$, $C_2=(C_2^1, C_2^2, . . . , C_2^n)$. Each category has at least a cover. Assume $C_i = \bigcup C_i^j$ and each C_i represents all covers of the ith category samples. We define POS of C_1, namely, $POS(C_1)$ by the difference of unions $\bigcup C_1^i - \bigcup C_2^j$, $NEG(C_1)$ by $\bigcup C_2^j - \bigcup C_1^i$ and $BND(C_1)$ by the rest, where $i=1,2,. . . , m$, $j=1,2,. . . , n$. That is to say, $POS(C_1)$ is equal to $NEG(C_2)$; $POS(C_2)$ is equal to $NEG(C_1)$; $BND(C_1)$ is equal to $BND(C_2)$.

3   Cost-Sensitive Three-Way Decisions Model Based on CCA

In this paper, a cost-sensitive three-way decisions model based on constructive covering algorithm (CCA) is introduced in this section. We propose a new way to get the thresholds of three-way decisions. We describe the process as Algorithm 1.

Algorithm 1:

Step 1: Formation of Covers. According to the constructive covering algorithm, we can obtain a set of cover C=$(C_1^1, C_1^2, \ldots, C_1^{m_1}, C_2^1, C_2^2, \ldots, C_2^{n_2})$. Each category has a cover at least. The radius of those covers are $\theta = (\theta_1^1, \theta_1^2, \ldots, \theta_1^{n_1}, \theta_2^1, \theta_2^2, \ldots, \theta_2^{n_2})$.

Step 2: Change the Radius θ.

We define the samples in $C_1^i (i=1,2,\ldots,n_1)$ belong to positive region and $C_2^j (j=1,2,\ldots,n_2)$ belong to negative region. We use following method to modify the radius of the covers. For covers of C_1, we regard the k-nearest distance between the center and the dissimilar point as radius ($k=0,1,2,3,4..$); for covers of C_2, we regard the t-nearest distance between the center and the dissimilar point as radius ($t=0,1,2,3,4..$). When $(k,t)=(0,0)$, the radius is the minimum radius which regards the max distance between the center and the similar points as the radius in the boundary that don't have any dissimilar points. Fig. 1 shows the change of radius.

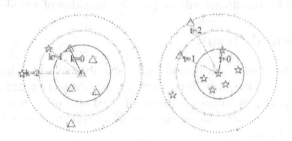

Fig. 1. Change of the Radius θ

Step 3: Compute average value of radius increase.

When the radius increase, the number of cover decrease. After increase the radius, we suppose the cover C=$(C_1^1, C_1^2, \ldots, C_1^{m_1}, C_2^1, C_2^2, \ldots, C_2^{m_2})$, and the radius $\theta = (\theta_{1k}^1, \theta_{1k}^2, \ldots, \theta_{1k}^{m_1}, \theta_{2t}^1, \theta_{2t}^2, \ldots, \theta_{2t}^{m_2})$. The average value of radius increase can be computed by following formula.

$$\triangle \theta_1 = \frac{\theta_1^1 + \theta_1^2 + \ldots + \theta_1^{n_1}}{n_1} - \frac{\theta_{1k}^1 + \theta_{1k}^2 + \ldots + \theta_{1k}^{m_1}}{m_1} \tag{2}$$

$$\triangle \theta_2 = \frac{\theta_2^1 + \theta_2^2 + \ldots + \theta_2^{n_2}}{n_2} - \frac{\theta_{2t}^1 + \theta_{2t}^2 + \ldots + \theta_{2t}^{m_2}}{m_2} \tag{3}$$

Step 4: Compute the ratio of radius increase.

The ratio of C_1' radius increase is R(C_1). The ratio of C_2' radius increase is R(C_2). The formulas are as following.

$$R(C_1) = \frac{\triangle \theta_1}{\frac{\theta_1^1 + \theta_1^2 + \ldots + \theta_1^{n_1}}{n_1}} \qquad R(C_2) = \frac{\triangle \theta_2}{\frac{\theta_2^1 + \theta_2^2 + \ldots + \theta_2^{n_2}}{n_2}} \tag{4}$$

Step 5: Compute the value of α and β.

The initial value of α is 1. The initial value of β is 0. We compute α and β according to the following formulas:

$$\alpha = 1 - R(C_1) \qquad \beta = 0 + R(C_2) \tag{5}$$

According to the above Algorithm, we can know that when k and t are different, the change of two covers are different, that is to say, the change of POS and NEG are different. The value of k and t are depended on the size of λ_{PN} and λ_{NP}. When $\lambda_{PN} < \lambda_{NP}$, in order to decrease loss, we need to increase the value of k and the value of t remains the same (or make the value of $k > t$). Then, the average radius value of C_1 will increase, and the area of POS will increase. Meanwhile the NEG remains the same, the value of α decrease, and $\beta = 0$. When $\lambda_{PN} > \lambda_{NP}$, in order to decrease loss, we need to keep the value k unchanged and increase the value of t(or make the value of $k < t$). Then, the average radius value of C_2 will increase, and the area of NEG will increase. Meanwhile the POS remains the same, the value of $\alpha = 1$, and β increase. When $\lambda_{PN} = \lambda_{NP}$, increase the value of k and t simultaneously or remain the value of k and t the same.

4   Experimental Result

Our experiments were performed on spambase and chess data set from UCI Machine Learning Repository(http://www.ics.uci.edu/mlearn/MLRepository.html). Spambase consists of 4601 instances and each instance is described by 58 attributes. Chess consists of 3196 instances and each instance is described by 36 attributes.

Firstly, we define some evaluation criteria.

Definition 1: The correct classification rate of samples in the cover (CRSC) is the ratio of the number of correct classification samples in the cover(CNSC) and all samples in cover(ASC). The samples in cover are the samples in POS and NEG. The formula is as follow.

$$CRSC = CNSC/ACS \tag{6}$$

Definition 2: The correct classification rate of samples (CRS) is the ratio of the number of correct classification samples (CNS) and all samples (AS). The formula is as follow.

$$CRS = CNS/AS \tag{7}$$

Definition 3: The correct classification rate of samples in positive region (CRSP) is the ratio of the number of correct classification samples in positive region (CNSP) and all samples in positive region (ASP). The formula is as follow.

$$CRSP = CNSP/ASP \tag{8}$$

Definition 4: The correct classification rate of samples in negative region (CRSN) is the ratio of the number of correct classification samples in negative region (CNSN) and all samples in negative region (ASN). The formula is as follow.

$$CRSN = CNSN/ASN \qquad (9)$$

In spambase, the losses of classify *legitimate* to *spam* is bigger than *spam* to *legitimate*, i.e., $\lambda_{NP}>\lambda_{PN}$. In chess, there is no actual meaning, we only assum $\lambda_{NP}<\lambda_{PN}$. According to the algorithm 1, in order to decrease loss, we need to decrease α in spambase and increase β in chess. In the three-way decision model based on CCA, the initial value of $\alpha=1$, $\beta=0$ when (k,t)=(0,0).

Table 1 and Table 2 show the accuracy and change of radius when t=0 on spambase and k=0 on chess respectively. D is average radius value of all covers, D_0 is average radius value of C_1, D_1 is average radius value of C_2. N_c denotes the number of the covers.

Table 1. The result of radius comparison and correct rate with t=0 on spambase

(k,t)	D	D_0	D_1	CRCS(%)	CRS(%)	N_C
(1,0)	0.124937	0.143770	0.109157	92.40	91.15	798
(2,0)	0.128826	0.155920	0.109146	91.58	90.70	742
(3,0)	0.127154	0.155880	0.108853	91.14	90.52	713
(4,0)	0.133110	0.175463	0.108991	90.90	90.04	685
(5,0)	0.135057	0.184366	0.109200	90.57	89.87	659
(6,0)	0.138107	0.198312	0.10888	90.12	89.35	651

Table 2. The result of radius comparison and correct rate with k=0 on chess

(k,t)	D	D_0	D_1	CRCS(%)	CRS(%)	N_C
(0,1)	0.346854	0.323234	0.380350	87.57	87.12	584
(0,2)	0.353722	0.323234	0.405312	85.36	85.05	545
(0,3)	0.356741	0.322775	0.420480	85.15	84.80	524
(0,4)	0.358964	0.323433	0.401010	84.66	84.20	503
(0,5)	0.359679	0.323047	0.438559	83.56	83.17	490
(0,6)	0.359363	0.323870	0.443968	84.05	83.89	488

When k increase, α will decrease. From Table 1 we can see that CRCS and CRS decrease (about 2%) with k increase. The reason is that there are more and more negative samples in C_1. The number of N_C decrease. If we need CRS(%)>90%, k increase from 1 to 4, the average radius of C_1 increases $\triangle\theta1=0.175463-0.143770=0.031693$ and the average radius of C_2 increases $\triangle\theta2=0$; the ratio of radius increases $R(C_1)=0.031693/0.143770=0.22044$, $R(C_2)=0$; and then when (k,t)=(4,0), we can compute $\alpha=1-R(C_1)=0.77956$, $\beta=R(C_2)=0$.

When t increase, α will increase. In Table 2, CRCS and CRS decrease (about 3%) with t increase. If we need CRS(%)>85%, we can compute $\beta=(0.420480-0.380350)/0.380350=0.105508$, $\alpha=0$ when (k,t)=(0,3).

Table 3 and Table 4 show the number of samples in three regions when t=0 on spambase and k=0 on chess respectively. P to P denotes the number of emails classified as legitimate, P to N denotes the number of emails classified as spam, P to B denotes the number of emails classified to boundary. N to P denotes the number of spam emails classified as legitimate, N to P denotes the number of spam emails classified as spam, N to B denotes the number of spam emails classified to boundary.

Table 3. The result of samples in three region with t=0 on spambase

(k,t)	PtoP	PtoN	PtoB	NtoP	NtoN	NtoB	CRSP(%)	CRSN(%)
(1,0)	253	11	13	21	147	12	92.34	93.04
(2,0)	257	11	10	25	145	10	91.13	92.92
(3,0)	262	9	7	29	143	8	90.03	94.08
(4,0)	262	9	6	31	141	8	89.42	94.00
(5,0)	264	9	5	33	140	7	88.89	93.96
(6,0)	265	8	4	32	142	6	89.83	94.67

Table 4. The result of samples in three region with k=0 on chess

(k,t)	PtoP	PtoN	PtoB	NtoP	NtoN	NtoB	CRSP(%)	CRSN(%)
(0,1)	136	21	8	16	131	4	82.42	84.52
(0,2)	131	28	6	17	132	2	79.39	85.16
(0,3)	129	31	6	15	134	3	78.18	86.45
(0,4)	128	33	4	14	135	2	77.58	87.10
(0,5)	124	36	5	14	135	2	77.15	87.10
(0,6)	124	37	4	12	138	1	77.15	89.03

From Table 3 we can see that with k increase, P to P and N to P increase, that is to say, the number of sample in C_1 is more and more. The area of NEG is not change, but the number of sample in NEG decrease. That is because each $C_1^i(i=1,2,...n_1)$ contains several negative samples, which makes P to N and N to N decrease, i.e., the number of sample in NEG decreases. With k increase, the radius of C_1 becomes big, and then the area of POS becomes big, which makes blank space samller, i.e., the area of BND becomes samll. We can see the number of samples in POS increase and the number of samples in BND decrease in Table 3. The loss of P to N is bigger than N to P, in order to reduce the loss of classification, we should reduce P to N. From Table 3, P to N decrease and N to P increase with the k increase. Meanwhile, CRSP is reduced and CRSN is increasing gradually. To the contrary, in Table 4, P to P , N to P decrease and P to N, N to N increase simultaneously. And the samples in BND decrease for the same reason.

5 Conclusions

In this paper, we proposed an effective method to compute thresholds α and β of three-way decisions model. The new model combined loss function with three-way decisions model based on constructive covering algorithm (CCA). We changed the radius of the cover according to the loss function with the purpose of reducing loss of classification, then computed the thresholds based on the modification.

In the experiment, with the radius increasing, the correct rate of classification decreases. We can compute the thresholds based on some certain conditions, such as the minimum accuracy we need to access, the maximum number of misclassification samples, and so on. Therefore, how to find appropriate conditions is our future research.

Acknowledgements. This work is partially supported by National Natural Science Foundation of China under Grant #61073117 and Grant #61175046, and supported by Natural Science Foundation of Anhui Province under Grant #1408085MF132, and supported by Provincial Natural Science Research Program of Higher Education Institutions of Anhui Province under Grant #KJ2013A016.

References

1. Yao, Y.Y.: Three-way decisions with probabilistic rough sets. Information Sciences 180(3), 341–353 (2010)
2. Yao, Y.Y.: The superiority of three-way decision in probabilistic rough set models. Information Sciences 181(6), 1080–1096 (2011)
3. Yao, Y.Y.: Three-Way Decision: An Interpretation of Rules in Rough Set Theory. In: Wen, P., Li, Y., Polkowski, L., Yao, Y., Tsumoto, S., Wang, G. (eds.) RSKT 2009. LNCS, vol. 5589, pp. 642–649. Springer, Heidelberg (2009)
4. Pauker, S.G., Kassirer, J.P.: The threshold approach to clinical decision making. The New England Journal of Medicine 302(20), 1109–1117 (1980)
5. Goudey, R.: Do statistical inferences allowing three alternative decision give better feedback for environmentally precautionary decision-making. Journal of Environmental Management 85(2), 338–344 (2007)
6. Abbas, A.R., Juan, L.: Supporting E-learning system with modified bayesian rough set model. In: Yu, W., He, H., Zhang, N. (eds.) ISNN 2009, Part II. LNCS, vol. 5552, pp. 192–200. Springer, Heidelberg (2009)
7. Zhou, B., Yao, Y.Y., Luo, J.G.: A three-way decision approach to email spam filtering. In: Farzindar, A., Kešelj, V. (eds.) Canadian AI 2010. LNCS, vol. 6085, pp. 28–39. Springer, Heidelberg (2010)
8. Forster, M.R.: Key concepts in model selection: performance and generalizability. Journal of Mathematical Psychology 4(1), 205–231 (2000)
9. Li, Y.Y., Zhang, C.C.: An information filtering model on the web and its application in jobAgent. Knowledge-Based Systems 13(5), 285–296 (2000)
10. Shen, L., Loh, H.L.: Applying rough sets to market timing decisions. Decision Support Systems 37(4), 583–597 (2004)

11. Yao, Y.Y., Wong, S.K.M., Lingras, P.: A decision-theoretic rough set model. In: Proceedings of the 5th International Symposium on Methodologies for Intelligent Systems, pp. 17–25. Springer, Heidelberg (1990)
12. Yao, Y.Y.: Two semantic issues in a probabilistic rough set model. Fundamenta Informaticae 108(3), 249–265 (2011)
13. Herbert, J.P., Yao, J.T.: Game-Theoretic Risk Analysis in Decision-Theoretic Rough Sets. In: Wang, G., Li, T., Grzymala-Busse, J.W., Miao, D., Skowron, A., Yao, Y. (eds.) RSKT 2008. LNCS (LNAI), vol. 5009, pp. 132–139. Springer, Heidelberg (2008)
14. Herbert, J.P., Yao, J.T.: Game-theoretic rough sets. Fundamenta Informaticae 108(3), 267–286 (2011)
15. Zhou, B., Yao, Y.Y., Yao, J.T.: Cost-sensitive three-way email spam filtering. Journal of Intelligent Information Systems, 1–27 (2014)
16. Zhang, Y.P., Xing, H., Zou, H.J., Zhao, S., Wang, X.Y.: A Three-Way Decisions Model Based on Constructive Covering Algorithm. In: Lingras, P., Wolski, M., Cornelis, C., Mitra, S., Wasilewski, P. (eds.) RSKT 2013. LNCS, vol. 8171, pp. 346–353. Springer, Heidelberg (2013)
17. Yao, Y.Y., Wong, S.: International Journal of Man-machine Studies. Journal of Man-Machine Studies 37(6), 793–809 (1992)

Feature Selection Based
on Confirmation-Theoretic Rough Sets

Bing Zhou[1] and Yiyu Yao[2]

[1] Department of Computer Science, Sam Houston State University
Huntsville, Texas, USA 77341
zhou@shsu.edu
[2] Department of Computer Science, University of Regina
Regina, Saskatchewan, Canada S4S 0A2
yyao@cs.uregina.ca

Abstract. As an important part of data preprocessing in machine learning and data mining, feature selection, also known as attribute reduction in rough set theory, is the process of choosing the most informative subset of features. Rough set theory has been used as such a tool with much success. The main objective of this paper is to propose a feature selection procedure based on a special group of probabilistic rough set models, called confirmation-theoretic rough set model(CTRS). Different from the existing attribute reduction methods, the definition of positive features is based on Bayesian confirmation measures. The proposed method is further divided into two categories based on the qualitative and quantitative nature of the underlying rough set models. This study provides new insights into the problem of attribute reduction.

Keywords: feature selection, attribute reduction, probabilistic rough set, confirmation-theoretic rough set.

1 Introduction

As an important part of data preprocessing in machine learning and data mining, feature selection is the process of choosing a most informative subset of features. It is used to break curse of dimensionality, reduce the amount of time and memory required by data mining algorithms, and allow data to be more easily visualized. Feature selection methods have been extensively studied, such as the brute-force approach that tries all possible feature subsets; the embedded approaches that occur naturally as part of the data mining algorithm. There are two basic categories of feature selection methods. Filter approaches select features before data mining algorithm is run; and the wrapper approaches use the data mining algorithm as a black box to find best subset of attributes. Examples of well known measures used in feature selection include information gain, best first search, and Genetic algorithms.

Rough set theory was introduced by Pawlak [5] in the early 1980s as a tool for analyzing data represented in an information table. It has been used for feature

C. Cornelis et al. (eds.): RSCTC 2014, LNAI 8536, pp. 181–188, 2014.

selection, also called attribute reduction, with much success. As an important concept of rough set theory, an attribute reduct is a subset of attributes that is jointly sufficient and individually necessary for preserving a particular property of the given information table. Attribute reduction has also been studied in generalized rough set models, e.g., probabilistic rough set models. This paper investigates attribute reduction in a special group of probabilistic rough set models, called confirmation-theoretic rough set model[17]. Different from the existing attribute reduction methods, the definition of positive features is based on Bayesian confirmation measures. The proposed method is further divided into two categories based on the qualitative and quantitative nature of the underlying rough set models. The qualitative measure does not consider the degree of support of certain feature to a decision class, whereas in the quantitative measure, we only select the features with degree of support beyond certain level.

2   Rough Set-Based Feature Selection

2.1   Feature Selection Based on Pawlak's Rough Set Model

In Pawlak's rough set model [5], information about a set of objects are represented in an information table with a finite set of attributes. Formally, an information table can be expressed as: $S = (U, At, \{V_a \mid a \in At\}, \{I_a \mid a \in At\})$, where U is a finite nonempty set of objects called universe; At is a finite nonempty set of attributes; V_a is a nonempty set of values for $a \in At$; $I_a : U \to V_a$ is an information function. An equivalence relation can be defined with respect to $A \subseteq At$, denoted as E. Two objects x and y in U are equivalent if and only if they have the same values on all attributes. For a subset $C \subseteq U$, the lower and upper approximations of C are defined by:

$$\underline{apr}(C) = \{x \in U \mid [x] \subseteq C\};$$
$$\overline{apr}(C) = \{x \in U \mid [x] \cap C \neq \emptyset\}. \tag{1}$$

Based on the rough set approximations of C defined by U/E, one can divide the universe U into three pair-wise disjoint regions: the positive region $\text{POS}(C)$ is the union of all the equivalence classes that is included in the decision class C; the boundary region $\text{BND}(C)$ is the union of all the equivalence classes that have a nonempty overlap with C; and the negative region $\text{NEG}(C)$ is the union of all equivalence classes that have an empty intersection with C:

$$\text{POS}(C) = \underline{apr}(C);$$
$$\text{NEG}(C) = U - (\text{POS}(C) \cup \text{BND}(C));$$
$$\text{BND}(C) = \overline{apr}(C) - \underline{apr}(C). \tag{2}$$

The basic idea of attribute reduction is to keep only those attributes that preserve the equivalence relation and, consequently, set approximation. Suppose the attribute set $At = A \cup \{D\}$, where $\{D\}$ is the decision attribute. An attribute set $A' \subseteq A$ is a attribute reduct of A with respect to D if it satisfies the following two conditions [11]:

$$\text{i)} \quad POS_{A'}(D) = POS_A(D);$$
$$\text{ii)} \quad \forall_{a \in A'}, POS_{A'-\{a\}}(D) \subset POS_A(D). \tag{3}$$

Conditions of equation (3) indicate that a reduct is i) jointly sufficient and ii) individually necessary for preserving the positive region of information table S.

2.2   Feature Selection Based on Probabilistic Rough Set Models

The positive and negative regions in Pawlak rough sets may be too restrictive to be practically useful in real applications. Attempts to use probabilistic information for approximations have been considered by many authors[9,14,16,17] to allow some tolerance of errors, in which the degrees of overlap between equivalence classes $[x]$ and a set C to be approximated are considered. A conditional probability is used to state the degree of overlapping and is defined as: $Pr(C \mid [x]) = \frac{|C \cap [x]|}{|[x]|}$, where $|\cdot|$ denotes the cardinality of a set, and the conditional probability is written as $Pr(C \mid [x])$ representing the probability that an object belongs to C given that the object is described by $[x]$.

Yao et al. [10,12] introduced decision-theoretic rough set (DTRS) model, in which a pair of thresholds α and β with $1 \geq \alpha > \beta \geq 0$ on the probability is used to define three probabilistic regions. The (α, β)-probabilistic positive, boundary and negative regions are defined by [12]:

$$POS_{(\alpha,\beta)}(C) = \{x \in U \mid Pr(C \mid [x]) \geq \alpha\},$$
$$BND_{(\alpha,\beta)}(C) = \{x \in U \mid \beta < Pr(C \mid [x]) < \alpha\},$$
$$NEG_{(\alpha,\beta)}(C) = \{x \in U \mid Pr(C \mid [x]) \leq \beta\}. \tag{4}$$

Similar to the study of a Pawlak reduct, a probabilistic attribute reduct can be defined by requiring that the (α, β)-probabilistic positive region of D is unchanged [11]:

$$\text{i)} \quad POS_{A'_{(\alpha,\beta)}}(D) = POS_{A_{(\alpha,\beta)}}(D);$$
$$\text{ii)} \quad \forall_{a \in A'}, POS_{A'-\{a\}_{(\alpha,\beta)}}(D) \subset POS_{A_{(\alpha,\beta)}}(D). \tag{5}$$

It may be considered as a generalization of Pawlak attribute reduct defined by $\alpha = 1$ and $\beta = 0$.

3   Confirmation-Theoretic Rough Set Models

Based on Bayes' theorem and Bayesian confirmation theory [1,2], alternative models of probabilistic rough sets have been proposed and studied [3,4,8,9]. In this section, we present a confirmation-theoretic framework to summarize the main results from these studies and show their differences from the (α, β)-probabilistic

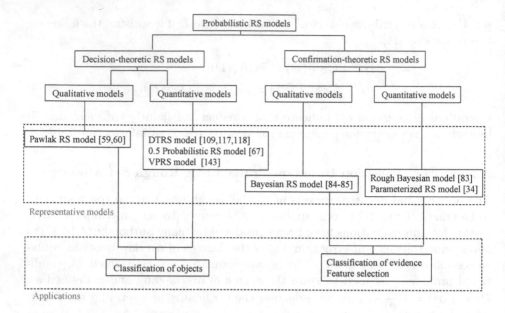

Fig. 1. Categorization of probabilistic rough set models

approximations. To differentiate these models from decision-theoretic models, we refer to them as (Bayesian) confirmation-theoretic rough set models. To differentiate the derived three regions from probabilistic approximation regions, we refer to them as (Bayesian) confirmation regions. Fig. 1 shows a categorization of the existing probabilistic rough set models and their applications.

3.1 Bayesian Rough Set Model and Rough Bayesian Model

Bayes' theorem [1] shows the relation between two conditional probabilities that are the reverse of each other. It expresses the conditional probability (or a posteriori probability) of an event H after E is observed in terms of a priori probability of H, probability of E, and the conditional probability of E given H. The Bayes' theorem is expressed as follows:

$$Pr(H|E) = \frac{Pr(E|H)}{Pr(E)} Pr(H), \tag{6}$$

where $Pr(H)$ is a priori probability of H that H happens or is true, $Pr(H|E)$ is the a posteriori probability that H happens after observing E, and $Pr(E|H)$ is the likelihood of H given E. Through Bayes' theorem, a difficulty to estimate probability $Pr(H|E)$ is expressed in terms of an easy to estimate likelihood $Pr(E|H)$. This makes Bayes' theorem particularly useful in data analysis and pattern classification.

Ślęzak [8] drew a natural correspondence between the fundamental notions of rough sets and Bayesian inference. The set to be approximated corresponds to a hypothesis and an equivalence class to a piece of evidence. Based on such a correspondence, Ślęzak and Ziarko [9] introduced a Bayesian rough set (BRS) model. A priori probability $Pr(C)$ is used to replace α and β in the (α, β)-probabilistic rough set model as a threshold for defining three regions:

$$\text{BPOS}(C) = \{x \in U \mid Pr(C \mid [x]) > Pr(C)\},$$
$$\text{BBND}(C) = \{x \in U \mid Pr(C \mid [x]) = Pr(C)\},$$
$$\text{BNEG}(C) = \{x \in U \mid Pr(C \mid [x]) < Pr(C)\}. \tag{7}$$

They also suggested to compare two likelihood functions $Pr([x] \mid C)$ and $Pr([x] \mid C^c)$ directly when neither a posteriori probability $Pr(C \mid [x])$ nor a priori probability $Pr(C)$ is derivable from data. That is,

$$\text{BPOS}(C) = \{x \in U \mid Pr([x] \mid C) > Pr([x] \mid C^c)\},$$
$$\text{BBND}(C) = \{x \in U \mid Pr([x] \mid C) = Pr([x] \mid C^c)\},$$
$$\text{BNEG}(C) = \{x \in U \mid Pr([x] \mid C) < Pr([x] \mid C^c)\}. \tag{8}$$

3.2 Parameterized Model Based on Bayesian Confirmation

Greco et al. [3] introduced a parameterized rough set model by considering a pair of thresholds on a Bayesian confirmation measure, in addition to a pair of thresholds on probability. The Bayesian confirmation measure is denoted by $c([x], C)$ which indicates the degree to which an equivalence class $[x]$ confirms the hypothesis C. Given a Bayesian confirmation measure $c([x], C)$ and a pair of thresholds (s, t) with $t < s$, three (α, β, s, t)-parameterized regions are defined by:

$$\text{PPOS}_{(\alpha,\beta,s,t)}(C) = \{x \in U \mid Pr(C \mid [x]) \geq \alpha \wedge c([x], C) \geq s\},$$
$$\text{PBND}_{(\alpha,\beta,s,t)}(C) = \{x \in U \mid (Pr(C \mid [x]) > \beta \vee c([x], C) > t) \wedge$$
$$(Pr(C \mid [x]) < \alpha \vee c([x], C) < s)\},$$

$$\text{PNEG}_{(\alpha,\beta,s,t)}(C) = \{x \in U \mid Pr(C \mid [x]) \leq \beta \wedge c([x], C) \leq t\}. \tag{9}$$

4 Feature Selection Based on Two Categories of CTRS Models

The three probabilistic regions defined in confirmation-theoretic rough set models provide a classification of attributes (features) of an information table. In machine learning and data mining, this is commonly referred to as the problem of feature selection. In rough set analysis, the problem is called attribute reduction, a selected set of attributes for rule induction is called a reduct. Generally speaking, an attribute reduct is a minimal subset of attributes whose induced rule sets have the same level of performance as the entire set of attributes, or a lower but satisfied level of performance.

Table 1. The training set

	s_1	s_2	s_3	s_4	$\cdots$	C
p_1	0	1	1	1		C
p_2	1	0	0	0		C^c
p_3	1	1	0	0		C
p_4	1	0	1	1		C
p_5	1	1	1	1		C^c
p_6	0	1	0	0		C^c
p_7	1	1	0	1		C

4.1 Feature Selection Based on Qualitative CTRS Models

Equation (7) and equation (8) in Section 3.1 provide qualitative measurements of attributes. Based on Bayes's theorem (equation (6)), we can proof that these two measures are equivalent. Table 1 is a training set representing the relationships between patients' symptoms to a certain disease C. The columns represent a set of symptoms $\{s_1, s_2, s_3, s_4 \ldots\}$ and the rows represent patients $\{p_1, p_2, p_3 \ldots\}$. For simplicity, I only use binary values where 1 represents symptom presents and 0 represents symptom not present. The decision attribute has two values, C indicates that a patient has disease and C^c (set complement) indicates that the patient does not have disease. At the first step, we calculate the a posteriori probability $Pr(C|s_i)$ of a patient having disease C given the symptom s_i. For example, for s_1, we have:

$$Pr(C \mid s_1) = \frac{|C \cap s_1|}{|s_1|} = \frac{|\{p_1, p_3, p_4, p_7\} \cap \{p_2, p_3, p_4, p_5, p_7\}|}{|\{p_2, p_3, p_4, p_5, p_7\}|} = \frac{3}{5}. \quad (10)$$

At the second step, we calculate the a priori probability $Pr(C)$ as:

$$Pr(C) = \frac{|C|}{|U|} = \frac{|\{p_1, p_3, p_4, p_7\}|}{|\{p_1, p_2, p_3, p_4, p_5, p_6, p_7\}|} = \frac{4}{7}. \quad (11)$$

At the third step, we look at the change between a posteriori probability and a priori probability, since $Pr(C \mid s_1) > Pr(C)$, s_1 should be classified into the positive features that supports decision C. Similarly, we can calculate the rest of symptoms and only select those in the positive regions for classification.

Note that the qualitative measurement did not consider the degree of support from certain feature to a decision class, we might end up with too many features in the positive region.

4.2 Feature Selection Based on Quantitative CTRS Models

Equation (9) in Section 3.2 provide quantitative measurements of attributes. We calculate the likelihood ratio $\frac{Pr(s_i|C)}{Pr(s_i|C^c)}$, where $Pr(s_i|C)$ indicates the chances of a patient having symptom s_i given that he/she has disease C, and $Pr(s_i|C^c)$

indicates the chances of a patient having symptom s_i given that he/she does not have disease C. For s_1, we have:

$$Pr(s_1 \mid C) = \frac{|C \cap s_1|}{|C|} = \frac{|\{p_1, p_3, p_4, p_7\} \cap \{p_2, p_3, p_4, p_5, p_7\}|}{|\{p_1, p_3, p_4, p_7\}|} = \frac{3}{4}. \quad (12)$$

Similarly, we can get,

$$Pr(s_1 \mid C^c) = \frac{|C^c \cap s_1|}{|C^c|} = \frac{|\{p_2, p_5, p_6\} \cap \{p_2, p_3, p_4, p_5, p_7\}|}{|\{p_2, p_5, p_6\}|} = \frac{2}{3}. \quad (13)$$

The likelihood ratio $\frac{Pr(s_i|C)}{Pr(s_i|C^c)} = \frac{9}{8}$. Assume the pair of thresholds (s', t') is set up as $(0.8, 0.2)$, since $\frac{9}{8} > 0.8$, s_1 should be classified into the positive region of features that supports decision C. Same procedure can be applied to calculate the rest of symptoms and only select those with the degree of support greater than 0.8 for classification.

In equation (9), a general form is used for quantitative measurement; any Bayesian confirmation measures $c([x], C)$ [3,4] can be used for feature selection. A pair of thresholds (s, t) is used. If $c([x], C) \geq s$, the features of x support C. If $c([x], C) \leq t$, the features of x are against C. If $t < c([x], C) < s$, the features of x are neutral to C and hence are not informative. Alternatively, the value of a Bayesian confirmation measure can be used to make pairwise comparisons between features. For example, if $c(s_1, C) > c(s_2, C)$, symptom s_1 provides a better indication for class C than symptom s_2, and s_1 will be selected over s_2.

4.3   Feature Selection Procedure Based on CTRS Models

Based on the above analysis, a feature selection procedure based on confirmation-theoretic rough set model can be summarized as follows.
1. Shuffle the data set and split into a training set and a testing set.
2. Let i vary among feature-set sizes: $i = (0, 1, 2, ..., n)$.
   Let fs_i = positive feature set of size i, where "positive" is measured by using qualitative measures described in Section 4.1 or quantitative measures described in Section 4.2 over the training set.
   End of loop of (i).
3. Output the feature set fs_i.
   The main differences between our approach and other existing attribute reduction methods are that we use Bayesian confirmation measures as criteria to choose positive features, whereas the definition of traditional attribute deduction is to choose the minimal set of attributes preserving the positive region.

5   Conclusions and Future Work

In this paper, a feature selection procedure based on a special group of probabilistic rough set models, called confirmation-theoretic rough set model, is proposed. Different to existing attribute reduction based on probabilistic rough set models, the definition of positive features is based on Bayesian confirmation measures. Instead of treating all probabilistic rough set models as the same in terms of their applications, we argue that confirmation-theoretic rough set models are well suited for feature selection and evaluation, whereas decision-theoretic rough set models are suitable for classification. The proposed feature selection method is further divided into two categories based on the qualitative and quantitative nature of their underlying rough set models. Note that the proposed method does not guarantee the selected feature set is minimal. Experimental study will be conducted as future work to compare with other feature selection methods.

References

1. Bayes, T., Price, R.: An essay towards solving a problem in the doctrine of chance. By the late Rev. Mr. Bayes, communicated by Mr. Price, in a letter to John Canton, M. A. and F. R. S. Philosophical Transactions of the Royal Society of London 53, 370–418 (1763)
2. Festa, R.: Bayesian Confirmation. In: Galavotti, M., Pagnini, A. (eds.) Experience, Reality, and Scientific Explanation, pp. 55–87. Kluwer Academic Publishers, Dordrecht (1999)
3. Greco, S., Pawlak, Z., Słowiński, R.: Bayesian confirmation measures within rough set approach. In: Tsumoto, S., Słowiński, R., Komorowski, J., Grzymała-Busse, J.W. (eds.) RSCTC 2004. LNCS (LNAI), vol. 3066, pp. 264–273. Springer, Heidelberg (2004)
4. Greco, S., Matarazzo, B., Słowiński, R.: Parameterized rough set model using rough membership and Bayesian confirmation measures. International Journal of Approximate Reasoning 49, 285–300 (2009)
5. Pawlak, Z.: Rough Sets. Theoretical Aspects of Reasoning about Data. Kluwer Academic Publishers, Dordrecht (1991)
6. Pawlak, Z., Skowron, A.: Rough membership functions. In: Yager, R.R., Fedrizzi, M., Kacprzyk, J. (eds.) Advances in the Dempster-Shafer Theory of Evidence, pp. 251–271. John Wiley and Sons, New York (1994)
7. Pawlak, Z., Wong, S.K.M., Ziarko, W.: Rough sets: probabilistic versus deterministic approach. International Journal of Man-Machine Studies 29, 81–95 (1988)
8. Ślęzak, D.: Rough Sets and Bayes Factor. In: Peters, J.F., Skowron, A. (eds.) Transactions on Rough Sets III. LNCS, vol. 3400, pp. 202–229. Springer, Heidelberg (2005)
9. Ślęzak, D., Ziarko, W.: Bayesian rough set model. In: Proceedings of FDM 2002, pp. 131–135 (2002)
10. Yao, Y.Y.: Decision-theoretic rough set models. In: Yao, J., Lingras, P., Wu, W.-Z., Szczuka, M.S., Cercone, N.J., Ślęzak, D. (eds.) RSKT 2007. LNCS (LNAI), vol. 4481, pp. 1–12. Springer, Heidelberg (2007)
11. Yao, Y.Y., Zhao, Y.: Attribute reduction in decision-theoretic rough set models. Information Sciences 178(17), 3356–3373 (2008)
12. Yao, Y.Y., Wong, S.K.M., Lingras, P.: A decision-theoretic rough set model. In: Ras, Z.W., Zemankova, M., Emrich, M.L. (eds.) Methodologies for Intelligent Systems 5, pp. 17–24. North-Holland, New York (1990)
13. Zhou, B., Yao, Y.Y.: Comparison of Two Models of Probabilistic Rough Sets. In: Lingras, P., Wolski, M., Cornelis, C., Mitra, S., Wasilewski, P. (eds.) RSKT 2013. LNCS, vol. 8171, pp. 121–132. Springer, Heidelberg (2013)

Fuzzy-Attributes
and a Method to Reduce Concept Lattices[*]

María Eugenia Cornejo[1], Jesús Medina-Moreno[2],
and Eloisa Ramírez-Poussa[1]

[1] Department of Statistic and O.R., University of Cádiz, Spain
{mariaeugenia.cornejo,eloisa.ramirez}@uca.es
[2] Department of Mathematics, University of Cádiz, Spain
jesus.medina@uca.es

Abstract. Reducing the size of the concept lattices is a fundamental problem in formal concept analysis. This paper presents several properties of useful fuzzy-attributes, in the general case of multi-adjoint concept lattices. Moreover, the use of these fuzzy-attributes provides a mechanism to reduce the size of concept lattices considering a subset of the original one and, therefore, without losing and modifying important information.

Keywords: Formal concept analysis, fuzzy sets, irreducible elements.

1 Introduction

In (fuzzy) Formal Concept Analysis (FCA) [14], the concept lattice obtained from usual relational database is huge and obtaining consequences from this is really difficult. Hence, looking for strategies to reduce the size of the obtained concept lattice, conserving the main information of the database, is very important.

There exist several mechanisms with this goal, however almost all of them modify the concept-forming operators and so, the original concepts, such as the use of hedges [1,8]. Other methodologies, change the original context (granular computing [7]) or consider a restrictive setting, for instance, they do not use fuzzy subsets of objects and attributes but, a crisp subset of objects and a fuzzy subset of attributes, as in [11].

The multi-adjoint concept lattice framework [12,13] is a general approach, in which the philosophy of the multi-adjoint paradigm was applied to the formal concept analysis. Adjoint triples [3] are used as basic operators to carry out the calculus in this framework and so, a general non-commutative environment can be considered. Moreover, different degrees of preference related to the set of objects and attributes can easily be established in this general concept lattice framework.

[*] Partially supported by the Spanish Science Ministry projects TIN2009-14562-C05-03 and TIN2012-39353-C04-04, and by Junta de Andalucía project P09-FQM-5233.

C. Cornelis et al. (eds.): RSCTC 2014, LNAI 8536, pp. 189–200, 2014.
© Springer International Publishing Switzerland 2014

In representation theory of fuzzy formal concept analysis, the fuzzy notions of attributes - fuzzy-attributes - which are fuzzy subsets of attributes, play an important role, for example, from the fuzzy-attributes the set of meet-irreducible elements of a concept lattice is obtained. In the general case of multi-adjoint concept lattices, this paper is a continuation of [4] which presents several properties of fuzzy-attributes and uses them in order to provide an original mechanism to reduce the size of concept lattices without losing and modifying important information.

The structure of the paper is as follows: preliminary notions and results, together with the multi-adjoint concept lattice framework, are introduced in Section 2; Section 4 presents irreducible α-cut concept lattices and several properties in order to reduce the size of multi-adjoint concept lattices. The paper finishes with several conclusions and future challenges.

2   Preliminaries

First of all, we recall several notions and results which are needed throughout the paper.

Definition 1. *Given a lattice $(L, \preceq)$, such that $\wedge, \vee$ are the meet and the join operators, and an element $x \in L$ verifying*

1. *If L has a top element $\top$, then $x \neq \top$.*
2. *If $x = y \wedge z$, then $x = y$ or $x = z$, for all $y, z \in L$.*

we call x meet-irreducible *($\wedge$-irreducible) element of L. Condition (2) is equivalent to*

2'. *If $x < y$ and $x < z$, then $x < y \wedge z$, for all $y, z \in L$.*

Hence, if x is $\wedge$-irreducible, then it cannot be represented as the infimum of strictly greatest elements. A join-irreducible *($\vee$-irreducible) element of L is defined dually.*

In a finite lattice, each element is equal to the infimum of meet-irreducible elements and the supremum of join-irreducible elements.

Other definitions and results about lattice theory, which will be used later, are the following.

Definition 2. *Let $(L, \preceq)$ be a lattice and $\varnothing \neq M \subseteq L$. Then $(M, \preceq)$ is a sublattice of $(L, \preceq)$, if for each $a, b \in M$ we have that $a \vee b \in M$ and $a \wedge b \in M$.*

Definition 3 ([2]). *A lattice $(L, \preceq)$ is called* distributive *if, for all $x, y, z \in L$,*

$$x \wedge (y \vee z) = (x \wedge y) \vee (x \wedge z)$$

Note that the above condition is equivalent to its dual: $x \vee (y \wedge z) = (x \vee y) \wedge (x \vee z)$, for all $x, y, z \in L$.

Fig. 1. Examples of no distributive lattices: $M3$ and $N5$ lattices

Theorem 1 ([2]). *A lattice which is not distributive contains one of the examples in Figure 1 as a sublattice.*

Lemma 1 ([2]). *In a distributive lattice, the decomposition of an element as a no-redundant meet of $\wedge$-irreducible elements is unique.*

The main goal of this paper is to reduce the size of multi-adjoint concept lattices. For that, we have considered the multi-adjoint concept lattice framework, since it is a general fuzzy setting which embeds other interesting frameworks and provides a great flexibility. Next, we recall this fuzzy concept lattice introduced in [13].

In the multi-adjoint concept lattice framework, considered operators in order to define the concept-forming operators are adjoint triples, which are generalizations of a triangular norm (t-norm) and its residuated implication [6].

Definition 4. *Let $(P_1, \leq_1)$, $(P_2, \leq_2)$, $(P_3, \leq_3)$ be posets and $\&\colon P_1 \times P_2 \to P_3$, $\swarrow\colon P_3 \times P_2 \to P_1$, $\nwarrow\colon P_3 \times P_1 \to P_2$ be mappings, then $(\&, \swarrow, \nwarrow)$ is an* adjoint triple *with respect to P_1, P_2, P_3 if:*

$$x \leq_1 z \swarrow y \quad \text{iff} \quad x \& y \leq_3 z \quad \text{iff} \quad y \leq_2 z \nwarrow x \tag{1}$$

where $x \in P_1$, $y \in P_2$ and $z \in P_3$. This condition is also called adjoint property.

Definition 5. *A* multi-adjoint frame *$\mathcal{L}$ is a tuple $(L_1, L_2, P, \&_1, \dots, \&_n)$ where $(L_1, \preceq_1)$ and $(L_2, \preceq_2)$ are complete lattices, $(P, \leq)$ is a poset and, for all $i = 1, \dots, n$, $(\&_i, \swarrow^i, \nwarrow_i)$ is an adjoint triple with respect to L_1, L_2, P.*

From a frame, a *multi-adjoint context* can be defined.

Definition 6. *Let $(L_1, L_2, P, \&_1, \dots, \&_n)$ be a multi-adjoint frame, a* context *is a tuple (A, B, R, σ) such that A and B are non-empty sets (usually interpreted as attributes and objects, respectively), R is a P-fuzzy relation $R\colon A \times B \to P$ and $\sigma\colon A \times B \to \{1, \dots, n\}$ is a mapping which associates any element in $A \times B$ with some particular adjoint triple in the frame.*

We will write L_2^B and L_1^A in order to represent the set of mappings $g\colon B \to L_2$, $f\colon A \to L_1$, respectively. On these sets a pointwise partial order can be considered from the partial orders in $(L_1, \preceq_1)$ and $(L_2, \preceq_2)$, which provides L_2^B

and L_1^A with the structure of complete lattice, that is, abusing notation, $(L_2^B, \preceq_2)$ and $(L_1^A, \preceq_1)$ are complete lattices where $\preceq_2$ and $\preceq_1$ are defined pointwise, that is, given $g_1, g_2 \in L_2^B$, $f_1, f_2 \in L_1^A$, $g_1 \preceq_2 g_2$ if and only if $g_1(b) \preceq_2 g_2(b)$, for all $b \in B$; and $f_1 \preceq_1 f_2$ if and only if $f_1(a) \preceq_1 f_2(a)$, for all $a \in A$.

Given a multi-adjoint frame and a context for that frame, the concept-forming operators are denoted as $\uparrow^\sigma : L_2^B \longrightarrow L_1^A$ and $\downarrow^\sigma : L_1^A \longrightarrow L_2^B$ are defined, for all $g \in L_2^B$, $f \in L_1^A$ and $a \in A$, $b \in B$, as

$$g^{\uparrow_\sigma}(a) = \inf\{R(a,b) \swarrow^{\sigma(a,b)} g(b) \mid b \in B\} \tag{2}$$

$$f^{\downarrow^\sigma}(b) = \inf\{R(a,b) \nwarrow_{\sigma(a,b)} f(a) \mid a \in A\} \tag{3}$$

and form a Galois connection [13]. In order to simplify the notation we will write $\uparrow$ and $\downarrow$ instead of $\uparrow^\sigma$ and $\downarrow^\sigma$, respectively.

A concept is defined as usual: a *multi-adjoint concept* is a pair $\langle g, f \rangle$ satisfying that $g \in L_2^B$, $f \in L_1^A$ and that $g^\uparrow = f$ and $f^\downarrow = g$; with $(\uparrow, \downarrow)$ being the Galois connection defined above.

Given $g \in L_2^B$ (resp. $f \in L_1^A$), *the generated concept from g (resp. f) is* $\langle g^{\uparrow\downarrow}, g^\uparrow \rangle$ (resp. $\langle f^\downarrow, f^{\downarrow\uparrow} \rangle$).

Finally, the definition of concept lattice in this framework is defined.

Definition 7. *The* multi-adjoint concept lattice *associated with a multi-adjoint frame* $(L_1, L_2, P, \&_1, \ldots, \&_n)$ *and a context* (A, B, R, σ) *is the set*

$$\mathcal{M} = \{\langle g, f \rangle \mid g \in L_2^B, f \in L_1^A \text{ and } g^\uparrow = f, f^\downarrow = g\}$$

in which the ordering is defined by $\langle g_1, f_1 \rangle \preceq \langle g_2, f_2 \rangle$ *if and only if* $g_1 \preceq_2 g_2$ *(equivalently* $f_2 \preceq_1 f_1$*).*

The ordering just defined above provides $\mathcal{M}$ with the structure of a complete lattice. The details can be seen in [13].

Now, a characterization of the $\wedge$-irreducible elements is extracted from [5]. Hereon, we will consider a multi-adjoint concept lattice $(\mathcal{M}, \preceq)$ associated with a multi-adjoint frame $(L_1, L_2, P, \&_1, \ldots, \&_n)$, a context (A, B, R, σ), an index set I, such that $A = \{a_i \mid i \in I\}$, and the following specific family of fuzzy subsets of L_1^A.

Definition 8. *For each* $a_i \in A$*, the fuzzy subsets of attributes* $\phi_{i,x} \in L_1^A$ *defined, for all* $x \in L_1$*, as*

$$\phi_{i,x}(a) = \begin{cases} x & \text{if } a = a_i \\ 0 & \text{if } a \neq a_i \end{cases}$$

will be called fuzzy-attributes. *The set of all fuzzy-attributes will be denoted as* $\Phi = \{\phi_{i,x} \mid a_i \in A, x \in L_1\}$.

Note that it is possible that there exist $\phi_{j,x_j}, \phi_{k,x_k}, \in \Phi$ and $i \in I$, such that $\phi_{j,x_j} = \phi_{i,x}$ and $\phi_{k,x_k} = \phi_{i,x'}$, with $x, x' \in L_1$ and $x \neq x'$.

Lemma 2. *For all* $f \in L_1^A$*, we have that* $f = \bigvee_{i \in I} \phi_{i,f(a_i)}$.

The following result characterizes the $\wedge$-irreducible elements of a multi-adjoint concept lattice.

Theorem 2. *The set of $\wedge$-irreducible elements of $\mathcal{M}$, $M_F(A)$, is formed by the pairs $\langle \phi^{\downarrow}_{i,x}, \phi^{\downarrow\uparrow}_{i,x} \rangle$ in $\mathcal{M}$, with $a_i \in A$ and $x \in L_1$, such that*

$$\phi^{\downarrow}_{i,x} \neq \bigwedge \{ \phi^{\downarrow}_{j,x_j} \mid \phi_{j,x_j} \in \Phi, \phi^{\downarrow}_{i,x} \prec_2 \phi^{\downarrow}_{j,x_j} \}$$

and $\phi^{\downarrow}_{i,x} \neq g_\top$, where $\top$ is the maximum element in L_2 and $g_\top \colon B \to L_2$ is the fuzzy subset defined as $g_\top(b) = \top$, for all $b \in B$.

Note that in a finite environment each concept in $\mathcal{M}$ is the infimum of $\wedge$-irreducible elements, hence $M_F(A)$ is like a base of $\mathcal{M}$. This idea will be used to reduce the size of $\mathcal{M}$, before that, several properties of these particular concepts need to be introduced.

3 Fuzzy-Attributes Generating Meet-Irreducible Elements

This section presents several properties about the meet-irreducible elements of a multi-adjoint concept lattice generated from a fuzzy-attribute. The first one introduces three technical properties.

Proposition 1. *Let g be an extension of a concept of $\mathcal{M}$ such that $g = \bigwedge_{j \in J} \phi^{\downarrow}_{j,x_j}$, where $\langle \phi^{\downarrow}_{j,x_j}, \phi^{\downarrow\uparrow}_{j,x_j} \rangle \in M_F(A)$. Then, the following properties hold:*

1. $g^{\uparrow\downarrow} = \bigwedge_{i \in I} \phi^{\downarrow}_{i,g^\uparrow(a_i)}$,
2. $x_j \preceq_1 g^\uparrow(a_j)$, *for all $j \in J$,*
3. $\phi^{\downarrow}_{j,g^\uparrow(a_j)} \preceq_2 \phi^{\downarrow}_{j,x_j}$, *for all $j \in J$.*

It is clear that $\phi^{\downarrow\uparrow}_{j,x_j}(a_j) = x_j$ does not hold in general, a counterexample is obtained from the concept C_{16} in Example 1, which satisfies that

$$C_{16} = \langle \phi^{\downarrow}_{2,0.5}, \phi^{\downarrow\uparrow}_{2,0.5} \rangle = \langle \phi^{\downarrow}_{2,0.4}, \phi^{\downarrow\uparrow}_{2,0.4} \rangle$$

Therefore, $0.4 \neq \phi^{\downarrow\uparrow}_{2,0.4} = 0.5$. However, this equality holds considering the values given by the intension of a fuzzy subset of objects, as the following proposition shows.

Proposition 2. *Let g be an extension of a concept of $\mathcal{M}$ such that $g = \bigwedge_{j \in J} \phi^{\downarrow}_{j,x_j}$, with $\langle \phi^{\downarrow}_{j,x_j}, \phi^{\downarrow\uparrow}_{j,x_j} \rangle \in M_F(A)$. Then*

$$\phi^{\downarrow\uparrow}_{j,g^\uparrow(a_j)}(a_j) = g^\uparrow(a_j)$$

for all $j \in J$.

Proposition 3. *Given an extension g of a concept of $\mathcal{M}$ such that $g = \bigwedge_{j \in J} \phi^{\downarrow}_{j,x_j}$, where $\langle \phi^{\downarrow}_{j,x_j}, \phi^{\downarrow\uparrow}_{j,x_j} \rangle \in M_F(A)$. The following equality holds.*

$$\bigwedge_{j \in J} \phi^{\downarrow}_{j,x_j} = \bigwedge_{j \in J} \phi^{\downarrow}_{j,g\uparrow(a_j)}$$

Applying the previous result to the particular fuzzy subset $\phi^{\downarrow}_{j,x_j}$, the following corollary is obtained.

Corollary 1. *The equality $\phi^{\downarrow}_{j,x_j} = \phi^{\downarrow}_{j,\phi^{\downarrow\uparrow}_{j,x_j}(a_j)}$ holds, for all $\langle \phi^{\downarrow}_{j,x_j}, \phi^{\downarrow\uparrow}_{j,x_j} \rangle \in M_F(A)$ and $j \in J$.*

Proposition 4. *Given an extension g of a concept of $\mathcal{M}$ such that $g = \bigwedge_{j \in J} \phi^{\downarrow}_{j,x_j}$, where $\langle \phi^{\downarrow}_{j,x_j}, \phi^{\downarrow\uparrow}_{j,x_j} \rangle \in M_F(A)$, we have*

$$g \preceq_2 \bigwedge_{t \in T} \phi^{\downarrow}_{t,g\uparrow(a_t)}$$

for all $T \subseteq J^c$, where J^c is the complement set of J.

The properties above are themselves important and in frameworks in which fuzzy-attributes are used, as the next one is.

4   Reducing the Size of Multi-Adjoint Concept Lattices

Decreasing the size of the concept lattices [1,7,9,10,11] is one of the most important problems in FCA. Nevertheless, several of the existing mechanisms modify the information given by the concepts. This section uses the previous characterization in order to provide a new procedure to reduce the size of the multi-adjoint concept lattices, without modifying the information given by the context, but beginning from the fuzzy-attributes that really can represent an attribute.

From Theorem 2 we have that every fuzzy-attribute $\phi_{i,x}$ associated with an attribute a_i, with $x \in L$, can be considered in the computation of the concept lattice, if this fuzzy-attribute generates a meet-irreducible element. For instance, if $L = \{0.0, 0.2, 0.5, 0.7, 1.0\}$, then $\phi_{i,0.2}$ could be considered, when the value for the attribute a_i is not representative, since $\alpha = 0.2$ is very small (although this depend on the context). Therefore, it could be more interesting to consider the fuzzy-attributes $\phi_{i,x}$ in which the value x exceeds a threshold α, proposed by an user or expert.

From the irreducible elements of $M_F(A)$ we will only consider the fuzzy-attributes with a considerable value. Hence, given a threshold α, we will only assume the fuzzy-attributes of each attribute a_i that provide to a_i a value greater than α, that is, we consider the following set of meet- irreducible elements of $(\mathcal{M}, \preceq)$:

$$M_F(A)_\alpha = \{\langle \phi^{\downarrow}_{i,x}, \phi^{\downarrow\uparrow}_{i,x} \rangle \in M_F(A) \mid \alpha \preceq_1 x\}$$

Hence, we only consider the concepts of $(\mathcal{M}, \preceq)$, which are obtained from the infimum of elements of $M_F(A)_\alpha$. Moreover, in order to obtain a complete lattice we also need to consider the greatest element in $(\mathcal{M}, \preceq)$, that is $\langle g_\top, g_\top^\uparrow \rangle$.

Definition 9. *Given* $\alpha \in L_1$, *the set* $\mathcal{M}_\alpha$, *defined as:*

$$\mathcal{M}_\alpha = \{\langle g, f \rangle \in \mathcal{M} \mid g = \bigwedge_{j \in J} \phi_{j,x_j}^\downarrow, \text{ with } \phi_{j,x_j} \in M_F(A)_\alpha\} \bigcup \{\langle g_\top, g_\top^\uparrow \rangle\}$$

is called irreducible α-cut of $\mathcal{M}$.

Attending to Definition 9, it is easy to check that if we increase the value of α, the size of the concept lattice of $\mathcal{M}_\alpha$ will be reduced.

The set, which is presented in Definition 9, with the ordering defined in $\mathcal{M}$, restricted to $\mathcal{M}_\alpha$, forms a lattice, indeed, this is a sublattice of the original one.

Theorem 3. *For each* $\alpha \in L_1$, *if* $\mathcal{M}_\alpha$ *is an irreducible* α-cut *of* $\mathcal{M}$, *then the pair* $(\mathcal{M}_\alpha, \preceq)$ *is a sublattice of* $(\mathcal{M}, \preceq)$.

Therefore, $(\mathcal{M}_\alpha, \preceq)$ is a concept lattice and, consequently, the following result holds.

Corollary 2. *Given* $(\mathcal{M}, \preceq)$ *and* $(\mathcal{M}_\alpha, \preceq)$, *we have that* $Ext(\mathcal{M}_\alpha) \subseteq Ext(\mathcal{M})$, $Int(\mathcal{M}_\alpha) \subseteq Int(\mathcal{M})$, *where* $Ext(\mathcal{M}_\alpha)$, $Ext(\mathcal{M})$, $Int(\mathcal{M}_\alpha)$ *and* $Int(\mathcal{M})$ *are the extension and intension sets of the concept lattices* $\mathcal{M}_\alpha$ *and* $\mathcal{M}$, *respectively.*

A similar procedure can be developed with respect to the join-irreducible elements. Now, more properties of these concept lattices are studied.

Proposition 5. *Given* $\alpha \in L_1$ *and* g *an extension of a concept of* $\mathcal{M}_\alpha$ *such that* $g = \bigwedge_{j \in J} \phi_{j,x_j}^\downarrow$, *with* $\langle \phi_{j,x_j}^\downarrow, \phi_{j,x_j}^{\downarrow\uparrow} \rangle \in M_F(A)_\alpha$. *Then* $\alpha \preceq_1 g^\uparrow(a_j)$, *for all* $j \in J$.

Proposition 6. *Given* $\alpha \in L_1$ *and* g *an extension of a concept of* $\mathcal{M}_\alpha$ *such that* $g = \bigwedge_{j \in J} \phi_{j,x_j}^\downarrow$, *with* $\langle \phi_{j,x_j}^\downarrow, \phi_{j,x_j}^{\downarrow\uparrow} \rangle \in M_F(A)_\alpha$, *the equality* $\phi_{j,g^\uparrow(a_j)}^{\downarrow\uparrow}(a_j) = g^\uparrow(a_j)$ *holds, for all* $j \in J$.

Note that, given $g \in L_2^B$, as $g^{\uparrow\downarrow}$ is the extension of a concept of $\mathcal{M}$, there exists a family of $\wedge$-irreducible elements in $\mathcal{M}$, $\langle \phi_{j,x_j}^\downarrow, \phi_{j,x_j}^{\downarrow\uparrow} \rangle_{j \in J} \subseteq M_F(A)$, such that $g^{\uparrow\downarrow} = \bigwedge_{j \in J} \phi_{j,x_j}^\downarrow$. The first conjecture we can propose is that the generated concept from g in $\mathcal{M}_\alpha$ is obtained considering only the $\wedge$-irreducible satisfying $\alpha \preceq_1 x_j$, that is:

$$g_\alpha = \bigwedge_{\substack{j \in J \\ \alpha \preceq_1 x_j}} \phi_{j,x_j}^\downarrow$$

However, this is not true in general. In Example 1 we have that a decomposition of the extension of the concept C_3 is $g^3 = \phi_{2,0.6}^\downarrow \wedge \phi_{1,0.7}^\downarrow$. Therefore, $g_{0.7}^3 = \bigwedge_{\substack{j \in J \\ \alpha \preceq_1 x_j}} \phi_{j,x_j}^\downarrow = \phi_{1,0.7}^\downarrow$, which clearly is a contradiction, since $C_3 \neq C_6$. The main reason is that this fuzzy subset has several decompositions in ∧-irreducible elements. In this case, C_3 has three decompositions: $C_6 \wedge C_{13}$, $C_6 \wedge C_{15}$ and $C_6 \wedge C_{16}$. In order to avoid this fact, distributive concept lattices should be considered. Hence, the following result is obtained.

Theorem 4. *Given a concept lattice* $\mathcal{M}$, *which is distributive,* $g \in L_2^B$ *and* $\alpha \in L_1$, *the least extension of a concept of* $\mathcal{M}_\alpha$ *that contains to* g *is*

$$g_\alpha = \bigwedge_{\substack{j \in J \\ \alpha \preceq_1 x_j}} \phi_{j,x_j}^\downarrow$$

where $g^{\uparrow\downarrow} = \bigwedge_{j \in J} \phi_{j,x_j}^\downarrow$ *with* $\langle \phi_{j,x_j}^\downarrow, \phi_{j,x_j}^{\downarrow\uparrow} \rangle_{j \in J} \subseteq M_F(A)$.

If the concept lattice $\mathcal{M}$ is not distributive, new strategies must be studied in order to compute the generated concept from a fuzzy subset of objects (attributes) in $\mathcal{M}_\alpha$.

The following example presents a particular multi-adjoint concept lattice and different irreducible α-cuts of it.

Example 1. The frame $\mathcal{L} = (L, \preceq, \&_P^*)$ is considered, where $L = [0,1]_{10}$ is a regular partition of $[0,1]$ in 10 pieces and $\&_P^*$ is the product conjunctor defined on L, see [3] for more details. In this framework, the context is (A, B, R, σ), where $A = \{a_1, a_2\}$, $B = \{b_1, b_2, b_3\}$, $R \colon A \times B \to L$ is given by Table 1, and σ is constant.

Table 1. Relation R of Example 1

R	b_1	b_2	b_3
a_1	0.2	0.7	0.5
a_2	0.5	1	0

The concept lattice $(\mathcal{M}, \preceq)$, associated with the framework and context considered, has 18 concepts listed below.

$$C_0 = \langle \{0.2/b_1, 0.7/b_2\}, \{1.0/a_1, 1.0/a_2\} \rangle$$
$$C_1 = \langle \{0.2/b_1, 0.8/b_2\}, \{0.8/a_1, 1.0/a_2\} \rangle$$
$$C_2 = \langle \{0.2/b_1, 0.7/b_2, 0.5/b_3\}, \{1.0/a_1\} \rangle$$
$$C_3 = \langle \{0.2/b_1, 1.0/b_2\}, \{0.7/a_1, 1.0/a_2\} \rangle$$
$$C_4 = \langle \{0.2/b_1, 0.8/b_2, 0.6/b_3\}, \{0.8/a_1\} \rangle$$
$$C_5 = \langle \{0.3/b_1, 1.0/b_2\}, \{0.6/a_1, 1.0/a_2\} \rangle$$
$$C_6 = \langle \{0.2/b_1, 1.0/b_2, 0.7/b_3\}, \{0.7/a_1\} \rangle$$
$$C_7 = \langle \{0.4/b_1, 1.0/b_2\}, \{0.5/a_1, 1.0/a_2\} \rangle$$
$$C_8 = \langle \{0.3/b_1, 1.0/b_2, 0.8/b_3\}, \{0.6/a_1\} \rangle$$
$$C_9 = \langle \{0.5/b_1, 1.0/b_2\}, \{0.4/a_1, 1.0/a_2\} \rangle$$
$$C_{10} = \langle \{0.4/b_1, 1.0/b_2, 1.0/b_3\}, \{0.5/a_1\} \rangle$$
$$C_{11} = \langle \{0.6/b_1, 1.0/b_2\}, \{0.3/a_1, 0.8/a_2\} \rangle$$

$$C_{12} = \langle \{0.5/b_1, 1.0/b_2, 1.0/b_3\}, \{0.4/a_1\} \rangle$$
$$C_{13} = \langle \{0.7/b_1, 1.0/b_2\}, \{0.2/a_1, 0.7/a_2\} \rangle$$
$$C_{14} = \langle \{0.6/b_1, 1.0/b_2, 1.0/b_3\}, \{0.3/a_1\} \rangle$$
$$C_{15} = \langle \{0.8/b_1, 1.0/b_2\}, \{0.2/a_1, 0.6/a_2\} \rangle$$
$$C_{16} = \langle \{1.0/b_1, 1.0/b_2\}, \{0.2/a_1, 0.5/a_2\} \rangle$$
$$C_{17} = \langle \{1.0/b_1, 1.0/b_2, 1.0/b_3\}, \{0.2/a_1\} \rangle$$

The Hasse diagram of this lattice is shown in the left side of Figure 2.

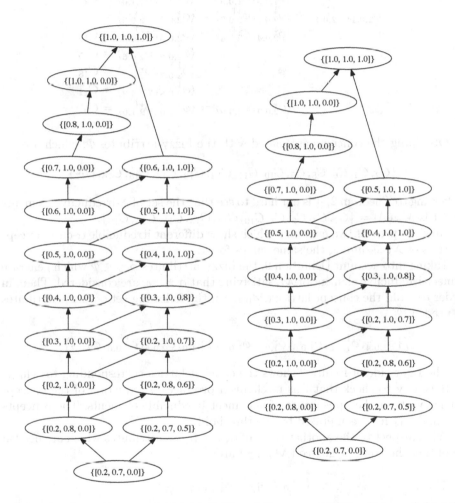

Fig. 2. The Hasse diagram of $(\mathcal{M}, \preceq)$ (left) and the concept lattice $(\mathcal{M}_{0.4}, \preceq)$ (right)

With respect to the fuzzy-attributes, clearly, $\langle \phi^{\downarrow}_{a,0.0}, \phi^{\downarrow\uparrow}_{a,0.0} \rangle = C_{17}$, for all $a \in A$. Moreover, $\langle \phi^{\downarrow}_{1,0.1}, \phi^{\downarrow\uparrow}_{1,0.1} \rangle$ and $\langle \phi^{\downarrow}_{1,0.2}, \phi^{\downarrow\uparrow}_{1,0.2} \rangle$ are C_{17}. The rest are

$$\langle \phi_{1,0.3}^{\downarrow}, \phi_{1,0.3}^{\downarrow\uparrow} \rangle = C_{14}$$
$$\langle \phi_{1,0.4}^{\downarrow}, \phi_{1,0.4}^{\downarrow\uparrow} \rangle = C_{12}$$
$$\langle \phi_{1,0.5}^{\downarrow}, \phi_{1,0.5}^{\downarrow\uparrow} \rangle = C_{10}$$
$$\langle \phi_{1,0.6}^{\downarrow}, \phi_{1,0.6}^{\downarrow\uparrow} \rangle = C_8$$
$$\langle \phi_{1,0.7}^{\downarrow}, \phi_{1,0.7}^{\downarrow\uparrow} \rangle = C_6$$
$$\langle \phi_{1,0.8}^{\downarrow}, \phi_{1,0.8}^{\downarrow\uparrow} \rangle = C_4$$
$$\langle \phi_{1,0.9}^{\downarrow}, \phi_{1,0.9}^{\downarrow\uparrow} \rangle = \langle \phi_{1,1.0}^{\downarrow}, \phi_{1,1.0}^{\downarrow\uparrow} \rangle = C_2$$
$$\langle \phi_{2,0.1}^{\downarrow}, \phi_{2,0.1}^{\downarrow\uparrow} \rangle = \langle \phi_{2,0.2}^{\downarrow}, \phi_{2,0.2}^{\downarrow\uparrow} \rangle = \langle \phi_{2,0.3}^{\downarrow}, \phi_{2,0.3}^{\downarrow\uparrow} \rangle =$$
$$= \langle \phi_{2,0.4}^{\downarrow}, \phi_{2,0.4}^{\downarrow\uparrow} \rangle = \langle \phi_{2,0.5}^{\downarrow}, \phi_{2,0.5}^{\downarrow\uparrow} \rangle = C_{16}$$
$$\langle \phi_{2,0.6}^{\downarrow}, \phi_{2,0.6}^{\downarrow\uparrow} \rangle = C_{15}$$
$$\langle \phi_{2,0.7}^{\downarrow}, \phi_{2,0.7}^{\downarrow\uparrow} \rangle = C_{13}$$
$$\langle \phi_{2,0.8}^{\downarrow}, \phi_{2,0.8}^{\downarrow\uparrow} \rangle = C_{11}$$
$$\langle \phi_{2,0.9}^{\downarrow}, \phi_{2,0.9}^{\downarrow\uparrow} \rangle = \langle \phi_{2,1.0}^{\downarrow}, \phi_{2,1.0}^{\downarrow\uparrow} \rangle = C_9$$

Obtaining the concepts associated with the fuzzy-attributes Φ, which are

$$\{C_2, C_4, C_6, C_8, C_9, C_{10}, C_{11}, C_{12}, C_{13}, C_{14}, C_{15}, C_{16}, C_{17}\}$$

Referring to Theorem 2, it is not hard to see that the set of $\wedge$-irreducible elements of $\mathcal{M}$ is $M_F(A) = \{C_2, C_4, C_6, C_8, C_{10}, C_{12}, C_{13}, C_{14}, C_{15}, C_{16}\}$.

Now, we present two examples which show different irreducible α-cuts concept lattices of $\mathcal{M}$ assuming the same values for α.

Taking into account Definition 9, the fuzzy-attributes $\phi_{i,x} \in \Phi$ which generate a meet-irreducible element of $\mathcal{M}$, satisfying that $\alpha \preceq_1 x$, are considered. Then, in order to build the concept lattice $(\mathcal{M}_{0.4}, \preceq)$, the following set of fuzzy-attributes is needed:

$$\{\phi_{1,0.4}^{\downarrow}, \phi_{1,0.5}^{\downarrow}, \phi_{1,0.6}^{\downarrow}, \phi_{1,0.7}^{\downarrow}, \phi_{1,0.8}^{\downarrow}, \phi_{1,1.0}^{\downarrow}, \phi_{2,0.5}^{\downarrow}, \phi_{2,0.6}^{\downarrow}, \phi_{2,0.7}^{\downarrow}\}$$

The right side of Figure 2 shows the concept lattice corresponding to $\mathcal{M}_{0.4}$.

It is easy to check that this mechanism produces a reduction of the concept lattice without the appearance of new meet-irreducible elements. The concepts C_{11} and C_{14} has been erased by the threshold.

With respect to the another value of α, the fuzzy-attributes which are needed to obtain the concept lattice $(\mathcal{M}_{0.7}, \preceq)$ are:

$$\{\phi_{1,0.7}^{\downarrow}, \phi_{1,0.8}^{\downarrow}, \phi_{1,1.0}^{\downarrow}, \phi_{2,0.7}^{\downarrow}\}$$

As result, the built concept lattice is presented in Figure 3. Note that irreducible 0.7-cut provides a concept lattice composed by 8 concepts, hence, a major reduction has been done.

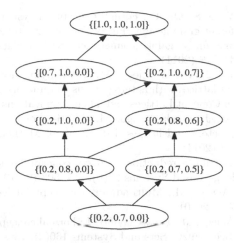

Fig. 3. Concept lattice $(\mathcal{M}_{0.7}, \preceq)$

5 Conclusions and Future Work

This paper has introduced several properties of the particular case of fuzzy subsets of attributes, which are called fuzzy-attributes. These mappings are used, for example, to characterize the meet-irreducible set of a concept lattice. Furthermore, considering an α-cut of the meet-irreducible elements, we have introduced a sublattice of the original concept lattice, which keep the main information obtained from the relational database, reducing the size and so, the complexity, of the concept lattice. This method is original and the comparison with other methods to reduce the size of a concept lattice will be studied in the future.

References

1. Bělohlávek, R., Vychodil, V.: Reducing the size of fuzzy concept lattices by hedges. In: The 2005 IEEE International Conference on Fuzzy Systems, pp. 663–668 (2005)
2. Birkhoff, G.: Lattice Theory, 3rd edn. American Mathematical Society, Providence (1967)
3. Cornejo, M., Medina, J., Ramírez, E.: A comparative study of adjoint triples. Fuzzy Sets and Systems 211, 1–14 (2013)
4. Cornejo, M., Medina, J., Ramírez, E.: Irreducible elements in multi-adjoint concept lattices. In: Intl Conference on Fuzzy Logic and Technology, EUSFLAT 2013, pp. 125–131 (2013)
5. Cornejo, M.E., Medina-Moreno, J., Ramírez, E.: On the classification of fuzzy-attributes in multi-adjoint concept lattices. In: Rojas, I., Joya, G., Cabestany, J. (eds.) IWANN 2013, Part II. LNCS, vol. 7903, pp. 266–277. Springer, Heidelberg (2013)
6. Hájek, P.: Metamathematics of Fuzzy Logic. Trends in Logic. Kluwer Academic (1998)

7. Kang, X., Li, D., Wang, S., Qu, K.: Formal concept analysis based on fuzzy granularity base for different granulations. Fuzzy Sets and Systems 203, 33–48 (2012)
8. Konecny, J.: Isotone fuzzy galois connections with hedges. Information Sciences 181(10), 1804–1817 (2011)
9. Konecny, J., Medina, J., Ojeda-Aciego, M.: Intensifying hedges and the size of multi-adjoint concept lattices with heterogeneous conjunctors. In: The 9th International Conference on Concept Lattices and Their Applications, pp. 245–256 (2012)
10. Konecny, J., Medina, J., Ojeda-Aciego, M.: Multi-adjoint concept lattices with heterogeneous conjunctors and hedges. In: Annals of Mathematics and Artificial Intelligence, pp. 1–17 (2014)
11. Li, L., Zhang, J.: Attribute reduction in fuzzy concept lattices based on the t-implication. Knowledge-Based Systems 23(6), 497–503 (2010)
12. Medina, J., Ojeda-Aciego, M.: Multi-adjoint t-concept lattices. Information Sciences 180(5), 712–725 (2010)
13. Medina, J., Ojeda-Aciego, M., Ruiz-Calviño, J.: Formal concept analysis via multi-adjoint concept lattices. Fuzzy Sets and Systems 160(2), 130–144 (2009)
14. Wille, R.: Restructuring lattice theory: an approach based on hierarchies of concepts. In: Rival, I. (ed.) Ordered Sets, pp. 445–470. Reidel (1982)

Monadic Formal Concept Analysis

Patrik Eklund[1], María Ángeles Galán García[2,*],
Jari Kortelainen[3], and Manuel Ojeda-Aciego[2,*]

[1] Umeå University, Department of Computing Science, Sweden
[2] Universidad de Málaga, Dept. Matemática Aplicada, Spain
[3] Mikkeli University of Applied Sciences, Department of Electrical Engineering
and Information Technology, Finland

Abstract. Formal Concept Analysis (FCA) as inherently relational can
be formalized and generalized by using categorical constructions. This
provides a categorical view of the relation between "object" and "attributes", which can be further extended to a more generalized view on
relations as morphisms in Kleisli categories of suitable monads. Structure of sets of "objects" and "attributes" can be provided e.g. by term
monads over particular signatures, and specific signatures drawn from
and developed within social and health care can be used to illuminate
the use of the categorical approach.

1 Introduction

In traditional FCA [15], a so called "context", or "formal context", is in the
end just a relation on sets, $I \subseteq G \times M$, often written as and said to be a triple
(G, M, I). Further, G is called the set of "objects" and M the set of "attributes".
However, the user of formal concept lattices is left with the burden to intuitively
explain what is really, if at all, meant by "object" and M the set of "attribute".
Sometimes "object" can be a name or a number, e.g., saying that 'Alice' as a
name has "attribute" 'old', or '92' as a number has "attribute' 'even'. Note,
however, we may also in another application say that 'Alice' as a name has
"attribute" '92', so it is not at all clear what the distinction between "object"
and "attribute" really is. Basically, in FCA, G and M are just plain sets, and
can be seen as objects in the category Set of sets and functions. Further, in
traditional FCA, the elements of those sets have no structure whatsoever. In
this paper we will formalize FCA categorically, thereby opening up possibilities
to give "object" and "attribute" more precise meanings, also going beyond just
using Set as the underlying category for FCA, and, needless to say, adopting
a much more generalized view on relations. In the simplest case, relations over
sets correspond precisely to morphisms in the Kleisli category of the ordinary
powerset monad (see e.g. [8] for basic notions related to monads and Kleisli
categories). It is not surprising that the notion of monad appears in this research
topic, since FCA is essentially built on the notion of Galois connection [1], a
kind of dual version of adjunction, and adjunctions are straightforwardly linked

---

* M.Á. Galán and M. Ojeda-Aciego has been partially suppported by Spanish project
TIN2012-39353-C04-01.

C. Cornelis et al. (eds.): RSCTC 2014, LNAI 8536, pp. 201–210, 2014.

to monads [11]. Other approaches related to both FCA and adjunctions/Galois connections can be seen in the development of the theory of generalized Chu mappings, within the category of contexts as objects and the so-called L-Chu correspondences as morphisms [10].

A so called "formal concept", or just a "concept" is a pair (A, B), with $A \subseteq G$ and $B \subseteq M$, such that $A = \{g \in G \mid gIm \text{ for all } m \in B\}$ and $B = \{m \in M \mid gIm \text{ for all } g \in A\}$. A lattice, the so called "formal concept lattice", is given for the set of all concepts by $(A_1, B_1) \leq (A_2, B_2)$ if and only if $A_1 \subseteq A_2$ (or, equivalently, $B_1 \supseteq B_2$).

Since there is no convention about how to use given names for objects and attributes in "informally constructed" names for formal concepts, combining names into names for concepts, or simply inventing the names otherwise, has become tradition within FCA. This, however, means that there is no terminological or ontology basis for FCA, but concepts themselves are seen as ontology objects. For applications e.g. in social and health care this is not recommendable since concepts related to disease and function must comply with classifications like WHO's ICD (International Classification of Diseases) and ICF (International Classification of Functions, Disabilities and Health). A similar observation can be made for description logics, which is successfully used for web ontology, but adopting success more than content for social and health care has turned out not to be as straightforward as expected e.g. by SNOMED. We may also note how rough set theory is related to these relational approaches, and a more formal intertwining of FCA, description logic and rough sets is certainly desirable, where some less strictly logical attempts have been made. Our approach to lative logic [6] using underlying categories has been shown to open up the possibility to strictly define description logic as being related to λ-calculus [4], and indeed not to first-order logic as frequently claimed. Further, rough sets have been shown to be 'monadic' in the Kleisli morphism sense [3].

2   Categorical Notions

In this paper the readers are assumed to be familiar with categorical terminology, however, we introduce some key definitions for convenience. Signatures and term monads over monoidal biclosed categories are given in [4]. These notions and constructions are given here over the category Set since traditional FCA is developed in Set. Naturally, the development in the current paper may be generalised to the Goguen category $\mathsf{Set}(\mathfrak{Q})$, where $\mathfrak{Q}$ is a quantale.

Let a category C and a set of sorts S be given. It is well-known that C_S is a category with objects[1] $X_S = (X_\mathbf{s})_{\mathbf{s} \in S}$ where each $X_\mathbf{s} \in \mathrm{Ob}(\mathsf{C})$. We have $f_S \colon X_S \to Y_S$ as morphisms, where $f_S = (f_\mathbf{s})_{\mathbf{s} \in S}$ and each $f_\mathbf{s} \in \hom_{\mathsf{Set}}(X_\mathbf{s}, Y_\mathbf{s})$. The composition of morphisms is defined by $f_S \circ g_S = (f_\mathbf{s} \circ g_\mathbf{s})_{\mathbf{s} \in S}$.

We may sometimes need to refer to an object $X_\mathbf{s} \in \mathrm{Ob}(\mathsf{C})$ when X_S is given in a form or another. We then need to define a functor $\mathsf{arg}^\mathbf{s} \colon \mathsf{C}_S \to \mathsf{C}$, which

---

[1] Note that a categorical object should obviously not be confused with an "object" in a formal context.

is given by $\arg^{\mathsf{s}} X_S = X_{\mathsf{s}}$ and $\arg^{\mathsf{s}} f_S = f_{\mathsf{s}}$. Note that a functor $\mathsf{F}\colon \mathsf{C} \to \mathsf{D}$ may be extended to a functor $\mathsf{F}_S\colon \mathsf{C}_S \to \mathsf{D}_S$ (the functor remains the same for all $\mathsf{s} \in S$). For example, the powerset functor $\mathsf{P}\colon \mathsf{Set} \to \mathsf{Set}$ and the many-valued powerset (based on the quantale $\mathfrak{Q}$) functor $\mathsf{Q}\colon \mathsf{Set} \to \mathsf{Set}$ both can be extended as functors on Set_S, and we write $\mathsf{P}_S = (\mathsf{P})_{\mathsf{s}\in S}$ and $\mathsf{Q}_S = (\mathsf{Q})_{\mathsf{s}\in S}$.

For any two functors $\mathsf{F}, \mathsf{G}\colon \mathsf{C} \to \mathsf{D}$ a natural transformation τ between F and G, denoted by $\tau\colon \mathsf{F} \to \mathsf{G}$, assigns for each C-object X a D-morphism $\tau_X\colon \mathsf{F}X \to \mathsf{G}X$ satisfying $\mathsf{G}f \circ \tau_X = \tau_Y \circ \mathsf{F}f$ for all $f \in \hom_{\mathsf{C}}(X,Y)$. Clearly, there may be natural transformations τ between functors $\mathsf{F}, \mathsf{G}\colon \mathsf{C}_S \to \mathsf{D}_S$ also such as $\tau\colon \mathsf{F} \to \mathsf{G}$, where $(\tau_{X_S})_S\colon (\mathsf{F}_{\mathsf{s}} X_S)_{\mathsf{s}\in S} \to (\mathsf{G}_{\mathsf{s}} X_S)_{\mathsf{s}\in S}$ satisfying $(\mathsf{G}_{\mathsf{s}} f_S)_{\mathsf{s}\in S} \circ (\tau_{X_S})_S = (\tau_{Y_S})_S \circ (\mathsf{F}_{\mathsf{s}} f_S)_{\mathsf{s}\in S}$.

Moreover, a monad $\mathbf{F}$ over a category C is a triple (F, η, μ), where $\mathsf{F}\colon \mathsf{C} \to \mathsf{C}$ is a (covariant) functor, and $\eta\colon \mathrm{id} \to \mathsf{F}$ and $\mu\colon \mathsf{F}\mathsf{F} \to \mathsf{F}$ are natural transformations satisfying $\mu \circ \mathsf{F}\mu = \mu \circ \mu\mathsf{F}$ and $\mu \circ \mathsf{F}\eta = \mu \circ \eta\mathsf{F} = \mathrm{id}_{\mathsf{F}}$. Note that we may have monads over C_S also.

2.1   Signatures and the Term Monad Construction

Using notations adopted in computer science, a many-sorted signature $\Sigma = (S, \Omega)$ over Set consists of a set S (of sorts) and a set Ω (of operators). Both S and Ω should be considered as objects in Set, however, S is used in this paper also as an index set since we have $S \cong \hom_{\mathsf{Set}}(\{\varnothing\}, S)$. Technically more precise description is given indeed in [4].

Intuitively, operators in $\Omega \cong \coprod_{\mathsf{s}\in S} \Omega_{\mathsf{s}}$ are written as $\omega\colon \mathsf{s}_1 \times \cdots \times \mathsf{s}_n \to \mathsf{s}$. This is seen by using a convenient notation $\Omega^{\mathbf{m}\to\mathsf{s}}$ for a set of operators $\omega\colon \mathsf{s}_1 \times \cdots \times \mathsf{s}_n \to \mathsf{s}$ as $\mathbf{m} = (\mathsf{s}_1, \ldots, \mathsf{s}_n) \in S^n$. We will consider $S^0 = \{\varnothing\}$ and $\hat{S} = \coprod_{n\in\mathbb{N}} S^n$ although, practically speaking, in an application there may be some maximum arity k such that always $\mathbf{m} \in \coprod_{n\le k} S^n$. Moreover, $\Omega^{\to\mathsf{s}}$ stands for the set of constants $\omega\colon \to \mathsf{s}$. With these notations we keep explicit track of sorts and arities and we have

$$\Omega_{\mathsf{s}} = \coprod_{\mathbf{m}\in\hat{S}} \Omega^{\mathbf{m}\to\mathsf{s}}.$$

For handling connections between signatures we have a category $\mathsf{Sign}_{\mathsf{Set}}$ with signatures Σ as its objects, where morphisms between signatures $\Sigma_1 = (S_1, \Omega_1)$ and $\Sigma_2 = (S_2, \Omega_2)$ are pairs (s, o) such that $\mathsf{s}\colon S_1 \to S_2$ and $\mathsf{o}\colon \Omega_1 \to \Omega_2$. Moreover, for each $\omega_1 \in \Omega_1^{\mathbf{m}\to\mathsf{s}}$, there exists a $\omega_2 \in \Omega_2^{\mathbf{n}\to\mathsf{s}(\mathsf{s})}$ such that $\mathsf{o}(\omega_1) = \omega_2$ and $\mathbf{n} = (\mathsf{s}(\mathsf{s}_1), \ldots, \mathsf{s}(\mathsf{s}_2))$. Composition is defined pairwise.

In our general term functor construction we have functors $\Psi_{\mathbf{m},\mathsf{s}}\colon \mathsf{Set}_S \to \mathsf{Set}$ such that

$$\Psi_{\mathbf{m},\mathsf{s}}((X_t)_{t\in S}) = \Omega^{\mathbf{m}\to\mathsf{s}} \times \prod_{i=1,\ldots,n} X_{\mathsf{s}_i}$$

as $\mathbf{m} = (\mathsf{s}_1, \ldots, \mathsf{s}_n)$ and action on morphisms is defined in a natural way. The inductive steps starts with $\mathsf{T}^1_{\Sigma,\mathsf{s}} = \coprod_{\mathbf{m}\in\hat{S}} \Psi_{\mathbf{m},\mathsf{s}}$. For $\iota > 1$ we proceed by $\mathsf{T}^{\iota}_{\Sigma,\mathsf{s}} X_S = \coprod_{\mathbf{m}\in\hat{S}} \Psi_{\mathbf{m},\mathsf{s}}(\mathsf{T}^{\iota-1}_{\Sigma,t} X_S \sqcup X_t)_{t\in S})$ and for morhisms $\mathsf{T}^{\iota}_{\Sigma,\mathsf{s}} f_S = \coprod_{\mathbf{m}\in\hat{S}} \Psi_{\mathbf{m},\mathsf{s}}(\mathsf{T}^{\iota-1}_{\Sigma,t} f_S \sqcup f_t)_{t\in S})$. We have now functors $\mathsf{T}^{\iota}_{\Sigma}\colon \mathsf{Set}_S \to \mathsf{Set}_S$ when assigning $\mathsf{T}^{\iota}_{\Sigma} X_S =$

$(T^{\iota}_{\Sigma,s} X_S)_{s \in S}$ and $T^{\iota}_{\Sigma} f_S = (T^{\iota}_{\Sigma,s} f_S)_{s \in S}$. Clearly, the system $(T^{\iota}_{\Sigma})_{\iota > 0}$ of endo-functors is an inductive system in a natural way, thus, there exist natural trans-formations $\Xi^{\iota+1}_{\iota} : T^{\iota}_{\Sigma} \to T^{\iota+1}_{\Sigma}$ of which each morphism is a canonical injection. There exists then an inductive limit F, and finally our term functor is given by $T_{\Sigma} = F \sqcup \mathrm{id}_{\mathrm{Set}_S}$, which can be extended to a monad.

2.2   Levels of Signatures

The three-level arrangement of signatures was presented in [4].

(i) Level one: The level of 'primitive and underlying' sorts and operations, with a many-sorted signature
$$\Sigma = (S, \Omega)$$

(ii) Level two: The level of 'type constructors', with a single-sorted signature
$$\lambda_{\Sigma} = (\{\iota\}, \{s :\to \iota \mid s \in S\} \cup \{\Rightarrow : \iota \times \iota \to \iota\})$$

(iii) Level three: The level in which we may construct 'λ-terms' based on the signature
$$\Sigma^{\lambda} = (S^{\lambda}, \Omega^{\lambda})$$
with $S^{\lambda} = T_{\lambda_{\Sigma}} \varnothing$, and $\Omega^{\lambda} = \mathfrak{M} \cup \{\mathrm{app}_{s,t} : (s \Rightarrow t) \times s \to t\}$, where

$$\mathfrak{M} = \{\omega^{\lambda}_{i_1,\ldots,i_n} :\to (s_{i_1} \Rightarrow \cdots \Rightarrow (s_{i_{n-1}} \Rightarrow (s_{i_n} \Rightarrow s)) \cdots) \mid$$
$$\mid \omega : s_1 \times \ldots \times s_n \to s \in \Omega, (i_1, \ldots, i_n) \text{ is a permutation of } (1, \ldots, n)\}$$

The natural numbers signature in levels is as follows

(i) Level one:
$$\mathrm{NAT} = (\{\mathrm{nat}\}, \{0 :\to \mathrm{nat}, \mathrm{succ} : \mathrm{nat} \to \mathrm{nat}\})$$

(ii) Level two:
$$\lambda_{\mathrm{NAT}} = (\{\iota\}, \{\mathrm{nat} :\to \iota, \Rightarrow : \iota \times \iota \to \iota\})$$

(iii) Level three:
$$\Sigma^{\lambda} = (T_{\lambda_{\mathrm{NAT}}} \varnothing, \Omega^{\lambda})$$
where

$$\Omega^{\lambda} = \{0^{\lambda} :\to \mathrm{nat}, \mathrm{succ}^{\lambda}_1 :\to (\mathrm{nat} \Rightarrow \mathrm{nat})\} \cup \{\mathrm{app}_{s,t} : (s \Rightarrow t) \times s \to t\}$$

Hierarchies of sets, or sets of sets, sets of sets of sets, and so on, can be modelled by the 'powerset' type constructor P : type $\to$ type on level two, i.e., intuitively thinking that the algebra P is a powerset functor. A signature $\Sigma_{\mathrm{DescriptionLogic}} = (S, \Omega)$ for description logic can then be provided as follows

(i) $S = \{\mathrm{concept}\}$, and we may add constants like $c_1, \ldots, c_n :\to \mathrm{concept}$.

(ii) We include a type constructor P : type $\rightarrow$ type into S_Ω, with an intuitive semantics of being the powerset functor, so that Pconcept is the constructed type for "powerconcept".

(iii) "Roles" are r :$\rightarrow$ (Pconcept $\Rightarrow$ PPconcept), and we need operators η :$\rightarrow$ (concept $\Rightarrow$ Pconcept) and μ :$\rightarrow$ (PPconcept $\Rightarrow$ Pconcept) in Ω', so that "∃r.x" can be defined as

$$\text{app}_{\text{PPconcept},\text{Pconcept}}(\mu, \text{app}_{\text{Pconcept},\text{PPconcept}}(r, x)).$$

term sets as appearing on different levels of signatures will be used in examples within categorical formal concept analysis.

3 Categorical Formal Concept Analysis

There is a number of options for the categorization of FCA. We start with a purely categorial extension of the relational view, also in a generalized relational setting, which makes no explicit enhancements of "object" and "attribute", and then we go further into specification of the content of "object" and "attribute".

3.1 Generalized Relations as Kleisli Morphisms

The most trivial categorical observation is that a *category of contexts* can be defined as the category SetRel of sets and relations, which is isomorphic to the Kleisli category Set$_\text{P}$, where $\mathbf{P} = (P, \eta, \mu)$ is the powerset monad over Set. Indeed, if (G, M, I) is a context, then we represent the relation $I \subset G \times M$ as a mapping $\iota : G \rightarrow PM$, where $\iota(g) = \{m \in M \mid gIm\}$. This mapping is the so called *Kleisli morphism*, i.e., a morphism in that Kleisli category. This morphisms indeed generates one of the derivation operators of FCA. The corresponding inverse relation I^{-1} is represented as $\iota^{-1}(m) = \{g \in G \mid mI^{-1}g\}$, i.e., providing the mapping $\iota^{-1} : M \rightarrow PG$.

A pair (A, B) in the context of (G, M, I) is now a formal concept if and only if

$$A = o(B)$$

and

$$B = \alpha(A),$$

where $\alpha : PG \rightarrow PM$ is given by

$$\alpha(A) = \{m \in M \mid A \subseteq \iota^{-1}(m)\},$$

and $o : PM \rightarrow PG$ by

$$o(B) = \{g \in G \mid B \subseteq \iota(g)\}.$$

Both mappings α and o have powersets as domain and range and are not in this sense Kleisli morphisms. However, the unit of the monad can be used to provide the equivalent conditions

$$A = \bigcup_{B \subseteq \iota(g)} \eta_G(g)$$

and

$$B = \bigcup_{A \subseteq \iota^{-1}(m)} \eta_M(m).$$

The multiplication of the monad can be used to introduce weaker conditions. In order to establish such weaker conditions, note that $A = o(B)$ means that $a \in A$ if and only if $B \subseteq \iota(a)$, and that $B = \alpha(A)$ means that $b \in B$ if and only if $A \subseteq \iota^{-1}(b)$. This then gives conditions

$$A \subseteq \bigcup_{b \in B} \iota^{-1}(b) = (\mu \circ P\iota^{-1})(B)$$

and

$$B \subseteq \bigcup_{a \in A} \iota(a) = (\mu \circ P\iota)(A),$$

which are implied by the formal concept conditions. These weaker conditions reveal that there are different kinds of possibilities to determine concepts in the given context.

Now note that in this categorical notation, there is no need to restrict to using the powerset functor only. Any monad $\mathbf{F} = (F, \eta, \mu)$ will act as a "context monad" for contexts (X, Y, ι), where $X, Y \in \mathrm{Ob}(\mathbf{Set})$, $\iota : X \to FY$, $\iota^{-1} : Y \to FX$, and either of the conditions above involving η or μ are fulfilled. We may then call such a pair (A, B) a "monadic concept", given the conditions adopted.

The ordering relation in formal concept lattice makes use of the $\subseteq$ relation which, for any set X, makes $(\mathsf{P}X, \subseteq)$ a partial order. The powerset monad is in fact a *partially ordered monad* [7], written as $(\mathsf{P}, \leq, \eta, \mu)$, where $\leq$ is $\subseteq$ in the partial order $(\mathsf{P}X, \leq)$. Monads, in particular over $\mathbf{Set}$, can often be extended to partially ordered monads $\mathbf{F} = (\mathsf{F}, \leq, \eta, \mu)$, so that F can represent a more deeply structured functor far beyond just the powerset monad. The fuzzy powerset monad and the filter monad [2] are typical examples. In these cases, a *concept lattice of monadic concepts* is still defined as $(A_1, B_1) \leq (A_2, B_2)$ if and only if $A_1 \leq A_2$ (or, equivalently, $B_1 \leq B_2$), but now a "concept" is not just a pair of sets. Composing contexts obviously creates a view on "composition of concept lattices", but is not treated in this paper.

These categorical formulations provide more content to the concept lattice, but still no new information concerning the specific explanation of "objects" and "attributes", as far as the underlying sets X and Y in (X, Y) are concerned. In the following subsections we will provide more elaborate examples on those X and Y, so that they are not just 'sets of points'.

3.2   Multi-Sorted Signatures and Terms

Let $\Sigma = (S, \Omega)$ be a multi-sorted signature on level one within the three-level arrangement of signatures. Whenever we have a term $t \in \mathsf{T}_{\Sigma^\lambda, \mathbf{s}} X_{S^\lambda}$, with $\mathbf{s} \in \mathsf{T}_{\lambda_\Sigma} \varnothing$, we write it shortly as $t :: \mathbf{s}$. Obviously, this creates a context

$$(\mathsf{T}_{\Sigma^\lambda, \mathbf{s}} X_{S^\lambda}, \mathsf{T}_{\lambda_\Sigma} \varnothing, ::)$$

where further enhancements of the relation :: can be extended to $::_\mathbf{F}$ given various choices of the corresponding monad $\mathbf{F}$. In this case "attributed object" would correspond to a some notion of "term of type", and broader view on "objects" would allow terms of any sort, so that the context is

$$(\bigcup_{\mathsf{s} \in \mathsf{T}_{\lambda_\Sigma} \varnothing} \mathsf{T}_{\Sigma^\lambda,\mathsf{s}} X_{S^\lambda}, \mathsf{T}_{\lambda_\Sigma} \varnothing, \kappa),$$

where κ would not necessarily have to comply with ::.

Similarly we may think of monadic contexts involving sorts only, in $(\mathsf{s} \in \mathsf{T}_{\lambda_\Sigma} \varnothing, \mathsf{s} \in \mathsf{T}_{\lambda_\Sigma} \varnothing, \iota_{\text{sorts}})$, or terms only, in $((\mathsf{T}_{\Sigma^\lambda,\mathsf{s}} X_{S^\lambda}, \mathsf{T}_{\Sigma^\lambda,\mathsf{s}} X_{S^\lambda}, \iota_{\text{terms}}))$. Further, we may involve *sentences* in the sense as formally produced by *sentence functors* [5], and from there on suggest contexts appearing like

$$(G_{\text{sentences}}, M_{\text{terms}}, \iota_{\text{sentences,terms}}),$$

$$(G_{\text{terms}}, M_{\text{sentences}}, \iota_{\text{terms,sentences}}),$$

and

$$(G_{\text{sentences}}, M_{\text{sentences}}, \iota_{\text{sentences,sentences}}).$$

Specific choices on the basic underlying signatures at level one will then determine most of the structure for these contexts and concepts. Again, F may take various 'generalized powerset' forms.

3.3 Some Further Generalized Approaches to Formal Concept Analysis

The logic of formal concepts can be viewed in various ways, and clearly depending on how we prefer to view concepts. Intuitively, concepts are 'statements' saying how "objects are attributed". The traditional view of the concept lattice is to see it as a lattice of propositions.

For the L-fuzzy powerset extension of FCA, the partially ordered monad $\mathbf{F}$ is the partially ordered L-fuzzy powerset monad $\mathbf{L} = (\mathsf{L}, \leq, \eta, \mu)$. Now, the definition of the derivation operators, is modified accordingly to the underlying framework of L-fuzzy powersets, and expressed in terms of the L-fuzzy subsethood relation. For instance, the derivation operator $f : \mathsf{L}G \to \mathsf{L}M$ is defined, for an L-set of objects $A : G \to L$, as the degree to which the attributes are satisfied with respect to the given L-fuzzy set of objects A; specifically, the result is the L-set of attributes $f(A) : M \to L$ defined by

$$f(A)(y) = \bigwedge_{x \in G} A(x) \to \iota(x)(y), \tag{1}$$

where $\to$ is the residuated implication of L, which is assumed to be a complete residuated lattice.

Even in the standard fuzzy case, where L is the real unit interval, there may be several candidates for the implication (there are lots of families of pairs formed by t-norms and their residuated mappings), and implication is essential when defining the subsethood relation and, hence, the derivation operators in a fuzzy setting.

The previous observation brings FCA into the realm of the multi-adjoint framework. Multi-adjoint FCA [14] starts on this generalization, by allowing to use a different implication for different subsets of objects/attributes, somehow considering different sorts within the sets of objects and attributes, and linking the approach to material introduced in Section 3.2. One of the initial advantages of the multi-adjoint approach was to obtain an easy and flexible way to formalize preferences; furthermore, it has proven to be a sufficiently general framework under which to interpret several other generalizations of FCA and related approaches [9,12].

Last but not least, there are options other than (1) to define the derivation operators, leading to concept lattices more easily interpretable in terms of roughness than fuzziness [13].

3.4   Health Contexts and Concepts Involving Classification for Disease, Drug and Functioning

WHO maintains and further develops a number of classifications, including classification for disease, drugs and functioning, where ICD (International Classification of Disease) and ICF (International Classification of Functioning, Disability and Health) are *reference classifications*, and ATC/DDD (Anatomical Therapeutic Chemicals Classification with Defined Daily Doses) is a *related classification*. Classifications and their codes are hierarchical, so that higher level codes are sets of lower level codes. Within these hierarchies there are hidden relations, that are indeed pointed out, but not formally related within the classifications. Arranging these classifications in signatures ICD, ATC and ICF enables to formalize such relations even as generalized relations.

Drugs for the nervous system is a typical example of a *main anatomical group* of drugs. This group is on the 1st ATC level and coded as 'N nervous system'. On 2nd level in this group there are e.g. 'N05 psycholeptics', and on 3rd level there are 'N05C hypnotics and sedatives' as an example of a pharmacological subgroup. Going to the chemical levels, the 4th level includes e.g. 'N05CD benzodiazepine derivatives', and the 5th level includes specific drugs like 'N05CD02 nitrazepam'.

The ATC signature could then on level two of the levels of signatures be arranged to include the ATC levels as 1st, 2nd, 3rd, 4th, 5th :→ type so that on level three we might say e.g. that "pharmacologic interventions" in general for a patient is a set of 3rd level items, whereas a set of "drug prescriptions" is a set of 5th level items. Clearly, there is lots of abuse of language in this contexts, even in the medical domain. However, a signature can help to make these notions more precise for a particular application context. We could have operators like PharmacologicIntervention :→ P(3rd) and DrugPrescriptions :→ P(5th) and we clearly "transformation" between the levels e.g. in form of $\phi^{5th\rightarrow3rd}$: 5th → 3rd, so that nitrazepam on 5th level can be formally treated also as 'sedative' on 3rd level.

Note also how *drug interactions*, e.g. according to the Swedish-Finnish SFINX model, can be syntactically described by the operator

$$\texttt{DrugDrugInteraction} : \texttt{5th} \times \texttt{5th} \to Q_{SFINX}$$

where Q_{SFINX} is the 5-level scale used in SFINX for describing the severity of interactions. Operations like

$$\texttt{DrugDrugInteraction}^{\texttt{ith,jth}} : \texttt{ith} \times \texttt{jth} \to Q_{SFINX}$$

$$\texttt{DrugSetOfDrugsInteraction}^{\texttt{ith,jth}} : \texttt{ith} \times \texttt{Pjth} \to Q_{SFINX}$$

$$\texttt{SetOfDrugsDrugSetOfDrugsInteraction}^{\texttt{ith,jth}} : \texttt{ith} \times \texttt{Pjth} \to Q_{SFINX}$$

can also be consider, where $\texttt{SetOfDrugsDrugSetOfDrugsInteraction}^{\texttt{5th,5th}}$ is a well recognized problem, for which there is still not a proper solution.

For ICF, similar encodings will be applied. For postural control we have factors like gait and muscle function, where, for coding purposes, the latter is to a larger extent identifiable with items in the ICF (WHO's International Classification of Function) classification. A typical example is

```
Muscle functions (ICF b730-b749)
  Muscle power functions (b730)
  ...
```

The ICF datatypes and its generic scale of quantifiers from `xxx.0 NO problem` to `xxx.4 COMPLETE problem` and including the `xxx.8 not specified` is suitable for modelling using quantales, where `xxx.8 not specified` will play the role of the unital. This can be formalized in a signature. Algebras must obviously be included in all these encodings, and for ICF it relates to the generic scale of quantifiers, and how that "8" can be situated in relation to the other quantifiers so as algebraically to form a quantale.

Generally, this view gives us several options for providing contexts involving diseases, functioning and drugs, respectively using sets of terms functors $\mathsf{T}_{\texttt{ICD}}$, $\mathsf{T}_{\texttt{ICF}}$ and $\mathsf{T}_{\texttt{ATC}}$, as introduced by ICD, ICF and ATC codes. Drug-drug interactions are then based on a context like $(\mathsf{T}_{\texttt{ATC}}X, \mathsf{T}_{\texttt{ATC}}X, \iota_{\texttt{ATC,ATC}})$, disease-functioning by $(\mathsf{T}_{\texttt{ICD}}X, \mathsf{T}_{\texttt{ICF}}Y, \iota_{\texttt{ICD,ICF}})$, and so on.

4   Conclusions

The current paper serves as our kick off to further elaboration of monadic FCA and its applications. We have presented an approach to formal concepts using Kleisli categories, including a syntactic view for contexts, which, on the one hand, opens up the structure and meaning of "objects" and "attributes", and, on the other hand, enables to adopt a generalized view of relations using Kleisli morphisms as substitutions. From practical point of view, social and health care is a typical application domain, where terminology and ontology e.g. for disorder and functioning need to be connected with various interventions and as evaluated using scales and qualifiers. In these application areas, it is important to clearly define the distinction between 'statement' and 'value', or, more broadly speaking, to provide a better understanding of the distinction between the use of logics and statistics, respectively.

References

1. Díaz, J., Medina, J., Ojeda-Aciego, M.: On basic conditions to generate multi-adjoint concept lattices via Galois connections. Intl Journal of General Systems 43(2), 149–161 (2014)
2. Eklund, P., Gähler, W.: Fuzzy filter functors and convergence. In: Rodabaugh, S.E., Klement, E.P., Höhle, U. (eds.) Applications of Category Theory to Fuzzy Subsets, pp. 109–136. Kluwer Academic Publishers (1992)
3. Eklund, P., Galán, M.Á.: Monads can be rough. In: Greco, S., Hata, Y., Hirano, S., Inuiguchi, M., Miyamoto, S., Nguyen, H.S., Słowiński, R. (eds.) RSCTC 2006. LNCS (LNAI), vol. 4259, pp. 77–84. Springer, Heidelberg (2006)
4. Eklund, P., Galán, M., Helgesson, R., Kortelainen, J.: Fuzzy terms. Fuzzy Sets and Systems (in press)
5. Eklund, P., Galán, M.Á., Helgesson, R., Kortelainen, J., Moreno, G., Vázquez, C.: Towards categorical fuzzy logic programming. In: Masulli, F. (ed.) WILF 2013. LNCS, vol. 8256, pp. 109–121. Springer, Heidelberg (2013)
6. Eklund, P., Höhle, U., Kortelainen, J.: The fundamentals of lative logic. In: LINZ 2014, 35th Linz Seminar on Fuzzy Set Theory (abstract) (2014)
7. Gähler, W.: General topology – the monadic case, examples, applications. Acta Math. Hungar. 88, 279–290 (2000)
8. Galán, M.: Categorical Unification. PhD thesis, Umeå University, Department of Computing Science (2004)
9. Konečný, J., Medina, J., Ojeda-Aciego, M.: Multi-adjoint concept lattices with heterogeneous conjunctors and hedges. Annals of Mathematics and Artificial Intelligence (accepted, 2014)
10. Krídlo, O., Ojeda-Aciego, M.: Linking L-Chu correspondences and completely lattice L-ordered sets. Annals of Mathematics and Artificial Intelligence (accepted, 2014)
11. Mac Lane, S.: Categories for the Working Mathematician. Graduate Texts in Mathematics. Springer (September 1998)
12. Madrid, N., Medina, J., Moreno, J., Ojeda-Aciego, M.: New links between mathematical morphology and fuzzy property-oriented concept lattices. In: IEEE Intl Conf. on Fuzzy Systems, FUZZ-IEEE 2014 (accepted, 2014)
13. Medina, J.: Multi-adjoint property-oriented and object-oriented concept lattices. Information Sciences 190, 95–106 (2012)
14. Medina, J., Ojeda-Aciego, M., Ruiz-Calviño, J.: Formal concept analysis via multi-adjoint concept lattices. Fuzzy Sets and Systems 160(2), 130–144 (2009)
15. Wille, R.: Restructuring lattice theory: An approach based on hierarchies of concepts. In: Rival, I. (ed.) Ordered Sets. NATO Advanced Study Institutes Series, vol. 83, pp. 445–470. Springer, Heidelberg (1982)

On Adjunctions between Fuzzy Preordered Sets: Necessary Conditions*

Francisca García-Pardo, Inma P. Cabrera,
Pablo Cordero, and Manuel Ojeda-Aciego

Universidad de Málaga, Spain
{fgarciap,ipcabrera,pcordero,aciego}@uma.es

Abstract. There exists a direct relation between fuzzy rough sets and fuzzy preorders. On the other hand, it is well known the existing parallelism between Formal Concept Analysis and Rough Set Theory. In both cases, Galois connections play a central role. In this work, we focus on adjunctions (also named isotone Galois connections) between fuzzy preordered sets; specifically, we study necessary conditions that have to be fulfilled in order such an adjunction to exist.

Keywords: Galois connection, Adjunction, Preorder, Fuzzy sets.

1 Introduction

Adjunctions, together with their antitone counterparts (also called Galois connections), have played an important role in computer science because its many applications, both theoretical and practical, and in mathematics because of its ability to link apparently very disparate worlds; this is why Denecke, Erné, and Wismath stated in their monograph [12] that *Galois connections provide the structure-preserving passage between two worlds of our imagination.*

Finding an adjunction (or Galois connection) between two fields is extremely useful, since it provides a strong link between both theories allowing for mutual synergistic advantages. The algebraic study of complexity of valued constraints, for instance, has been studied in terms of establishing a Galois connection [10].

This work is focused on the study of adjunctions between fuzzy (pre-)ordered structures. Both research topics are related to, on the one hand, the theory of formal concept analysis (FCA) and, on the other hand, to rough set theory. For instance, in[22] Pawlak's information systems are studied in terms of Galois connections and functional dependencies; there are also papers which develop rough extensions of FCA by using rough Galois connections, see for instance [25]; there are works which study whether certain extensions of the upper and lower approximation operators form a Galois connection [11].

There is a number of papers which study Galois connections from the abstract algebraic standpoint [1,2,8,9,14,15,20] and also focusing on its applications [12,13,24,26,27,28,29]. In previous works [18,19], the authors studied the

---

* Partially supported by the Spanish Science Ministry projects TIN12-39353-C04-01 and TIN11-28084.

C. Cornelis et al. (eds.): RSCTC 2014, LNAI 8536, pp. 211–221, 2014.

problem of defining a right adjoint for a mapping $f\colon (A, \leq_A) \to B$ from a partially (pre)ordered set A to an unstructured set B. The natural extension of that approach is to consider a fuzzy preordered set (A, ρ_A).

In this paper, we start the study of conditions which guarantee the existence of adjunctions between sets with a fuzzy preorder. Specifically, we provide here a set of necessary conditions for an adjunction exists between (A, ρ_A) and (B, ρ_B).

2   Preliminary Definitions and Results

The most usual underlying structure for considering fuzzy extensions of Galois connections is that of residuated lattice, $\mathbb{L} = (L, \vee, \wedge, \top, \bot, \otimes, \to)$. An $\mathbb{L}$-fuzzy set is a mapping from the universe set to the membership values structure $X\colon U \to L$ where $X(u)$ means the degree in which u belongs to X. Given X and Y two $\mathbb{L}$-fuzzy sets, X is said to be included in Y, denoted as $X \subseteq Y$, if $X(u) \leq Y(u)$ for all $u \in U$.

An $\mathbb{L}$-fuzzy binary relation on U is an $\mathbb{L}$-fuzzy subset of $U \times U$, that is $\rho_U\colon U \times U \to L$, and it is said to be:

- *Reflexive* if $\rho_U(a, a) = \top$ for all $a \in U$.
- *Transitive* if $\rho_U(a, b) \otimes \rho_U(b, c) \leq \rho_U(a, c)$ for all $a, b, c \in U$.
- *Symmetric* if $\rho_U(a, b) = \rho_U(b, a)$ for all $a, b \in U$.
- *Antisymmetric* if $\rho_U(a, b) = \rho_U(b, a) = \top$ implies $a = b$, for all $a, b \in U$.

Definition 1 (Fuzzy poset)
An $\mathbb{L}$-fuzzy partially ordered set is a pair $\mathbb{U} = (U, \rho_U)$ in which ρ_U is a reflexive, antisymmetric and transitive $\mathbb{L}$-fuzzy relation on U.

A crisp ordering can be given in U by $a \leq_U b$ if and only if $\rho_U(a, b) = \top$.

From now on, when no confusion arises, we will omit the prefix "$\mathbb{L}$-".

Definition 2. *For every element $a \in U$, the extension to the fuzzy setting of the notions of upset and downset of the element a are defined by $a^\uparrow, a^\downarrow\colon U \to L$ where $a^\downarrow(u) = \rho_U(u, a)$ and $a^\uparrow(u) = \rho_U(a, u)$ for all $u \in U$.*

An element $a \in U$ is a maximum for a fuzzy set X if $X(a) = \top$ and $X \subseteq a^\downarrow$. The definition of minimum is similar.

Note that maximum and minimum elements are necessarily unique, because of antisymmetry.

Definition 3. *Let $\mathbb{A} = (A, \rho_A)$ and $\mathbb{B} = (B, \rho_B)$ be fuzzy ordered sets.*

1. *A mapping $f\colon A \to B$ is said to be* isotone *if $\rho_A(a_1, a_2) \leq \rho_B(f(a_1), f(a_2))$ for each $a_1, a_2 \in A$.*
2. *Moreover, a mapping $f\colon A \to A$ is said to be* inflationary *if $\rho_A(a, f(a)) = \top$ for all $a \in A$. Similarly. a mapping f is* deflationary *if $\rho_A(f(a), a) = \top$ for all $a \in A$.*

Definition 4 (Fuzzy adjunction). *Let* $\mathbb{A} = (A, \rho_A)$, $\mathbb{B} = (B, \rho_B)$ *be fuzzy posets, and two mappings* $f \colon A \to B$ *and* $g \colon B \to A$. *The pair* (f, g) *forms an adjunction between* A *and* B, *denoted* $(f, g) \colon \mathbb{A} \leftrightharpoons \mathbb{B}$ *if, for all* $a \in A$ *and* $b \in B$, *the equality* $\rho_A(a, g(b)) = \rho_B(f(a), b)$ *holds.*

Notation 1. *From now on, we will use the following notation, for a mapping* $f \colon A \to B$ *and a fuzzy subset* Y *of* B, *the fuzzy set* $f^{-1}(Y)$ *is defined as* $f^{-1}(Y)(a) = Y(f(a))$, *for all* $a \in A$.

Finally, we recall the following theorem which states different equivalent forms to define a fuzzy adjunction.

Theorem 1 ([16]). *Let* $\mathbb{A} = (A, \rho_A)$, $\mathbb{B} = (B, \rho_B)$ *be fuzzy posets, and two mappings* $f \colon A \to B$ *and* $g \colon B \to A$. *The following conditions are equivalent:*

1. $(f, g) \colon \mathbb{A} \leftrightharpoons \mathbb{B}$.
2. f *and* g *are isotone,* $g \circ f$ *is inflationary, and* $f \circ g$ *is deflationary.*
3. $f(a)^{\uparrow} = g^{-1}(a^{\uparrow})$ *for all* $a \in A$.
4. $g(b)^{\downarrow} = f^{-1}(b^{\downarrow})$ *for all* $b \in B$.
5. f *is isotone and* $g(b) = \max f^{-1}(b^{\downarrow})$ *for all* $b \in B$.
6. g *is isotone and* $f(a) = \min g^{-1}(a^{\uparrow})$ *for each* $a \in A$.

Theorem 2 ([17]). *Let* (A, ρ_A) *be a fuzzy poset and a mapping* $f \colon A \longrightarrow B$. *Let* A_f *be the quotient set over the kernel relation* $a \equiv_f b \iff f(a) = f(b)$. *Then, there exists a fuzzy order* ρ_B *in* B *and a map* $g \colon B \longrightarrow A$ *such that* $A \leftrightharpoons B$ *if and only if the following conditions hold:*

1. *There exists* $\max[a]_f$ *for all* $a \in A$.
2. $\rho_A(a_1, a_2) \leq \rho_A(\max[a_1]_f, \max[a_2]_f)$, *for all* $a_1, a_2 \in A$.

3   Building Adjunctions between Fuzzy Preordered Sets

In this section we start the generalization of Theorem 2 above to the framework of fuzzy preordered sets.

The construction will follow that given in [19] as much as possible. Therefore, we need to define a suitable fuzzy version of the p-kernel relation.

Firstly, we need to set the corresponding fuzzy notion of transitive closure of a fuzzy relation, and this is done in the definition below:

Definition 5. *Given a fuzzy relation* $S \colon U \times U \to L$, *for all* $n \in \mathbb{N}$, *the iterations* $S^n \colon U \times U \to L$ *are recursively defined by the base case* $S^1 = S$ *and, then,*

$$S^n(a, b) = \bigvee_{x \in U} \left(S^{n-1}(a, x) \otimes S(x, b) \right)$$

The transitive closure of S *is a fuzzy relation* $S^{tr} \colon U \times U \to L$ *defined by*

$$S^{tr}(a, b) = \bigvee_{n=1}^{\infty} S^n(a, b)$$

The relation $\approx_A$ allows for gettting rid of the absence of antisymmetry, by linking together elements which are 'almost coincident'; formally, the relation $\approx_A$ is defined on a fuzzy preordered set (A, ρ_A) as follows:

$$(a_1 \approx_A a_2) = \rho_A(a_1, a_2) \otimes \rho_A(a_2, a_1) \qquad \text{for } a_1, a_2 \in A$$

The kernel equivalence relation $\equiv_f$ associated to a mapping $f \colon A \to B$ is defined as follows for $a_1, a_2 \in A$:

$$(a_1 \equiv_f a_2) = \begin{cases} \perp & \text{if } f(a_1) \neq f(a_2) \\ \top & \text{if } f(a_1) = f(a_2) \end{cases}$$

Definition 6. *Let* $\mathbb{A} = (A, \rho_A)$ *be a fuzzy preordered set, and* $f \colon A \to B$ *a mapping. The* fuzzy p-kernel *relation* $\cong_A$ *is the fuzzy equivalence relation obtained as the transitive closure of the union of the relations* $\approx_A$ *and* $\equiv_f$.

Notice that the fuzzy equivalence classes $[a]_{\cong_A} : A \to L$ are fuzzy sets defined as

$$[a]_{\cong_A}(x) = (x \cong_A a) \tag{1}$$

The notion of maximum or minimum element of a fuzzy subset X of a fuzzy preordered set is the same as in Definition 2. There is an important difference which justifies the introduction of special terminology in this context: due to the absence of antisymmetry, there exists a crisp set of maxima (resp. minima) for X, which is not necessarily a singleton, which we will denote p-max(X) (resp., p-min(X)).

The following theorem states the different equivalent characterizations of the notion of adjunction between fuzzy preordered sets. As expected, the general structure of the definitions is preserved, but those concerning the actual definition of the adjoints have to be modified by using the notions of p-maximum and p-minimum.

Theorem 3 ([16]). *Let* $\mathbb{A} = (A, \rho_A), \mathbb{B} = (B, \rho_B)$ *be two fuzzy preordered sets, and* $f \colon \mathbb{A} \to \mathbb{B}$ *and* $g \colon \mathbb{B} \to \mathbb{A}$ *be two mappings. The following statements are equivalent:*

1. $(f, g) : \mathbb{A} \leftrightharpoons \mathbb{B}$.
2. f *and* g *are isotone, and* $g \circ f$ *is inflationary,* $f \circ g$ *is deflationary.*
3. $f(a)^\uparrow = g^{-1}(a^\uparrow)$ *for all* $a \in A$.
4. $g(b)^\downarrow = f^{-1}(b^\downarrow)$ *for all* $b \in B$.
5. f *is isotone and* $g(b) \in$ p-max $f^{-1}(b^\downarrow)$ *for all* $b \in B$.
6. g *is isotone and* $f(a) \in$ p-min $g^{-1}(a^\uparrow)$ *for all* $a \in A$.

The following definitions recall the notion of Hoare ordering between crisp subsets, and then introduces an alternative statement in the subsequent lemma:

Definition 7. *Consider a fuzzy preordered set* (A, ρ_A), *and* C, D *crisp subsets of* A, *we define the following relations*

$$- (C \sqsubseteq_W D) = \bigvee_{c \in C} \bigvee_{d \in D} \rho_A(c, d)$$

$$- (C \sqsubseteq_H D) = \bigwedge_{c \in C} \bigvee_{d \in D} \rho_A(c, d)$$

$$- (C \sqsubseteq_S D) = \bigwedge_{c \in C} \bigwedge_{d \in D} \rho_A(c, d)$$

Lemma 1. *Consider a fuzzy preordered set (A, ρ_A), and $X, Y \subseteq A$ such that* p-min $X \neq \varnothing \neq$ p-min Y, *then*

$$\big(\text{p-min } X \sqsubseteq_W \text{p-min } Y\big) = \big(\text{p-min } X \sqsubseteq_H \text{p-min } Y\big) = \big(\text{p-min } X \sqsubseteq_S \text{p-min } Y\big)$$

and their value coincides with $\rho_A(x, y)$ for any $x \in$ p-min X and $y \in$ p-min Y

Proof. Firstly, notice that if $u_1, u_2 \in$ p-min X, then $\rho_A(u_1, u_2) = \top$, by the definition of p-min X.

Secondly, $\rho_A(x_1, y_1) = \rho_A(x_2, y_2)$ for all $x_1, x_2 \in$ p-min X, $y_1, y_2 \in$ p-min Y. Indeed, $\rho_A(x_1, y_1) \geq \rho_A(x_1, x_2) \otimes \rho_A(x_2, y_1) = \top \otimes \rho_A(x_2, y_1) \geq \rho_A(x_2, y_2) \otimes \rho_A(y_2, y_1) = \rho_A(x_2, y_2)$. Analogously, $\rho_A(x_2, y_2) \geq \rho_A(x_1, y_1)$. □

We can now state the main contribution of this work: some necessary conditions for the existence of fuzzy adjunctions between fuzzy preordered sets. The result obtained resembles that in the crisp case [19]:

Theorem 4. *Given fuzzy preordered sets $\mathbb{A} = (A, \rho_A)$ and $\mathbb{B} = (B, \rho_B)$, and mappings $f : A \to B$ and $g : B \to A$ such that $(f, g) : \mathbb{A} \leftrightharpoons \mathbb{B}$ then*

1. $gf(A) \subseteq \bigcup_{a \in A} \text{p-max}[a]_{\cong_A}$

2. p-min$(UB[a]_{\cong_A} \cap gf(A)) \neq \varnothing$, *for all $a \in A$.*

3. $\rho_A(a_1, a_2) \leq \Big(\text{p-min}(UB[a_1]_{\cong_A} \cap gf(A)) \sqsubseteq \text{p-min}(UB[a_2]_{\cong_A} \cap gf(A))\Big)$ *for all $a_1, a_2 \in A$.*

Proof. 1. Consider $a \in A$, and let us show that $gf(a) \in$ p-max$[gf(a)]_{\cong_A}$.

By definition of p-maximum element of a fuzzy set, we have to prove that it is an element of its core, and also an upper bound. To begin with, it is straightforward that $[gf(a)]_{\cong_A}(gf(a)) = \top$, therefore we have just to prove the inclusion $[gf(a)]_{\cong_A} \subseteq (gf(a))^{\downarrow}$ between fuzzy sets, that is, we have to prove $[gf(a)]_{\cong_A}(u) \leq \rho_A(u, gf(a))$ for all $u \in A$.

Recall that relation $\cong_A$ has been defined as the transitive closure of the join $\approx_A \cup \equiv_f$, which we will denote R hereafter. Specifically, by using the definition of transitive closure (Defn. 5) and properties of the supremum, we will prove by induction that any iteration R^n satisfies the following inequality:

$$gf(a) R^n u \leq \rho_A(u, gf(a)) \qquad \forall u \in A \tag{2}$$

- For $n = 1$ and $u \in A$, let us prove the inequality by using the definition of the relations involved we obtain

$$
\begin{aligned}
gf(a)Ru &= (gf(a) \approx_A u) \vee (gf(a) \equiv_f u) \\
&= (\rho_A(gf(a), u) \otimes \rho_A(u, gf(a)) \vee (gf(a) \equiv_f u) \\
&\leq \rho_A(u, gf(a)) \vee (gf(a) \equiv_f u)
\end{aligned}
$$

Depending on the value of $gf(a) \equiv_f u$, which is a crisp relation, there are just two possible cases to consider, and both are straightforward:
If $(gf(a) \equiv_f u) = \bot$, there is nothing to prove, as the previous inequality collapses to inequality (2).
If $(gf(a) \equiv_f u) = \top$, inequality (2) degenerates to a tautology since the upper bound turns out to be $\top$. In effect, we have $fgf(a) = f(u)$ by definition of the kernel relation $\equiv_f$, in addition, using the hypothesis $(f, g) : \mathbb{A} \leftrightharpoons \mathbb{B}$, we have that

$$
\begin{aligned}
\rho_A(u, gf(a)) &= \rho_B(f(u), f(a)) \\
&= \rho_B(fgf(a), f(a)) = \rho_A(gf(a), gf(a)) = \top
\end{aligned}
$$

- Assume inequality (2) holds for $n - 1$. By definition of the n-th iteration of a fuzzy relation, and the induction hypothesis, we have that

$$
\begin{aligned}
gf(a)R^n u &= \bigvee_{x \in A} \left(gf(a)R^{n-1}x \otimes xRu \right) \\
&\leq \bigvee_{x \in A} \left(\rho_A(x, gf(a)) \otimes ((x \approx_A u) \vee (x \equiv_f u)) \right) \\
&= \bigvee_{x \in A} \left(\rho_A(x, gf(a)) \otimes ((\rho_A(x, u) \otimes \rho_A(u, x)) \vee (x \equiv_f u)) \right) \\
&\leq \bigvee_{x \in A} \left(\rho_A(x, gf(a)) \otimes (\rho_A(u, x) \vee (x \equiv_f u)) \right).
\end{aligned}
$$

Now, similarly to case $n = 1$, for every disjunct above there are two cases depending on the outcome of the kernel relation:
If $(x \equiv_f u) = \bot$, by commutativity of $\otimes$ and transitivity of ρ_A, then the corresponding disjunct simplifies to $\rho_A(u, gf(a))$.
If $(x \equiv_f u) = \top$, then the disjunct simplifies to $\rho_A(x, gf(a))$; but, moreover, using the fact that $f(x) = f(u)$ and the hypothesis $(f, g) : \mathbb{A} \leftrightharpoons \mathbb{B}$, we have that

$$
\rho_A(x, gf(a)) = \rho_B(f(x), f(a)) = \rho_B(f(u), f(a)) = \rho_A(u, gf(a))
$$

Summarizing, inequation (2) holds for all n and, by definition of the transitive closure, we have $[gf(a)]_{\approx_A}(u) \leq \rho_A(u, gf(a))$ for all $u \in A$.

2. Note that the set of upper bounds and the image involved in this condition are crisp sets. Specifically, we will prove that $gf(a)$ belongs to the intersection p-min$(UB[a]_{\approx_A} \cap g(f(A)))$.

To begin with, we have to check that $gf(a) \in UB[a]_{\cong_A} \cap gf(A)$. As it is obvious that $gf(a) \in gf(A)$, we have just to show $gf(a) \in UB[a]_{\cong_A}$, that is, $gf(a)$ is an upper bound of the fuzzy set $[a]_{\cong_A}$. We have to prove that $(a \cong_A u) \leq \rho_A(u, gf(a))$ holds for all $u \in A$. Again, by using the definition of $\cong_A$ as transitive closure, and properties of the supremum, it is sufficient to show that

$$aR^n u \leq \rho_A(u, gf(a)) \qquad \forall u \in A \tag{3}$$

From now on, the proof follows the line of the previous item.

- For $n = 1$, and $u \in A$, we have that

$$\begin{aligned} aRu &= (a \approx_A u) \vee (a \equiv_f u) \\ &= (\rho_A(a, u) \otimes \rho_A(u, a)) \vee (a \equiv_f u) \\ &\leq \rho_A(u, a) \vee (a \equiv_f u). \end{aligned}$$

Considering the two possible values of $a \equiv_f u$:
If $(a \equiv_f u) = \bot$, by monotonicity of f and the adjunction property, we have that

$$\rho_A(u, a) \leq \rho_B(f(u), f(a)) = \rho_A(u, gf(a)).$$

If $(a \equiv_f u) = \top$, inequality (3) once again degenerates to a tautology. Specifically, using $f(a) = f(u)$ and the adjunction property, we have

$$\rho_A(u, gf(a)) = \rho_B(f(u), f(a)) = \rho_B(f(a), f(a)) = \top$$

- Assume the inequality (3) holds for $n - 1$, and let us prove it for n. For this, consider $x \in A$,

$$\begin{aligned} aR^n u &= \bigvee_{x \in A} aR^{n-1}x \otimes xRu \\ &\leq \bigvee_{x \in A} \rho_A(x, gf(a)) \otimes ((x \approx_A u) \vee (x \equiv_f u)) \\ &= \bigvee_{x \in A} \rho_A(x, gf(a)) \otimes ((\rho_A(x, u) \otimes \rho_A(u, x)) \vee (x \equiv_f u)) \\ &\leq \bigvee_{x \in A} \rho_A(x, gf(a)) \otimes (\rho_A(u, x) \vee (x \equiv_f u)). \end{aligned}$$

Once again, we reason on each disjunct separately, considering the possible results of $x \equiv_f a$, and using monotonicity of f and the hypothesis $(f, g) : \mathbb{A} \leftrightarrows \mathbb{B}$ when necessary:
If $(x \equiv_f u) = \bot$, then the result follows by commutativity of $\otimes$ and transitivity of ρ_A.
If $(x \equiv_f u) = \top$, from $f(x) = f(u)$, then we have

$$\rho_A(x, gf(a)) = \rho_B(f(x), f(a)) = \rho_B(f(u), f(a)) = \rho_A(u, gf(a))$$

Summarizing, we have proved that $gf(a)$ is an upper bound of the fuzzy set $[a]_{\cong_A}$.

Finally, for the minimality, we have to check that $\rho_A(gf(a), x) = \top$ for all $x \in UB[a]_{\cong_A} \cap g(f(A)))$.

Consider $x \in UB[a]_{\cong_A} \cap g(f(A))$; then there exists $a_1 \in A$ such that $x = gf(a_1)$ and $(a \cong_A u) \leq \rho_A(u, x)$ for all $u \in A$. Particularly, considering $u = a$ and using the monotonicity of g and the adjunction property, we have that,

$$\top = (a \cong_A a) \leq \rho_A(a, x) = \rho_A(a, gf(a_1))$$
$$= \rho_B(f(a), f(a_1))$$
$$\leq \rho_A(gf(a), gf(a_1)) = \rho_A(gf(a), x).$$

3. Consider $a_1, a_2 \in A$, as f and g are isotone maps, then we have

$$\rho_A(a_1, a_2) \leq \rho_A(g(f(a_1)), g(f(a_2)))$$

From the inequality above, we directly obtain the required condition

$$\rho_A(a_1, a_2) \leq \left(\text{p-min}(UB[a_1]_{\cong_A} \cap g(f(A))) \sqsubseteq \text{p-min}(UB[a_2]_{\cong_A} \cap g(f(A))) \right)$$

since we have just proved above that $g(f(a)) \in \text{p-min}(UB[a]_{\cong_A} \cap gf(A))$ for all $a \in A$. $\square$

Corollary 1. *Let* $\mathbb{A} = (A, \rho_A)$ *be a fuzzy preordered set, let* B *be an unstructured set and* $f \colon A \to B$ *be a mapping. If* f *is the left adjoint for an adjunction, then there exists a subset* $S \subseteq A$ *such that*

(1) $S \subseteq \bigcup_{a \in A} \text{p-max}[a]_{\cong_A}$

(2) $\text{p-min}(UB[a]_{\cong_A} \cap S) \neq \varnothing$, *for all* $a \in A$.

(3) $\rho_A(a_1, a_2) \leq \left(\text{p-min}(UB[a_1]_{\cong_A} \cap S) \sqsubseteq \text{p-min}(UB[a_2]_{\cong_A} \cap S) \right)$ *for all* $a_1, a_2 \in A$.

It is worth to notice that the necessary conditions obtained above closely follow the characterizations one obtained in the crisp case for existence of adjunctions between preordered sets. Specifically, in [19], it was proved that given any (crisp) preordered set $\mathbb{A} = (A, \leq_A)$ and a mapping $f \colon \mathbb{A} \to B$, there exists a preorder $\mathbb{B} = (B, \leq_B)$ and $g \colon B \to A$ such that (f, g) forms a crisp adjunction between $\mathbb{A}$ and $\mathbb{B}$ *if and only if* there exists a subset S of A such that the following conditions hold:

(1) $S \subseteq \bigcup_{a \in A} \text{p-max}[a]_{\cong_A}$

(2) $\text{p-min}(UB[a]_{\cong_A} \cap S) \neq \varnothing$, for all $a \in A$.

(3) If $a_1 \leq_A a_2$, then $\left(\text{p-min}(UB[a_1]_{\cong_A} \cap S) \sqsubseteq \text{p-min}(UB[a_2]_{\cong_A} \cap S) \right)$, for $a_1, a_2 \in A$.

Obviously, although in this paper we have just proved one implication (the necessary conditions), as the obtained results are exactly the corresponding fuzzy translation of the crisp one, it seems likely that the converse should hold as well.

In order to provide some clue about the significance of the obtained conditions, it is worth to recall the characterization of the existence of adjunctions from a crisp poset to an unstructured set, which somehow unifies some well-known facts about adjunctions in a categorical sense, i.e. if g is a right adjoint then it preserves limits.

In [18] it was proved that given a poset $(A, \leq_A)$ and a map $f \colon A \to B$, there exists an ordering $\leq_B$ in B and a map $g \colon B \to A$ such that (f, g) is a crisp adjunction between posets from $(A, \leq_A)$ to $(B, \leq_B)$ if and only if

(i) There exists $\max([a]_{\equiv_f})$ for all $a \in A$.
(ii) $a_1 \leq_A a_2$ implies $\max([a_1]_{\equiv_f}) \leq_A \max([a_2]_{\equiv_f})$, for all $a_1, a_2 \in A$

where $\equiv_f$ is the kernel relation wrt f.

These two conditions are closely related to the different characterizations of the notion of adjunction, as stated in Theorem 1 (items 5 and 6); specifically, condition (i) above states that if $b \in B$ and $f(a) = b$, then necessarily $g(b) = \max([a]_{\equiv_f})$, whereas condition (ii) is related to the isotonicity of both f and g.

In some sense, the necessary conditions (1), (2), (3) obtained in Corollary 1 reflect the considerations given in the previous paragraph, but the different underlying ordered structure leads to a different formalization. Formally, condition (i) above is split into (1) and (2), since in a preordered setting, if $b \in B$ and $f(a) = b$, $g(b)$ needs not be in the same class as a but being maximum in its class (1). However, the latter condition is too weak and (2) provides exactly the remaining requirements needed in order to adequately reproduce the desired properties for g. Now, condition (3) it just the rephrasing of (ii) in terms of the properties described in (2).

4   Future Work

We have provided a set of necessary conditions for the existence of right adjunction to a mapping $f \colon (A, \rho_A) \to B$. The immediate future task is to study the other implication in order to find a set of necessary and sufficient conditions so that it is possible to define a fuzzy preorder on B such that f is a left adjoint.

Several papers on *fuzzy* Galois connections have been written since its introduction in [1]; consider for instance [3,14,23] for some recent ones. Another source of future work will be to study possible generalizations of the previously obtained results to the existence of fuzzy adjunctions within appropriate structures, and study the potential relationship to other approaches based on adequate versions of fuzzy closure systems [21].

Last but not least, in the recent years there has been some interesting developments on the study of both fuzzy partial orders and fuzzy preorders, see [4,5,6,7] for instance. In these works, it is noticed that versions antisymmetry and reflexivity commonly used are too strong and, as a consequence, the resulting fuzzy partial orders are very close to the classical case. Accordingly, another line of future work will be the adaptation of the current results to these alternative weaker definitions.

References

1. Bělohlávek, R.: Fuzzy Galois connections. Mathematical Logic Quarterly 45(4), 497–504 (1999)
2. Bělohlávek, R.: Lattices of fixed points of fuzzy Galois connections. Mathematical Logic Quartely 47(1), 111–116 (2001)
3. Bělohlávek, R., Osička, P.: Triadic fuzzy Galois connections as ordinary connections. In: IEEE Intl Conf. on Fuzzy Systems (2012)
4. Bodenhofer, U.: A similarity-based generalization of fuzzy orderings preserving the classical axioms. International Journal of Uncertainty, Fuzziness and Knowledge-Based Systems 8(5), 593–610 (2000)
5. Bodenhofer, U.: Representations and constructions of similarity-based fuzzy orderings. Fuzzy Sets and Systems 137(1), 113–136 (2003)
6. Bodenhofer, U., Klawonn, F.: A formal study of linearity axioms for fuzzy orderings. Fuzzy Sets and Systems 145(3), 323–354 (2004)
7. Bodenhofer, U., De Baets, B., Fodor, J.: A compendium of fuzzy weak orders: Representations and constructions. Fuzzy Sets and Systems 158(8), 811–829 (2007)
8. Börner, F.: Basics of Galois connections. In: Creignou, N., Kolaitis, P.G., Vollmer, H. (eds.) Complexity of Constraints. LNCS, vol. 5250, pp. 38–67. Springer, Heidelberg (2008)
9. Castellini, G., Koslowski, J., Strecker, G.E.: Categorical closure operators via Galois connections. Mathematical Research 67, 72–72 (1992)
10. Cohen, D., Creed, P., Jeavons, P., Živný, S.: An algebraic theory of complexity for valued constraints: Establishing a Galois connection. In: Murlak, F., Sankowski, P. (eds.) MFCS 2011. LNCS, vol. 6907, pp. 231–242. Springer, Heidelberg (2011)
11. Csajbók, Z., Mihálydeák, T.: Partial approximative set theory: generalization of the rough set theory. Intl J. of Computer Information Systems and Industrial Management Applications 4, 437–444 (2012)
12. Denecke, K., Erné, M., Wismath, S.L.: Galois connections and applications, vol. 565. Springer (2004)
13. Djouadi, Y., Prade, H.: Interval-valued fuzzy Galois connections: Algebraic requirements and concept lattice construction. Fundamenta Informaticae 99(2), 169–186 (2010)
14. Frascella, A.: Fuzzy Galois connections under weak conditions. Fuzzy Sets and Systems 172(1), 33–50 (2011)
15. García, J.G., Mardones-Pérez, I., de Prada-Vicente, M.A., Zhang, D.: Fuzzy Galois connections categorically. Math. Log. Q. 56(2), 131–147 (2010)
16. García-Pardo, F., Cabrera, I.P., Cordero, P., Ojeda-Aciego, M.: On Galois connections and Soft Computing. In: Rojas, I., Joya, G., Cabestany, J. (eds.) IWANN 2013, Part II. LNCS, vol. 7903, pp. 224–235. Springer, Heidelberg (2013)
17. García-Pardo, F., Cabrera, I.P., Cordero, P., Ojeda-Aciego, M.: On the construction of fuzzy Galois connections. In: Proc. of XVII Spanish Conference on Fuzzy Logic and Technology, pp. 99–102 (2014)

18. García-Pardo, F., Cabrera, I.P., Cordero, P., Ojeda-Aciego, M., Rodríguez, F.J.: Generating isotone Galois connections on an unstructured codomain. In: Proc. of Information Processing and Management of Uncertainty in Knowlegde-Based Systems (IPMU) (to appear, 2014)
19. García-Pardo, F., Cabrera, I.P., Cordero, P., Ojeda-Aciego, M., Rodríguez-Sanchez, F.J.: On the Existence of Isotone Galois Connections between Preorders. In: Glodeanu, C.V., Kaytoue, M., Sacarea, C. (eds.) ICFCA 2014. LNCS, vol. 8478, pp. 67–79. Springer, Heidelberg (2014)
20. Georgescu, G., Popescu, A.: Non-commutative fuzzy Galois connections. Soft Computing 7(7), 458–467 (2003)
21. Guo, L., Zhang, G.-Q., Li, Q.: Fuzzy closure systems on L-ordered sets. Mathematical Logic Quarterly 57(3), 281–291 (2011)
22. Järvinen, J.: Pawlak's information systems in terms of Galois connections and functional dependencies. Fundamenta Informaticae 75, 315–330 (2007)
23. Konecny, J.: Isotone fuzzy Galois connections with hedges. Information Sciences 181(10), 1804–1817 (2011)
24. Kuznetsov, S.: Galois connections in data analysis: Contributions from the soviet era and modern russian research. In: Ganter, B., Stumme, G., Wille, R. (eds.) Formal Concept Analysis. LNCS (LNAI), vol. 3626, pp. 196–225. Springer, Heidelberg (2005)
25. Li, F., Liu, Z.: Concewpt lattice based on the rough sets. Intl J. of Advanced Intelligence 1, 141–151 (2009)
26. Melton, A., Schmidt, D.A., Strecker, G.E.: Galois connections and computer science applications. In: Poigné, A., Pitt, D.H., Rydeheard, D.E., Abramsky, S. (eds.) Category Theory and Computer Programming. LNCS, vol. 240, pp. 299–312. Springer, Heidelberg (1986)
27. Mu, S.-C., Oliveira, J.: Programming from Galois connections. Journal of Logic and Algebraic Programming 81(6), 680–704 (2012)
28. Propp, J.: A Galois connection in the social network. Mathematics Magazine 85(1), 34–36 (2012)
29. Wolski, M.: Galois connections and data analysis. Fundamenta Informaticae 60, 401–415 (2004)

A New Consensus Tool in Digital Libraries

Francisco Javier Cabrerizo[1], Ignacio Javier Pérez[2],
Juan Antonio Morente-Molinera[3], Raquel Ureña[3],
and Enrique Herrera-Viedma[3]

[1] Dept. of Software Engineering and Computer Systems,
Distance Learning University of Spain (UNED), 28040 Madrid, Spain
`cabrerizo@issi.uned.es`
[2] Dept. of Computer Science and Engineering, University of Cádiz,
Cádiz 11002, Spain
`ignaciojavier.perez@uca.es`
[3] Dept. of Computer Science and Artificial Intelligence, University of Granada,
Granada 18071, Spain
`{jamoren,raquel,viedma}@decsai.ugr.es`

Abstract. Libraries represent a focal point of academic life and as such serve also a societal purpose of bringing together people around common themes. This purpose is nowadays enhanced and facilitated by the so-called digital libraries. For this reason, it is necessary to develop tools for helping users to reach decision with a high level of conensus in these virtual environments. The aim of this contribution is to present a tool for reaching consensus in order to minimize the main problems that these virtual environments presents (difficulty of establishing trust relations, low and intermittent participation rates, and so on) while incorporating the the benefits that they offers (rich and diverse knowledge due to a large number of users, real-time comunication, and so on). To do so, the fuzzy linguistic modelling is used to represent the users' opinions.

Keywords: Consensus, linguistic information, digital library.

1 Introduction

Libraries form an essential part of academic institutions, enabling and facilitating the exchange and growth of information, knowledge and culture among teachers, students and the general public [1]. This purpose is nowadays enhanced and facilitated by the use of technology and, in recent times, by the so-called digital libraries [2,3].

Digital libraries are the logical extension of physical libraries in an electronic information society, offering new levels of access to broader audiences of users. Digital libraries enable users to access human knowledge at anytime and anywhere, in a friendly multimodal way, by overcoming barriers of distancia, language and culture, and by using multiple network-connected devices. Therefore, the decisions about important issues in digital libraries have to be made by their own users.

C. Cornelis et al. (eds.): RSCTC 2014, LNAI 8536, pp. 222–231, 2014.

This situation can be seen as a group decision making problem. Group decision making (GDM) is a situation faced when individuals collectively make a choice from a suitable set of alternatives. This decision is no longer attributable to any single individual who is a member of the group. This is because all the individuals and social group processes such as social influence contribute to the outcome [4,5].

One the one hand, as the natural language is the standard representation of those concepts that humans use for communication, it seems natural that users use words instead of numerical values to provide their opinions. The *linguistic approach* is an approximate technique that represents qualitative aspects as linguistic values by means of *linguistic variables*, that is, variables whose values are not numbers but words or sentences in a natural or artificial language [6].

On the other hand, it is clear that involving a very large number of individuals in a decision process is a difficult task but, with the appearance of new electronic technologies, we are in the beginning of a new stage where traditional decision models may leave some space to a more direct participation of the "webizens". In fact, Web 2.0 represents a paradigm shift in how people use the Web as nowadays; everyone can actively contribute content on-line. However, the challenge is to develop more sophisticated Web 2.0 applications with better "participation architectures" that allow sharing data to their users, trusting users as co-developers, harnessing collective intelligence, etc., [7]. They should be able to overcome the inherent problems of the Web 2.0 Communities as [8,9]: (i) large user base, (ii) heterogeneity in the users, which present different backgrounds and use different expression domains, (iii) the low and intermittent participation rates, (iv) the dynamism of the Web 2.0 frameworks, e.g. the group of users could vary over time, and (v) difficulties of establishing trust relations.

In this paper we propose a new tool for reaching consensus in digital libraries by assuming fuzzy linguistic information to represent the users' opinions. As many traditional consensus approaches, it implements an iterative process in which the users of the digital library interact in order to reach a consensus solution on a particular problem. As digital libraries present several different inherent characteristics that are not present in usual decision making problems, this consensus tool incorporates some different modules to tackle them.

To do so, the paper is organized as follows. In Section 2 some of the most important characteristics of digital libraries and of group decision making problems under fuzzy linguistic preference relations are presented. In Section 3 the new linguistic consensus tool helping users of the digital libraries to obtain consensus solutions is described. Section 4 deals with a real world application of the tool. Finally, in Section 5 we point out our concluding remarks.

2 Preliminaries

In this section we present some important information about digital libraries and Web 2.0, and some generalities on GDM problems.

2.1   Digital Libraries and Web 2.0

The term Web 2.0 [10] was coined to describe the trends and business models that survived the technology sector market crash of the 1990s. He noted that the companies, which had survived the collapse, seemed to have some things in common: they were collaborative in nature, interactive, dynamic, and users created the content in these sites as much as they consumed it.

Developing the idea of Web 2.0 in the library context, the concept of digital library 2.0 emerges. Digital library 2.0 can be seen as a reaction from librarians to the increasingly relevant developments in information and communications technology (i.e., Web 2.0 and social software) and to an environment that is saturated with information available through more easily accessible channels. This reaction comes in the form of increased openness and trust toward library users, and in the development of new communication channels and services that are in tune with social developments. Digital library 2.0 has multiple facets reflecting the typical means of user participation that Web 2.0 enables. These facets include blogging, tagging, social bookmarking, social networking, podcasting and so on.

New Web 2.0 technologies have provided a new framework in which virtual communities can be created in order to collaborate, communicate, and share information and resources, and so on. This very recent kind of communities allows people from all over the globe to meet other individuals that share some of their interests. Among the different activities that the users of digital library 2.0 usually perform we can cite: (i) generate on-line contents and documents, which is greatly improved with the diversity and knowledge of the involved people, (ii) provide recommendations about different products and services, and (iii) participate in discussions and forums.

Apart from the obvious advantage of meeting new people with similar interests, digital libraries communities present some characteristics that make them different from other more usual kinds of organizations.:

- *Large user base.* Digital libraries communities usually have a large user base [8]. This can be seen from a double perspective. On the one hand, the total knowledge that a large user base implies is usually greater and more diverse than in a small community. This can be seen as a clear advantage: making decisions is usually performed better when there is a rich knowledge on the evaluated subject. On the other hand, managing a large and diverse amount of opinions in order to extract and use that knowledge might be a difficult task: for example, some of the users might not find easy to use typical numerical preference representation formats and thus, linguistic ones should be implemented.
- *Heterogeneous user base.* Not only the user base in digital library communities is large, but it is usually heterogeneous. This fact implies that we cannot easily assume that all the individuals may find easy to use the tools that are being developed and introduced in the websites. A clear example is the use of numerical ratings: some users may find difficult to express their

preferences about a set of alternatives using numerical ratings and thus, it may be interesting to provide tools which can deal with natural language or linguistic assessments.

– *Low participation and contribution rates.* Although many digital library communities have a quite large user base, many of those users do not directly participate in the community activities. Moreover, encouraging them to do so can be difficult [9]. Many of the users of a digital library community are mere spectators that make use of the produced resources but they do not (and is not willing to) contribute themselves with additional resources. This can be a serious issue when making decisions if only a small subset of the users contribute to a decision and it does not reflect the overall opinion of the community.

– *Intermittent contributions.* Partially due to the fast communication possibilities and due to a very diverse involvement of the different members, it is a common issue that some of them might not be able to collaborate during a whole decision process, but only in part of it. This phenomenon is well known in web communities: new members are continuously incorporated to the community and existing users leave it or temporarily cease in their contributions.

– *Real time communication.* The technologies that support digital library communities allow near real time communication among its members. This fact let us create models that in traditional scenarios would be quite impractical. For example, in a referendum, it is not easy at all to make a second round if there has been a problem in the first one due to the high amount of resources that it requires.

– *Difficulty of establishing trust relations.* As the main communication way, digital library communities use electronic devices and, in the majority of the cases, the members of the community do not know each other personally, it might be difficult to trust other members to, for example, delegate votes. This fact implies that it might be necessary to implement control mechanisms to avoid a malicious user taking advantage of others.

2.2   GDM Problems under Fuzzy Linguistic Information

A GDM situation consists of a problem to solve, a solution set of possible alternatives, $X = \{x_1, x_2, \ldots, x_n\}$, $(n \geq 2)$, and a group of two or more experts, $E = \{e_1, e_2, \ldots, e_m\}$, $(m \geq 2)$, who express their opinions about the set of alternatives to achieve a common solution.

One of the problems in this field is to find the best way to represent the information. There are situations in which the information cannot be assessed precisely in a quantitative form but may be in a qualitative one. For example, when attempting to qualify phenomena related to human perception, we are often led to use words in natural language instead of numerical values, e.g. when evaluating quality of a football player, terms like "good", "medium" or "bad" can be used.

The ordinal fuzzy linguistic approach [11] is a tool based on the concept of linguistic variable [12,13,14] to deal with qualitative assessments. It is a very useful kind of fuzzy linguistic approach because its use simplifies the processes of computing with words as well as linguistic representation aspects of problems. It is defined by considering a finite and totally ordered label set $S = \{s_i\}$, $i \in \{0, \ldots, g\}$, in the usual sense, i.e., $s_i \geq s_j$ if $i \geq j$, and with odd cardinality (usually 7 or 9 labels). The midterm represents an assessment of "approximately 0.5", and the rest of the terms are placed symmetrically around it. The semantics of the label set is established from the ordered structure of the label set by considering that each label for the pair (s_i, s_{g-i}) is equally informative [25]. For example, we can use the following set of seven labels to represent linguistic information: $S = \{s_0 = N, s_1 = VL, s_2 = L, s_3 = M, s_4 = H, s_5 = VH, s_6 = P\}$, where $N =$ Null, $VL =$ Very Low, $L =$ Low, $M =$ Medium, $H =$ High, VH = Very High and $P =$ Perfect. Using this approach, it is possible to define automatic and symbolic aggregation operators of linguistic information, as for example the LOWA operator [4].

In such a way, we assume that the experts give their preferences by using fuzzy linguistic preference relations (FLPR). A FLPR P^h given by an expert e_h is a fuzzy set defined on the product set $X \times X$, that is characterized by a linguistic membership function: $\mu P^h : X \times X \rightarrow S$, where the value $\mu P^h(x_i, x_j) = p_{ij}^h$ is interpreted as the linguistic preference degree of the alternative x_i over x_j for the expert e_h.

Usually, to solve a GDM problem, two processes are considered [4,15,16]:

- *Consensus process.* This process refers to how to obtain the maximum degree of agreement among the experts on the solution alternatives [5]. Usually, this process is guided by the figure of a moderator and it is carried out before the selection process. Clearly, it is preferable that the experts reach a high degree of consensus on the solution set of alternatives before obtaining the final solution.
- *Selection process.* This process describes how to obtain the solution set of alternatives from the opinions on the alternatives given by the experts. It consists of two phases: aggregation and exploitation. The aggregation phase defines a collective opinion according to the preferences provided by the experts. The exploitation phase transforms the global information about the alternatives into a global ranking.

Initially, in the consensus tool proposed in this contribution, we consider that in any nontrivial GDM problem the experts disagree in their opinions so that decision making has to be viewed as an iterative process composed by several discussion rounds, in which experts are expected to modify their preferences according to the advice given by the moderator. This means that agreement is obtained only after some rounds of consultation. In each round, we calculate the consensus measures and check the current agreement existing among experts.

Normally, to achieve consensus among the experts, it is necessary to provide the whole group of experts with some advice (feedback information) on how far the group is from consensus, what are the most controversial issues (alternatives), whose preferences are in the highest disagreement with the rest of the group, how their change would influence the consensus degree, and so on. In such a way, the moderator carries out three main tasks: (i) to compute the consensus measures, (ii) to check the level of agreement and (iii) to produce some advice for those experts that should change their minds.

3   A Tool for Reaching Consensus in Digital Libraries

There have been some attempts to model consensus processes that use new web environments to solve decision problems. For example, in [17] a web based consensus support system for GDM was presented. This system was prepared to be incorporated into GDM processes in which the experts would interact using a simple web platform. Therefore, it could not be used to deal with Web 2.0 decision frameworks in which we find large number of individuals with intermittent contributions and low participation rates. Recently, Alonso et al. [18] presented a theoretical model to overcome all these characteristic.

In this section we show a new tool that implements as a basis the theoretical model presented in [18]. It is specifically adapted to deal with digital library's features in order to increase the consensus level of the library users when making a decision on a set of alternatives. Some of the properties of the presented tool are:

- It does not require the existence of a moderator. It is the own application who acts as virtual moderator.
- It allows working in highly dynamical environments where participation and contribution rates change.
- It uses linguistic information to model user's preferences and trust relations.
- It allows weighting the contributions of each user according to some degree of expertise (staff, students and professors' opinions could have different weights).
- It implements a feedback module to help experts to change their preferences about the alternatives (the virtual moderator provides recommendations to join the expert's opinions).
- It provides a delegation scheme based on trust that allows minimizing communications and easing the computation of solutions.
- It implements a trust checking procedure to avoid some of the difficulties that the delegation scheme could introduce in the consensus reaching model.

Its operation implies the implementation of several different modules that are applied sequentially in each consensus round:

- *Initialization module.* This module serves as an entry point for the experts that are going to participate in the consensus process. Thus, this module presents the different alternatives $X = \{x1, \ldots, x_n\}$ in the problem to the experts. Once they know the feasible alternatives, each expert $e_h \in E$ is asked to provide a fuzzy linguistic preference relation P^h that represents his/her opinions about the alternatives.

- *Neighbors computation module.* For each participating expert e_h a set of neighbors (experts with similar opinions) is computed along with a global current preference relation. This information is presented to the experts. To calculate the neighborhoods we use a distance measure defined in [18]. The global current preference relation is computed by aggregating all the individual FLPRs using the LOWA operator [4].
- *Delegation module.* Another important mechanism that has been widely used in decision processes where lots of individuals are involved is delegation [19]. In fact, classical democratic systems rely on delegation in order to simplify the decision making processes: as not all the individuals are involved in the decision process (some of them delegate on others), the final decisions are usually achieved faster and in a simpler way. Therefore, this module allows each expert to delegate in other experts (presumably from his computed neighborhood, with similar opinions). It creates a kind of trust network that allows experts to leave (temporally or not) the decision process but maintaining part of his influence in the problem. This delegation mechanism is introduced to soften the intermittent contributions problem (because an expert who cannot continue the resolution process may choose to delegate to other experts instead of just leaving the process) and to decrease the number of preference relations involved in the problem.
- *Feedback module.* To ease the update of the preferences of the experts that have not delegated (in order to achieve a greater level of consensus) the system will provide several easy to follow feedback rules to the experts. The users will then update their preferences.
- *Consensus checking module.* The system will check the consensus status by computing different consensus measures [16]. If the consensus degree is high enough the consensus phase ends and the selection one is applied.
- *Trust checking module.* This module is carried out if the consensus measure is not high enough. It is introduced in order to avoid some of the problems that can be derived of the characteristics of digital library communities: the difficulties of establishing real trust relations. It is not difficult to imagine a scenario where some experts delegate to another that shares a common point of view on the decision that has to be made and, in a certain consensus round, this expert decides to drastically change his/her preferences, probably not reflecting the opinions of the experts that delegated to him anymore. To avoid this kind of situations this module will compare the last preference relation expressed by expert eh with the last preference relations of the experts that delegated to him/her (direct or indirectly). This comparison is made by applying a distance operator.

It is worth noting that, although the presented tool manages a series of more or less complex interactions, it is based on existing mechanisms (delegation and feedback reinforcement), which are used in real decision making problems. Moreover, the computations of all variables are quite straightforward and thus, the computational complexity of the tools is low. As it happens in real decision making problems, some of the modules of the tool as the delegation or change of preferences steps can slow down the resolution process. However, as those steps do work only during a fixed amount of time for each round this does not represent a noticeable problem. In addition, as the experts are not forced to provide their preferences, to delegate or even to change their opinions, those steps would not interfere with the whole resolution time, even if an expert "misses" one of the consensus iterations.

4   Application

Although the main goal of this contribution is to present a theoretical model which could be adapted to deal with similar GDM problems, in order to test and analyze the presented tool, in this section, we describe the use of the system and its applicability to one of the most important European digital library communities: the Spanish Open University (UNED) academic digital library.

With more than 260,000 scholars, UNED is, in terms of the number of students, the largest university of Spain and the second largest in Europe, next to the Open University in the UK. At this moment, the UNED library provides a wide range of 2.0 functionalities. Thus, it is a suitable scenario where our system can be applied perfectly.

In such a vast environment, where thousands of users interact it has been necessary to introduce new tools to avoid conflicts and increase the consensus of the main decisions. As the UNED digital library covers a large variety of controversial issues, this kind of tools can help to reach better decisions.

The consensus system proposed in this contribution suits some of those situations. For example, lets imagine a particular conflictive issue as the funding distribution problem. This problem can be modeled like a GDM problem in which the experts are the digital library users (students, professors, staff and so on), who has to compare the different alternatives to assign the money. We model the alternatives as the library resources (technology resources, science resources, medical resources, humanities resources and so on).

When we apply our tool, we allow choosing a solution of consensus among the alternatives in which:

- Every user that is willing to participate can do it (thus increasing the level of confidence in the final decision making).
- New users may incorporate in the middle of the consensus process.
- Participating users will not be forced to finish the consensus process, as they may choose to delegate to other users.
- Some users may have higher weight than others (for example, registered users, or library staff).
- The consensus status may be reached faster than using traditional discussion mechanism (due to the incorporation of the feedback mechanism).
- The preferences of the users are given in a linguistic way increasing their understandability.

Therefore, the final chosen option for the different resources would be a much justified one than if we apply some much more traditional models as direct and simple voting mechanisms. After using the proposed application, the system will show to the general manager the computed ranking of these resources in order to manage the funding distribution.

The mission of incorporating the new tool to an already constructed digital library web community as the UNED digital library is not a difficult one. In fact, it can be incorporated as a separate tab like polls and forums.

To do so, this tool incorporates two different interfaces, the first one for the library managers and the website administrator and a second one in which the rest of users could express their preferences. The former allows managers to define the problem (creating a description of what is the decision that has to be carried out), to define the available alternatives and other details as the linguistic set that is going to be used. On the other hand, the user interface, that allows them to provide their preferences or delegate, is implemented as a client that communicates with a server where all the required processes are carried out. Finally, the computed information is presented in the website and it is updated in real time as well as the resolution process is carried out.

5   Conclusions

In this contribution we have presented a new tool for reaching consensus that has been specially designed to be applied in digital libraries. This tool makes use of existing mechanism that are applied in real decision making situations: it uses fuzzy linguistic preference relations for the expression and management of experts preferences and it has been designed to manage a large users base by means of a delegation scheme. This delegation scheme is based in a particular kind of trust network created from linguistic trust evaluations given by the experts that simplifies the computations and the time needed to obtain the users preferences. Moreover, this delegation scheme also solves the intermittent contribution problem, which is present in almost any on-line community (that is, many of the users will not continuously collaborate but will do it from time to time). The model also incorporates a feedback mechanism to help the experts in changing their preferences in order to quickly obtain a high level of consensus.

The tool also allows incorporating new experts to the consensus process. It is able to handle some of the dynamic properties that real digital library communities have. Finally, it incorporates a trust check mechanism that allows detecting some abnormal situations in which an expert may try to take advantage of others by drastically changing his opinion and benefiting from the trust that the other experts might have deposited in him in previous consensus rounds.

Acknowledgments. This work has been developed with the financing of FEDER funds in FUZZYLING-II Project TIN2010-17876, the Andalusian Excellence Projects TIC-05299 and TIC-5991, and "Proyecto de Investigación del Plan de Promoción de la Investigación UNED 2011 (2011/PUNED/0003)".

References

1. Herrera-Viedma, E., López-Gijón, J.: Libraries' social role in the information age. Science 339(6126), 1382 (2013)
2. Heradio, R., Fernandez-Amoros, D., Cabrerizo, F.J., Herrera-Viedma, E.: A review of quality evaluation of digital libraries based on users perceptions. Journal of Information Science 38(3), 269–283 (2012)
3. Heradio, R., Cabrerizo, F.J., Fernandez-Amoros, D., Herrera, M., Herrera-Viedma, E.: A fuzzy linguistic model to evaluate the quality of Library 2.0 functionalities. International Journal of Information Management 33(4), 642–654 (2013)
4. Herrera, F., Herrera-Viedma, E., Verdegay, J.L.: Direct approach processes in group decision making using linguistic OWA operators. Fuzzy Sets and Systems 79, 175–190 (1996)
5. Herrera-Viedma, E., Cabrerizo, F.J., Kacprzyk, J., Pedrycz, W.: A review of soft consensus models in a fuzzy environment. Information Fusion 17, 4–13 (2014)
6. Roubens, M.: Fuzzy sets and decision analysis. Fuzzy Sets and Systems 90, 199–206 (1997)
7. Ling, K.: E-commerce technology: back to a prominent future. IEEE Internet Computing 12(1), 60–65 (2008)
8. Baym, N.: Cybersociety 2.0. In: The Emergence of On-line Community, pp. 35–68 (1998)
9. Ling, K., Beenen, G., Ludford, P., Wang, X., Chang, K., Li, X., Cosley, D., Frankowski, D., Terveen, L., Rashid, A.M., Resnick, P., Kraut, R.: Using social psychology to motivate contributions to online communities. Journal of Computer-Mediated Communication 10(4) (2005)
10. O'Reilly, T.: What is web 2.0: design patterns and business models for the next generation of software. Communications & Strategies 65, 17–37 (2007)
11. Herrera, F., Herrera-Viedma, E., Verdegay, J.L.: A model of consensus in group decision making under linguistic assessments. Fuzzy Sets and Systems 78(1), 73–87 (1996)
12. Zadeh, L.A.: The concept of a linguistic variable and its applications to approximate reasoning. Part I. Information Sciences 8(3), 199–249 (1975)
13. Zadeh, L.A.: The concept of a linguistic variable and its applications to approximate reasoning. Part II. Information Sciences 8(4), 301–357 (1975)
14. Zadeh, L.A.: The concept of a linguistic variable and its applications to approximate reasoning. Part III. Information Sciences 9(1), 43–80 (1975)
15. Kacprzyk, J., Fedrizzi, M.: Multiperson decision making models using fuzzy sets and possibility theory. Kluwer, Dortdrecht (1990)
16. Cabrerizo, F.J., Moreno, J.M., Perez, I.J., Herrera-Viedma, E.: Analyzing consensus approaches in fuzzy GDM: advantages and drawbacks. Soft Computing 14(5), 451–463 (2009)
17. Alonso, S., Herrera-Viedma, E., Chiclana, F., Herrera, F.: A web based consensus support system for group decision making problems and incomplete preferencess. Information Sciences 180(23), 4477–4495 (2010)
18. Alonso, S., Perez, I.J., Cabrerizo, F.J., Herrera-Viedma, E.: A linguistic consensus model for web 2.0 communities. Applied Soft Computing 13(1), 149–157 (2013)
19. Harstad, B.: Strategic delegation and voting rules. Journal of Public Economics 94(1-2), 102–113 (2010)

An Implementation of a Linguistic Multi-Criteria Decision Making Model: An Application to Tourism

Ramón Alberto Carrasco[1], María Francisca Blasco[1], and Enrique Herrera-Viedma[2]

[1] Department of Marketing and Market Research, Complutense University, Spain
[2] Department of Computer Science and Artificial Intelligence, University of Granada, Spain
ramoncar@ucm.es, fblasco@emp.ucm.es, viedma@decsai.ugr.es

Abstract. In tourism there are attempts to standardize the service quality evaluation, such as the SERVQUAL instrument, which is a five-item scale consisting of: tangibles, reliability, responsiveness, assurance and empathy. However, this scale is not commonly used in the most popular tourism websites. In this context, we present an implementation, using IBM SPSS Modeler, of a linguistic *multi-criteria decision making* model to integrate the hotel guests' opinions included in the WWW and expressed on other dimensions (or attributes) in order to obtain a SERVQUAL scale evaluation value of service quality. As a particular case study, we show an application example using TripAdvisor website.

1   Introduction

The fuzzy linguistic approach is a tool intended for modeling qualitative information in a problem. It is based on the concept of linguistic variable and has been satisfactorily used in *multi-criteria decision making* (MCDM) problems [3]. The 2-tuple fuzzy linguistic approach [2] is a model of information representation that carries out processes of "computing with words" without the loss of information.

The SERVQUAL scale [6] is a survey instrument which claims to measure the service quality in any type of service organization. In [1] we have presented a model to integrate the hotel guests' opinions included in several websites (and expressed in other different dimension or attributes) in order to get a SERVQUAL overall evaluation value of service quality by means two *linguistic multi-criteria decision making* (LMCDM) processes based on the 2-tuple fuzzy linguistic approach.

In this paper we present a general SPSS Modeler [5] implementation of this 2-tuple LMCDM model. This makes able to be widely applied at a practical level and not only at a theoretical one. SPSS Modeler© is a robust data mining software, produced by IBM©. It has a visual interface which allows users to leverage data mining algorithms without programming.

The rest of the paper is organized as follows: Section 2 revises the preliminary concepts, i.e. the 2-tuple linguistic modeling and the SERVQUAL scale. Section 3 presents the implementation of the LMCDM model using IBM SPSS Modeler. Section 4 shows an application example using a two-stage LMCDM model integrating customers' opinions collected from TripAdvisor [7] website. Finally, we point out some concluding remarks and future work.

C. Cornelis et al. (eds.): RSCTC 2014, LNAI 8536, pp. 232–239, 2014.

2   Preliminaries

2.1   The 2-Tuple Fuzzy Linguistic Approach

Let $S = \{s_0,\ldots,s_T\}$ be a linguistic term set with odd cardinality, where the mid-term represents a indifference value and the rest of terms are symmetric with respect to it. We assume that the semantics of labels is given by means of triangular membership functions and consider all terms distributed on a scale on which a total order is defined, i.e. $s_i \leq s_j \Leftrightarrow i < j$. In this fuzzy linguistic context, if a symbolic method aggregating linguistic information obtains a value $b \in [0,T]$, and $b \notin \{0,\ldots,T\}$, then an approximation function is used to express the result in S.

DEFINITION 1 [2]. *Let b be the result of an aggregation of the indexes of a set of labels assessed in a linguistic term set S, i.e. the result of a symbolic aggregation operation, $b \in [0,T]$. Let $i = round(b)$ and $\alpha = b - i$ be two values, such that $i \in [0,T]$ and $\alpha \in [-0.5,0.5)$, then α is called a Symbolic Translation.*

The 2-tuple fuzzy linguistic approach [2] is developed from the concept of symbolic translation by representing the linguistic information by means of 2-tuple (s_i, α_i), $s_i \in S$ and $\alpha_i \in [-0.5,0.5)$, where s_i represents the information linguistic label, and α_i is a numerical value expressing the value of the translation from the original result b to the closest index label, i, in the linguistic term set S. This model defines a set of transformation functions between numeric values and 2-tuple:

DEFINITION 2 [2]. *Let $S = \{s_1,\ldots,s_T\}$ be a linguistic term set and $b \in [0,T]$ a value representing the result of a symbolic aggregation operation, then the 2-tuple that expresses the equivalent information to b is obtained with the following function:*

$$\Delta: [0,T] \rightarrow S \times [-0.5,0.5)$$
$$\Delta(b) = (s_i, \alpha), \text{ with } s_i, \ i = round(b) \text{ and } \alpha = b - i, \alpha \in [-0.5,0.5) \tag{1}$$

where round(·) is the usual round operation, s_i has the closest index label to b and α is the value of the symbolic translation.

For all Δ, there exists Δ^{-1}, defined as $\Delta^{-1}(s_i, \alpha) = i + \alpha$. $\tag{2}$

Below, we describe the aggregation operators which we use in our model:

DEFINITION 3 [4]. *Let $A = \{(l_1, \alpha_1),\ldots, (l_n, \alpha_n)\}$ be a set of linguistic 2-tuple and $W = \{(w_1, \alpha_1^w),\ldots, (w_n, \alpha_n^w)\}$ be their linguistic 2-tuple associated weights. The 2-tuple linguistic weighted average $\bar{A}^w$ is:*

$$\bar{A}^w [((l_1, \alpha_1),(w_1, \alpha_1^w)),\ldots, ((l_n, \alpha_n),(w_n, \alpha_n^w))] = \Delta\left(\frac{\sum_{i=1}^{n} \beta_i \cdot \beta_{wi}}{\sum_{i=1}^{n} \beta_{wi}} \right), \tag{3}$$

with $\beta_i = \Delta^{-1}(l_i, \alpha_i)$ and $\beta_{wi} = \Delta^{-1}(w_i, \alpha_i^w)$.

2.2   The SERVQUAL Scale

Below, we will explain the five resultant scales proposed for SERVQUAL [6] and their adaptation to hotel guests' perceptions [1]:

- *Tangibles:* It makes reference to the appearance of the physical facilities, equipment, personnel, and communication materials.
- *Reliability:* This is the ability to perform the promised service dependably and accurately. Customers generally place heavy emphasis on the image, sanitary condition, safety and privacy of the hotel.
- *Responsiveness:* Willingness to help customers and provide prompt service. A courteous and friendly attitude by the service personnel makes the consumer feel respected, and definitely enhances the customer's appraisal of the hotel.
- *Assurance:* Knowledge and courtesy of employees and their ability to inspire trust and confidence. The price level is usually one of the most important factors that will influence the evaluation result by customers.
- *Empathy:* Caring and individualized attention that the firm provides its customers. If the hotel is located in a remote district, whether the hotel provides a tourist route suggestion, convenient traffic routes, or a shuttle bus to pick up customers will influence customers' desire to go to the hotel.

3   Implementing a LMCDM Using IBM SPSS Modeler

In a LMCDM [3] process, the goal consists in searching the best alternatives of the set $id\_alternative = \{id\_alternative_1,\ldots, id\_alternative_n\}$ according to the linguistic assessments provided by a group of experts, $id\_criterion = \{id\_criterion_1,\ldots, id\_criterion_m\}$ with respect to a set of evaluation criteria. In our model, we assume that these assessments are weighted by the self-rated expertise level set $weight = \{weight_1,\ldots, weight_m\}$. We also assume that we have p decision problems, one for each dimension (attribute or set of attributes of the problem domain) value. Therefore, we define the assessments set as $assessments = \{assessments_{dij}\}$ $\forall i, j, d, i \in \{1,\ldots, n\}$, $j \in \{1,\ldots, m\}, d \in \{1,\ldots, p\}$.

In order to obtain an easy linguistic interpretability and the high precision of the model results, we assume that all the information provided by the experts is in the 2-tuple form [2]. In our system to represent the 2-tuple values with a single attribute (string data type), we denote the pair (s_i, α), $s_i \in S$, with "$s_i\ sign(\alpha)\ abs(\alpha)$", e.g. $(s_0, -0.1)$ is denoted by the string "s_0 -0.1" and $(s_0, 0)$ by "s_0".

Our goal is to design a SPSS IBM Modeler Super Node that solves all the problems that are under the previous approach. These Super Nodes (symbolized by ☆) are similar to a procedure with inputs (labeled with *From Stream*, for example in Fig. 2) and/or outputs values (labeled with *To Stream,* see Fig. 2).

First, we need to store information about the symmetric and uniformly distributed domain $S = \{s_0,\ldots, s_T\}$, $T = 4$: $s_0 = Strongly\ Disagree = SD$, $s_1 = Disagree = D$, $s_2 = Neutral = N$, $s_3 = Agree = A$, and $s_4 = Strongly\ Agree = SA$. For this purpose we use a metadata called 2T-LABELS (see Table 1 and Fig. 1).

Table 1. Example of metadata 2T-LABELS

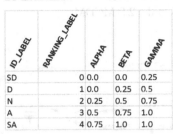

ID_LABEL	RANKING_LABEL	ALPHA	BETA	GAMMA
SD	0	0.0	0.0	0.25
D	1	0.0	0.25	0.5
N	2	0.25	0.5	0.75
A	3	0.5	0.75	1.0
SA	4	0.75	1.0	1.0

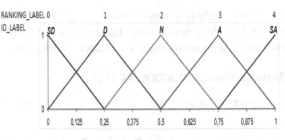

Fig. 1. Representation of the metadata showed in Table 1

Below we show implementations made to solve this problem:

Super Node 2T-MCDM (Fig. 2)

```
Inputs: ∀d ∈ {1,..., p}
```

$$
\begin{array}{c}
id\_criterion_1 \quad \cdots \quad id\_criterion_m \\
weight_1 \quad \cdots \quad weight_m
\end{array}
$$

$$
\begin{array}{c}
id\_alternative_1 \\
... \\
id\_alternative_n
\end{array}
\begin{bmatrix}
assessment_{d11} & \cdots & assessment_{dm1} \\
\vdots & \ddots & \vdots \\
assessment_{d1n} & \cdots & assessment_{dmn}
\end{bmatrix}
$$

```
Outputs: ∀d ∈ {1,..., p}
```

$$
\begin{array}{c}
id\_dimension_1 \quad \cdots \quad id\_dimension_p
\end{array}
$$

$$
\begin{array}{c}
id\_alternative_1 \\
... \\
id\_alternative_n
\end{array}
\begin{bmatrix}
performance\_value\_2t_{11} & \cdots & performance\_value\_2t_{p1} \\
\vdots & \ddots & \vdots \\
performance\_value\_2t_{1n} & \cdots & performance\_value\_2t_{pn}
\end{bmatrix}
$$

$$
\begin{array}{c}
id\_dimension_1 \quad \cdots \quad id\_dimension_p
\end{array}
$$

$$
\begin{array}{c}
id\_alternative_1 \\
... \\
id\_alternative_n
\end{array}
\begin{bmatrix}
consensus\_value\_2t_{11} & \cdots & consensus\_value\_2t_{p1} \\
\vdots & \ddots & \vdots \\
consensus\_value\_2t_{1n} & \cdots & consensus\_value\_2t_{pn}
\end{bmatrix}
$$

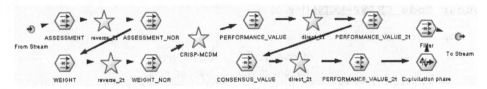

Fig. 2. 2T-MCDM: IBM SPSS Super Node for the MCDM calculation

This node from the 2-tuple assessment values (for each alternative and dimension) and the 2-tuple weight values (for each criterion) obtain the performance and consensus 2-tuple values (for each alternative and dimension) result of the LMCDM. The performance value is obtained using the 2-tuple linguistic weighted average showed in Eq. 3. The consensus value is obtained using the weighted Euclidean distance between this performance value and the corresponding assessments. The basic idea of this node is to convert the input 2-tuple values to crisp values and then perform a crisp MCDM process. After this process, the results are converted to 2-tuple values. Following, we explain the rest of Super Nodes contained in the 2T-MCDM:

Super Node reverse 2t (Fig. 3)

```
Input: label_2t. Output: reverse_label_2t
```

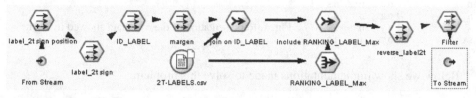

Fig. 3. reverse_2t: IBM SPSS Super Node for Δ^{-1} calculation

This node implements the Δ^{-1} function (Eq. 2), i.e. the output is $\Delta^{-1}(label\_2t)$. For this purpose the node uses the metadata 2T-LABELS showed in Table 1.

Super Node direct 2t (Fig. 4)

```
Input: reverse_label_2t. Output: label_2t
```

Fig. 4. direct_2t: IBM SPSS Super Node for Δ calculation

This node implements the Δ function (Eq. 1), i.e. the output is $\Delta(reverse\_label\_2t)$ using the metadata 2T-LABELS (see Table 1).

Super Node CRISP-MCDM (Fig. 5)

inputs: $\forall d \in \{1,\ldots, p\}$

$$
\begin{array}{c}
 \begin{array}{ccc} id\_criterion_1 & \ldots & id\_criterion_m \\ weight\_nor_1 & \ldots & weight\_nor_m \end{array} \\
\begin{array}{c} id\_alternative_1 \\ \ldots \\ id\_alternative_n \end{array}
\begin{bmatrix} assessment\_nor_{d11} & \cdots & assessment\_nor_{dm1} \\ \vdots & \ddots & \vdots \\ assessment\_nor_{d1n} & \cdots & assessment\_nor_{dmn} \end{bmatrix}
\end{array}
$$

outputs: $\forall d \in \{1,\ldots, p\}$

$$
\begin{array}{c}
 \begin{array}{ccc} id\_dimension_1 & \ldots & id\_dimension_p \end{array} \\
\begin{array}{c} id\_alternative_1 \\ \ldots \\ id\_alternative_n \end{array}
\begin{bmatrix} performance\_value_{11} & \cdots & performance\_value_{p1} \\ \vdots & \ddots & \vdots \\ performance\_value_{1n} & \cdots & performance\_value_{pn} \end{bmatrix}
\end{array}
$$

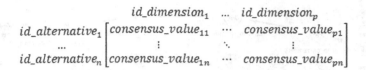

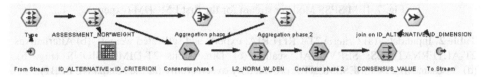

Fig. 5. CRISP-MCDM: IBM SPSS Super Node for a crisp MCDM calculation

This node performs a MCDM process which manages only crisp values. Decision processes are composed by two phases: *Aggregation* that combines the expert preferences. *Exploitation* that obtains a solution set of alternatives for the decision problem. Also it obtains the values of consensus using a Euclidean distance.

4   Application Example

In [1] we have presented a two-stage LMCDM to integrate the hotel guests' opinions included in several websites in order to get a SERVQUAL evaluation value of service quality. In this section, we present a SPSS IBM Modeler implementation of this model, using the Super Node presented in the previous section, with an application example using customers' opinions about the high-end hotels located in Granada (Spain), which were collected from TripAdvisor [7] during the year 2013. In this website customers write reviews on the following dimensions: *sleep quality*, *location*, *rooms*, *service*, *value* and *cleanliness* (see Fig. 6) using a linguistic five scale which can be modeled with $S = \{s_0, \ldots, s_T\}$, $T = 4$: $s_0 = Terrible = SD$, $s_1 = Poor = D$, $s_2 = Average = N$, $s_3 = Very Good = A$, and $s_4 = Excellent = SA$ (see Table and Fig. 1).

Fig. 6. Evaluation form of a hotel in TripAdvisor website

Following, we explain the implemented two-stage LMCDM model:

1. LMCDM process addressed by the information provided by hotel experts
In this step, we have counted on the collaboration of five experts. The objective is to obtain the linguistic importance of the input attributes for each SERVQUAL scale. The stream to solve this phase is very simple (Fig. 7). It collects the input information provided by experts (Table 2) and then executes the node 2T-MCDM obtaining the output data (Table 3) with an acceptable level of consensus among experts.

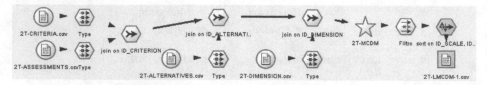

Fig. 7. IBM SPSS Modeler stream for the first LMCDM (experts)

Table 2. Input data: (a) Criteria 2T-CRITERIA: experts and self-rated weight. (b) Alternatives 2T-ALTERNATIVES: SERVQUAL scales. (c) Dimensions 2T-DIMENSIONS (Fig. 6). (d) Assessments 2T-ASSESSMENTS: provided by experts (only for the alternative PZB1).

ID_CRITERION	DES_CRITERION	WEIGHT
E1	EXPERT 1	SA
E2	EXPERT 2	SA
E3	EXPERT 3	SA
E4	EXPERT 4	SA
E5	EXPERT 5	A

(a)

ID_ALTERNATIVE	DES_ALTERNATIVE
PZB1	TANGIBLES
PZB2	RELIABILITY
PZB3	RESPONSIVENESS
PZB4	ASSURANCE
PZB5	EMPATHY

(b)

ID_DIMENSION	DES_DIMENSION
D1	Sleep quality
D2	Location
D3	Rooms
D4	Service
D5	Value
D6	Cleanliness

(c)

ID_DIMENSION	ID_ALTERNATIVE	ID_CRITERION	ASSESSMENT
D1	PZB1	E1	SA
D1	PZB1	E2	SA
D1	PZB1	E3	SA
D1	PZB1	E4	SA
D1	PZB1	E5	A
D2	PZB1	E1	SA
D2	PZB1	E2	SA
D2	PZB1	E3	SA
D2	PZB1	E4	SA
D2	PZB1	E5	A
D3	PZB1	E1	SA
D3	PZB1	E2	A
D3	PZB1	E3	SA
D3	PZB1	E4	SA
D3	PZB1	E5	SA
D6	PZB1	E1	SA
D6	PZB1	E2	A
D6	PZB1	E3	SA
D6	PZB1	E4	SA
D6	PZB1	E5	A

(d)

Table 3. Output data 2T-LMCDM-1: Importance of the dimensions (Fig. 6) for each scale

ID_SCALE	DES_SCALE	ID_DIMENSION	DES_DIMENSION	PERFORMANCE_VALUE_2t	CONSENSUS_VALUE_2t
PZB1	TANGIBLES	D1	Sleep quality	SA-0.039474	SA-0.080998
PZB1	TANGIBLES	D2	Location	SA-0.039474	SA-0.080998
PZB1	TANGIBLES	D3	Rooms	SA-0.052632	SA-0.101383
PZB1	TANGIBLES	D6	Cleanliness	SA-0.092105	SA-0.116442
PZB2	RELIABILITY	D3	Rooms	A-0.052632	SA-0.116442
PZB2	RELIABILITY	D4	Service	SA-0.092105	SA-0.116442
PZB2	RELIABILITY	D6	Cleanliness	A+0.052632	SA-0.101383
PZB3	RESPONSIVENESS	D1	Sleep quality	N+0.052632	SA-0.101383
PZB3	RESPONSIVENESS	D4	Service	A+0.052632	SA-0.101383
PZB4	ASSURANCE	D1	Sleep quality	A+0.052632	SA-0.101383
PZB4	ASSURANCE	D4	Service	A+0.052632	SA-0.101383
PZB4	ASSURANCE	D5	Value	SA-0.105263	SA-0.121647
PZB4	ASSURANCE	D6	Cleanliness	N+0.105263	A+0.047234
PZB5	EMPATHY	D4	Service	A+0.013158	A+0.108157

2. LMCDM process addressed by the information provided by hotel guests

In order to get the SERVQUAL evaluation value we use the linguistic importance for each SERVQUAL scale, obtained in the previous step (Table 3) playing the role of *weight* criteria. The alternatives are the SERVQUAL scales. In addition we also use the TripAdvisor hotel guests' (*id_criterion*) opinions from 2T-QUESTIONNAIRE as *assessments* (Table 4a) on the dimensions (*id_dimension*) showed in Fig. 6. The stream to solve this phase is showed in Fig. 8. It collects the input sources and then executes the Super Node 2T-MCDM obtaining the evaluation of the hotels (Table 4b).

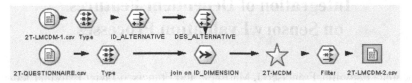

Fig. 8. IBM SPSS Modeler stream for the second LMCDM (hotel guests' opinions)

Table 4. (a) Input data 2T-QUESTIONNAIRE: TripAdvisor hotel guests' opinions (only a few rows). (b) Output data 2T-LMCDM-2: SERVQUAL evaluation for all hotels.

ID_OPINION	ID_DIMENSION	ASSESSMENT_DIMENSION
333	D4	A
334	D4	A
335	D4	A
336	D4	A
337	D4	D
338	D4	N
339	D4	SA

(a)

ID_ALTERNATIVE	DES_ALTERNATIVE	PERFORMANCE_VALUE_2t	CONSENSUS_VALUE_2t
PZB1	TANGIBLES	A+0.020467	A-0.023722
PZB2	RELIABILITY	A-0.005434	A-0.022190
PZB3	RESPONSIVENESS	A-0.016990	A-0.008856
PZB4	ASSURANCE	A+0.029214	A+0.000078
PZB5	EMPATHY	A-0.038889	A-0.033657

(b)

5   Concluding Remarks and Future Work

We have presented a general IBM SPSS Modeler [5] implementation of a 2-tuple LMCDM model. This makes able to be widely applied at a practical level on several types of problems. Thus, we have implemented the model [1] for integrating the opinions expressed by hotel guests in the TripAdvisor website [7], in order to obtain the overall value of service quality under the SERVQUAL instrument perspective. We are currently focusing on the development of an entire opinion aggregation architecture which will use the implementation presented in this paper.

Acknowledgements. This work has been developed with the financing of FEDER funds in FUZZYLING-II Project TIN2010-17876, the Andalusian Excellence Projects TIC-05299 and TIC-5991.

References

1. Carrasco, R.A., Villar, P., Hornos, M., Herrera-Viedma, E.: A Linguistic Multi-Criteria Decision Making Model Applied to Hotel Service Quality Evaluation from Web Data Sources. Int. J. Intell. Syst. 27(7), 704–731 (2012)
2. Herrera, F., Martínez, L.: A 2-tuple fuzzy linguistic representation model for computing with words. IEEE Transactions on Fuzzy Systems 8(6), 746–752 (2000)
3. Herrera, F., Herrera-Viedma, E.: Linguistic decision analysis: Steps for solving decision problems under linguistic information. Fuzzy Sets and Systems 115(10), 67–82 (2000)
4. Herrera-Viedma, E., Herrera, F., Chiclana, F., Luque, M.: Some issues on consistency of fuzzy preference relations. European Journal of Operational Research 154(1), 98–109 (2004)
5. IBM SPSS Modeler. URL (consulted 2014),
   http://www.ibm.com/software/analytics/spss/products/modeler/
6. Parasuraman, A., Zeithaml, V.A., Berry, L.L.: SERVQUAL: A multiple-item scale for measuring consumer perceptions of service quality. Journal of Retailing 64, 12–40 (1988)
7. TripAdvisor. Branded sites alone make up the most popular and largest travel community in the world. URL (consulted 2014), http://www.tripadvisor.es

Integration of Dependent Features
on Sensory Evaluation Processes

Macarena Espinilla, Francisco J. Martínez, and Francisco Javier Estrella Liébana

Department of Computer Science, University of Jaén, Jaén 23071, Spain
{mestevez}@ujaen.es

Abstract. The aim of a sensory evaluation process is to compute the global value of each evaluated product by means of an evaluator set, according to a set of sensory features. Several sensory evaluation models have been proposed which use classical aggregation operators to summary the sensory information, assuming independent sensory features, i.e, there is not interaction among them. However, the sensory information is perceived by the set of human senses and, depending on the evaluated product, its sensory features may be dependent and present interaction among them. In this contribution, we present the integration of dependent sensory features in sensory evaluation processes. To do so, we propose the use of the fuzzy measure in conjunction with the Choquet integral to deal with this dependence, extending a sensory evaluation model proposed in the literature. This sensory evaluation model has the advantage that offers linguistic terms to handle the uncertainty and imprecision involved in evaluation sensory processes. Finally, an illustrative example of a sensory evaluation process with dependent sensory features is shown.

Keywords: Sensory evaluation, decision analysis, sensory information, linguistic information, interaction, dependence.

1 Introduction

Evaluation processes are key in quality inspection, marketing and other fields in industrial companies. In these processes, it is very common that a group of evaluators assess a set of evaluated product, according to a set of criteria in order to obtain a global value of each evaluated product. To achieve this aim, some evaluation models are based on decision analysis methods due to the fact that these methods offer a simple and rational analysis that can be adapted in the evaluation context.

The sensory evaluation is an evaluation discipline that is carried out to evoke, measure, analyze, and interpret reactions of the sensory features of products [3]. This evaluation discipline has an important impact on many industrial areas such as comestibles, cosmetic and textile [16].

In the literature, several sensory evaluation models [6, 11–13, 15] have been proposed. These evaluation models assume that the multiple sensory features are completely independent, without presenting interaction among them. However, sensory features are perceived by the set of human senses *sight, smell, taste, touch* and *hearing* and, depending on the evaluated product, its sensory features may not be independent. For

C. Cornelis et al. (eds.): RSCTC 2014, LNAI 8536, pp. 240–249, 2014.

example, the *texture* and *appearance* are sensory features that are evaluated in clothing fabrics and there is a dependence between them. Another example are fruits where, usually, the sensory feature of *taste* can interact with other sensory features [16].

Therefore, in each sensory evaluation process is necessary to analyze each sensory feature and its relationships or dependence among them, i.e., to consider the interaction among sensory features. Furthermore, this interaction should be managed when the set of assessments is aggregated [10, 16] to obtain successful results that model the reality of sensory evaluation processes. Thereby, in order to manage the interaction among sensory features, sensory evaluation models should be extended.

In this contribution, we propose the use of fuzzy measures [18] in conjunction with fuzzy integrals [17] to capture the interaction among sensory features and to manage this interaction to compute the global value of each evaluated product in the sensory evaluation process. To do so, we propose the use of the Choquet integral [1] that is a fuzzy integral as well as a useful tool to model the dependence or interaction of criteria in several applications [4].

The information involved in sensory processes is perceived by the human senses and always involves imprecision and uncertainty that has a non-probabilistic nature [11]. For this reason, we propose to extend a sensory evaluation model that uses the fuzzy linguistic approach [21] to model and manage such an uncertainty by means of linguistic variables. The use of linguistic information in sensory evaluation processes involves Computing with Words (CWW) processes in which the objects of computation are words or sentences from a natural language and the results are also expressed in a linguistic expression domain [8]. Therefore, CWW processes are carried out in our proposal, considering the interaction among aggregated arguments. Finally, we show a case study to illustrate the usefulness and effectiveness of the fuzzy measures with the Choquet integral in a sensory evaluation process with linguistic information for fruit jam samples.

The rest of the contribution is organized as follows: Section 2 reviews the CWW processes in the context of dependent aggregation as well as the linguistic sensory evaluation model that will be extended. In Section 3, the integration of dependent features in the linguistic sensory evaluation model is presented. In Section 4, an illustrative example of the extended linguistic evaluation model is shown. In the last section, we give the conclusions.

2   Preliminaries

In this section, CWW processes with dependent arguments are reviewed by means of the 2-tuple linguistic model, which is used to represent the linguistic information in the linguistic sensory evaluation model that will be reviewed later.

2.1   CWW with Presence of Dependence

In this section, we review the 2-tuple linguistic model and dependent aggregation operators for linguistic 2-tuples, these will be used in our proposal to capture the interaction among sensory features and to carry out CWW processes, considering such interaction.

2-Tuple Linguistic Model. The 2-tuple linguistic model has been successfully applied in different fields such as sustainable energy [5], recommender systems [14], quality of service [7], performance appraisal [4], etc. This model represents the information by means of a pair of values (s, α), where s is a linguistic term with syntax and semantics and α is a numerical value that represents the value of the *symbolic translation*. The symbolic translation is a numerical value assessed in $[-0.5, 0.5)$ that supports the difference of information between a counting of information β assessed in the interval of granularity $[0, g]$ of the linguistic term set S and the closest value in $S = \{s_0, \ldots, s_g\}$ which indicates the index of the closest linguistic term in S.

This model defined a set of functions to facilitate the computational processes with linguistic 2-tuples [9].

Definition 1 *[9]. Let $S = \{s_0, \ldots, s_g\}$ be a set of linguistic terms. The 2-tuple set associated with S is defined as $\langle S \rangle = S \times [-0.5, 0.5)$. The function $\Delta_S : [0, g] \longrightarrow \langle S \rangle$ is given by,*

$$\Delta_S(\beta) = (s_i, \alpha), \quad with \quad \begin{cases} i = round\,(\beta), \\ \alpha = \beta - i, \end{cases} \tag{1}$$

where $round(\cdot)$ assigns to β the integer number $i \in \{0, 1, \ldots, g\}$ closest to β.

Proposition 1 *Let $S = \{s_0, \ldots, s_g\}$ be a linguistic term set and (s_i, α) be a linguistic 2-tuple. There is always a function $\Delta_S^{-1} = i + \alpha$ such that, from a linguistic 2-tuple, it returns its equivalent numerical value $\beta \in [0, g]$.*

Remark 1 *The conversion of a linguistic term into linguistic 2-tuple consists of adding a value 0 as symbolic translation. $H : S \rightarrow \langle S \rangle$ that allows us to transform a linguistic term s_i into a linguistic 2-tuple $(s_i, 0)$.*

The 2-tuple linguistic representation model has a linguistic computational model associated based on Δ_S^{-1} and Δ_S that accomplishes CWW processes in a precise way. Different 2-tuple linguistic aggregation operators have been proposed [9] that consist of obtaining a linguistic 2-tuple value that summarizes a set of linguistic 2-tuples. The 2-tuple ordered weighted averaging (OWA) aggregation operator that will be used in our case study is defined as follows:

Definition 2 *[9] Let $S = \{(s_1, \alpha_1), (s_2, \alpha_2), \ldots, (s_n, \alpha_n)\}$ be a set of 2-tuples and $w = (w_1, w_2, \ldots, w_n) \in [0, 1]^n$ be the weighting vector of S such that $\sum_{i=1}^{n} w_i = 1$. The 2-tuple ordered weighted averaging aggregation operators for linguistic 2-tuples is defined as:*

$$2TOWA_w(S) = \Delta_S \left(\sum_{i=1}^{n} w_i \Delta_S^{-1}(s_{\sigma(i)}, \alpha_{\sigma(i)}) \right),$$

with σ a permutation on $\{1, \ldots, n\}$ such that $(s_{\sigma(1)}, \alpha_{\sigma(1)}) \geq \ldots \geq (s_{\sigma(n)}, \alpha_{\sigma(n)})$.

Dependent Aggregation Operators. Choquet integral-based aggregation operators [1] consider the dependence of the aggregated arguments in order to deal with the interaction among them. These aggregation operators require fuzzy measures [18] in order to represent the interaction among arguments. Following, the fuzzy measures and the Choquet integral for linguistic 2-tuples are defined.

Definition 3 *[18]. Let* $N = \{1, \ldots, m\}$ *be a set of* m *arguments. A* fuzzy measure *is a set function* $\mu : 2^N \to [0,1]$ *that satisfies the following conditions:* $\mu(\emptyset) = 0$, $\mu(N) = 1$ *and* $\mu(S) \leq \mu(T)$ *whenever* $S \subseteq T$ *(*μ *is monotonic)*

To represent set functions, for a small m, it is convenient to arrange their values into an array. For example, the fuzzy measure for $m = 3$ is represented as follows:

$$\begin{array}{ccc} & \mu(f_1, f_2, f_3) & \\ \mu(f_1, f_2) & \mu(f_1, f_3) & \mu(f_2, f_3) \\ \mu(f_1) & \mu(f_2) & \mu(f_3) \\ & \mu(\emptyset) & \end{array} \tag{2}$$

Definition 4 *[20]. Let* μ *be a fuzzy measures on* $X = \{x_1, x_2, \ldots, x_n\}$ *and a set of linguistic 2-tuples* $S = \{(s_1, \alpha_1), (s_2, \alpha_2), \ldots, (s_n, \alpha_n)\}$. *The 2-tuple Choquet integral* $(2TCI)$ *for linguistic 2-tuples is defined as:*

$$2TCI_\mu(S) = \Delta_S \left(\sum_{i=1}^{n} w_i \Delta_S^{-1}(s_{\sigma(i)}, \alpha_{\sigma(i)}) \right)$$

where $w_i = \mu(H_{\sigma(i)}) - \mu(H_{\sigma(i-1)})$, *with* σ *a permutation on* $\{1, \ldots, n\}$ *such that* $(s_{\sigma(1)}, \alpha_{\sigma(1)}) \geq (s_{\sigma(2)}, \alpha_{\sigma(2)}) \geq \ldots \geq (s_{\sigma(n)}, \alpha_{\sigma(n)})$ *and* $x_{\sigma(i)}$ *is the attribute corresponding to* $(s_{\sigma(i)}, \alpha_{\sigma(i)})$. *With the convention* $H_{\sigma(0)} = \emptyset$ *and* $H_{\sigma(i)} = \{x_{\sigma(1)}, \ldots, x_{\sigma(i)}\}$, *for* $i \geq 1$. *Obviously* $w_i \geq 0$ *and* $\sum_{i=1}^{n} w_i = 1$.

By using the Choquet integral, in [20] some aggregation operators for linguistic 2-tuples was introduced, including the 2-tuple correlated averaging operator, the 2-tuple correlated geometric operator and the generalized 2-tuple correlated averaging operator.

2.2   Linguistic Sensory Evaluation Model

In this section, we briefly review the linguistic sensory model based on linguistic 2-tuples [11] that offers linguistic terms to handle the uncertainty and imprecision involved in evaluation sensory processes, providing linguistic results.

The linguistic sensory evaluation model adapts the common decision resolution scheme proposed in [2] and consists of three phases (see Figure 1) that are reviewed in the following subsections.

Evaluation Framework. This phase defines the structure and the set of elements in the sensory evaluation process that these are:

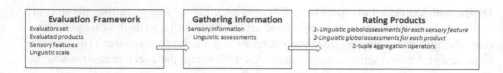

Fig. 1. Linguistic sensory evaluation model with independent sensory features

- $C = \{c_k : k = 1, ..., m\}$ is the evaluator panel.
- $X = \{x_j : j = 1, ..., n\}$ is the set of evaluated products.
- $F = \{f_i : i = 1, ..., h\}$ is the set of sensory features that identify each evaluated product.
- $S = \{s_l : l = 0, ..., g\}$ is the linguistic scale in which evaluators' assessments will be expressed.

Gathering Information. In this phase, each evaluator $c_k \in C$ expresses his/her assessment value of each evaluated product $x_j \in X$ by means of a linguistic assessment vector: $U_j^k = \{u_{1j}^k, u_{2j}^k, \ldots, u_{nj}^k\}$. This linguistic information is transformed into linguistic 2-tuples, using the *Remark* 1.

Rating Products. The aim of the sensory evaluation process is to compute a global value of the set of evaluated products, according to the sensory information gathered in the previous phase. A key issue in this process is to carry out CWW processes, aggregating the sensory information in an appropriate way by means of aggregation operators for linguistic 2-tuples. To do so, this phase consists of two steps.

1. *Computing a global value for each sensory feature*: first, it is computed a global linguistic 2-tuple, (u_{jk}, α), for each sensory feature f_k, of the evaluated product x_j, using an aggregation operator for linguistic 2-tuples $2TAO_1$.

$$(u_{jk}, \alpha) = 2TAO_1((u_{jk}^1, \alpha_1), \ldots, (u_{jk}^n, \alpha_n))$$

2. *Computing a global value for each evaluated product*: the final aim of the rating process is to obtain a global value, (u_j, α), for each evaluated product, x_j according to its global values for the set of sensory features. To do so, this step aggregates the global linguistic 2-tuple values for each feature sensory, (u_{jk}, α), using an aggregation operator for linguistic 2-tuples $2TAO_2$, assuming that the set of sensory features is independent.

$$(u_j, \alpha) = 2TAO_2((u_{j1}, \alpha_1), \ldots, (u_{jh}, \alpha_h))$$

Until now, sensory evaluation models have been used classical aggregation operators for linguistic 2-tuples as arithmetic average, weighted average or median average [6, 11–13] to compute the global value for each evaluated product, without assuming the interaction among sensory features.

3   Integration of Dependent Features on Sensory Evaluation Process

In this section, we present the management of dependent sensory features in sensory evaluation processes. To do so, we extend the reviewed linguistic sensory evaluation model in Section 2.2 in order to capture and model the interaction among sensory features and compute global assessments, taking into account this interaction.

In order to achieve the aim of this contribution, the fuzzy measures and the Choquet Integral are used to manage the interaction among sensory features. So, these will be used in the phase of *Evaluation Framework* as well as in the step of computing global value for each evaluated product of *Rating Products*. These phases are described bellow and illustrated in Figure 2 in the extended linguistic sensory evaluation model

Fig. 2. Extended linguistic sensory evaluation model to manage dependent sensory features

Extending the Evaluation Framework. In the evaluation framework, it is necessary to analyze each sensory feature and its relationships or depende among them. To do so, it is necessary to define in the evaluation framework the fuzzy measure associated with the set of sensory features $F = \{f_i : i = 1, ..., h\}$:

- $\mu : 2^F \longrightarrow [0, 1]$ are the fuzzy measures that represent the dependence among the set of sensory features.

In order to clarify the use of the fuzzy measures in the sensory evaluation process, three examples are illustrated, considering a sensory evaluation process in which three sensory features are evaluated: $F = \{f_1, f_2, f_3\}$.

Example 1. *Let μ be the fuzzy measure on F given by Eq. (3), these fuzzy measures represent interaction among f_1 and f_2 due to the fact that $\mu(f_1, f_2) = 0.6 > \mu(f_1) + \mu(f_2) = 0.5$. The sensory feature f_3 is independent respect to f_1 and f_2 because $\mu(f_1, f_3) = \mu(f_1) + \mu(f_3) = 0.7$ and $\mu(f_2, f_3) = \mu(f_2) + \mu(f_3) = 0.8$.*

$$
\begin{matrix}
1 \\
0.6 \ 0.7 \ 0.8 \\
0.2 \ 0.3 \ 0.5 \\
0
\end{matrix}
\tag{3}
$$

Example 2. *Let μ be the fuzzy measure on F given by Eq. (4) that is a symmetric additive fuzzy measure, these sensory features are independent and have the same* *weight in the evaluation process due to the fact that the same cardinalities of the corresponding subsets have the same value in* μ: $\mu(f_1) = \mu(f_2) = \mu(f_3) = 1/3$ *and* $\mu(f_1, f_2) = \mu(f_1, f_3) = \mu(f_2, f_3) = 2/3$.

$$
\begin{array}{ccc}
& 1 & \\
2/3 & 2/3 & 2/3 \\
1/3 & 1/3 & 1/3 \\
& 0 &
\end{array}
\tag{4}
$$

Example 3. *Let* μ *be the fuzzy measure on F given by Eq. (5) that is an additive fuzzy measure, these features sensory are independent, there is not interaction among them. However, the sensory features not have the same weight in the set because* $\mu(f_3) = 0.5 > \mu(f_2) = 0.3 > \mu(f_3) = 0.2$.

$$
\begin{array}{ccc}
& 1 & \\
0.5 & 0.7 & 0.8 \\
0.2 & 0.3 & 0.5 \\
& 0 &
\end{array}
\tag{5}
$$

Extending the Rating Products. In this phase, it is carried out CWW processes, aggregating the sensory information and considering the interaction among sensory features. To do so, the second step is extended in order to compute the linguistic global value (u_j, α), for each evaluated product, according to the interaction as well as the global values of the set of sensory features computed previously in the firs step.

- *Computing a global value for each evaluated product*: In order to manage the interaction among sensory features when the set of global values are aggregated, it is necessary to use an aggregation operator that can deal with the fuzzy measure defined in the evaluation framework. Therefore, *Choquet integral-based aggregation operators* for linguistic 2-tuples, which consider the fuzzy measure, are used to manage the interaction among sensory features and to aggregate the linguistic information.

$$(u_j, \alpha) = 2TAO_\mu((u_{j1}, \alpha_1), ..., (u_{jh}, \alpha_h))$$

It is noteworthy that the interaction of the set of sensory features is captured by fuzzy measures. Therefore, if the fuzzy measure does not capture the interaction, the Choquet integral does not consider such interaction. Thereby, depending on the properties of fuzzy measures, the Choquet Integral can include classical aggregation operators, for example *weighted means*. So, the Choquet integral with respect to an additive fuzzy measure μ is the weighted arithmetic mean with the weights $w_i = \mu(i)$. With respect to a symmetric additive fuzzy measure, the Choquet integral is the arithmetic mean and the values of μ are given by $\mu(A) = |A|/n$.

4   Dependent Sensory Features of Fruit Jam. Case Study

In this section, we present a case study to illustrate the usefulness and effectiveness of the integration of dependent sensory features in the sensory evaluation process of fruit jam samples.

4.1 Evaluation Framework

The evaluation framework includes a set of twenty evaluators, $C = \{c_1, \ldots, c_{20}\}$, who assess two samples of fruit jam, $X = \{x_1, x_2\}$. Each fruit jam sample is characterized by three sensory features $f = \{f_1, f_2, f_3\}$ which are respectively: *taste, smell* and *texture*. In this case study, evaluators express their assessments about the set of sensory features, using the linguistic term set S that is illustrated in the Figure 3.

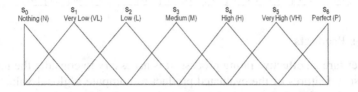

Fig. 3. Linguistic term set

Furthermore, an expert in the company provides the interaction among the set of sensory features by means of fuzzy measures that are shown in Eq. (6):

$$
\begin{array}{l}
1 \\
0.6\ 0.5\ 0.4 \\
0.2\ 0.2\ 0.2 \\
0
\end{array}
\tag{6}
$$

In this sensory evaluation process, the three independent sensory features are associated with the same weight $\mu(f_1) = 0.2$, $\mu(f_2) = 0.2$ and $\mu(f_3) = 0.2$. However, the company establishes that the sensory feature of *taste* is more important in coalition with the other two sensory features, *smell* and *texture*. Due to the fact that $\mu(f_1, f_2) = 0.6 > \mu(f_1) + \mu(f_2) = 0.4$ and $\mu(f_1, f_3) = 0.5 > \mu(f_1) + \mu(f_3) = 0.4$. Furthermore, it is more important the interaction among *taste* and *smell* than *taste* and *texture* because $\mu(f_1, f_2) = 0.6 > \mu(f_1, f_3) = 0.5$. Finally, the sensory features of *smell* and *taste* are independent, since their fuzzy measures are additive, $\mu(f_2, f_3) = 0.4 = \mu(f_2) + \mu(f_3) = 0.4$.

4.2 Gathering Information

Once the evaluation framework has been defined in the previous phase, the information must be gathered. The evaluator set provides their assessments by using linguistic assessment vectors. Therefore, each evaluator c_k provides his/her assessments about the evaluated product x_j according to each sensory feature f_i.

In this case study, the evaluator set provides all assessments, but if the provided information is incomplete, some decision-making models could be used to manage it. The linguistic information gathered by the evaluator set is shown in Table 1, this information is transformed into linguistic 2-tuples, using the *Remark 1*.

Table 1. Assessments about x_1 and x_2 provided by the evaluators

x_1	c_1	c_2	c_3	c_4	c_5	c_6	c_7	c_8	c_9	c_{10}	c_{11}	c_{12}	c_{13}	c_{14}	c_{15}	c_{16}	c_{17}	c_{18}	c_{19}	c_{20}
f_1	VH	P	P	P	P	H	H	H	H	H	L	M	M	M	M	VL	VL	VL	VL	L
f_2	P	P	P	P	P	VH	VH	VH	P	P	VH	VH	VH	VH	VH	L	VH	VH	VH	VH
f_3	P	P	P	P	P	P	P	P	P	P	VH	VH	VH	VH	VH	L	VH	VH	VH	VH

x_2	c_1	c_2	c_3	c_4	c_5	c_6	c_7	c_8	c_9	c_{10}	c_{11}	c_{12}	c_{13}	c_{14}	c_{15}	c_{16}	c_{17}	c_{18}	c_{19}	c_{20}
f_1	L	M	M	M	H	H	H	VH	VH	VH	N	N	VL	VL	VL	VL	VL	L	L	L
f_2	H	VH	VH	P	P	P	P	P	P	P	L	L	M	M	M	M	M	M	H	H
f_3	VH	VH	VH	P	P	P	P	P	P	P	VL	VL	L	L	M	H	H	H	VH	VH

4.3 Rating Products

In order to ensure an effective rating process, it is necessary to consider the interaction among sensory features of the evaluated product to compute a global value for each evaluated product. The description of the rating process is described as follows:

Computing a Global Value for Each Sensory Feature. In the first step of this process, the *2-tuple OWA operator* is applied, which requires an *weighting vector*. In this case study, the linguistic quantifier *"Most"* [19] is used to obtain the weight vector that is: $w = (0, 0, 0, 0, 0, .1, .1, .1, .1, .1, .1, .1, .1, .1, .1, .1, 0, 0, 0, 0)$. Global values for each sensory feature are shown in Table 2.

Table 2. Global values for each sensory feature

	x_1	x_2
f_1	$(s_3, -0.2) = (M, -0.2)$	$(s_2, -0.4) = (L, -0.4)$
f_2	$(s_5, 0.06) = (VH, 0.06)$	$(s_4, -0.03) = (H, -0.03)$
f_3	$(s_5, 0.26) = (VH, 0.26)$	$(s_4, 0.26) = (H, 0.26)$

Computing a Global Value for Each Product. Aggregating the global values for each sensory feature of each sample x_j is obtained its global value. Considering the interaction among sensory features, the Choquet Integral is used with the fuzzy measure defined in the evaluation framework. The computed linguistic global assessments are $x_1 = (s_4, 0.3) = (H, 0.3)$ and $x_2 = (s_3, 0.04) = (H, 0.04)$.

5 Conclusions

In this paper, we have presented the integration of dependent sensory features in a linguistic sensory evaluation model to capture the interaction among sensory features and to compute the global assessment of each evaluated product, considering such interaction. To do so, we have proposed the use of fuzzy measure to model the dependence among sensory features and the Choquet integral aggregation operator for linguistic 2-tuples to obtain the linguistic global assessment of each evaluated product. Finally, an illustrative example of a sensory evaluation process with interaction among some sensory features has been presented.

Acknowledgments. This contribution has been supported by research projects TIN2012-31263 and AGR-6487.

References

1. Choquet, G.: Theory of capacities, vol. 5, pp. 131–295. Annales de l'institut Fourier (1953)
2. Clemen, R.T.: Making Hard Decisions. An Introduction to Decision Analisys. Duxbury Press (1995)
3. Dijksterhuis, G.B.: Multivariate Data Analysis in Sensory and Consumer Science, Food and Nutrition. Press Inc., Trumbull (1997)
4. Espinilla, M., de Andrés, R., Martínez, F.J., Martínez, L.: A 360-degree performance appraisal model dealing with heterogeneous information and dependent criteria. Information Sciences 222, 459–471 (2013)
5. Espinilla, M., Palomares, I., Martínez, L., Ruan, D.: A comparative study of heterogeneous decision analysis approaches applied to sustainable energy evaluation. International Journal on Uncertainty, Fuzziness and Knowledge-Based Systems 20(suppl. 1), 159–174 (2012)
6. Estrella, F.J., Espinilla, M., Martínez, L.: Fuzzy linguistic olive oil sensory evaluation model based on unbalanced linguistic scales. Journal of Multiple-Valued Logic and Soft Computing 22, 501–520 (2014)
7. Gramajo, S., Martínez, L.: A linguistic decision support model for QoS priorities in networking. Knowledge-Based Systems 32(1), 65–75 (2012)
8. Herrera, F., Alonso, S., Chiclana, F., Herrera-Viedma, E.: Computing with words in decision making: foundations, trends and prospects. Fuzzy Optimization and Decision Making 8, 337–364 (2009)
9. Herrera, F., Martínez, L.: A 2-tuple fuzzy linguistic representation model for computing with words. IEEE Transactions on Fuzzy Systems 8(6), 746–752 (2000)
10. Marichal, J.-L.: An axiomatic approach of the discrete choquet integral as a tool to aggregate interacting criteria. IEEE Transactions on Fuzzy Systems 8(6), 800–807 (2000)
11. Martínez, L.: Sensory evaluation based on linguistic decision analysis. International Journal of Approximate Reasoning 44(2), 148–164 (2007)
12. Martínez, L., Espinilla, M., Liu, J., Pérez, L.G., Sánchez, P.J.: An evaluation model with unbalanced linguistic information applied to olive oil sensory evaluation. Journal of Multiple-Valued Logic and Soft Computing 15(2-3), 229–251 (2009)
13. Martínez, L., Espinilla, M., Pérez, L.G.: A linguistic multigranular sensory evaluation model for olive oil. International Journal of Computational Intelligence Systems 1(2), 148–158 (2008)
14. Rodríguez, R.M., Espinilla, M., Sánchez, P.J., Martínez, L.: Using linguistic incomplete preference relations to cold start recommendations. Internet Research 20(3), 296–315 (2010)
15. Ruan, D., Zeng, X. (eds.): Intelligent Sensory Evaluation: Methodologies and Applications. Springer (2004)
16. Stone, H., Sidel, J.L.: Sensory Evaluation Practice. Academic Press Inc., San Diego (1993)
17. Torra, V., Narukawa, Y.: Modeling decisions: Information fusion and aggregation operators. Cognitive Technologies 13 (2007)
18. Wang, Z., Klir, G.: Fuzzy measure theory. Plenum Press, New York (1992)
19. Yager, R.R.: On ordered weighted averaging operators in multicriteria decision making. IEEE Transactions on Systems, Man, and Cybernetics 18, 183–190 (1988)
20. Yang, W., Chen, Z.: New aggregation operators based on the choquet integral and 2-tuple linguistic information. Expert Systems with Applications 39(3), 2662–2668 (2012)
21. Zadeh, L.A.: The concept of a linguistic variable and its applications to approximate reasoning. Information Sciences, Part I, II, III, 8,8,9:199–249,301–357,43–80 (1975)

Aiding in the Treatment of Low Back Pain by a Fuzzy Linguistic Web System

Bernabé Esteban[1], Álvaro Tejeda-Lorente[2], Carlos Porcel[3],
José Antonio Moral-Muñoz[4], and Enrique Herrera-Viedma[5]

[1] Dept. of Physical Therapy, University of Granada, Granada, Spain, 18071
[2] Dept. of Computer Science and Artificial Intelligence, University of Granada, Granada, Spain, 18071
[3] Dept. of Computer Science, University of Jaen, Jaen, Spain, 23071
[4] Dept. of Librarianship and Information Science, University of Granada, Granada, Spain, 18071
[5] Dept. of Computer Science and Artificial Intelligence, University of Granada, Granada Spain, 18071
{bernabe,jamoral}@ugr.es, {atejeda,viedma}@decsai.ugr.es,
cporcel@ujaen.es

Abstract. Low back pain affects a large proportion of the adult population at some point in their lives and has a major economic and social impact. To soften this impact, one possible solution is to make use of recommender systems, which have already been introduced in several health fields. In this paper, we present TPLUFIB-WEB, a novel fuzzy linguistic Web system that uses a recommender system to provide personalized exercises to patients with low back pain problems and to offer recommendations for their prevention. This system may be useful to reduce the economic impact of low back pain, help professionals to assist patients, and inform users on low back pain prevention measures. A strong part of TPLUFIB-WEB is that it satisfies the Web quality standards proposed by the Health On the Net Foundation (HON), Official College of Physicians of Barcelona, and Health Quality Agency of the Andalusian Regional Government, endorsing the health information provided and warranting the trust of users.

Keywords: Low back pain, health care, recommender systems, fuzzy linguistic modeling, quality evaluation.

1 Introduction

Low back pain is a painful and economically costly syndrome that affects two-thirds of adults in developed societies at some point in their lives [1]. It is almost always a self-limiting episode of pain, with a tendency to spontaneous and complete improvement, although there is frequently a transition from acute to chronic disease [2]. Low back pain has an enormous social and economic impact and is a leading cause of absenteeism in all professions. Physical exercise has proven effective to protect against low back pain and promote recovery from processes that can transform into chronic pain, reducing the number of days off work and helping in the treatment of psychological components of this condition [3].

C. Cornelis et al. (eds.): RSCTC 2014, LNAI 8536, pp. 250–261, 2014.
© Springer International Publishing Switzerland 2014

Recently developed Information and Communication Technology (ICT) applications in healthcare have demonstrated potential for addressing different challenges, including: the development of personalized medicine, i.e., the tailoring of medical decisions, practices, and/or products to individual patients [4], the reduction of healthcare costs [5], and the universalization of health, i.e., the accessibility of care to all citizens, regardless of their resources or place of residence [6]. *Recommender Systems* (RSs) are one ICT application that may be useful in the healthcare field [7]. RSs assist users in their decision making processes to find relevant information. They could be seen as a Decision Support System, where the solution alternatives are the information items to be recommended and the criteria to satisfy are the user preferences and needs [8,9]. RSs offer a personalized approach, because each user can be treated in a different way. They may be useful in the diagnosis of chronic disease, offering a prediction of the disease risk to support the selection of appropriate medical advice for patients [10]. Thus, in the field of physiotherapy, RSs may help to achieve an effective personalization of recommended exercises.

The aim of this article is to present a fuzzy linguistic Web system, designated TPLUFIB-WEB[1], for individuals with low back pain, providing them with appropriate exercises and information. The major innovations and contributions of the system include:

1. The provision of personalized exercises by using a recommender system.
2. The ability to use it in any place and at any time, yielding savings in travel and staffing costs.
3. Its user-friendly nature, designed for individuals with minimal skills and using fuzzy linguistic modeling to improve the representation of user preferences and facilitate user-system interactions [11,12].
4. The reliability of the information offered and the selection of exercises, endorsed by a team of experts in physiotherapy from the School of Health Sciences of the University of Granada. We emphasize that the aim was not to develop new exercises or treatments for low back pain but rather to incorporate clinically validated proposals [3,13], including preventive strategies, in a Web tool to facilitate their use by individuals at any time anywhere.

The utilization of the Internet to seek medical information has increased sharply over recent years. Figure 1 shows the Web search interest in "low back pain" worldwide since 2004 according to the "Google Trends" tool[2]. The maximum search interest is scored as 100, and the interest was 70 by June 2013. As depicted in Figure 2, the search interest in the Spanish term "*lumbalgia*" in the same month was also very high (90).

The number of physiotherapists per 100 000 inhabitants in Spain is low in comparison to other European countries (see the report listed at: http://www.pordata. pt/en/Europe/Physiotherapists+per+100+thousand+inhabitants-1925), supporting the need for complementary tele-rehabilitation systems to assess low back

---

[1] Accessible in: http://sci2s.ugr.es/sapluweb/

[2] http://www.google.com/trends/

pain. The enormous number of health recommendations available on the Web is cause of concern to the user, who needs to be sure of their provenance and reliability. For this reason, measures were taken to guarantee the quality and reliability of the data in our Web system. Thus, TPLUFIB-WEB satisfies the requirements of the World Wide Web Consortium (W3C) Web Accessibility Initiative[3] and of health accreditation bodies, i.e., the Health On the Net Foundation (HON)[4], Official College of Physicians of Barcelona[5] and Health Quality Agency of the Andalusian Regional Government[6].

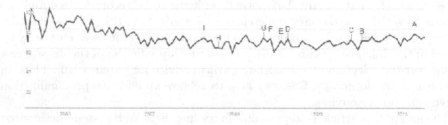

Fig. 1. Google trends for "low back pain"

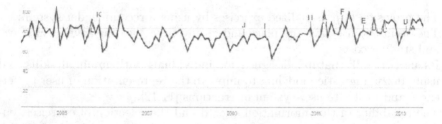

Fig. 2. Google trends for "lumbalgia"

The paper is organized as follows: Section 2 describes preliminary information pertaining to recomender systems and the fuzzy linguistic modeling; Section 3 presents the new Web system, TPLUFIB-WEB; Section 4 addresses the validation of the system, and section 5 offers conclusions based on the study findings.

2   Preliminaries

2.1   Recommender Systems

RSs are systems that produce individualized recommendations as output or have the effect of guiding the user in a personalized manner towards appropriate tasks

---

[3] http://www.w3.org
[4] http://www.healthonnet.org/
[5] http://wma.comb.es/es/home.php
[6] http://www.juntadeandalucia.es/agenciadecalidadsanitaria

among a wide range of possible options [7]. In order to provide personalized recommendations, system requires knowledge about users, such as ratings provided
of already explored items [7,14]. To maintain available this knowledge it implies
system should keep also user profiles that contain also users preferences and ne-
cessities. Nevertheless, the way system acquires this information depends on the
recommendation scheme used. The system could obtain the information about
users either in an *implicit* way, that is analyzing their behavior, or *explicitly*
requiring user to specify their preferences.

One of the most popular method used to obtain recommendations is the collaborative approach [7]. In this approach the recommendations for a user are
based on the ratings provided by other users similar to this user. Another method
more simple but not less important is the content-based approach [7]. This approach recommends items to a user by matching the content of the item and the
user's past experience with similar items, ignoring data from other users. These
aproaches work with the set of historic ratings, which are provided by the users
when they experience an item or update a previous rating.

Each technique has its advantages and disadvantages, according to the setting. However, a hybrid approach can also be adopted to compensate for their
weaknesses and benefit from their strengths [7,15,9].

2.2   Fuzzy Linguistic Modeling

The fuzzy linguistic modeling is a tool based on the concept of *linguistic variable*
[16] which has given very good results for modeling qualitative information in
many problems in which quantitative information can not be assessed precisely
[17].

The 2-tuple FLM [18] is a continuous model of representation of information
that allows to reduce the loss of information typical of other fuzzy linguistic
approaches (classical and ordinal, see [16]).

Let $S = \{s_0, ..., s_g\}$ be a linguistic term set with odd cardinality. We assume
that the semantics of labels is given by means of triangular membership functions and consider all terms distributed on a scale on which a total order is
defined. In this fuzzy linguistic context, if a symbolic method aggregating lin-
guistic information obtains a value $\beta \in [0, g]$, and $\beta \notin \{0, ..., g\}$, β is represented
by means of 2-tuples (s_i, α_i), where $s_i \in S$ represents the linguistic label of the
information, and α_i is a numerical value expressing the value of the translation
from the original result β to the closest index label, i, in the linguistic term set
$(s_i \in S)$.

This model defines a set of transformation functions between numeric values
and 2-tuples $\Delta(\beta) = (s_i, \alpha)$ and $\Delta^{-1}(s_i, \alpha) = \beta \in [0, g]$ [18].

In order to establish the computational model a negation, comparison and
aggregation operators are defined. Using functions Δ and Δ^{-1} that transform,
without loss of information, numerical values into linguistic 2-tuples and vicev-
ersa, any of the existing aggregation operators (i.e. arithmetic mean, weighted
average operator or linguistic weighted average operator) can be easily extended
for dealing with linguistic 2-tuples [18].

In any fuzzy linguistic approach, an important parameter to determine is the "granularity of uncertainty", i.e., the cardinality of the linguistic term set S. When different experts have different uncertainty degrees on the phenomenon or when an expert has to assess different concepts, then several linguistic term sets with a different granularity of uncertainty are necessary [19]. In [19] a multi-granular 2-tuple FLM based on the concept of linguistic hierarchy is proposed.

A *Linguistic Hierarchy, LH,* is a set of levels $l(t,n(t))$, where each level t is a linguistic term set with different granularity $n(t)$ from the remaining of levels of the hierarchy. The levels are ordered according to their granularity, i.e., a level $t+1$ provides a linguistic refinement of the previous level t. We can define a level from its predecessor level as: $l(t,n(t)) \to l(t+1, 2 \cdot n(t) - 1)$. A graphical example of a linguistic hierarchy is shown in Figure 3. Using this LH, the linguistic terms in each level are the following:

- $S^3 = \{a_0 = Null = N, a_1 = Medium = M, a_2 = Total = T\}$.
- $S^5 = \{b_0 = None = N, b_1 = Low = L, b_2 = Medium = M, b_3 = High = H, b_4 = Total = T\}$
- $S^9 = \{c_0 = None = N, c_1 = Very\_Low = VL, c_2 = Low = L, c_3 = More\_Less\_Low = MLL, c_4 = Medium = M, c_5 = More\_Less\_High = MLH,$
  $c_6 = High = H, c_7 = Very\_High - VH, c_8 = Total = T\}$

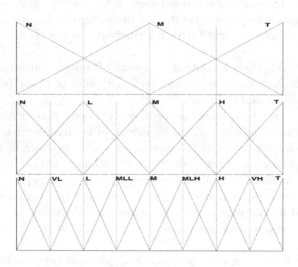

Fig. 3. Linguistic Hierarchy of 3, 5 and 9 labels

In [19] a family of transformation functions between labels from different levels was introduced. This family of transformation functions is bijective. This result guarantees that the transformations between levels of a linguistic hierarchy are carried out without loss of information.

3   TPLUFIB-WEB: A Web System to Help in the Treatment of Low Back Pain Problems

TPLUFIB-WEB is accessible at: http://sci2s.ugr.es/sapluweb/. Figure 4 shows the system structure. The system structure has three main components (see Figure 4):

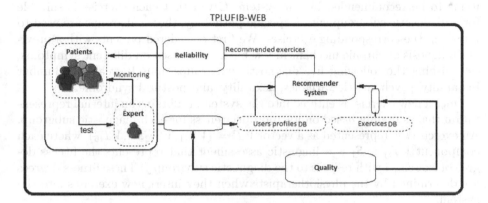

Fig. 4. Operating scheme

1. *A multimedia database of exercises* for recommendation to patients according to their pathology.
2. *A database of patient profiles* that stores the characteristics of each patient, not only the internal representation of their diagnostics but also their personal evaluations obtained after user-system interaction.
3. *A personalized method for generating exercise recommendations* that implements the hybrid recommendation policy based on information from the multimedia and patient profile databases.

Different sets of linguistic labels $(S_1, S_2, ...)$ are used to represent the different concepts necessary for the system activity; we selected the LH presented in section 2.2. The different concepts assessed in the system are the following:

- The *membership degree* of patient diseases with respect to each of the defined diagnostic subgroups, which is labeled in S_1.
- The predicted *degree of relevance* of exercise for a patient, which is labeled in S_2.
- The *degree of similarity* between the diseases of two patients or between exercises, which is labeled in S_3.
- The *degree of satisfaction* with a recommended exercise expressed by a patient, which is labeled in S_4.

We use the set with 5 labels to represent the degrees of membership and satisfaction $(S_1 = S^5$ and $S_4 = S^5)$ and 9 labels to represent the degrees of predicted relevance $(S_2 = S^9)$ and similarity $(S_3 = S^9)$.

3.1   Multimedia Database

A multimedia database was developed that contained exercises for all possible pathologies. Exercises can be exchanged among different subgroups in the construction of a customized program for each patient. Instruction videos were recorded for reproduction on computers and mobile devices. It is very important to obtain an adequate representation of exercises, because these are the items to be recommended by our system. Given that each exercise is suitable for a diagnostic subgroup with a specific pathology, these subgroups are used to represent the corresponding exercises. We first considered patients with a previous diagnosis of chronic mechanical low back pain based on different symptoms, establishing the following five diagnostic subgroups: muscle weakness, lumbar instability, psychometric variables, flexibility, and postural syndrome.

Once a new exercise is entered into the system, it obtains an internal representation that is mainly based on its appropriateness for each diagnostic subgroup. An exercise i is represented as a vector $VT_i = (VT_{i1}, VT_{i2}, ..., VT_{i5})$, where each component $VT_{ij} \in S_1$ is a linguistic assessment that represents the fitness degree of exercise i with respect to the diagnostic subgroup j. These fitness degrees are determined by the physiotherapists when they insert new exercises into the system.

For instance, suppose that a new exercise x, specially designed to improve the lumbar instability and postural syndrome, is inserted. Then, the physiotherapist would select for it the diagnostic sugroups 2 and 5 with a membership degree "Total" because they are in the positions 2 and 5 respectively; these membership degrees belong to the label set S_1, i.e. in our proposal the set S^5, with labels $b_0, b_1, ..., b_4$. The rest of the disgnostic subgroup have a membership degree with a value of "None". So, x is represented as: $VT_x = ((b_0, 0), (b_4, 0), (b_0, 0), (b_0, 0), (b_4, 0))$.

3.2   Patient Profiles Database

The patient profiles database stores the patients' pathological conditions, which are used to personalize the exercises. The results of a series of tests undergone by patients [13] are analyzed by experts to establish the pathology used to represent their respective profiles. The representation of the pathologies is also based on the same features as those applied for representation of the exercises. After obtaining the test results, the experts assess the membership of the patient's pathology in each one of the five diagnosis subgroups. A patient i is represented as a vector $VP_i = (VP_{i1}, VP_{i2}, ..., VP_{i5})$, where each component $VP_{ij} \in S_1$ is a linguistic assessment (i.e., a 2-tuple) that represents the fitness degree of i for each subgroup j.

For instance, suppose a patient p whose pathology is muscle weakness and some flexibility. Then, the physiotherapist would select for it the diagnostic sugroups 1 and 3 with a membership degree "Total" and "Medium" respectively, because they are in the positions 1 and 3; these membership degrees belong to the label set S_1, i.e. in our proposal the set S^5, with labels $b_0, b_1, ..., b_4$. The rest of the disgnostic subgroup have a membership degree with a value of "None". So, p is represented as: $VT_p = ((b_4, 0), (b_0, 0), (b_2, 0), (b_0, 0), (b_0, 0))$.

3.3 Method of Generating Recommendations of Exercises

TPLUFIB-WEB is based on a hybrid recommendation strategy, which switches between a content-based and a collaborative approach. The former approach is applied when a new exercise is entered into the system and the latter when a new patient is registered or when previous recommendations to a patient are updated, whenever the system has received sufficient ratings. We rely on a matching process by similarity measures among vectors. Particularly, we use the standard cosine measure, but defined in a linguistic context:

$$\sigma_l(V_1, V_2) = \Delta(g \times \frac{\sum_{k=1}^{n}(\Delta^{-1}(v_{1k}, \alpha_{v1k}) \times \Delta^{-1}(v_{2k}, \alpha_{v2k}))}{\sqrt{\sum_{k=1}^{n}(\Delta^{-1}(v_{1k}, \alpha_{v1k}))^2} \times \sqrt{\sum_{k=1}^{n}(\Delta^{-1}(v_{2k}, \alpha_{v2k}))^2}})(1)$$

with $\sigma_l(V_1, V_2) \in S_3 \times [-0.5, 0.5]$, and where g is the granularity of the term set used to express the similarity degree, i.e. S_3, n is the number of terms used to define the vectors (i.e. the number of diagnosis subgroups that have been considered) and (v_{ik}, α_{vik}) is the 2-tuple linguistic value of the diagnostic subgroup k in the exercise or patient vector V_i (label of S_1).

When a new exercise i is entered into the system, a content-based approach is used to know if it could be appropriate for a patinent p, as follows:

1. Compute $\sigma_l(VT_i, VP_p) \in S_3$. As $S_3 = S^9$, exercise i is considered appropriate for patient p if $\sigma_l(VT_i, VP_p) > (s_4^9, 0)$
2. If exercise i is considered appropriate for patient p, then the system recommends i to p with an estimated relevance degree $i(p) \in S_2 \times [-0.5, 0.5]$, which is obtained as follows:
   (a) Look for all exercises stored in the system that were previously assessed by p.
   (b) To aggregate all the ratings of p over these exercises, weigthed by the similarity between i and each of the exercises. To do that we use the linguistic weighted average operator [18].

As mentioned above, TPLUFIB-WEB also applies a collaborative approach to generate recommendations. The number of ratings rises with the increase in patients using the system, thereby allowing a collaborative approach to be adopted. Moreover, when new patients are entered into the system, they receive recommendations about existing exercises that may be of interest to them. Because these patients have not yet evaluated any exercise, the collaborative approach is used to generate these recommendations. To estimate (when no ratings are yet scored) or upgrade the relevance of a exercise i for a patient p following the collaborative approach:

1. Compute $\sigma_l(VT_i, VP_p) \in S_3$. If $S_3 = S^9$, i is considered appropriate for p if $\sigma_l(VT_i, VP_p) > (s_4^9, 0)$.

2. If i is appropriate, the set of patients $\aleph_p$ with a similar pathology to that of p, usually called *nearest neighbors*, is identified. This is done by calculating $\sigma_l(VP_p, VP_y) \in S_3$, between p and the vectors of all patients already in the system (VP_y, $y = 1..n$ where n is the number of patients). Because $S_3 = S^9$, patient y is considered a nearest neighbor to p if $\sigma_l(VP_p, VP_y) > (s_4^9, 0)$.

3. Retrieve the exercises positively rated by the nearest neighbors of p.

4. Each exercise i of recovered in the previous step is recommended to p with a predicted relevance degree $i(p) \in S_2 \times [-0.5, 0.5]$, computed as the aggregation of all the ratings, weighted by the similarity between e and their nearest neighbors. To do that we use the linguistic weighted average operator [18].

3.4 Feedback Phase

When patients have completed the recommended exercises, they are asked to assess the relevance of these recommendations in order to update their patient profiles. TPLUFIB-WEB receives the user feedback in this way. Patients communicate their linguistic evaluation judgements to the system, $rc \in S_4$, indicating their satisfaction with the recommendations (higher values of rc = greater satisfaction). Future recommendations are strenghened by taking account of patients' ratings, and the user-system interaction required is minimal in order to facilitate the sending of this important information.

4   Validating TPLUFIB-WEB

As previously stated, all the exercises recommended by TPLUFIB-WEB have already been approved by physiotherapists [3,13]. Furthermore, it is not our intention to validate the performance of the recommendation system in a strict sense. We have focused on the quality of TPLUFIB-WEB and the confidence that it inspires.

TPLUFIB-WEB satisfies the following quality criteria:

1. *Reliability of information provided.* The health Web underwent an accreditation process to ensure compliance with ethical codes and user rights and satisfactory fulfillment of quality standards. To date, the quality of the system has been accredited by the following:
   - Health on the net foundation, HONcode[7], certifying that the website was reviewed by the HONcode Team at a given date and complies with the eight principles of this code.
   - The Official College of Physicians of Barcelona (COMB)[8], a non-profit organization started in 1999 to provide benchmarks for reliability and service and improve the quality of health information on the Internet.

---

[7] http://www.healthonnet.org/
[8] http://wma.comb.es/es/home.php

2. *Quality of the website.* The system complies with the protocols laid down by the Health Quality Agency of the Andalusian Regional Goverment[9], designed to guarantee the reliability of the information and paying special attention to the protection and rights of patients. Accordingly, TPLUFIB-WEB is governed by very strict rules and fulfills the requirements of the World Wide Web Consortium (W3C) Web Accessibility Initiative[10], including compliance with XHTML 1.0 and CSS standards to facilitate use of the website on all types of device/platform.

3. *Usability.* Evaluation of the user-friendliness of the system is based on the responses of TPLUFIB-WEB users themselves to a questionnaire hosted on the home page during the trial period (one month). In that period, 64 individuals completed the survey, which comprises ten items. The first six questions are related to their understanding of the information by patients. The next three questions regard their ability and efficiency in using the website. The last item asked for a global evaluation of the health website, on a scale of *0 to 10*.

   The results demonstrate that the website is very positively perceived by its users. The patients were able to understand the received information and perform the exercises themselves. The usability and efficiency of the website was rated as "*Very Good* or *Good*" by 95% of the responders, and the patients evaluated the website with an average global score of 8.84 out of 10.

5   Concluding Remarks

This study presents a fuzzy linguistic Web tool named TPLUFIB-WEB, which incorporates a recommender system to provide personalized exercises to patients with low back pain. A physiotherapist establishes the pathology of a patient after evaluating the results of different tests, which are used to generate the recommendations. The website also provides patients with advice for handling future problems. The main benefits of this system deal with the personalization and the possibility of following the exercises anywhere and at anytime, potentially contributing to the reduction in the economic impact of low back pain. We have applied TPLUFIB-WEB in a real environment, and the experimental results demonstrate that acceptance of the system by users and patients is very high and that it may be able to achieve major costs savings for national health systems and patients by enhancing the effectiveness of each health professional involved.

Further research is warranted to explore other ICT applications in healthcare, especially in areas in which the physical presence of the health professionals is not wholly necessary and minimal supervision is adequate. There is also a need to improve the proposed recommendation approach, investigating new methodologies for the generation of recommendations.

---

[9] http://www.juntadeandalucia.es/agenciadecalidadsanitaria
[10] http://www.w3.org

Acknowledgments. This study received funds from National Projects TIN2010-17876 and Regional Projects P09-TIC-5299 and P10-TIC-5991.

References

1. Ehrlich, G.E.: Low back pain. Bulletin of the World Health Organization 81(9), 671–676 (2003)
2. Burton, A.K., McClune, T.D., Clarke, R.D., Main, C.J.: Long-term follow-up of patients with low back pain attending for manipulative care: outcomes and predictors. Manual Therapy 9(1), 30–35 (2004)
3. Cohen, I., Rainville, J.: Aggressive exercise as treatment for chronic low back pain. Sports Medicine 32(1), 75–82 (2002)
4. Vizirianakis, I.S.: Nanomedicine and personalized medicine toward the application of pharmacotyping in clinical practice to improve drug-delivery outcomes. Nanomedicine: Nanotechnology, Biology and Medicine 7, 11–17 (2011)
5. Al-Shorbaji, N.: Health and medical informatics. Technical paper, World Health Organization, RA/HIS, Regional Office for the Eastern Mediterranean (2001)
6. Santana, S., Lausen, B., Bujnowska-Fedak, M., Chronaki, C., Prokosch, H.U., Wynn, R.: Informed citizen and empowered citizen in health: results from an european survey. BMC Family Practice 12, 20 (2011)
7. Burke, R.: Hybrid web recommender systems. In: Brusilovsky, P., Kobsa, A., Nejdl, W. (eds.) Adaptive Web 2007. LNCS, vol. 4321, pp. 377–408. Springer, Heidelberg (2007)
8. Porcel, C., Herrera-Viedma, E.: Dealing with incomplete information in a fuzzy linguistic recommender system to disseminate information in university digital libraries. Knowledge-Based Systems 23, 32–39 (2010)
9. Tejeda-Lorente, A., Porcel, C., Peis, E., Sanz, R., Herrera-Viedma, E.: A quality based recommender system to disseminate information in a university digital library. Information Sciences 261(52-69) (2014)
10. Hussein, A., Omar, W., Li, X., Ati, M.: Efficient chronic disease diagnosis prediction and recommendation system. In: IEEE EMBS Conference on Biomedical Engineering and Sciences (IECBES), Malaysia, pp. 17–19 (2012)
11. Alonso, S., Pérez, I., Cabrerizo, F., Herrera-Viedma, E.: A linguistic consensus model for web 2.0 communities. Applied Soft Computing 13(1), 149–157 (2013)
12. Herrera, F., Alonso, S., Chiclana, F., Herrera-Viedma, E.: Computing with words in decision making: Foundations, trends and prospects. Fuzzy Optimization and Decision Making 8(4), 337–364 (2009)
13. Palacín-Marín, F., Esteban-Moreno, B., Olea, N., Herrera-Viedma, E., Arroyo-Morales, M.: Agreement between telerehabilitation and face-to-face clinical outcome measurements for low back pain in primary care. Spine 38(11), 947–952 (2013)
14. Porcel, C., Moreno, J., Herrera-Viedma, E.: A multi-disciplinar recommender system to advice research resources in university digital libraries. Expert Systems with Applications 36(10), 12520–12528 (2009)
15. Porcel, C., Tejeda-Lorente, A., Martínez, M.A., Herrera-Viedma, E.: A hybrid recommender system for the selective dissemination of research resources in a technology transfer office. Information Sciences 184(1), 1–19 (2012)
16. Zadeh, L.: The concept of a linguistic variable and its applications to approximate reasoning. Part I, Information Sciences 8, 199–249 (1975), Part II, Information Sciences 8, 301–357 (1975), Part III, Information Sciences 9, 43–80 (1975)

17. Herrera-Viedma, E., López-Herrera, A., Luque, M., Porcel, C.: A fuzzy linguistic IRS model based on a 2-tuple fuzzy linguistic approach. International Journal of Uncertainty, Fuzziness and Knowledge-based Systems 15(2), 225–250 (2007)
18. Herrera, F., Martínez, L.: A 2-tuple fuzzy linguistic representation model for computing with words. IEEE Transactions on Fuzzy Systems 8(6), 746–752 (2000)
19. Herrera, F., Martínez, L.: A model based on linguistic 2-tuples for dealing with multigranularity hierarchical linguistic contexts in multiexpert decision-making. IEEE Transactions on Systems, Man and Cybernetics. Part B: Cybernetics 31(2), 227–234 (2001)

On Distances for Cooperative Games and Non-additive Measures with Communication Situations

Vicenç Torra[1] and Yasuo Narukawa[2]

[1] IIIA, Institut d'Investigació en Intel·ligència Artificial -
CSIC, Consejo Superior de Investigaciones Científicas
Campus UAB s/n, 08193 Bellaterra, Catalonia, Spain
[2] Toho Gakuen,
3-1-10, Naka, Kunitachi, Tokyo, 186-0004, Japan

Abstract. Non-additive (fuzzy) measures also known as cooperative games or capacities are set functions that can be used to evaluate subsets of a reference set. In order to evaluate their similarities and differences, we can consider distances between pairs of measures.

Games have been extended to communication situations in which besides of the game there is a graph that establishes which sets are feasible (which coalitions are possible, which individuals can cooperate).

In this paper we consider the problem of defining a distance for pairs of measures when not all sets are feasible.

1 Introduction

Non-additive measures, also known as fuzzy measures, cooperative games and capacities are set functions that are monotonic. They can be used in applications that range from decision making to data fusion. They are successful applications in computer vision, database integration, risk assessment and game theory.

The literature presents different studies on these measures. For example, some indices have been defined to evaluate the power of one of the individuals in a coallition or their interaction. In decision making and information fusion, the measures are often combined with integrals. Choquet [3] and Sugeno [11] integrals are two examples of fuzzy integrals that permit us to integrate a function with respect to a non-additive measure. See e.g. [12] for examples and definitions.

An important aspect in all applications is how to determine the appropriate measure for a given system. For this purpose, different approaches have been considered for measure identification. Some are based on heuristic techniques, other in machine learning techniques (either supervised or unsupervised ones).

In some cases, several non-additive measures can be obtained for the same problem due e.g. to the execution of the same algorithm with different data sets. In this case it is of relevance to consider the similarity between these measures obtained. For example, in [1] several non-additive measures were generated in a reidentification problem for risk assessment.

C. Cornelis et al. (eds.): RSCTC 2014, LNAI 8536, pp. 262–269, 2014.

Up to our knowledge, the problem of defining distances for non-additive measures has not been much explored. [13,8] considered some distances for non-additive measures when the reference set is not finite. In [4] a distance was introduced in the case that the measure is finite. It is based on the Euclidean distance.

In this paper we further explore the problem of distance definition.

More specifically, we focus in the case that not all coalitions are possible. That is, following [5,2] we consider the case that for some reason not all the subsets of the reference set are feasible. Using the terminology of cooperative games, this means that some subsets cannot cooperate.

A typical example is when the reference set is the set of political parties in a parliament. The measure evaluates the worth of a certain coalition. However, because some incompatibilities between parties, not all coalitions between them are possible.

Another example is when the reference set corresponds to the nodes of a network. In this case it may happen that there are nodes that are not reachable from other nodes (i.e., the graph is disconnected) maybe because part of the network is down. In this case, the measure (e.g., the worth of a subset of nodes) has to take into account these isolated nodes.

The structure of the paper is as follows. In Section 2 we review some reliminaries. In Section 3 we introduce our new definitions. We prove some properties of the distances. The paper finishes with some conclusions.

2   Preliminaries

In this section we review the concepts that are needed in the rest of the paper. We begin defining non-additive measure. As stated above, non-additive measure is also known with the name fuzzy measure, and cooperative game.

The measure is defined over a reference set X. In the area of (multicriteria) decision making, X usually corresponds to the set of criteria, and the measure $\mu(A)$ for $A \subseteq X$ corresponds to the importance of the set of criteria A. One interpretation in game theory is that A is a coalition of individuals or parties from X. Then, $\mu(A)$ is the power of the coalition A in e.g. a parliament.

Definition 1. *A non-additive measure μ on a set X is a set function μ : $\wp(X) \to [0,1]$ satisfying the following axioms:*

(i) $\mu(\emptyset) = 0$ (boundary conditions)
(ii) $A \subseteq B$ implies $\mu(A) \le \mu(B)$ (monotonicity)

In the area of fuzzy measures it is usual to require $\mu(X) = 1$. Unless we state otherwise, we will presume that $\mu(X) = 1$.

In this definition, we can evaluate the measure for any subset of X. Let us now consider a communication network. We will consider that the elements of the reference set are the ones that define the network. Then, the network will establish which communications between pairs of elements of X are feasible.

So, in other words, the network will be used to restrict the possible coalitions. Definitions below follow [5]. See also [2].

Given X, we represent a communication network by an undirected graph $G = (X, E)$ where X is the set of nodes of the graph (in our case, the elements of the reference set X) and E is the set of edges of the graph. As usual, $E \subset X \times X$.

Given X, the graph $G = (X, E)$, and a set of elements S, we define the graph $G(S) = (S, E(S))$ where $E(S) = \{(n_1, n_2) \in E | n_1 \in S, n_2 \in S\}$. We call $G(S)$ the subgraph induced from the underlying graph G and S.

A sequence of different nodes $n_1, \ldots, n_m$ where $(n_i, n_{i+1}) \in E$ is called a path (with lenght $m - 1$). If there is a path between n_i and n_j we say that n_j is reachable from n_i in graph G, and we write $n_i \sim_G n_j$.

The relation $\sim_G$ is an equivalence relation on X. X/E is defined in terms of this equivalence relation $X/E := X/ \sim_G$.

Definition 2. *[5] Let X be a reference set, μ a non-additive measure on X, and $G = (X, E)$ a graph on X. Then, the triplet (X, μ, E) is called a communication situation.*

Definition 3. *[5] Let X be a reference set, μ a non-additive measure on X, and $G = (X, E)$ a graph on X which define a communication situation (X, μ, E). Then, a coalition $S \subseteq X$ is said to be feasible if S is connected in G.*

Taking into account the communication network, not all sets $S \subseteq X$ are feasible. From the point of view of coalitions, this means that not all of them are possible. That is, when the members of a set S cannot communicate with each other because there is no path between some of the members, the value of the set is different from $\mu(S)$. The next definition assigns a value to the set taking into account the feasible sets (the feasible coalitions).

Definition 4. *[5] Let X be a reference set, μ a non-additive measure on X, and $G = (X, E)$ a graph on X which defines a communication situation (X, μ, E). Then, the network restricted measure (X, μ^E) associated with (X, μ, E) is defined by*

$$\mu^E(S) = \sum_{T \in S/E} \mu(T)$$

for all $S \subseteq X$.

Lemma 1. *[5] When the graph is complete (i.e., $E = X \times X$), $\mu^E = \mu$.*

This lemma follows naturally from the fact that $S/E = \{S\}$ when the graph is complete.

It is relevant here to underline that having a measure μ such that $\mu(X) = 1$, this does not imply that $\mu^E(X) = 1$. Note that we may have either $\mu^E(X) < 1$ or $\mu^E(X) > 1$. To see this, let us consider a reference set X, and the graph $G = (X, E)$ with $E = \emptyset$. Now, if we define μ_1 such that $\sum_{x \in X} \mu_1(x) < 1$ we will have $\mu_1^E(X) < 1$ while if we define μ_2 such that $\sum_{x \in X} \mu_2(x) > 1$ we will have $\mu_2^E > 1$.

Because of that we introduce a new definition of a network restricted measure. The difference is that now we normalize the measure so that the measure for the whole reference set is one.

Definition 5. *Let X be a reference set, μ a non-additive measure on X, and $G = (X, E)$ a graph on X which defines a communication situation (X, μ, E). Then, the normalized network restricted measure (X, μ^E) associated with (X, μ, E) is defined by*

$$\mu_N^E(S) = \frac{\sum_{T \in S/E} \mu(T)}{\sum_{T \in X/E} \mu(T)}$$

for all $S \subseteq X$.

2.1 Shapley and Banzhaf Values

Different indices have been defined in the literature for non-additive measures. The Shapley and Banzhaf are used in the context of coalitions to evaluate the strength or power of an individual or a party. See e.g. [2,12] for details.

We review here the Shapley and the Banzhaf index.

Definition 6. *Given a reference set X and a non-additive measure μ on X, the Shapley value of μ for $x_i \in X$, denoted by $\varphi_{x_i}(\mu)$, is defined as follows:*

$$\varphi_{x_i}(\mu) := \sum_{S \subseteq X \setminus \{x_i\}} \frac{|S|!}{(N - |S| - 1)! N!} \left(\mu(S \cup \{x_i\}) - \mu(S) \right) \tag{1}$$

Definition 7. *Given a reference set X and a non-additive measure μ on X; then,*

1. *The* unnormalized *(or nonstandardized or absolute) Banzhaf index of μ for x_i is defined by*

    $$\beta'_{x_i}(\mu) := \frac{\sum_{S \subseteq X} \left(\mu(S) - \mu(S \setminus \{x_i\}) \right)}{2^{N-1}}.$$

2. *The* Penrose index *(or normalized Banzhaf index or relative Banzhaf index) of μ for x_i is defined by*

    $$\beta_{x_i}(\mu) := \frac{\sum_{S \subseteq X} \left(\mu(S) - \mu(S \setminus \{x_i\}) \right)}{\sum_{i=1}^{N} \sum_{S \subseteq X} \left(\mu(S) - \mu(S \setminus \{x_i\}) \right)}.$$

These indices are defined for a non-additive measure μ. Let us consider the case in which we have a communication situation (X, μ, E). Myerson [10] defined the following value for this situation.

Definition 8. *[10] Let X be a reference set, μ a non-additive measure on X, and $G = (X, E)$ a graph on X which defines a communication situation (X, μ, E). Then, the Myerson value for (X, μ, E) is denoted by $\Phi(X, \mu, E)$ and it is defined by:*

$$\Phi(X, \mu, E) := \phi(\mu^E)$$

where $\phi(\mu^E)$ is the Shapley value of (X, μ^E).

Lemma 2. *When G is the complete graph, $\Phi(X, \mu, E) = \phi(\mu)$.*

This follows from Lemma 1. We will use the same definition but using the normalized measure μ_N^E instead of μ^E. We will denote it by Φ_N.

Lemma 3. *Let X be a reference set, and let μ be a non-additive measure such that $\mu(X) = K$; then, the Shapley value and the Myerson value is positive and such that $\sum_{x \in X} \phi(x) = K$.*

In our case that we require $\mu(X) = 1$, we have that the Shapley value is such that $\sum_{x \in X} \phi(x) = 1$.

3 Distance on Non-additive Measures

In [4], the Euclidean distance was used for pairs of (generalized) Shapley values. That is, given two vectors of Shapley values ϕ_1 and ϕ_2 the following distance was considered:

$$d(\phi_1, \phi_2) = \sqrt{\sum_{x \in X} (\phi_1(x) - \phi_2(x))^2}.$$

In the next definition we propose to use this expression to define a distance between pairs of non-additive measures. This can be naturally extended for non-additive measures in a communication situation using the Myerson value.

Definition 9. *Let X be a reference set and μ_1 and μ_2 to non-additive measures on this set. Then, we define a distance between μ_1 and μ_2 as follows:*

$$d(\mu_1, \mu_2) = \sqrt{\sum_{x \in X} (\phi_1(x) - \phi_2(x))^2}$$

where ϕ_1 and ϕ_2 are, respectively, the Shapley values of μ_1 and μ_2.

Definition 10. *Let X be a reference set, μ_1 and μ_2 non-additive measures on X, and $G = (X, E)$ a graph on X which defines a communication situation (X, μ, E). Then, we define a distance between μ_1 and μ_2 on the communication situation (X, μ, E) as follows:*

$$d(\mu_1, \mu_2) = \sqrt{\sum_{x \in X} (\Phi_N(X, \mu_1, E)(x) - \Phi_N(X, \mu_2, E)(x))^2}$$

where Φ_N is the Myerson value with the normalized network restricted measure.

Lemma 4. *Definitions 9 and 10 define a distance (i.e., they satisfy positiveness, symmetry, and triangular inequality).*

Proof. Positiveness and symmetry follow from the definition. Triangular inequality follows from the triangular inequality of the Euclidean distance. Note that for the measures, the triangular inequality means:

$$d(\mu_x, \mu_y) + d(\mu_y, \mu_z) \geq d(\mu_x, \mu_z)$$

which according to the definition of the distance corresponds to

$$d(x, y) + d(y, z) \geq d(x, z)$$

where x, y, and z are the Shapley value or the Myerson value of the measures μ_x, μ_y, and μ_z. As the last equation holds for the Euclidean distance, the triangular inequality holds for the distances above.   □

According to Lemma 3, when $\mu(X) = 1$, the Shapley value is positive and such that $\sum_{x \in X} \phi(x) = 1$. So, Φ_N can be seen as a probability distribution. In this framework, it is possible to consider distances and expressions for probability distributions, and all the properties for these distances will hold for the measure. In particular, we can consider e.g. f-divergence, Hellinger distance, KL-divergence, and other standard distances and similarity measures for probabilities. Taking advantage of this fact, we define below the Hellinger distance.

Definition 11. *Let X be a reference set, μ_1 and μ_2 non-additive measures on X, and $G = (X, E)$ a graph on X which defines a communication situation (X, μ, E). Then, we define the Hellinger distance between μ_1 and μ_2 on the communication situation (X, μ, E) as follows:*

$$d(\mu_1, \mu_2) = \frac{1}{2} \sqrt{\sum_{x \in X} \left(\sqrt{\Phi_N(X, \mu_1, E)(x)} - \sqrt{\Phi_N(X, \mu_2, E)(x)} \right)^2}$$

where Φ_N is the Myerson value.

Lemma 5. *The Hellinger distance in Definition 11 satisfies positiveness, symmetry and triangular inequality.*

We have defined some distances and shown that they satisfy the basic properties of distances. These definitions were based on the Shapley index. Analogous definitions can be given with the Banzhaf index. The proof of their properties is also analogous to the proofs given here.

3.1   Normalization

In Definition 5 we have introduced the normalized network restricted measure. The normalization was done so that $\mu_N^E(X) = 1$ when $\mu(X) = 1$ as this is not the case with μ^E.

Note that this is not the only possible way to normalize the measure. An alternative way is to define the network restricted measure using the maximum (or a t-conorm) instead of using the summatory. Finally, the normalization would just to fix $\mu(X) = 1$.

The definition of the network restricted measure using the maximum would permit us to have a measure that is monotonic with respect to the links included in the set of edges E. That is, if we define $\mu_\wedge^E(S) = \sum_{T \in S/E} \mu(T)$ then for all graphs $G = (X, E)$ on X, and all $(i, j) \in E$,

$$\mu_\wedge^E(A) \geq \mu_\wedge^{E \setminus (i,j)}(A)$$

for all $A \subseteq X$.

This property does not hold for other measures.

Any of these normalizations would lead to a measure whose Shapley value is a probability distributions and, thus, the expressions above would apply.

4   Summary and Future Work

In this paper we have introduced distances for non-additive measures when there is a communication situation and not all coalitions are possible. The distance between two non-additive measures is defined in terms of the Shapley value of their normalized network restricted measures. We need to introduce the normalized network restricted measures so that the Shapley value of the measure is a probability distribution.

As future work we will consider additional properties of the distance. One of them is to know whether the distance is monotonic decreasing with respect to removing links to the network from the complete graph.

Then, we will also consider the case of defining distances when there is a fuzzy communication structure in the line of [7] and [6].

Finally, another line for future research is to consider the study of distances taking into account the index introduced in [9] (a Myerson value for complete coalition systems). The authors study an extension of the Myerson value for complete coalitions.

Acknowledgements. Partial support by the Spanish MEC projects ARES (CONSOLIDER INGENIO 2010 CSD2007-00004) and COPRIVACY (TIN2011-27076-C03-03) is acknowledged.

References

1. Abril, D., Navarro-Arribas, G., Torra, V.: Choquet Integral for Record Linkage. Annals of Operations Research 195, 97–110 (2012)
2. Bilbao, J.M.: Cooperative Games on Combinatorial Structures. Kluwer Academic Publishers (2000)
3. Choquet, G.: Theory of capacities. Ann. Inst. Fourier 5, 131–295 (1953)
4. De, A., Diakonikolas, I., Servedio, R.A.: The inverse Shapley value problem. Electronic Colloquium on Computational Complexity, Report No. 181 (2012)

5. Fujimoto, K.: Cooperative game as non-additive measure. In: Torra, V., Narukawa, Y., Sugeno, M. (eds.) Non-Additive Measures. Studies in Fuzziness and Soft Computing, vol. 310, pp. 131–171. Springer, Heidelberg (2014)
6. Gallego, I., Fernández, J.R., Jiménez-Losada, A., Ordoñez, M.: A Banzhaf value for games with fuzzy communication structure: Computing the power of the political groups in the European Parliament. Fuzzy Sets and Systems (in press, 2014)
7. Jiménez-Losada, A., Fernández, J.R., Ordóñez, M.: Myerson values for games with fuzzy communication structure. Fuzzy sets and systems 213, 74–90 (2013)
8. Kawabe, J.: Metrizability of the Lévy topology on the space of nonadditive measures on metric spaces. Fuzzy sets and systems 204, 93–105 (2012)
9. Kajii, A., Kojima, H., Ui, T.: The Myerson Value for Complete Coalition Systems. Institute for Mathematical Sciences (IMS) preprint series # 2006-25 (previous version with title "A Refinement of the Myerson Value") (2006)
10. Myerson, R.: Graphs and cooperation in games. Mathematics of Operations Research 2, 225–229 (1977)
11. Sugeno, M.: Theory of Fuzzy Integrals and its Applications, Ph. D. Dissertation, Tokyo Institute of Technology, Tokyo, Japan (1974)
12. Torra, V., Narukawa, Y.: Modeling decisions: information fusion and aggregation operators. Springer (2007)
13. Torra, V., Narukawa, Y., Sugeno, M.: Carlson, On the f-divergence for non-additive measures (submitted)

TSK-0 Fuzzy Rule-Based Systems
for High-Dimensional Problems Using
the Apriori Principle for Rule Generation

Javier Cózar, Luis de la Ossa, and José A. Gámez

University of Castilla-La Mancha, Albacete, Spain
{javier.cozar,luis.delaossa,jose.gamez}@uclm.es

Abstract. Algorithms which learn Linguistic Fuzzy Rule-Based Systems from data usually start up from the definition of the linguistic variables, generate a set of candidate rules and, afterwards, search a subset of them through a metaheuristic technique. In high-dimensional datasets the number of candidate rules is intractable, and a preselection is a must. This work adapts an existing preselection algorithm for Fuzzy Asociation Rule-Based Classification Systems to deal with TSK-0 LFRBSs. Experimental results show a good behaviour of the adaptation allowing to build precise and simple models for high-dimensional problems.

Keywords: linguistic fuzzy modeling, machine learning, high-dimensional, Takagi-Sugeno-Kang.

1 Introduction

Fuzzy Rule-Based Systems (FRBSs) have been used to solve different classification and prediction problems [1,2,4] due to their interpretability for the end user. These systems can be designed ad-hoc by an expert, or generated automatically from a data set. In the last case, there are many algorithms which consider fixed fuzzy partitions of the variables and derive only the rule base [1,3,6]. In general, such algorithms consider a set of candidate fuzzy rules and try to find the optimal subset.

When dealing with relatively high-dimensional data sets, the set of candidate rules grows exponentially. Some works deal with this problem by generating a subset with only the most promising candidate rules i.e., those which cover a high number of instances and maximize some metric relative to accuracy or precision. For example, in [4] the authors choose a set of rules based on their support and individual performance for classification. After that, a genetic algorithm carries out a rule subset selection and derive the final fuzzy system. In this paper, we adapt this strategy to design an algorithm for regresion. In particular, we aim to learn TSK-0 Linguistic Fuzzy Rule-Based Systems (LFRBSs) in high-dimensional problems.

This paper is divided into four sections besides this introduction. In Section 2, we describe TSK-0 LFRBSs. The following section explains our adaptation to

C. Cornelis et al. (eds.): RSCTC 2014, LNAI 8536, pp. 270–279, 2014.
© Springer International Publishing Switzerland 2014

the aforementioned strategy to derive only the most promising candidate rules from data, and the method used to determine the final rule set. Section 4 includes the set up of experimentation and results. And finally, in Section 5, we sumarize the conclusions and expose some future work.

2 TSK-0 Linguistic Fuzzy Rule-Based Systems

In TSK FRBSs, the antecedent of each rule consists of a number of predicates with the form X is F, where X is a variable of the problem, and F is a fuzzy set defined over the domain of X. If we use a finite partition of linguistic labels instead of fuzzy sets (X is A) and we relate each label to a particular fuzzy set F, the interpretability grows up. This type of systems are called Linguistic Fuzzy Rule-Based Systems (LFRBSs) [13]. On the other hand, the consequent of a TSK rule is formed by a polynomial function of the input variables, $P_S(X_1, \ldots, X_n)$. The order of a TSK FRBS refers to the degree of the polynomial. Thus, a TSK-0 system means that P_s are polynomial of degree 0. Having said that, a fuzzy rule R_s is represented as:

$$R_s : \text{If } X_1 \text{ is } A_1^s \text{ and } \ldots \text{ and } X_n \text{ is } A_n^s \text{ then } Y = b_s$$

Given the set of rules $\mathcal{RB}$, the output produced by a TSK-0 FRBS when processing an instance $e_l = (x_1^l, \ldots, x_n^l, y^l)$ is a weighted average of the individual outputs generated by each rule $R_s \in \mathcal{RB}$:

$$\hat{y}^l = \frac{\sum_{R_s \in \mathcal{RB}} h_s^l b_s}{\sum_{R_s \in \mathcal{RB}} h_s^l} \tag{1}$$

where $h_s^l(e_l) = T(A_1^s(x_1^l), \ldots, A_n^s(x_n^l))$ is the compatibility degree of the instance e_l with the rule R_s, T is a T-norm [1], and $P_s(x_1^l, \ldots, x_n^l)$ is the value of the polynomial of the rule R_s for the input instance e_l.

3 Learning TSK-0 Fuzzy Rules in High-Dimensional Problems

As we use linguistic variables for the ancededents our algorithm takes as starting point two elements:

- A data set $\mathcal{E} = \{e_1, \ldots, e_l, \ldots, e_N\}$, where $e_l = (x_1^l, \ldots, x_n^l, y^l)$, x_i^l is the input part of the instance and y^l is the output[2].
- The data base, which contains the definition of the linguistic variables (their domains, fuzzy partitions and fuzzy terms).

---

[1] In this work we use *min* as a T-Norm.
[2] In this work we only consider one output variable, but this number can be greater than one.

In general, rule derivation methods proceed in two main stages. Firstly, they generate a set of candidate rules $\mathcal{RB}$ (only the antecedent part), such that every rule $R_s \in \mathcal{RB}$ fires at least one instance $e_l \in \mathcal{E}$. Afterwards, they choose a subset of $\mathcal{RB}$ and fix the consequent of each selected rule. In order to do that, they usually consider metrics based on the error. In this work, we have used the Mean Squared Error. Let be $\hat{y}_o^l$ the output of the LFRBS which contains the set of rules $\mathcal{RB}_o$ when processing the instance e_l. Then, $\mathrm{MSE}_o(\mathcal{E})$ is obtained as follows:

$$\mathrm{MSE}_o(\mathcal{E}) = \sum_{l=1}^{N} \frac{(\hat{y}_o^l - y^l)^2}{N}$$

3.1 Candidate Rule Extraction

This task, divided into two stages, is an adaptation of the method described in [4] for fuzzy association rule extraction in classification problems. First of all, it carries out a fuzzy association rule extraction [5] to obtain all the frequent item sets. After that, the process derives fuzzy rules from the previous item sets (they correspond to the antecedent part) and carries out a prescreening process to reduce the amount of fuzzy rules. Next, both parts will be explained in detail.

Rule Extraction. An item set A is a set of predicates (X_i is A_i) which does not contain two predicates involving the same variable X_i. Therefore, it represents the antecedent of a linguistic fuzzy rule. In order to quantify the support in the context of regression, we have adapted the formula used in [4] for Fuzzy Rule-Based Classification Systems (FRBCSs) [9]. Let $\mathcal{E}$ be the data set, A a fuzzy item set, e_l an instance of the data set $\mathcal{E}$ and $\mu_A(e_l)$ the compatibility degree of e_l with the predicates from the item set A. We define the support of A as:

$$\mathrm{Support}(A) = \frac{\sum_{e_l \in \mathcal{E}} \mu_A(e_l)}{N}$$

Frequent item sets are those whose support is greater than a given threshold (minimum support). Following the apriori principle [10], if a fuzzy item set is not frequent, all the item sets derived from it –by adding a fuzzy predicate– are not frequent either, so there is no need to calculate their support. This is important, since time and space complexity reduction are improved by a factor of 10 according to [5].

In order to obtain frequent item sets efficiently, they are listed by means of a tree. In such a tree, the root (level 0) represents the empty item set, and a child of a node represents an item set which contains the same items of the father plus one more. Therefore, in the level l_{th} there are item sets formed by exactly l predicates. If the item set in a node is not frequent, there is no need to expand it. We can see a general structure of the tree in Figure 1.

Once the search tree has been built, we derive the corresponding candidate fuzzy rule set, $\mathcal{RB}_{ap}$, from the item sets.

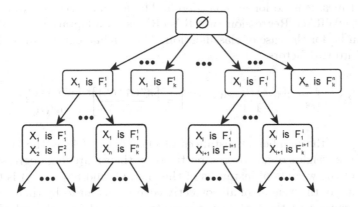

Fig. 1. General structure of the search tree

Prescreening. In high-dimensional problems, the amount of candidate rules generated by the previous method can still be huge. In this stage we select the most promising rules, making the search space tractable by the algorithm which selects rules and finds the consequents (in the next stage).

The process of prescreening constists of an iterative rule selection, as it is described in Algorithm 1.

> **Input:** $\mathcal{RB}_{ap}, k_t, \mathcal{E}$
> **Output:** $\mathcal{RB}_o$
> $\mathcal{E}' \leftarrow \mathcal{E}$;
> $c_l = 0, \forall e_l \in \mathcal{E}'$;
> **while** $\exists e_l \in \mathcal{E}' \mid c_l < k_t$ **do**
> $\quad R_s = \arg\max_{R_t} \text{wWRAccR}(R_t) \mid R_t \in \mathcal{RB}_{ap}$;
> $\quad \mathcal{RB}_{ap} = \mathcal{RB}_{ap} - \{R_s\}$;
> $\quad \mathcal{RB}_o = \mathcal{RB}_o \cup \{R_s\}$;
> $\quad c_l = c_l + 1, \forall e_l \in \mathcal{E}'^s$;
> $\quad \mathcal{E}' \leftarrow \mathcal{E}' - \{c_l\} \mid c_l \geq k_t, \forall e_l \in \mathcal{E}'^s$;
> **end**

Algorithm 1: Prescreening algorithm

On each step, one rule (the best according to some criterium) is selected from $\mathcal{RB}_{ap}$. This is repeated until all the instances of the data set are covered at least k_t times (by different selected rules). The criterium to select the best rule is based on a pattern weighting sheme. Every example e_l has a weight $w(e_l, i) = \frac{1}{i+1}$, such that i is the number of selected rules which cover e_l. When an instance has been covered k_t times, it is no longer considered. The formula used to evaluate each rule, called wWRAcc Regression (wWRAccR), is an adaptation of wWRAcc" described in [4] for the case of classification. The original expression wWRAcc" is divided into two factors:

$$\text{wWRAcc"}(R_s) = \left[f_a^{wWRAcc''} \right] \cdot \left[f_b^{wWRAcc''} \right] = \left[\frac{n''(A \cdot C_j)}{n'(C_j)} \right] \cdot \left[\frac{n''(A \cdot C_j)}{n''(A)} - \frac{n(C_j)}{N} \right],$$

where $n(C_j)$ is the number of instances of class C_j; $n''(A)$ is the sum of the product of the weights of all covered patterns by their matching degrees; $n'(C_j)$ is the sum of the weights of instances of the class C_j and $n''(A \cdot C_j)$ is the sum of the product of the weights of all correctly covered instances by their matching degrees. In some way, the first factor tries to reward those rules which cover, with a high degree, highly-weighted instances, whereas the second one aims to select accurate rules.

As in the case of regression y_l is a real number, and a rule consequent is a polynomial of degree 0 (again a real number), the adaptation of $f_b^{wWRAcc''}$ is not straightforward. Based on the results shown in [12], the standard deviation of the instances (the output values) which fire a particular rule seems to be a good quality indicator for that rule. Given a rule R_s and the set of instances covered by it $\mathcal{E}^s$, we use the standard deviation of $\mathcal{E}^s$, $sd(\mathcal{E}^s)$, to determine if a rule is better than other in terms of precission. In order to give both factors $f_a^{wWRAccR}$ and $f_b^{wWRAccR}$ similar importance, as their domains are not the same, we normalize them in the range $[0-1]$. As this adaptation is not simple, and in order to test if its behaviour is correct or not, we have considered two variations for the formula which include and does not include the second factor (wWRAccR and wWRAccR' respectively).

Let h_s^l be the compatibility degree of the rule R_s with the instance e_l, $\mathcal{E}^s$ the instances from the data set covered by the rule R_s and $sd(\mathcal{E}^s)$ the standard deviation of the class variable Y for the examples covered by R_s. Then:

$$\text{wWRAccR}(R_s) = \left[f_a^{wWRAccR} \right] = \left[\sum_{l=1}^{N} \frac{w(e_l, i) \cdot h_s^l}{w(e_l, i)} \right],$$

$$\text{wWRAccR}'(R_s) = \left[\frac{f_a^{wWRAccR'} - \min_a^{wWRAccR'}}{\max_a^{wWRAccR'} - \min_a^{wWRAccR'}} \right] \cdot \left[\frac{f_b^{wWRAccR'} - \min_b^{wWRAccR'}}{\max_b^{wWRAccR'} - \min_b^{wWRAccR'}} \right],$$

where $f_a^{wWRAccR'} = \sum_{l=1}^{N} \frac{w(e_l,i) \cdot h_s^l}{w(e_l,i)}$, $f_b^{wWRAccR'} = sd(\mathcal{E}^s)$, $\min_a^{wWRAccR'}$ is the minimum value for the first factor for all the rules, $\max_a^{wWRAccR'}$ is the maximum value for the first factor for all the rules, and $\min_b^{wWRAccR'}$ and $\max_b^{wWRAccR'}$ the minimum and maximum values for the second factor.

3.2   Rulebase Generation

In this stage, we fix the consequent for every single rule $R_s \in \mathcal{RB}_o$. We have focused this problem as a linear regression one, and solved by minimizing the squared error in the prediction using the Least Squares method [8]. As it fixes the consequent of all rules at once, cooperation among rules [7] is considered.

The output when processing an example e_l, given a rulebase $\mathcal{RB}_o$, is computed as shown in Equation 1. This operation can be vectorized as:

$$[\hat{y}_o^l] = [c_1^l \cdots c_{\mathcal{RB}_o}^l] \begin{bmatrix} b_1 \\ \vdots \\ b_{|\mathcal{RB}_o|} \end{bmatrix}$$

where c_s^l is:

$$c_s^l = \frac{h_s^l}{\sum_{\mathcal{RB}_o} h_s^l}$$

Using all the N examples in the data set, the previous expression turns into the following:

$$Y = C \cdot B = \begin{bmatrix} y_1 \\ \vdots \\ y_N \end{bmatrix} = \begin{bmatrix} c_1^l & \cdots & c_{|\mathcal{RB}_o|}^1 \\ \vdots & \vdots & \vdots \\ c_N^l & \cdots & c_{|\mathcal{RB}_o|}^N \end{bmatrix} \begin{bmatrix} b_1 \\ \vdots \\ b_{|\mathcal{RB}_o|} \end{bmatrix}$$

The matrix of consequents B can be obtained as $B = C^{-1} \cdot Y$. In order to compute C^{-1} we have used the Moore-Penrose pseudoinverse built by the Singular Value Decomposition method [11].

4   Experiments

In order to test the proposed algorithms, we have used five data sets for regression from the KEEL repository[3]. Two of them (*pole* and *puma32h*) have been generated artificially, while the rest are problems from the real world. The data sets *treasury* and *mortgage* have 15 features, *compactiv* has 21, *pole* 26 and *puma32h* has 32 variables.

We have tested the two alternatives designed for the adaptation of the formula for the prescreening stage (wWRAccR and wWRAccR'). Both have been executed using the described method for the rulebase generation (the Least Squares method).

For all the configurations the number of labels to model each fuzzy variable has been set to 5, and symmetrical triangular partitions have been used. We have set the maximum number of antecedents per rule to 3. The *min support*

---

[3] http://www.keel.es/

has been established to 0.05 and $k_t = 20$. These parameters have been experimentally chosen in order to achieve a tractable number of candidate rules but also good systems in terms of error. Results have been averaged over 30 independent executions for all the data sets. For each one, the training set has been set randomly selecting 80% of the instances, taking as seed the number of execution. The remaining instances have been left for testing. As we run 30 executions per problem and prescreening expression, we can suppose that results are normally distributed, so we have used a parametrical statistical test (t-Student) to compare the results obtained with both approaches (wWRAccR or wWRAccR'), setting the confidence level of $\alpha = 0.05$.

The results are shown in Table 1. The first column shows the name of the dataset, the second is the prescreening expression used in the learning process, the two following columns are the number of candidate rules generated by the rule extraction and prescreening stages, the next two are the training and test error expressed in RMSE (root of MSE), and the last column is the mean average of the number of antecedents by rule. We can see an improvement in terms of test error for all the data sets, except for pole, when using wWRAccR'. In Table 2 the p-values for the comparison between both approaches are shown for each dataset, setting the alternative hypothesis as wWRAccR test errors are greater than wWRAccR'.

In relation to the number of generated rules, we have noticed an increase in the case of wWRAccR', as well as the mean number of antecedents per rule. Having seen that, we wonder if the improvement in precision is due to the quality of the selected rules or if it is just an effect of having more rules for the Least Squares method. In order to analyze that, we have re-executed the previous experimentation for the wWRAccR' configuration, but changing the stop criteria for the prescreening method in order to obtain the same number of candidate rules as wWRAccR. In this case, it will stop iterating when $|\mathcal{RB}_o|$ is equal to the number of generated rules for the same configuration but using wWRAccR for the prescreening expression.

Table 1. Comparison between the two prescreening expressions

Problem	Prescr. Exp.	$|\mathcal{RB}_{ap}|$	$|\mathcal{RB}_o|$	Tra.err	Test err.	#Ant.
compactiv	wWRAccR	8269.3	41.2	7.3230	7.4438	1.9
compactiv	wWRAccR'	8269.3	51.0	6.7967	**6.9010**	2.1
mortgage	wWRAccR	8104.3	51.1	0.5256	0.6631	1.7
mortgage	wWRAccR'	8104.3	66.9	0.2987	**0.4037**	2.0
pole	wWRAccR	9766.8	32.4	35.2445	**35.2378**	1.4
pole	wWRAccR'	9766.8	34.1	36.5824	36.6168	1.6
puma32h	wWRAccR	4624.0	51.4	0.0294	0.0296	1.0
puma32h	wWRAccR'	4624.0	52.9	0.0232	**0.0233**	1.0
treasury	wWRAccR	8153.1	51.0	0.7340	0.8625	1.7
treasury	wWRAccR'	8153.1	73.0	0.3748	**0.5442**	2.0

Table 2. Statistical test (t-Student) for comparison between wWRAccR and wWRAccR' test error

	compactiv	mortgage	pole	puma32h	treasury
p-value	$8.70 \cdot 10^{-6}$	$5.46 \cdot 10^{-8}$	$1.00 \cdot 10^0$	$4.30 \cdot 10^{-14}$	$5.76 \cdot 10^{-11}$

As it is shown in Table 3, when the number of generated rules is the same, wWRAccR' improves the results in terms of test error for 3 out of 5 problems. Table 4 shows the p-values for the comparison between both approaches for each dataset. As we can see, the precision for the wWRAccR' expression decreases for all the data sets, so having more rules for the Least Squares method leads to an improvement in test errors. However, results are still statistically better for the problems mortgage, puma32h and treasury.

Table 3. Comparison between the two prescreening expressions when the number of candidate rules is forced to be the same

Problem	Prescr. Exp.	$|\mathcal{RB}_{ap}|$	$|\mathcal{RB}_o|$	Tra.err	Test err.	#Ant.
compactiv	wWRAccR	8269.3	41.2	7.3230	7.4438	1.9
compactiv	wWRAccR'	8269.3	41.2	7.2728	**7.3880**	2.1
mortgage	wWRAccR	8104.3	51.1	0.5256	0.6631	1.7
mortgage	wWRAccR'	8104.3	51.1	0.3702	**0.4733**	1.9
pole	wWRAccR	9766.8	32.4	35.2445	**35.2378**	1.4
pole	wWRAccR'	9766.8	32.4	36.8963	36.9311	1.6
puma32h	wWRAccR	4624.0	51.4	0.0294	0.0296	1.0
puma32h	wWRAccR'	4624.0	51.4	0.0237	**0.0238**	1.0
treasury	wWRAccR	8153.1	51.0	0.7340	0.8625	1.7
treasury	wWRAccR'	8153.1	51.0	0.4826	**0.5695**	1.8

Table 4. Statistical test (t-Student) for comparison between wWRAccR and wWRAccR' test error when $|\mathcal{RB}_o|^{\text{wWRAccR'}} = |\mathcal{RB}_o|^{\text{wWRAccR}}$

	compactiv	mortgage	pole	puma32h	treasury
p-value	$3.43 \cdot 10^{-1}$	$3.36 \cdot 10^{-5}$	$1.00 \cdot 10^0$	$3.52 \cdot 10^{-12}$	$3.88 \cdot 10^{-10}$

5   Conclusions

In this work we have adapted the candidate rule generation process exposed in [4] from classification to modeling by using TSK-0 LFRBSs. Afterwards, we apply the Least Squares method to fix the consequent value for all the candidate rules. Results show that we can deal with relatively high-dimensional problems and build systems with a small number of simple rules (which have, on average, a maximum of 2.1 antecedents per rule).

We have noticed, in most cases, an improvement in terms of test error when the standard deviation is used, even when the number of candidate rules generated by the prescreening stage is forced to be the same. Therefore, the use of the standard deviation as a metric for rule's quality seems to perform well.

However, we also have noticed that using wWRAccR' for the data set pole does not lead to an improvement in any of the two experiments. We found out that so many variables are unequally distributed in this problem, being the most part of the data in one extreme of the variable domain. As we use symmetrical triangular partitions for each fuzzy variable, many candidate rules covers the most part of the data set. Hence, for both expressions used in the prescreening stage, the factor which takes into account the number of covered instances by this rules is far better, and significantly reduce the impact of the second factor in case of wWRaccR'. As a consequence, the set of candidate rules is mostly the same when using any of the two expressions.

In future works, we plan to extend this paper using TSK-1 LFRBSs, and test them with more and larger data sets. We also are attempting to improve the way fuzzy sets are generated to outperform the prediction capacity of the LFRBSs, exploiting the expressiveness of TSK-1 fuzzy rules.

Acknowledgements. This work has been partially funded by FEDER funds and the Spanish Government (MINECO) through project TIN2010-20900-C04-03. Javier Cózar is also funded by the MICINN grant FPU12/05102.

References

1. Cordón, O., Herrera, F.: A proposal for improving the accuracy of linguistic modeling. IEEE Transactions on Fuzzy Systems 8(3), 335–344 (2000)
2. Nozaki, K., Ishibuchi, H., Tanaka, H.: A simple but powerful heuristic method for generating fuzzy rules from numerical data. Fuzzy Sets and Systems 86, 251–270 (1997)
3. Wang, L.X., Mendel, J.M.: Generating fuzzy rules by learning from examples. IEEE Transactions on Systems, Man, and Cybernetics 15(1), 116–132 (1985)
4. Alcala-Fdez, J., Alcala, R., Herrera, F.: A fuzzy association rule-based classification model for high-dimensional problems with genetic rule selection and lateral tuning. IEEE Transactions on Fuzzy Systems 19(5), 857–872 (2011)
5. Kavsek, B., Lavrac, N.: APRIORI-SD: Adapting association rule learning to subgroup discovery. Appl. Artif. Intell. 20(7), 543–583 (2006)
6. Cózar, J., delaOssa, L., Gámez, J.A.: Learning TSK-0 linguistic fuzzy systems by means of Local Search Algorithms. Applied Soft Computing (2014), doi:10.1016/j.asoc.2014.03.003 (advance access published March 18, 2014)
7. Casillas, J., Cordon, O., Herrera, F.: COR: A methodology to improve ad hoc data-driven linguistic rule learning methods by including cooperation among rules. IEEE Transactions on Systems, Man and Cybernetics 32(4), 526–537 (2002)

8. Nozaki, K., Ishibuchi, H., Tanaka, H.: A simple but powerful heuristic method for generating fuzzy rules from numerical data. Fuzzy Sets and Systems 86, 251–270 (1997)
9. Ishibuchi, H., Nakashima, T., Nii, M.: Classification and Modeling with Linguistic Information Granules: Advanced Approaches to Linguistic Data Mining. Springer, Berlin (2005)
10. Agrawal, R., Imielinski, T., Swami, A.: Mining association rules between sets of items in large databases. SIGMOD Rec. 22(2), 207–216 (1993)
11. Klema, V., Laub, A.: The singular value decomposition: Its computation and some applications. IEEE Transactions on Automatic Control 25(2), 164–176 (1980)
12. Cózar, J., de la Ossa, L., Gámez, J.: Using Apriori Algorithm + Standard Deviation to Improve the Scalability of TSK-0 Rule Learning Algorithms. In: XV Conferencia de la Asociación Española Para la Inteligencia Artificial (CAEPIA), pp. 99–108 (2013)
13. Zadeh, L.: The concept of a linguistic variable and its application to approximate reasoning. Information Science 8, 199–249 (1975)

Vehicle Accident Severity Rules Mining
Using Fuzzy Granular Decision Tree

Hamid Kiavarz Moghaddam and Xin Wang

Department of Geomatics Engineering, Schulich School of Engineering, University of Calgary,
2500 University Dr. N.W. Calgary, Alberta, Canada, T2N 1N4
{hkiavarz,xcwang}@ucalgary.ca

Abstract. Road accident is a disaster that vocalizes a major cause of disability, untimely death and the loss of human lives. Therefore, investigating the condition of road accidents for prediction and prevention purposes on highways is significant. In this paper, we propose a new fuzzy granular decision tree to generate the road accident rules applying the discrete and continuous data stored in accident databases. Among all critical factors in the occurrence of traffic accidents, environmental factors and road design (geometry) are considered in this study. This method establishes an optimized fuzzy granular decision tree with the minimum redundancy and road accident severity classification using fuzzy reasoning. California highways were considered as the case study to examine the proposed approach. The experimental results demonstrate that the proposed method is approximately 16% more accurate than the fuzzy ID3 method with less redundancy in constructing the decision tree.

Keywords: Fuzzy granular decision tree, ID3, fuzzy reasoning, Fuzzy granular entropy.

1   Introduction

Road traffic accident is a social and public health challenge as it always comes with injuries and fatalities [1]. The World Health Organization reports that road collisions, as the ninth leading cause of death in 2004, will be ranked as the fifth factor in 2030 [1]. It estimates over 1 million people are killed each year in road collisions, which is equal to 2.1% of the annual global mortality, resulting in an estimated social cost of $518 billion[2]. In Canada, approximately 3,000 people are killed every year on the roads [3]. The previous traffic safety studies show that, in most accident cases, the occurrences of traffic accidents are rarely random in space and time. In this regard, identifying the risk factors that significantly influence the severity of traffic accidents, and discovering the relationships between such factors and collision severity is an important research topic. Large amount of the data regarding road collision such as the collision attributes, road condition and environmental attributes, road geometry and conditions has been accumulated over the time,. It is still a challenge to analyze and extract rules from diverse historic collision data from a large database. Data mining is a suitable solution to help decision makers recognize the rules of vehicular collision severity of large databases[4] .

C. Cornelis et al. (eds.): RSCTC 2014, LNAI 8536, pp. 280–287, 2014.
© Springer International Publishing Switzerland 2014

Decision tree is a data mining method that can generate understandable rules [5]. It provides a hierarchical representation of the data and decision path to create logical rules. ID3 is a classical method to generate decision trees. Based on the literature review, there are some problems in traditional decision tree. A major problem in the traditional decision trees such as ID3 and C4.5 is the large number of branches that causes duplication and repetition of subtrees within the tree [6] dealing with large number of attributes. Both repetition and duplication make redundancy in the decision tree methods. In that case, the tree needs to be pruned while maintaining the accuracy of the tree. Also, another issue is to prevent over-fitting [7]. The other problem is related to splitting a node. An attribute chosen based on only information about only this node causes redundant attributes at different levels. This problem also creates redundancy in the decision tree. Besides, the vehicle collision events' database contains two different kinds of attributes: discrete and continuous. The entropy in traditional algorithms considers only discrete attributes and not continuous data such as spatial measures that are some of the important factors in vehicle incidents.

To overcome above mentioned problems, this paper proposes the fuzzy granular decision tree (FGDT). The main contributions of this paper are listed as follows: 1) the fuzzy granular entropy is proposed to measure the degree of disorder or uncertainty of objects in each granular with respect to both discrete and numerical data. 2) To demonstrate the potential of fuzzy granular computing against other existing problem solving approaches, the fuzzy granular computing is applied in a case study. This method considers the environmental factors and spatial characteristics of accident locations to make a decision to categorize the road vehicle collision events in to three classes of PDO[1], Injury and Fatal. 3) The inference process of fuzzy reasoning system are proposed to use fuzzy membership functions input and the fuzzy granular decision tree rules which are if-then linguistic rules, whose antecedents and consequents are composed of fuzzy statements.

The rest of the paper is organized as follows: Constructing the fuzzy granular decision tree for vehicle collisions rule extraction is discussed in the Section 2. The implementation and case study are also elaborated in Section 3. Section 4 describes the assessment of fuzzy granular decision tree and fuzzy ID3 methods. Section 5 concludes the paper with the results from this study.

2   Fuzzy Granular Decision Tree for Vehicle Collisions Severity Rules Extraction

This section introduces the proposed method in the framework of fuzzy granular computing. As discussed earlier, in constructing the decision tree in vehicle collision database, this paper proposes a fuzzy granular decision tree (FGDT) to support the discrete and continuous data by defining the membership functions and fuzzification of data in the database. While other decision tree methods such as ID3 consider only discrete attributes; the fuzzy granular decision tree which is an extension of classical decision tree perceives both discrete and continuous attributes.

Also, to overcome the over-fitting problem in conventional decision tree methods, the FGDT chooses an attribute-value in favor of all nodes at the same level when

---

[1] Property Damage Only.

splitting a node. But in the conventional methods, the attribute is solely chosen based on the information about this node; not any other nodes at the same level. Thus, in the conventional decision tree, different nodes at the same level may use different attributes, and the same attribute with all possible values may be used at different levels which causes the over-fitting issue. The vehicle collision event rules' mining is employed to demonstrate the potential of fuzzy granular decision tree in solving the mentioned issues. The traffic accident is usually caused by human, vehicle, environmental factors, roadway design and some spatial factors [8]. Due to the lack of sufficient human and vehicle historical damage data, this research attempts to impose the environmental, roadway design and spatial factors to test the proposed methodology. Fuzzy Granular Decision Tree (FGDT) is the proposed decision tree in this paper which is a generalization of the classical decision tree. First, all the data in the training data set are fuzzified in the form of membership functions. Then, the fuzzy granular entropy is calculated for each object in the data set. Next, according to the calculated fuzzy granular entropy and generality and redundancy criteria, the fuzzy granular decision tree is constructed. The last step expresses the decision method of the final classification that is done by training and checking data using by fuzzy rules based system. In the following, the FGDT will be discussed step by step

2.1   Fuzzification of Data

This paper attempts to use the fuzzy concepts to calculate and construct the decision tree, substituting the training data with the fuzzy expression and forming the fuzzy granular decision tree. The function of fuzzy membership functions for each attribute is very significant in the creation of fuzzy granular decision tree. As such, various functions are tested, and appropriate function for each factor is determined. Triangular and trapezoidal functions (with a maximum equal to 1 and the minimum equal to 0) are widely-applied membership functions. This research uses triangular and trapezoidal membership functions because of their simplicity, learning capability, and the short amount of time required for designing the system. Based on the collision data in fuzzy membership function, the fuzzification of collision data is applied to the database by the defined membership functions.

2.2   Fuzzy Granular Conditional Entropy

Conditional entropy is the most commonly used measure for selecting attribute-value in the construction of the decision tree for classification. Many decision tree algorithms such as ID3 and common granular decision tree require data with a discrete value. Discretization of a continuous variable is not easy, particularly to determine the boundary of each interval. As an example the distance from a collision to an intersection can be named. Therefore, the fuzzy concept in the process of granular decision tree is substituted with the data set with the fuzzy expression and form the fuzzy granular decision tree method. Based on this concept, the fuzzy granular entropy is proposed to employ the continuous and discrete values in decision tree construction. The fuzzy granular conditional entropy is introduced based on defined membership values of each object in each granular because of the data fuzzy expression. Equation 1 specifies the Fuzzy Granular Conditional Entropy with the given granular universe S.

$$\text{Fuzzy Granular Conditional Entropy}(a = v) = -\sum_{i=1}^{c} \frac{\sum_{j}^{N} \mu_{ij}}{S} log_2 \frac{\sum_{j}^{N} \mu_{ij}}{S} \qquad (1)$$

where μ_{ij} is the membership value of the jth granule to the ith class. This equation is defined based on the concept of applying the value of the membership function of each factor of collision events rather than using the crisp values. The summation of the membership values of granular in the specific class is designed as a numerator of Equation 1 and the summation of membership values of all granular in a specific formula is calculated as the denominator of this equation. The calculated entropy presents the entropy of the granular S of formula $a = v$ related to training data. In the fuzzy granular conditional entropy the membership function (which belongs to formula $a = v$ for each granule) are involved to calculate the entropy.

2.3   Road Vehicle Collision Fuzzy Granular Decision Tree

To create a road collision fuzzy granular decision tree with minimum uncertainty, the subset of formula (attribute-value) with the highest values of coverage, confidence, generality and minimum granular fuzzy entropy should be selected as a node of the tree. Then the road collision rules are generated automatically based on training data of the case study. Constructing fuzzy granular decision tree involves applying concept of generality which represents the presence of a granular rather than the other granular in universe. Also, employing the notion of fuzzy granular entropy which measures the homogeneity of each granular and decrease the redundancy by selecting those granular which have the minimum redundancy rather than the other granular to cover the universe. The proposed decision tree automatically selects more appropriate nodes based on measurement of the redundancy by counting the repetitive object of granular and universe in each step to select the node with minimum redundancy and maximum coverage of universe granular. Furthermore, this process can recognize which granule is more appropriate at the end of each level to be broken down first until it reaches the granular which objects will be the subset of final classes. This granular is called non-active granular. After creating the fuzzy granular decision tree, the rules are extracted from it. The rules that are extracted from fuzzy granular decision tree have description structure based on IF-Then phrases called linguistic rules. To extract the rules, we have used a simple approach by following a path through the tree to one of the leaves. This path starts from the root of the tree to a leaf and establishes conditions, in terms of specifying the final class.

2.4   Reasoning with Fuzzy Granular Decision Tree

The process of FRBS is started from a given input to output using a set of fuzzy if-then linguistic rules which are generated from FGDC, whose antecedents and consequents are compound of fuzzy statements, related by the concepts of fuzzy implication and the compositional rule of inference [9]. The fuzzy reasoning is applied to determine the final classes of collision events in four steps. The last step called defuzzification is used to convert fuzzy value to the final crisp classes of PDO, Injury and Fatal value. This crisp number is obtained in a process known as the defuzzification. Centroid defuzzification method denotes a point representing the center of gravity of the aggregated fuzzy set A, on the interval [a,b] which can be calculated using [10]:

$$z_{COG} = \frac{\int_a^b z \cdot \mu_A(z)dz}{\int_a^b \mu_A(z)dz} \tag{2}$$

where z_{COG} is the crisp output, $\mu_A(Z)$ is the aggregated membership function, and z is the output variable. In the case study section, there is an example of FRBS based on real data of collision events.

3   Case Study in California State

The data which is used in this study includes 1837 collision event points which are recorded in the databases of years 2009 and 2010 belonging to 13 counties of California State. In the dataset, 1004 points were considered as training data and the other 833 event points are taken as testing data. Table 1 demonstrates the sample of the first level of fuzzy granular decision tree measures so that the universe is equal to {O1,O2,O3, O4, O5, O6, O7, O8, O9, O10, O10, O11, O12, O13, O14, O15, O16}. The seven granules of formulas Road Lighting = Day Light, Dist to Intersection = Far, Weather = Clear, Dist to Intersection = Very Near, Collision Time = Morning, Road Radius = Small and Road Slope = Low have the minimum values of entropy. They are sorted in order of their value of entropy. The formula Road Lighting = Day Light with granule = {O1, O4, O7, O10, O13} is chosen as the first node of fuzzy granular decision tree due to its least entropy. These other six granules cannot include the universe thereby; they are not a covering solution to reduce the redundancy. The algorithm will search and analyze other granules in order to find a set of granules that cover the whole universe. The algorithm considers the non-redundant covering and removes those six adding candidates since they cannot form a non-redundant covering.

Table 1. The Road Collision Information Table

Objects	Weather	Surface	Lighting	Time	Radius (m)	Slope (%)	Distance from Intersection(m)	Distance from Population Centers(m)	Severity
O_1	Clear	Dry	Day-Light	11	800.00	3.00	360.00	1700.00	PDO
O_2	Clear	Dry	Dusky/Dark	10	850.00	7.00	180.00	2800.00	PDO
O_3	Clear	Dry	Dusky/Dark	20	900.00	8.00	280.00	1000.00	Injury
O_4	Clear	Not Dry	Day-Light	9	300.00	4.00	200.00	1900.00	Injury
O_5	Clear	Not Dry	Dusky/Dark	13	550.00	11.00	90.00	900.00	Injury
O_6	Clear	Not Dry	Dusky/Dark	21	880.00	3.50	170.00	1800.00	Injury
O_7	Raining	Not Dry	Day-Light	12	890.00	4.00	420.00	1100.00	PDO
O_8	Raining	Not Dry	Dusky/Dark	14	350.00	10.00	220.00	750.00	Injury
O_9	Raining	Not Dry	Dusky/Dark	15	350.00	18.00	75.00	420.00	Fatal
O_{10}	Fog	Dry	Day-Light	11	770.00	6.00	320.00	1450.00	PDO
O_{11}	Fog	Dry	Dusky/Dark	10	500.00	9.00	240.00	350.00	Injury
O_{12}	Fog	Dry	Dusky/Dark	21	780.00	13.00	130.00	980.00	Injury
O_{13}	Fog	Not Dry	Day-Light	10	1500.00	4.00	170.00	1200.00	PDO
O_{14}	Fog	Not Dry	Dusky/Dark	20	450.00	3.00	110.00	650.00	Fatal
O_{15}	Fog	Not Dry	Dusky/Dark	22	600.00	7.00	110.00	650.00	Fatal
O_{16}	Fog	Not Dry	Dusky/Dark	21	850.00	14.00	110.00	650.00	Fatal

As a consequence, it will not choose these granules even if other measures are in favor of this granule. If many objects in a candidate granule are already in granular decision tree, this granule will not be chosen. The formula of Road Lighting = Dusk-Down with the granular of {O2,O3,O5,O6,O8,O9,O11,O12,O14,O15,O16} is considered the most suitable granule to cover the universe, and will be chosen accordingly. It can be verified that the union of two chosen formula granules {O1,O4,O7,O10,O13},{O2,O3,O5,O6,O8,O9,O11,O12,O14,O15,O16}   satisfied covering universe with no redundancy. Obviously, the objects in Road Lighting = Day Light and Lighting = Dusk-Down do not belong to the same decision classes because they are active nodes, therefore, further granulation to this granule will be conducted in order to find smaller definable granules.

4   Reasoning with Fuzzy Granular Decision Tree

The first step is the fuzzification interface which is transforming a crisp data into fuzzy sets. The rule evaluation is the next step in which the strengths of rules are computed based on the extracted rules and inputs. Then, they should be applied to antecedents of the fuzzy rules. In this study, the minimum (AND) fuzzy value is applied as the strengths of rules. Fig.1 illustrates the rules evaluation step of some training data as a sample of fuzzy rule evaluation. The third step is the aggregation of the rules' outputs which is the process of unification of the outputs of all rules. The last step is the defuzzification that translates the fuzzy rule which in turn, translates the results back to the real world values.

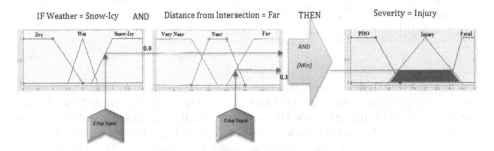

Fig. 1. The rule evaluation of extracted rules

5   Accuracy Assessment of Fuzzy Granular Decision Tree and Fuzzy ID3 Collisions Methods

With implementing FGDT and fuzzy ID3methods, 22 rules are extracted from granular, and 25 rules are extracted from fuzzy ID3, which are applied to 833 road collision events in California. To assess the accuracy of the results, the error matrix is proposed to be used in classification accuracy assessment [11]. Overall accuracy, validity (producer accuracy) and reliability (or user accuracy which are represented in Equation 3, 4 and 5, respectively), are derived from the error matrix for both fuzzy granular and fuzzy ID3 resulted classes. Table 2 represents the overall, producer and user accuracy.

$$OveralAccuracy = O.A. = \frac{\sum_{i=1}^{c} a_{ii}}{N} \tag{3}$$

$$\Pr oducer.Accuracy = P.A._j = \frac{a_{jj}}{\sum_{i=1}^{r} a_{ij}} \tag{4}$$

$$User.Accuracy = U.A._j = \frac{a_{ii}}{\sum_{j=1}^{r} a_{ij}} \tag{5}$$

where a_{ii} is number of collision events in the class i in row, which are classified by classifiers and class j in column which are labeled in reality.

Table 2. The O.A,O P.A. U.A. values of Granular Algorithm

	Fuzzy Granular Decision Tree			Fuzzy ID3		
	O.A.	**P.A.**	**U.A.**	**O.A.**	**P.A.**	**U.A.**
Class1 = PDO	68.7%	85%	62%	52%	77%	58%
Class2 = Injury		33%	62%		26%	52%
Class3 = Fatal		60%	79%		48%	27%

The results in Table 2 proves that near 68% of the collision events in testing data are in complete agreement using the fuzzy granular decision tree while 52% of them are consistent with that of fuzzy ID3 decision. As the calculated reliability of the fuzzy granular method results shows, it can be expected that 62% of all the classified collision events PDO and Injury class are indeed the same as PDO and Injury on checking data. Also, 79% of fuzzy collision events as Fatal are indeed in a class of Fatal on the checking data. However, the validity shows that the fuzzy granular classified 85% of all PDO events, 33% of all Injury events and 60% of all Fatal events in the database. The calculated user accuracy of the fuzzy ID3 method shows 52% of all the classified collision events as PDO are the same as PDO class in checking data; Moreover, 52% of all classified collision events as Injury are indeed Injury on the checking data and 27% of all the classified collision events as Fatal matches with Fatal class in checking data. It can be concluded that the overall accuracy, reliability and accuracy of each individual class on fuzzy granular method, are higher than those of the fuzzy ID3 method.

6   Conclusion

This paper proposed a new approach to extract rules for predicting vehicular road collisions. The results indicate that the fuzzy granular decision tree finds the most suitable granules defined by an attribute-value pair which is selected considering the continuous values in the database. Moreover, using the fuzzy data and fuzzy entropy efficiently impacts the performance of the learning by involving the discrete and continuous values in the database. This leads to more accurate results in comparison with classical decision trees including fuzzy ID3. This research attempted to compare the proposed method with decision tree method and not other classification approaches. As a future work, the proposed method can be compared with other classification methods such as C-RT, Naïve Bayes, Rnd Tree, MLN, ADA-boost and SVM.

References

1. Organization, W.H.: WHO global status report on road safety 2013: supporting a decade of action. World Health Organization (2013)
2. Peden, M., et al.: World report on road traffic injury prevention. World Health Organization Geneva (2004)
3. Wang, J.: A traffic accident risk mapping framework (2012)
4. Shanthi, S., Ramani, R.G.: Classification of Vehicle Collision Patterns in Road Accidents using Data Mining Algorithms. International Journal of Computer Applications 35 (2011)
5. Peng, Y., Flach, P.A.: Soft discretization to enhance the continuous decision tree induction. In: Proceedings of ECML/PKDD-2001 Workshop IDDM 2001, Freiburg, Germany (2001)
6. Han, J., Kamber, M., Pci, J.: Data mining: concepts and techniques. Morgan Kaufmann (2006)
7. Osei-Bryson, K.-M.: Post-pruning in decision tree induction using multiple performance measures. Computers & Operations Research 34(11), 3331–3345 (2007)
8. Geurts, K., et al.: Profiling of high-frequency accident locations by use of association rules. Transportation Research Record: Journal of the Transportation Research Board 1840(1), 123–130 (2003)
9. Roisenberg, M., Schoeninger, C., da Silva, R.R.: A hybrid fuzzy-probabilistic system for risk analysis in petroleum exploration prospects. Expert Systems with Applications 36(3), 6282–6294 (2009)
10. Runkler, T.A.: Selection of appropriate defuzzification methods using application specific properties. IEEE Transactions on Fuzzy Systems 5(1), 72–79 (1997)
11. Zhao, Y., Yao, Y., Yan, M.: ICS: An interactive classification system. In: Kobti, Z., Wu, D. (eds.) Canadian AI 2007. LNCS (LNAI), vol. 4509, pp. 134–145. Springer, Heidelberg (2007)

Using Imprecise Probabilities to Extract Decision Rules via Decision Trees for Analysis of Traffic Accidents

Griselda López[1], Laura Garach[1], Joaquín Abellán[2],
Javier G. Castellano[2], and Carlos J. Mantas[2]

[1] Department of Civil Engineering, University of Granada, Spain
{griselda,lgarach}@ugr.es
[2] Department of Computer Science and A.I., University of Granada, Spain
{jabellan,fjgc,cmantas}@decsai.ugr.es

Abstract. The main aim of this study is focused on the extraction or obtaining of important decision rules (DRs) using decision trees (DTs) from traffic accidents' data. These decision rules identify patterns related with the severity of the accident. In this work, we have incorporated a new split criterion to built decision trees in a method named *Information Root Node Variation* (IRNV) used for extracting these DRs. It will be shown that, with the adding of this criterion, the information obtained from the method is improved trough new and different decision rules, some of them use different variables than the ones obtained with the original method.

Keywords: IRNV, Imprecise Info-Gain, decision rules, traffic accident severity.

1 Introduction

DTs have been widely applied in the road safety research, being the CART method developed in [1] the most used. One of the reasons for use DTs in order to analyze traffic accident′ severity is that the structure of a DT permits easily the extraction of Decision Rules (DRs). These DRs provide a direct way to describe the relationships between the accident attributes and can be used by safety analysts to understand the events leading up to an accident and identify the variables that determine how serious it will be. However, the extraction of knowledge with DRs obtained from a DT is very limited, and some important pattern cannot be found using only one DT.

In order to extract all the knowledge from a particular dataset, the IRNV method used in [2] is applied in this study. The main characteristic of this method is that different DTs are built by varying the root node. The useful rules could be used by road safety analyst to establish specific measures of performance.

It has been shown that the new split criterion of Abellán and Moral: *Imprecise Info-Gain* [3], noted as IIG, based on imprecise probabilities and uncertainty measures, has a different performing than the ones from the classic split criteria [4]. The handling of the imprecision is a key part of the difference above mentioned. As a logical consequence of this, we have incorporated this criterion in the method of IRNV, adding it as a third criterion in that method. We considered that the new split criterion

C. Cornelis et al. (eds.): RSCTC 2014, LNAI 8536, pp. 288–298, 2014.
© Springer International Publishing Switzerland 2014

could increase the information obtained from data because it has a different performing. In this paper, we will prove this assertion in an experimental study. In addition it
the first practical application (in particular, studding traffic accident), that has been
realized using this split criterion.

The paper is organized as follows: In Section 2, the data used to carry on this study
is presented, and the methodology used is described. In Section 3, the outcomes obtained with the extended method are detailed and analyzed. Finally, the last section is
devoted to the conclusions.

2   Materials and Methods

2.1   Data Description

The data used in this study comes from the Directorate General of Traffic [5]. Only
traffic accidents that occurred on rural two-lane highways for the province of Granada
(Spain) were analyzed. The period of study was seven years (2003-2009). The data set
was first checked out for questionable data, and those which were found to be unrea-
listic were screened out. The accidents analyzed involved 1 vehicle and they did not
occur on intersections. Therefore the data set used to conduct the study contains 1801
accidents.

The class variable is the severity of the accidents (SEV). It was defined according
to the level of injury for the worst injured occupant (following previous studies such
as [6,7], [8]). With the original classification of accidents by severity, there are 149
fatal, 723 serious and 929 slight ones. Since the different categories of the variable
severity are not balanced and this issue affects the overall accuracy of the model [8],
the class variable was re-coded in two levels: SI - accidents with slightly injured
(929); and KSI - accidents with killed or seriously injured (872).

Nineteen variables (see description in Table 1) were used with the class variable
(SEV) in an attempt to identify the important patterns of an accident related with their
severity. The choice of the variables and their categorization were mainly guided by
previous studies ([7,9], [10]).

The dataset includes variables describing the conditions that contributed to the accident and injury severity:

— Injury severity variables: number of injuries and severity level of injuries
— Roadway information: safety barriers, pavement width, lane width, shoulder type,
  paved shoulder, road markings and sight distance.
— Context information: atmospheric factors and lighting.
— Accident information: causes, day, hour, month, occupant involved, type of acci-
  dent.
— Vehicle information: type of vehicle.
— Driver information: age and gender.

Table 1. Description of the set of variables in the dataset

NUM.	VARIABLE: CODE	VALUES: CODE	TOTAL	SEVERITY %SI	%KSI
1	Accident type: ACT	Fixed objects collision: **CO**	19	76.47	23.53
		Collision with pedestrian: **CP**	152	33.33	66.67
		Other (collision with animals, etc.): **OT**	32	68.57	31.43
		Rollover (carriage without collision): **RO**	118	61.86	38.14
		Run off road (with or without collision): **ROR**	1480	51.77	48.23
2	Age: AGE	≤ 20: **≤ 20**	219	52.73	47.27
		[21-27]: **[21-27]**	492	50	50
		[28-60]: **[28-60]**	948	51.76	48.24
		≥ 61: **≥ 61**	110	59.68	40.32
		Unknown: **UN**	32	27.59	72.41
3	Atmospheric factors: ATF	Good weather: **GW**	1540	50.58	49.42
		Heavy rain: **HR**	43	63.16	36.84
		Light rain: **LR**	161	58.75	41.25
		Other: **O**	57	51.06	48.94
4	Safety barriers: BAR	No: **N**	1740	48.3	54.7
		Yes: **Y**	61	53.6	46.4
5	Cause: CAU	Driver characteristics: **DC**	1471	48.99	51.01
		Combination of factors: **CO**	262	61.16	38.84
		Other: **OT**	29	72.73	27.27
		Road characteristics: **RC**	24	84	16
		Vehicle characteristics: **VC**	15	63.64	36.36
6	Day: DAY	Working day after weekend or public holiday: **APH**	131	57.62	42.38
		Working day before weekend or public holiday: **BPH**	286	52.26	47.74
		On a weekend or public holiday: **PH**	532	50.36	49.64
		Regular working day: **WD**	852	51.05	48.95
7	Lane width: LAW	< 3,25 m: **THI**	503	46.87	53.13
		[3,25-3,75] m: **MED**	1264	53.2	46.8
		> 3,75 m: **WID**	34	58.54	41.46
8	Lighting: LIG	Daylight: **DAY**	958	55.49	44.51
		Dusk: **DU**	103	54.29	45.71
		Insufficient (night-time): **IL**	131	51.15	48.85
		Sufficient (night-time): **SL**	66	59.72	48.28
		Without lighting (night-time): **WL**	543	43.1	56.9
9	Month: MON	Autumn: **AUT**	412	53.07	46.93
		Spring: **SPR**	440	53.64	46.36
		Summer: **SUM**	479	51.63	48.37
		Winter: **WIN**	470	47.92	52.08
10	Number of injuries: NOI	1 injury: **[1]**	1233	53.43	46.57
		> 1 injury: **[>1]**	568	47.35	52.65
11	Occupants involved: OI	1 occupant: **[1]**	1171	51.2	48.8
		2 occupants: **[2]**	374	51.48	48.52
		> 2 occupants: **[>2]**	256	53.71	46.29
12	Paved shoulder: SHT	No: **N**	309	49.35	50.65
		Non existent or impassable: **NE**	580	50.89	49.11
		Yes: **Y**	912	52.74	47.26

Table 1. (*continued*)

13	Pavement width: PAW	[6-7] m: **MED**	530	53.19	46.81
		< 6 m: **THI**	282	45.56	54.44
		> 7 m: **WID**	989	52.27	47.73
14	Pavement markings: ROM	Does not exist or was deleted: **DME**	168	52.35	47.65
		Separate margins of roadway: **DMR**	180	48.31	51.69
		Separate lanes and define road margins: **SLD**	1368	52.23	47.77
		Separate lanes only: **SLO**	85	46.59	53.41
15	Gender: SEX	Female: **F**	286	62.18	37.82
		Male: **M**	1513	49.61	50.39
		Unknown: **UN**	2	75	25
16	Shoulder type: SHW	< 1.5 m: **THI**	699	52.54	47.46
		[1.5-2.5] m: **MED**	898	50.28	49.72
		Non existent or impassable: **NE**	204	50.57	49.43
17	Sight distance: SID	Atmospheric: **ATM**	30	67.5	32.5
		Building: **BU**	6	36.36	63.64
		Other: **OT**	12	50	50
		Topography: **TOP**	420	49.39	50.61
		Vegetation: **VEG**	13	50	50
		Without restriction: **WR**	1320	51.94	48.06
18	Time: TIM	[00:00-05:59]: **[0-6)**	340	48.06	51.94
		[06:00-11:59]: **[6-12)**	380	58.73	41.27
		[12:00-17:59]: **[12-18)**	591	52.77	47.23
		[18:00-23:59]: **[18-24)**	490	47.22	52.78
19	Vehicle type: VEH	Cars: **CAR**	1287	47.1	52.9
		Trucks: **TRU**	78	53.8	46.2
		Motorbikes and motorcycles: **MOT**	385	35.6	64.4
		Other: **OT**	51	50.6	49.4

2.2 Decision Trees, Split Criteria and IRNV Method

A decision tree (DT) is a structure that can be used in classification and regression tasks. If the class variable, i.e. the variable under study, has a finite set of possible values, the task is named as classification; in other case, is named as regression.

Within a decision tree, each node represents an attribute variable or feature (a characteristic of each item in the dataset) and each branch represents one of the values or states of this variable. A tree leaf specifies the expected value of the class variable. Associated to each node is the most informative variable, according a split criterion (the criterion to branching), which has not already been selected in the path from the root to this node. If the information about the class variable is not improved or there are no more features to choose, a leaf node is added with the most probable class value for the partition of the dataset associated to that node. A DT can be interpreted as a compact set of rules about the class variable.

A key part of the procedure to build a DT is the split criterion. In the literature we can see many works focused on the use of classic split criteria (see [4]). The most used ones are the Information Gain (IG), Information Gain Ratio (IGR) and the Gini Index (GInf). IG and IGR were presented in [11,12] and GInf in [1].

The Imprecise Info Gain (IIG) was presented in [3] and has a different performing than the classic ones (see [4]). It is based on the use of imprecise probabilities and uncertainty measures. This criterion can be defined as follows: in a classification problem, let C be the class variable, $\{X_1,...,X_m\}$ the set of features, and X a feature; then

$$IIG(C,X) = H^*(K(C)) - \sum_i P(X = x_i) H^*(K(C \mid X = x_i)),$$ (1)

where $K(C)$ and $K(C|X=x_i)$ are the convex sets of probability distributions obtained via the Imprecise Dirichlet Model for the C and $(C|X=x_i)$ variables respectively [3] and the function $H^*(K(Z))$ is the maximum Shannon's entropy function of all the probability distributions that belong to the set K. This set K on a variable Z with values belong to $\{z_1,...,z_k\}$ is defined as

$$K(Z) = \left\{ p \mid p(z_j) \in \left[\frac{n_{z_j}}{N+s}, \frac{n_{z_j}+s}{N+s} \right], j = 1,...,k \right\}$$ (2)

with nz_j as the frequency of the set of values $(Z=z_j)$ in the dataset, N the sample size and s a given hyperparameter. To calculate H^* for $s=1$, the simple procedure of [13] can be used.

A procedure to build DTs can be explained of the following way: Each node No in a DT produces a partition D of the dataset (for the root node the entire dataset is considered). Also, each node N has associated a list "Γ" of labels of features (features that are not in the path from the root node to No). A recursive and simple procedure to build a DT can be expressed by the algorithm shown in Figure 1.

Procedure BuildTree (No, Γ)

1. *If $\Gamma = \Phi$, then **Exit***
2. *Let D be the partition associated with node No*
3. *Compute the value of the maximum gain of information for a feature on D (using a split criterion: SC)*
   $\delta= max\ SC(C,X)$
4. *If δ is lower than or equal to 0 then **Exit***
5. *Else*
   6. *Let X_t be the variable for which the maximum δ is attained*
   7. *Remove X_t from Γ*
   8. *Assing X_t to the node No*
   9. *For each possible value x_t of X_t*
   10. *Add a node N_t*
   11. *Make No_t a child of No*
   12. *Call BuilTree (No_t, Γ)*

Fig. 1. Algorithm to build a DT

Each Exit state in the above procedure corresponds to a leaf node. Here, the most probable value of the class variable, associated with the corresponding partition, is selected.

The method called Information Root Node Variation (IRNV), to extract Decision Rules (DRs), is based on using different trees obtained by varying the root node. In this method, if there are m features, and RXi is the feature that occupies position i in importance (gain of information via a split criterion); RXi is used as the root to build DTi (i=1,...,m). We use the simple method for building trees explained above, none-theless now the root node is selected directly for each tree (the rest of the building procedure remains the same). Thus, we obtain m trees and m rule sets (RS), DTi and RSi (i=1,...,m), respectively. Each RSi is checked in the test set to obtain the final rule set. The entire procedure is carried out using GInf and IGR criteria.

The process of the method can be explained via the following scheme (for more details see [2]):

1. Select a split criterion (SC) for building trees.
2. Build DT_i using RX_i, as the root node, and SC (i=1,...,m.)
3. Extract RS_i from each DT_i.
4. Check RS_i in the corresponding TEST set → Selection of rules from RS_i.
5. Extract the final rule set obtained by using the SC.
6. Change of SC and go back to step 2.
7. Join the final rule sets obtained using GInf and IGR.

In the original method of the IRNV, IGR and GInf are used as SC. In this work, we incorporate the IIG split criterion in the IRNV method, i.e. we also use the IIG crite-rion in the point 6 of the above scheme. Hence, we build 3m DTs in the method to extract DRs.

The new split criterion is more complex than the classic ones. The computational complexity of the method for building DTs via the IIG criterion is analyzed in [14]. It has a complexity of order $O(N^2m)$, with N the sample size of the data set and m the number of features.

2.3   Significant Decision Rules

A DR conforms a logical-conditional structure of the type "IF (X) → THEN (Y)", where A is the antecedent of the rule (in our case, a set of statuses of several attribute variables); and B is the consequent (in our case, it is only one state of the class vari-able).

Each rule starts at the root node, and each variable that intervenes in tree division makes an IF of the rule, which ends in leaf nodes with a value of THEN (associated with the state resulting from the leaf node). The resulting state is the status of the class variable that shows the highest number of cases in the leaf node analyzed.

A priori, the number of rules can be identified with the number of terminal nodes in the tree. Then, specific parameter and minimum thresholds are used to extract sig-nificant rules ([2], [7], [10]):

— *Support* (S) is the percentage of the dataset where "A & B" appear. Minimum threshold is S≥ 0.6%.

— *Population* (Po) is the percentage of the dataset where "A" appears. Minimum threshold is Po≥1%.

— *Probability* (P) is the percentage of cases in which the rule is accurate (i.e. P=S/Po expressed as percentage). Minimum threshold is P≥60%.

The parameters' thresholds for Po, C, and S are normally selected depending on the nature of the data (balanced or unbalanced), significant interest in fatal crashes (rare events), and sample size (small or large datasets). As this work uses the same data as in [9], we also use the same thresholds.

Due to the large number of patterns considered, DTs can suffer from an extreme risk of Type-1 error, that is, of finding patterns that appear due to chance alone to satisfy constraints on the sample data [15]. To reduce this error and following other authors ([16] [7][8][10]) the rules extracted on the training set (with the minimum parameters) are validate using the test set.

In addition the rules are formed by four variables as much because DTs with only four levels of proof are built. The main reason for including only four levels of proof is that simple and understandable rules (from safety point of view) are needed by Administration and Authorities. If the rules have more variables could not be useful. Previous studies such as Montella et al. [10] and Montella et al. [17] in which DTs and Association Rules are obtained to study traffic accidents, use the same levels of proof.

3   Results

In the first step, the dataset was randomly split into two different sets: training (70%) and test (30%) as in [6], [8]. Thus, 1,260 accidents formed the training set with the following severity distribution: 646- KSI, and 614-SI.

Next, using the training set, the IRNV method is applied. The different DTs varying the root node were built using Weka platform [18]. The procedures for building the DTs based on Imprecise Info-Gain and the root node variation procedure were implemented using the method proposed in [13]. DTs were built with four levels of proof; previous studies such as [2], [10,17] use the same number of levels. This number of levels allows us to find useful and understandable rules by the safety analysts.

Table 2 provides the number of DTs, the root nodes and the number of rules in the different steps of IRNV method obtained with the IIG criterion.

Table 2. Number of rules obtained with IRNV using the IIG criterion

DTs	DT$_1$	DT$_2$	DT$_3$	DT$_4$	DT$_5$	DT$_6$	DT$_7$	DT$_8$	DT$_9$	DT$_{10}$	DT$_{11}$	DT$_{12}$	DT$_{13}$	DT$_{14}$	DT$_{15}$	DT$_{16}$	DT$_{17}$	DT$_{18}$	DT$_{19}$	Total
R.N	ACT	LIG	SEX	CAU	VEH	ATF	PAW	TIM	AGE	NOI	DAY	SID	LAW	MON	OI	ROM	SHW	BAR	SHT	
R.T.	12	19	13	15	6	9	14	24	16	13	17	20	14	18	19	13	16	9	11	278
R.V.	5	3	4	4	4	5	4	8	7	8	4	8	9	6	7	5	7	3	6	107

R.N. Root node; R.T. Rules Training; R.V. Rules validated.

Next, we comment the results with IIG split criterion. The root node in DT_1 is the variable ACT. In this tree, 12 rules can be extracted from the training set. With all the trees, 278 rules are obtained. In addition, the variable that generates the highest number of rules when they are used as root node is TIM with 24 rules (see Table 2). When the rules are validated using the test set, the number of rules decreases to 107. Highlighted that, all DTs generate valid DRs (verify the minimum threshold fixed for the parameters S, Po and P). Remark that when we use the variable LAW as root node it is generates the highest number of valid rules (9 rules).

In its original form, IRNV method is applied with two different split criterions GInf (based on the Gini Index) and IGR (info gain ratio). The method obtains for the training set 227 rules with GInf and 174 with IGR; however applying the IRNV method with the IIG criterion a bigger set of rules is obtained (278 rules).

As the same way, for the test set with GInf and IGR, a minor number of rules are validated, 78 and 81 rules respectively (see [2]); whereas with the IIG split criterion we obtain 107 validated rules.

Overall of the rules identified with the IIG split criterion, 21 rules (of 107) also were identified with GInf, and only 6 rules (of 107) are shared with IGR. Then, using this split criterion in the IRNV method, new and interesting information from the same dataset has been obtained.

Regarding with the parameters, in the rules extracted with the three split criteria, confidence ranges between 60% and 100% are presented. The rule with the biggest support (27%) is obtained with GInf, for IIG the maximum support is 17.6%, whereas with IGR the values of support are minor, being the maximum support 8.9%. About the population parameter, the biggest value is attained for GInf (with 41.3%), following for the IIG (28%) and the last is IGR with 13%.

The IIG criterion identifies more different variables in the rules that the criterion GInf or IGR.

The following new statuses of variables in the rules are identified (they appear in significant rules) when the IIG split criterion is applied instead of the criterion GInf:

ROM=SLO, ROM=DMR, VEH=TRU and LIG=DU.

Also, when the IIG split criterion is applied instead of the criterion IGR the following new statuses of variables are identified:

ROM=SLO; ROM=DMR, VEH=TRU, MON=SUM, PAW=THI and AGE $\geq$ 61

The distribution of the 107 rules obtained with the new criterion is the following: 50 patterns for SI accidents and 57 patterns for KSI accidents. Due to the large number of patterns obtained, only rules with S>4% are extracted on Table 3. The support is a parameter that combines confidence and population (a support higher than 4% implies that the rule is met by at least 50 accidents in the sample under study).

For KSI rules, the maximum value for confidence is 78.57% (rules 1 and 3). Support ranking between 4.92 and 4.37; and population ranges between 6.35 and 5.56.

To reduce run-off-road accidents is one of the priorities of the Spanish Road Safety Strategy 2011-2020 [5]. Rules 1 and 3 identify one of the most important concerns for road safety in Spain: run-off-road accidents and they are related with motorcycles.

In addition these patterns shown that the presence of shoulders affects the severity of the accident.

Rule 2 shows a problem related with the lighting condition and lane width. [7,9] also pointed out that KSI accidents are associated with roadways with no lighting.

For SI rules parameter of confidence varies from 60.92% (rule 7) to 68.67% (rule 6). Support varies from 4.21 (rule 7) to 17.62 (rule 10). And population ranges from 28.10% (rule 10) to 6.35% (rule 11).

Most of the SI rules shown patterns for run-off-road accidents, in which, the type of vehicle involved is a car. These patterns have one parameter related with the road: medium pavement (rules 4, 8 and 12) or lane (rule 10), shoulders minor than 1.5 m (rule 13) or paved shoulder (rule 14). So, one safety measure in order to reduce the severity is the improvement of the shoulders.

Rule 7 identifies run-off-road accidents with car for young drivers when the type of day is a working day. This pattern could be showing a particular problem related to the inexperience of young drivers in this kind of roads.

Rules 6, 9 and 11 show similar SI patterns in which vehicle involved is a car. Two of them occur at the morning time (rules 6 and 9), for working days with good weather conditions (rule 6) or during a day on roads without sight distance restrained (rule 9). In rule 11 accidents also occurs on working days, in summer with only one occupant involved.

Then, from safety point of view accidents which occur on working days, with good weather conditions or during a day have a minor severity.

Table 3. DRs ordered by severity

NUM	ANTECEDENT	CONSE-QUENT	Po (%)	S (%)	C (%)
1	NOI=[1];VEH=MOT; ACT=ROR;SHT=Y	KSI	5.56	4.37	78.57
2	SID=WR;LIG=WL; LAW=THI;SEX=M	KSI	6.35	4.92	77.50
3	OI=[1];VEH=MOT; ACT=ROR;SHT=Y	KSI	5.56	4.37	78.57
4	ACT=ROR;VEH=CAR; PAW=MED;ROM=SLD	SI	14.21	9.13	64.25
5	ATF=GW;LIG=DAY; SEX=F;OI=[1]	SI	6.59	4.44	67.47
6	TIM=(6-12);VEH=CAR; DAY=WD;ATF=GW	SI	6.59	4.52	68.67
7	AGE=(20-27);ACT=ROR; DAY=WD;VEH=CAR	SI	6.90	4.21	60.92
8	NOI=[1];VEH=CAR; ACT=ROR;PAW=MED	SI	11.27	7.54	66.90
9	SID=WR;LIG=DAY; VEH=CAR;TIM=(6-12)	SI	9.37	6.35	67.80
10	LAW=MED;VEH=CAR; ACT=ROR;NOI=[1]	SI	28.10	17.62	62.71

Table 3. (*continued*)

11	MON=SUM;DAY=WD; VEH=CAR;OI=[1]	SI	6.35	4.60	72.50
12	OI=[1];VEH=CAR; ACT=ROR;PAW=MED	SI	10.56	6.98	66.17
13	SHW=THI;ACT=ROR; VEH=CAR;NOI=[1]	SI	17.46	10.79	61.82
14	SHT=Y;ACT=ROR; VEH=CAR;NOI=[1]	SI	20.48	13.02	63.57

4   Conclusions

We have incorporated a new split criterion based on imprecise probabilities and uncertainty measures on a method to obtain Decision Rules called IRNV. This method uses only classic split criteria. We have showed that the incorporation of this new split criterion can complete the information extracted for data in different ways.

With the new split criterion in the procedure of the IRNV, more rules than with the classic criteria are generated. Finally, 170 new validated rules have been obtained.

From safety point of view interested results have been presented using the new split criterion: KSI rules for the motorcyclists' run-off-road accidents are obtained. In additions, the rules obtained imply others important variables: roadways with no lighting and paved shoulder. The first is associated with KSI accidents. The presence of shoulders affects the severity of the accident Therefore, some countermeasures that could be applied by the Administration in order to reduce the severity of these accidents are: remove dangerous obstacles in the roadside and adopt improved safety barrier designs and new crash test procedures to protect Powered Two Wheelers (PTWs).

Also, it is stressed the need for studying the conditions in the environment of two-lane rural highways (i.e. shoulders or lighting, etc.), because they have a substantial impact on crash severity; it is found particular factors related with a minor level of severity (such as working days, good weather conditions or lighting day).

As a future work, from the safety point of view, specific studies analyzing the main factors that affect the severity of motorcycled will be realized. And the used of the proposed method for studies other datasets (i.e., other infrastructure, roads and countries) or specific dangerous locations (i.e., black points, or intersections).

References

1. Breiman, L., Friedman, J., Olshen, R., Stone, C.: Classification and Regression Trees. Chapman & Hall, Belmont (1984)
2. Abellán, J., De Oña, J., López, G.: Analysis of traffic accident severity using decision rules via decision trees. Expert Systems with Application 40, 6047–6054 (2013)
3. Abellán, J., Moral, S.: Building classification trees using the total uncertainty criterion. International Journal of Intelligent Systems 18(12), 1215–1225 (2003)

4. Mantas, C.J., Abellán, J.: Analysis and extension of decision trees based on imprecise probabilities: Application on noisy data. Expert Systems with Applications 41, 2514–2525 (2014)

5. DGT. Spanish Road Safety Strategy 2011-2020. Traffic General Directorate, Madrid, 222 p. (2011)

6. Chang, L.Y., Wang, H.W.: Analysis of traffic injury severity: an application of non-parametric classification tree techniques. Accident Analysis and Prevention 38, 1019–1027 (2006)

7. De Oña, J., López, G., Abellán, J.: Extracting decision rules from police accident reports through decision trees. Accident Analysis and Prevention 50, 1151–1160 (2013)

8. Kashani, A., Mohaymany, A.: Analysis of the traffic injury severity on two-lane, two-way rural roads based on classification tree models. Safety Science 49, 1314–1320 (2011)

9. De Oña, J., López, G., Mujalli, R.O., Calvo, F.J.: Analysis of traffic accidents on rural highways using Latent Class Clustering and Bayesian Networks. Accident Analysis and Prevention 51, 1–10 (2013)

10. Montella, A., Aria, M., D'Ambrosio, A., Mauriello, F.: Data Mining Techniques for Exploratory Analysis of Pedestrian Crashes. Transportation Research Record 2237, 107–116 (2011)

11. Quinlan, J.R.: Induction of decision trees. Machine Learning 1, 81–106 (1986)

12. Quinlan, J.R.: Programs for machine learning. Morgan Kaufmann series in Machine Learning (1993)

13. Abellán, J., Masegosa, A.: An ensemble method using credal decision trees. European Journal of Operational Research 205(1), 218–226 (2010)

14. Abellán, J., Moral, S.: Upper entropy of credal sets. Applications to credal classification, International Journal of Approximate Reasoning 39(2-3), 235–255 (2005)

15. Webb, G.I.: Discovering significant patterns. Machine Learning 68, 1–33 (2007)

16. Chang, L.Y., Chien, J.T.: Analysis of driver injury severity in truck involved accidents using a non-parametric classification tree model. Safety Science 51, 17–22 (2013)

17. Montella, A., Aria, M., D'Ambrosio, A., Mauriello, F.: Analysis of powered two-wheeler crashes in Italy by classification trees and rules discovery. Accident Analysis and Prevention 49, 58–72 (2012)

18. Witten, I.H., Frank, E.: Data Mining: Practical Machine Learning Tools and Techniques, 2nd edn. Morgan Kaufmann, San Francisco (2005)

Selecting the Most Informative Inputs in Modelling Problems with Vague Data Applied to the Search of Informative Code Metrics for Continuous Assessment in Computer Science Online Courses

José Otero[1], Maria Del Rosario Suárez[1], Ana Palacios[2],
Inés Couso[3], and Luciano Sánchez[1]

[1] Universidad de Oviedo, Departamento de Informática, Edificio Departamental
Oeste, Campus de Viesques, Gijón 33204, Asturias, España
[2] Universidad de Granada, Departamento de Ciencias de la Computación e
Inteligencia Artificial, ETSIIyT, Granada 18071, Granada, España
[3] Universidad de Oviedo, Departamento de Estadística e I.O. y D. M. EPI Gijón,
Edificio Principal, Campus de Viesques, Gijón 33204, Asturias, España

Abstract. Sorting a set of inputs for relevance in modeling problems
may be ambiguous if the data is vague. A general extension procedure
is proposed in this paper that allows applying different deterministic
or random feature selection algorithms to fuzzy data. This extension is
based on a model of the relevance of a feature as a possibility distribution. The possibilistic relevances are ordered with the help of a fuzzy
ranking. A practical problem where the most informative software metrics are searched for in an automatic grading problem is solved with this
technique.

Keywords: Low Quality Data, Vague Data, Genetic Fuzzy Systems,
Feature Selection.

1 Introduction

Learning Management Systems or Content Management Systems allow students
and teachers to interact via lectures, assignments, exams or gradings. Open on-
line courses take advantage of these resources, but tracking students and taking
examination are nonetheless time consuming tasks, thus there is demand for intelligent techniques that help the instructor to manage large groups of students.
In particular, procedures are sought that automate the grading process, under-
stood as taking standardized measurements of varying levels of achievement in
a course [2].

Many different automatic grading systems exist in the context of Computer
Science online courses. For instance, in [17] a semi-automated system for task
submission and grading is proposed, but the grading itself must be done manually by the teacher. The *WebToTeach* system [2], on the contrary, is able to check

C. Cornelis et al. (eds.): RSCTC 2014, LNAI 8536, pp. 299–308, 2014.

submitted source code automatically. Similar to this, and focused on programming, the methods in [15] or [11] achieve an automatic grading by comparing the output of each student program with the output of a correct program. There is no measurement of the internals of the source code, which it is labelled as correct if the output is correct, regardless of the solution strategy. The *AutoLEP* system [23] is more recent. One of the salient points of the last work is a procedure to compare any implementation of an algorithm against a single model. Furthermore, in [22] a methodology is presented that accomplishes automatic grading by testing the program results against a predefined set of inputs, and also by formally verifying the source code or by measuring the similarities between the control flow graph and the teacher's solution. The parameters of a linear model are found that averages the influence of the three techniques in order to match teacher's and automatic grading in a corpus of manually graded exercises. Finally, in [14], software metrics are used to measure the properties of the students' programs and a fuzzy rule-based system is used to determine how close the programs submitted by students and the solutions provided by the teacher are, partially achieving an automatic grading. On the whole, these approaches pay particular attention to exam grading, that comprises comparing the outputs of student programs to that of a correct program for the problem at hand and also checking certain aspects about the internals of the source code (i.e., code style, documentation, etc.)

However, for the most part, the purpose of the student when following an open online course is not to obtain a certificate but to acquire a knowledge. In this respect, from the instructor's side it is important that early corrective actions can be taken when learning difficulties are detected and therefore a continuous assessment of the student must be carried out. Since the evolution of each student should not be tracked down to a single exam, incremental measurements should be taken, and the students must upload many different assignments. This introduces a new source of uncertainty into this problem, as the number of assignments is variable, and some of them might be incomplete or missing. Furthermore, if software metrics are used to assess the quality of the assignments, not all of them will be equally informative. Because of the mentioned reasons, in this paper:

- A method is proposed for computing a fuzzy aggregated value that summarizes the metrics of the different source files that are related to the same programming concept.
- The relevance of the different metrics is assessed with an extension of a crisp feature selection algorithm to fuzzy data.
- A learning fuzzy system that can extract if-then rules from fuzzy data is used to build the rule based system that performs the grading.

This paper is organized as follows: in Section 2, the method for combining the values of a metric over a set of different source files, and a method for ranking the importance of the fuzzy aggregated values are described. In Section 3, a case study with actual data collected in classroom lectures in 2013 and 2014 is provided. Section 4 concludes the paper and highlights future research lines.

2   Feature Selection for Regression with Vague Data

As mentioned, the grading process is intended to determine the level of achievement of each programming concept, which in turn is assessed by means of a set of source code files written by the students. The metrics of all files in these sets are jointly considered. Given that these sets are of different sizes for different students and some of its elements may be missing, a robust combination method is needed. The proposed combination will be based on the assumption that the application of a software metric to a given source code can be assimilated to the process of measuring the value of an observable variable that provides partial information about an unobservable variable, which in this case is the degree of assessment of a given programming concept. It is remarked that the information provided by different observations may be in conflict. The conversion of a set of observations or items into an estimation of a non-observable or latent variable that can be fed into a model has been solved in different ways in other contexts. For instance, there exists models in marketing where sets of items are preprocessed and aggregated into a characteristic value [8]. The most commonly used aggregation operator is the mean, although many different functions may be used instead.

In [20], however, a different approach was used: it was assumed that there exists a true value for the latent variable, but also that this value cannot be precised further than a set that contains it. The same idea will be adopted in this paper. This method is compatible with a possibilistic view of the uncertainty, where fuzzy sets are used for describing partial knowledge about the data. This interpretation is grounded in the result that the contour function of a possibility distribution is a fuzzy set [12], and α-cuts of fuzzy sets are linked to confidence intervals about the unknown value of the feature with significance levels $1 - \alpha$ (see reference [10]). This last property supports the use of intervals or fuzzy data for modelling uncertain data: a fuzzy set is searched for such that their α-cuts are confidence intervals with degree $1 - \alpha$ of the expected value of the observation error. In this paper, bootstrap estimates of these confidence intervals have been used, that are stacked to form the fuzzy membership functions describing the aggregated value of the metrics. In the following of this section, a method for ranking the importance of the fuzzy aggregated metrics in relation to the grading problem is presented that allows aplying an arbitrary deterministic or random feature selection algorithm to this problem.

2.1   Random Feature Selection Algorithms Extended to Vague Data

In the following, the grades and fuzzy aggregated metrics will be regarded as random and fuzzy random variables, respectively. A fuzzy random variable will be regarded as a nested family of random sets,

$$(\Lambda_\alpha)_{\alpha \in (0,1)}, \tag{1}$$

each one associated to a confidence level $1 - \alpha$ [9]. A random set is a mapping where the images of the outcomes of the random experiment are crisp sets.

A random variable X is a selection of a random set Γ when the image of any outcome by X is contained in the image of the same outcome by Γ. For a random variable $X : \Omega \to \mathbf{R}$ and a random set $\Gamma : \Omega \to \mathcal{P}(\mathbf{R})$, X is a selection of Γ (written $X \in S(\Gamma)$) when

$$X(\omega) \in \Gamma(\omega) \quad \text{for all } \omega \in \Omega. \tag{2}$$

In turn, a random set can be viewed as a family of random variables (its selections.)

Let be $M + 1$ paired samples $(X_1^k, X_2^k, \ldots, X_N^k)$, and $(Y_1, Y_2, \ldots, Y_N)$, with $k = 1, \ldots, M$, from $M + 1$ standard random variables $X^1, X^2, \ldots, X^M$ and Y (in this particular case, M is the number of metrics and N is the number of students). It will be assumed that all universes of discourse are finite. Let be assumed that a feature selection algorithm is a random mapping between the $M + 1$ paired samples and a permutation σ of $\{1, \ldots, M\}$ that sorts the metrics according to their relevance:

$$\sigma(X_1^1, X_2^1, \ldots, X_N^M, Y_1, Y_2, \ldots, Y_N, \omega) = (\sigma_1, \ldots, \sigma_M)(\omega) \tag{3}$$

where $p_{ik} = P(\sigma_i = k) = P(\omega | \sigma_i(\omega) = k)$ with $i, k = 1 \ldots, M$, is the probability that the k-th random variable X^k is ranked as the i-th most relevant feature. If the feature selection criterion is deterministic (for instance, a correlation or mutual information-based criterion [3]) then $p_{ik} \in \{0, 1\}$. In other cases, successive launches of the feature selection algorithm over the same sample will produce different permutations (think for instance in random forest feature importance measures [18]).

Now let be $M + 1$ fuzzy paired samples $(\widetilde{X}_1^k, \widetilde{X}_2^k, \ldots, \widetilde{X}_N^k)$, and an also paired crisp sample $(Y_1, Y_2, \ldots, Y_N)$ from $M+1$ fuzzy random variables $\widetilde{X}^1, \widetilde{X}^2, \ldots, \widetilde{X}^M$ and the random variable Y. Let the list of fuzzy numbers $\widetilde{\sigma} = (\widetilde{\sigma}_1, \ldots, \widetilde{\sigma}_M)$ be defined as

$$\mu_{\widetilde{\sigma}_i}(k) = \sup\{\alpha \mid P(\sigma_i(X_1^1, X_2^1, \ldots, X_N^M, Y_1, Y_2, \ldots, Y_N) = k) \geq \epsilon$$
$$X_i^k \in S([\widetilde{X}_i^k]_\alpha), \ i, k = 1, \ldots, M\} \tag{4}$$

for a given small value ϵ. It will be shown later in this paper that each fuzzy number $\widetilde{\sigma}$ models our incomplete knowledge about the possible ranks of each fuzzy aggregated metric $\widetilde{X}^k$; these metrics will be ordered according to a ranking between fuzzy numbers. In the next section a detailed practical case is worked.

3 Case Study

Fourty six volunteering students from the first course of an Engineering Degree in Computer Science at Oviedo University, Spain, participated in this study. The Python programming language was used. Students were allowed to upload as many source code files as they wished, ranging from none to more than a solution for each problem. 800 files were uploaded. Seven programming concepts

Table 1. Relevant metrics

Programming concept	Description of the metric	Rank 99%	Rank 80%
Conditional	COCOMO SLOC [4]	1 ± 0	11 ± 10
Conditional	Number of tokens	2 ± 0	8.5 ± 7.5
Conditional	Code ratio	4 ± 1	$25 + 24$
File I/O	Number of characters	4 ± 1	11 ± 8
Conditional	Number of lines	7 ± 1	21 ± 19
Functions	Number of characters	4 ± 1	43 ± 37
Conditional	Number of keywords	7.5 ± 1.5	36 ± 33
Conditional	Number of comments	17 ± 6	48 ± 44
File I/O	Ratio of comments	17 ± 6	48 ± 44
File I/O	McCabe Complexity [16]	17 ± 9	30 ± 24
File I/O	Number of blocks	17 ± 9	31 ± 25

were studied: Standard I/O, Conditionals, While loop, For loop, Functions, File I/O and Lists. The evaluation of the students comprised both theoretical and practice skills, with two exams each, at the midterm and at the end of the term. The uploaded exercises were not part of the exams and had no impact on the final grading. 23 software metrics and properties were measured for each source file[1] [1], thus the feature selection stage has to choose between 161 different combinations of programming concept and software metric.

The feature selection algorithm to be extended is based on the random forest feature importance measures [18]. A fuzzy rank (see [5]) was used to sort the fuzzy rankings. In Figure 3, right part, an example is given with the shapes of the membership functions of the fuzzy ranks of the metrics "COCOMO SLOC" [4] and "McCabe Complexity" [16]. The most relevant metrics (this is the subset for which the best model attained a minimum error, details are given later in this section), and two α-cuts of their fuzzy ranks, are shown in Table 1

Interestingly enough, only two programming concepts (Conditional and File I/O) made into this list. The most informative metric was COCOMO SLOC, followed by other indicators related to the fact that the best students seem to have better coding style and produce a larger code base with a better documentation.

In Figure 1, the rank of the most relevant metrics, according to the proposed algorithm, are graphically displayed. Supports (dashed lines), modal points (bars), fuzzy ranks (abscissa) and crisp ranks (ordinate of the squares) of the 50 most relevant metrics are displayed. Observe that those metrics whose square is plotted below the diagonal line ocupy a more relevant position under the fuzzy rank than they were assigned by the crisp feature selection algorithm. Squares over the diagonal line, on the contrary, are assigned more weight by the crisp algorithm than they are with the fuzzy extension.

From a methodological point of view the proposed technique is robust and the available information is better exploited with the combination of the fuzzy feature selection and a genetic fuzzy system (the NMIC modelling algorithm

---

[1] http://www.webappsec.org/

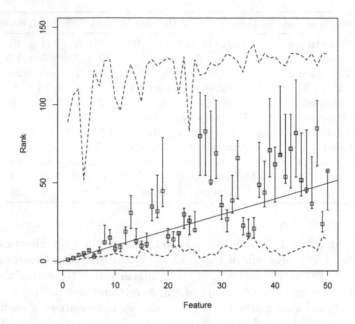

Fig. 1. Supports (dashed lines), modal points (bars), fuzzy ranks (abscissa) and crisp ranks (ordinate of the squares) of the 50 most relevant metrics. Those metrics whose square is plotted below the diagonal line ocupy a more relevant position under the fuzzy rank.

for low quality data [19]) than it is with standard feature selection and model learning algorithms. To prove this fact, regression trees [6], neural networks [13], support vector machines [21], random forests [7] and the NMIC algorithm were launched over subsets sweeping the range between 10 and 20 metrics, found by both the extended fuzzy feature selection algorithm and the original crisp version operating on the centerpoints of the aggregated data. In Table 2 these results are jointly displayed. In Figure 2, test errors corresponding to the selection of the most relevant variables with random forest feature importance measures (applied to the centerpoints of the fuzzy data) are drawn with dashed lines. The proposed extension of the same feature selection method to fuzzy data, followed by a learning with the same centerpoints for Regression Trees, Neural Networks, Support Vector Machines and Random Forest, but the whole fuzzy data for NMIC, are drawn with solid lines. Observe that, if the number of features associated to the lowest test error is chosen as a quality index, the proposed extension improved the accuracy of the grading system in 4 of 5 cases (all but the Regression Tree, with incidentally attained the worst results).

The combination of the NMIC algorithm with fuzzy data was consistently better in all cases (statistically relevant results, according to Friedman/Wilcoxon tests, p-value better than 0.05). In the left part of Figure 3 a set of boxplots is

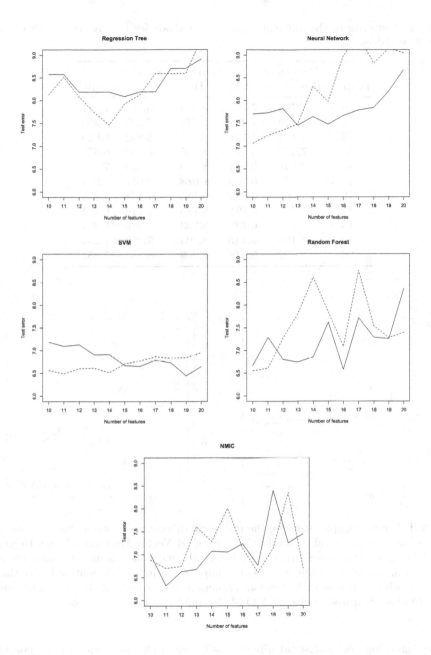

Fig. 2. Test error as a function of the number of features for feature subsets of sizes 10 to 20. Test errors after a random forest feature importance-based selection, applied to the centerpoints of the fuzzy data, are drawn with dashed lines. The extension of this method to fuzzy data, followed by a learning with the same centerpoints for Regression Trees, Neural Networks, Support Vector Machines and Random Forest, but the whole fuzzy data for NMIC, are drawn with solid lines.

Table 2. Test error or the different regression methods for feature sets ranging from 10 to 20 variables

Features	Multilayer Perceptron	SVM	Regression Tree	Random Forest	NMIC
10	7.703	7.185	8.574	6.671	7.011
11	7.729	7.093	8.574	7.285	**6.321**
12	7.823	7.128	8.185	6.802	6.629
13	**7.451**	6.911	8.185	6.742	6.678
14	7.641	6.914	8.185	6.854	7.073
15	7.472	6.670	**8.084**	7.617	7.056
16	7.661	6.652	8.185	**6.576**	7.236
17	7.785	6.791	8.185	7.716	6.764
18	7.838	6.728	8.703	7.285	8.399
19	8.195	**6.445**	8.703	7.256	7.256
20	8.672	6.648	8.909	8.357	7.451

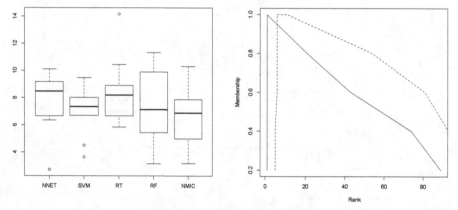

Fig. 3. Left part: Boxplot showing the statistical differences between the test error of the combination of Neural Networks (NN), Support Vector Machines (SVM), Regression Trees (RT), Random Forests (RF) and NMIC with the feature set computed as described in this paper. NMIC exploits the imprecision in the information better than the alternatives. Right part: Membership functions of the ranks of the first (solid) and 10-th (dashed) features, i.e. COCOMO SLOC and McCabe complexity.

drawn, showing the statistical differences between the test error of the combination of Neural Networks (NN), Support Vector Machines (SVM), Regression Trees (RT), Random Forests (RF) and NMIC with the feature set computed as described in this paper. The last graphic is intended to shown that NMIC exploits the imprecision in the information better than the alternatives, demonstrating that the fuzzy aggregation loses less information than the alternatives and also that the proposed method is able to exploit this extra information.

4   Concluding Remarks and Future Work

A method for ranking software tests according to their relevance in an automatic grading systems has been proposed. The main innovation of the new method lies in the development of a set of techniques that can make use of a fuzzy aggregation of the information contained in a variable number of exercises about the same learning subject.

From a methodological point of view, the new algorithm is a solid alternative. The combination of a learning algorithm for vague data and the extended feature selection proposed in this paper was shown to make a better use of the imprecision in the information than any of the alternatives, demonstrating that the fuzzy aggregation keeps valuable information and also that the proposed method is able to exploit this. On the other hand, from the point of view of the automated grading techniques, it has been found that the most informative metrics are some measures of the cost and complexity of the code, followed by indicators related to the code size and quality of the documentation. However, there is still a margin for improving this knowledge, as the number of students participating in ths study was small and further work is needed to build a larger corpus of hand-graded assignments.

Acknowledgements. This work was supported by the Spanish Ministerio de Economía y Competitividad under Project TIN2011-24302, including funding from the European Regional Development Fund.

References

1. Abdellatief, M., et al.: A mapping study to investigate component-based software system metrics. Journal of Systems and Software 86(3), 587–603 (2013)
2. Arnow, D., Barshay, O.: On-line programming examinations using Web to teach. In: Proceedings of the 4th Annual SIGCSE/SIGCUE ITiCSE Conference on Innovation and Technology in Computer Science Education (ITiCSE 1999), pp. 21–24 (1999)
3. Battiti, R.: Using mutual information for selecting features in supervised neural net learning. IEEE Trans. Neural Nets. 5(4) (July 1994)
4. Boehm, B., et al.: Cost models for future software life cycle processes: COCOMO 2.0. Annals of Software Engineering 1(1), 57–94 (1995)
5. Bortolan, G., Degani, R.: A review of some methods for ranking fuzzy subsets. Fuzzy Sets and Systems 15(1) (1985)
6. Breiman, L., Friedman, L.J., Olshen, A., Stone, C.: Classification and Regression Trees. Wadsworth (1984)
7. Breiman, L.: Random Forests. Machine Learning 45(1), 5–32 (2001)
8. Casillas, J., Martinez-Lopez, F., Martinez, F.: Fuzzy association rules for estimating consumer behaviour models and their application to explaining trust in internet shopping. Fuzzy Economic Review IX(2), 3–26 (2004)
9. Couso, I., Montes, S., Gil, P.: The necessity of the strong alpha-cuts of a fuzzy set. International Journal of Uncertainty, Fuzziness and Knowledge-Based Systems 9(2), 249–262 (2001)

10. Couso, I., Sanchez, L.: Higher order models for fuzzy random variables. Fuzzy Sets and Systems 159(3), 237–258 (2008)
11. Cheang, B., Kurnia, A., Lim, A., Oon, W.: On automated grading of programming assignments in an academic institution. Comput. Educ. 41(2), 121–131 (2003)
12. Dubois, D., Prade, H.: Fuzzy sets - a convenient fiction for modeling vagueness and possibility. IEEE Transactions on Fuzzy Systems 2(1), 16–21 (1994)
13. Haykin, S.: Neural Networks: A Comprehensive Foundation, 2nd edn. Prentice Hall (1998)
14. Jurado, F., Redondo, M., Ortega, M.: Using fuzzy logic applied to software metrics and test cases to assess programming assignments and give advice. J. Netw. Comput. Appl. 35(2) (2012)
15. Kurnia, A., Lim, A., Cheang, B.: Online judge. Comput. Educ. 36(4), 299–315 (2001)
16. McCabe, T.: A complexity measure. IEEE Trans. on Software Engineering 2(4), 308–320 (1976)
17. Reek, K.A.: A software infrastructure to support introductory computer science courses. In: Klee, K.J. (ed.) Proceedings of the Twenty-Seventh SIGCSE Technical Symposium on Computer Science Education (SIGCSE 1996), pp. 125–129. ACM, New York (1996)
18. Saeys, Y., Abeel, T., Van de Peer, Y.: Robust Feature Selection Using Ensemble Feature Selection Techniques. In: Daelemans, W., Goethals, B., Morik, K. (eds.) ECML PKDD 2008, Part II. LNCS (LNAI), vol. 5212, pp. 313–325. Springer, Heidelberg (2008)
19. Sanchez, L., Otero, J., Couso, I.: Obtaining linguistic fuzzy rule-based regression models from imprecise data with multiobjective genetic algorithms. Soft Comput. 13(5), 467–479 (2008)
20. Sanchez, L., Couso, I., Casillas, J.: Genetic learning of fuzzy rules on low quality data. Fuzzy Sets and Systems 160(17), 2524–2552 (2009)
21. Smola, A.J., Schölkopf, B.: A tutorial on support vector regression. Statistics and Computing 14(3), 199–222 (2004)
22. Vujosevic-Janicica, M., Nikolica, M., Tosica, D., Kuncak, V.: Software verification and graph similarity for automated evaluation of students assignments. Information and Software Technology 55(6), 1004–1016 (2013)
23. Wang, T., Su, X., Ma, P., Wang, Y., Wang, K.: Ability-training-oriented automated assessment in introductory programming course. Comput. Educ. 56(1), 220–226 (2011)

Classification of a Sequence of Objects
with the Fuzzy Decoding Method

Andrey V. Savchenko[1] and Lyudmila V. Savchenko[2]

[1] National Research University Higher School of Economics, Nizhny Novgorod, Russia
avsavchenko@hse.ru
[2] Nizhny Novgorod State Linguistic University, Russia
LyudmilaSavchenko@yandex.ru

Abstract. The problem of recognition of a sequence of objects (e.g., video-based image recognition, phoneme recognition) is explored. The generalization of the fuzzy phonetic decoding method is proposed by assuming the distribution of the classified object to be of exponential type. Its preliminary phase includes association of each model object with the fuzzy set of model classes with grades of membership defined as the confusion probabilities estimated with the Kullback-Leibler divergence between model distributions. At first, each object (e.g., frame) in a classified sequence is put in correspondence with the fuzzy set which grades are defined as the posterior probabilities. Next, this fuzzy set is intersected with the fuzzy set corresponding to the nearest neighbor. Finally, the arithmetic mean of these fuzzy intersections is assigned to the decision for the whole sequence. In this paper we propose not to limit the method's usage with the Kullback-Leibler discrimination and to estimate the grades of membership of models and query objects based on an arbitrary distance with appropriate scale factor. The experimental results in the problem of isolated Russian vowel phonemes and words recognition for state-of-the-art measures of similarity are presented. It is shown that the correct choice of the scale parameter can significantly increase the recognition accuracy.

Keywords: Classification, sequence of objects, fuzzy sets, phoneme recognition, fuzzy decoding method, Kullback-Leibler discrimination, Mel-frequency cepstral coefficients.

1 Introduction

The problem of recognition of a set of objects is quite acute [1] though it usually appears as a part of complex object or speech recognition algorithms [2]. For instance, conventional speech recognition system includes the extraction of regular segments extracted by any phoneme segmentation method, dividing them into a sequence of partially-overlapped frames, classification of each frame with available phonetic database (acoustic model) and further processing of frame's recognition result (Dynamic Time Warping or Hidden Markov Models, language models, etc) [3]. The efficiency of speech recognition is largely determined by the quality of phoneme recognition [3, 4].

C. Cornelis et al. (eds.): RSCTC 2014, LNAI 8536, pp. 309–318, 2014.

The latter is exactly the task of recognition of a sequence of objects (frames). Another important example is a still-to-still approach to a video-based image recognition [5]. According to this approach, at first, detected object is tracked and recognized in each video frame. Next, recognition results for each frame are aggregated by a simple voting (SV) procedure or more complex algorithms to provide final solution [2].

This paper seeks the way to improve the quality of conventional SV scheme by the usage of the fuzzy set theory [6, 7]. Following the ideas of our fuzzy phonetic decoding method (FPD) [4], we define each class (i.e., phoneme label, person in face recognition) as a fuzzy set of all available model objects. In comparison with conventional class definition as a crisp set of models corresponding to this class, our approach allows to take into account the closeness of models from different classes. The distance between each frame and every model X_r should be approximately equal to the distance between the model from correct class and X_r. In order to verify the frame's recognition result, we use the fuzzy intersection of sets corresponding to this frame and to the nearest neighbor model. Thus, the main difference between the currently proposed fuzzy decoding (FD) method and the original FPD is as follows:

1) The FD method is a generalization of the FPD to a problem of statistical classification of the sequence of object observations (e.g., still-to-still video recognition). Hence the FD is not restricted to the use only in the field of phoneme recognition.

2) The FD method is able to be combined with the nearest neighbor rule with an *arbitrary* distance, while the original FPD was designed *only* for its application with comparison of power spectral density (PSD) speech features with the Kullback-Leibler (KL) minimum information discrimination principle [8]. It is known from the speech recognition theory [3, 9], other methods (first of all, comparison of the mel-frequency cepstral coefficients (MFCC)) are characterized with better accuracy and performance. Our current experimental study shows that the proposed FD method can increase the recognition accuracy even for MFCC features compared with conventional Euclidean metric.

The rest of the paper is organized as follows: Section 2 presents the SV solution of the problem of statistical classification of objects' sequence by assuming the objects distribution to be of exponential family. In Section 3, we introduce our generalization of the FPD, namely, the fuzzy decoding method, and show the possibility of its usage with an arbitrary distance. In Section 4, we present the experimental results of the proposed method in the isolated vowel phonemes and words recognition task for Russian language [4, 10]. Finally, concluding comments are given in Section 5.

2   Classification of a Sequence of Objects

Let the input sequence $\{X(t)\}, t = \overline{1, T}$ of $T \geq 1$ objects be specified. For example, each object $X(t)$ may be a feature vector of speech frame (in automatic speech recognition) or an image frame (in video-based object recognition task). The problem is to assign the sequence $\{X(t)\}$ to one of $R>1$ classes. We suppose that different *observations* of only one object is presented in this sequence (i.e., object detection and tracking procedures were performed preliminarily, e.g., the sequence $\{X(t)\}$ contains several serial speech frames of one phone). Each r-th class is given with the model training set X_r. In this paper we define this recognition task in terms of statistical classification. It is assumed that objects in each class are identically distributed and all distributions are of multivariate exponential type $f_\theta(X)$ generated by the fixed (for all classes) function $f_0(X)$ with K-dimensional parameter vector θ [11]

$$f_\theta(X) = \exp\left(\tau(\theta)\cdot\hat{\theta}(X)\right)\cdot f_0(X)/M(\tau),$$

$$M(\tau) = \int \exp\left(\tau(\theta)\cdot\hat{\theta}(X)\right)\cdot f_0(X)dX \tag{1}$$

where $\hat{\theta}(X)$ is an estimation of parameter θ using available data (random sample) X, and $\tau(\theta)$ is a normalizing function (K-dimensional parameter vector) defined by the following equation if the parameter estimation $\hat{\theta}(X)$ is unbiased (see [8] for details):

$$\int \hat{\theta}(X)\cdot f_\theta(X)dX \equiv \frac{d}{d\tau}\ln M(\tau) = \theta. \tag{2}$$

Each class is determined by its parameter vector θ_r. This assumption about exponential family of each class $f_{\hat{\theta}(X_r)}(X)$ in which parameter θ_r of r-th class is estimated by using the observed (given) sample X_r, covers wide range of known distributions (polynomial, normal, etc.). In such case an optimal Bayesian solution for classification of observation $X(t)$ is obtained with the nearest neighbor rule with the KL discrimination [8]

$$\nu(t) = \arg\min_{r=1,R} \hat{I}\left(*: f_{\hat{\theta}(X_r)}; X(t)\right), \tag{3}$$

where

$$\hat{I}\left(*: f_{\hat{\theta}(X_r)}; X(t)\right) = \int f_{\hat{\theta}(X(t))}(X)\cdot\ln\frac{f_{\hat{\theta}(X(t))}(X)}{f_{\hat{\theta}(X_r)}(X)}dX$$

The final solution of a sequence recognition task is obtained in favor of r^* class by SV [2, 4]:

$$r^* = \arg\max_{r=1,R} \mu_r, \tag{4}$$

where

$$\mu_r = \sum_{t=1}^{T} \delta(v(t) - r) \qquad (5)$$

and $\delta(\cdot)$ is the discrete Dirac delta function.

3   Fuzzy Decoding Method

Unfortunately, if several model classes are quite close to each other, the accuracy of straightforward SV (3)-(5) is rather low. To improve the quality we propose to follow our FPD method [4] and to define each class as the fuzzy set of models. Namely, the j-th ($j = \overline{1,R}$) class is represented not only by a model X_j, but by a fuzzy set $\left\{\left(X_r, \mu_r^{(j)}\right)\right\}$, where the grade of membership $\mu_r^{(j)}$ is defined as the confusion probability $P\left(X_r | X_j\right)$ of marking j-th class as an object of the r-th class (i.e., the distance between the object from j-th class and X_r is minimal). This probability can be estimated on the basis of known asymptotic properties of the KL divergence [8]. Namely, the double KL distance between objects from j-th and r-th classes is asymptotically distributed as a non-central chi square with $(K-1)$ degrees of freedom and noncentrality parameter $2 \cdot \hat{I}\left(*: f_{\hat{\theta}(X_r)}; X_j\right)$. As the number of parameters K is usually quite large, the central limit theorem can be applied and the KL discrimination is asymptotically normally distributed. Hence, the probability $P\left(X_r | X_j\right)$ can be estimated with the known distribution of independent minimum normal variables [13]:

$$P\left(X_r | X_j\right) = \frac{1}{\sqrt{2\pi}} \int_{-\infty}^{+\infty} \exp\left(-t^2/2\right) dt \prod_{\substack{i=1 \\ i \neq r}}^{R} \left(1 - \right.$$

$$\left. - \Phi\left(\frac{t \cdot \sqrt{8\hat{I}\left(*: f_{\hat{\theta}(X_r)}; X_j\right) + K - 1} + 2\left(\hat{I}\left(*: f_{\hat{\theta}(X_r)}; X_j\right) - \hat{I}\left(*: f_{\hat{\theta}(X_i)}; X_j\right)\right)}{\sqrt{8\hat{I}\left(*: f_{\hat{\theta}(X_i)}; X_j\right) + K - 1}}\right)\right) \qquad (6)$$

where $\Phi(\cdot)$ is a cumulative distribution function of $N(0;1)$, and any numeric method is used for integration.

Next, each query object $X(t)$ is associated with the fuzzy set $\left\{\left(X_r, \mu_r(X(t))\right)\right\}$, where the grade of membership $\mu_r(X(t))$ is defined as the r-th posterior probability

$P\left(X_r|X(t)\right)$ [2]. If all classes are assumed to be equiprobable, the latter probability is estimated from the known equation [11, 14]

$$P\left(X_r|X(t)\right)=\frac{\exp\left(-\hat{l}\left(*:f_{\hat{\theta}(X_r)};X(t)\right)\right)}{\sum\limits_{i=1}^{R}\exp\left(-\hat{l}\left(*:f_{\hat{\theta}(X_i)};X(t)\right)\right)},\tag{7}$$

According to the asymptotic properties of the KL divergence [8], the distance between the object $X(t)$ from j-th class and X_r is approximately equal to $\hat{l}\left(*:f_{\hat{\theta}(X_r)};X_j\right)$. In such case, conditional probabilities $P\left(X_r|X(t)\right)$ and $P\left(X_r|X_j\right)$ are approximately identical for all $r\in\{1,...,R\}$. Thus, to verify the correctness of the nearest neighbor class $v(t)$ (3), the decision $\{\!\{X_r,\mu_{r;t}\}\!\}$ for the frame $X(t)$ is a fuzzy intersection of sets $\left\{\left(X_r,\mu_r^{(j)}\right)\right\}$ and $\{\!\{X_r,\mu_r(X(t))\}\!\}$

$$\mu_{r;t}=\min\left(\mu_r^{(v(t))},\mu_r(X(t))\right),\tag{8}$$

where $\mu_r^{(v(t))}$ is determined by (6) after substitution $j=v(t)$. Final decision is taken in favor of one of the model minimal speech units on the basis of all $\mu_{r;t}$ (8). In this work, a decision is made by SV (4), but μ_r is defined as follows

$$\mu_r=\frac{1}{T}\sum_{t=1}^{T}\mu_{r;t},\tag{9}$$

The proposed classification method (3), (4), (6)-(9) is a generalization of the FPD in the problem of recognition of a sequence of objects. As one can notice, it includes only calculation of the KL divergence in (3), (6) and (7). Hence, our final proposal is to replace this discrimination by the similar to (3) nearest neighbor rule with arbitrary distance $\rho(\cdot)$:

$$v(t)=\arg\min_{r=\overline{1,R}}\rho\left(X(t),X_r\right),\tag{10}$$

Equations (6) and (7) should be slightly modified to add appropriate smoothing factor $\alpha=const>0$. The latter can be found experimentally. In such case, equation (7) becomes equivalent to the known output of the probabilistic neural network [15]

$$\mu_r(X(t)) = \frac{\exp(-\alpha \cdot \rho(X(t), X_r))}{\sum\limits_{i=1}^{R} \exp(-\alpha \cdot \rho(X(t), X_i))}, \tag{11}$$

By using the same idea, conditional probability (6) is estimated with

$$\mu_r^{(j)} = \frac{1}{\sqrt{2\pi}} \int\limits_{-\infty}^{+\infty} \exp(-t^2/2) \prod\limits_{\substack{i=1 \\ i \neq r}}^{R} \left(1 - \right.$$

$$\left. -\Phi\left(\frac{t \cdot \sqrt{8\alpha \cdot \rho(X_j, X_r) + K - 1} + 2\alpha \cdot (\rho(X_j, X_r) - \rho(X_j, X_i))}{\sqrt{8\alpha \cdot \rho(X_j, X_i) + K - 1}}\right)\right) dt \tag{12}$$

Thus, in this section we proposed the fuzzy decoding method (8)-(12) in the recognition of a sequence of objects problem. Our method is based on the nearest neighbor rule (10) and may be used with arbitrary distance to increase the accuracy of conventional SV (4), (5), (10). The next section experimentally supports this claim.

4   Experimental Results

In this section we examine the usage of the proposed FD method in the typical problem of set of objects classification, namely, vowel phonemes recognition for Russian language. To implement our FD method we use the following state-of-the-art similarity measures and speech features:

1) 12 MFCCs features + their first derivatives (totally, $K=24$ parameters) compared with the Euclidean distance [3].
2) Autoregression (AR) estimates of the speech signal PSD [3] compared with:
a) the Itakura-Saito (IS) divergence [16] (equivalent with a constant factor to the KL information discrimination for Gaussian signals [17])

$$\rho_{IS}(X_1, X_2) = \frac{1}{F} \sum\limits_{f=1}^{F} \left(\frac{G_1(f)}{G_2(f)} - \ln\frac{G_1(f)}{G_2(f)} - 1\right),$$

where $G_1(f)$ and $G_2(f)$ are the PSDs of signals X_1 and X_2, respectively, F is the speech sampling rate.
b) Spectral distortion (SD) which is the known equivalent to the linear prediction coding cepstral coefficients' comparison in Euclidean space [3]

$$\rho_{SD}(X_1, X_2) = \frac{1}{F} \sum\limits_{f=1}^{F} (\ln G_1(f) - \ln G_2(f))^2.$$

The model database was filled with $R=10$ Russian vowels pronounced by each of 5 speakers (3 men and 2 women) in isolated mode. These models were used to recognize the vowel in isolated syllables and words produced by the same speaker, i.e., we test a speaker-dependent mode of speech recognition [3]. The following parameters were chosen: sampling frequency $F=8$ kHz, AR-model order 20 (i.e., $K=20$ for the PSD features), signal to noise ratio is equal to 30 dB, utterances are divided into 45-ms frames with 30-ms overlap. The range of the smoothing parameter variation is $\alpha \in \{0.005, 0.01, 0.05, 0.1, 0.5, 1, 1.5\}$. All tests are performed in a modern laptop (4 core i7, 6 Gb RAM).

In the first experiment every speaker produced 1000 isolated vowels (100 for each class). The dependence of the average recognition accuracy for various metrics on α is shown in Fig. 1. To compare the computing efficiency of the proposed FD method with the SV we summarized the recognition results for the best (in terms of achieved accuracy) value of α in Table 1.

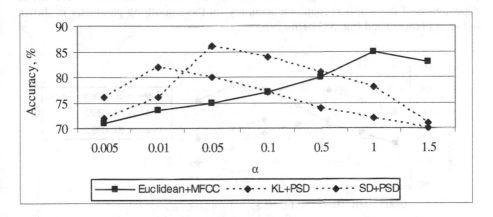

Fig. 1. Dependence of the phoneme recognition accuracy on α

Table 1. The best phoneme recognition experimental results

Distance/features	The best obtained α	Time, ms		Accuracy, %	
		SV	FD	SV	FD
Euclidean+MFCC	1	0.7±0.02	1.0±0.01	80±1.7	85±1.4
IS+PSD	0.05	3.5±0.05	5.5±0.04	81.5±1.9	86±1.4
SD+PSD	0.01	1.8±0.03	2.2±0.04	77±1.7	82±1.5

Based on these results, it is possible to draw the following conclusions. First, experimentally obtained parameter α can significantly increase the accuracy of phonemes recognition. For example, the SD's accuracy for the best obtained smoothing parameter α=0.01 is 10% higher than the accuracy for α=1. Second, the conventional approach to speech recognition (Euclidean discrimination with MFCC features) is expectedly characterized with the best computing efficiency. Third, though the FD's performance is a bit higher than the SV's performance, the proposed FD method's error rate is usually much lower than the SV's error.

The second experiment is devoted to isolated words recognition. Two vocabularies were used, namely, 1) the list of 1832 Russian towns with corresponding regions; and 2) the list of 1913 drugs (hereinafter "Pharmacy"). All speakers pronounced every word from all vocabularies twice in isolated syllable mode to simplify the recognition procedure [18]. The phonetic database of $R=10$ speaker-dependent vowels from the first experiment was used. The algorithm is quite straightforward [19, 20]. Syllables are extracted with simple amplitude detector and vowels are recognized in each syllable [21] by described SV and PD methods. Finally, each word is associated with the mean of grades of memberships μ_r (5), (9) of syllables contained in this word. The

word with the highest mean of grades is put into solution. To compare our results with the state-of-the-art approach, we used CMU Pocketsphinx 0.8 [22] with the MLLR speaker adaptation [3] to recognize vowels in each syllable. Posterior probabilities evaluated by Sphinx for each syllable are aggregated with SV (similar to (4), (5)). The dependence of the words recognition accuracy on the smoothing parameter is shown in Fig. 2 (Towns) and Fig. 3 (Pharmacy).

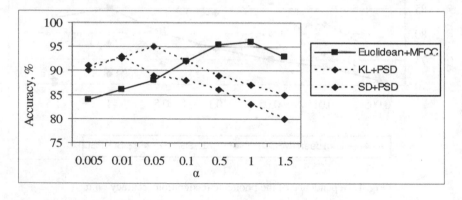

Fig. 2. Dependence of the words recognition accuracy α, vocabulary Towns

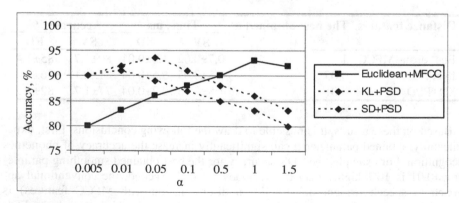

Fig. 3. Dependence of the words recognition accuracy α, vocabulary Pharmacy

Comparing these figures with the results of the first experiment (Fig. 1), one can notice that the best smoothing parameter is quite stable to the choice of distance and features. Namely, the best accuracy of words recognition is achieved with the same values of α as for the phoneme recognition task (Table 1). The average accuracy for the best α is summarized in Table 2. Though the standard deviation in each case seems to be quite high, these results are statistically significant as we used quite large test databases. The high variation may be explained by the fact, that all speakers cannot reach the same error rate due to their physical abilities.

Table 2. Isolated words recognition accuracy, %

Distance/features	Towns		Pharmacy	
	SV	FD	SV	FD
Euclidean+MFCC	92±3.4	96±2.9	89.5±2.2	93±2.0
IS+PSD	91.5±3.2	95±3.0	90±2.0	93.5±1.9
SD+PSD	88.5±2.7	93±2.4	87±2.9	91±2.8
CMU Pocketsphinx	90.5±2.3	-	89.4±3.0	-

The most significant conclusion here is the superiority of the proposed FD method over conventional SV in terms of achieved accuracy. These experiments support the fact that our method may be successfully applied not only with the IS divergence (equivalent to the KL discrimination), but with various measures of similarity. For instance, the best recognition accuracy is achieved with state-of-the-art MFCC features comparison in Euclidean space. It is also remarkable that the error rate of the SV with MFCC is lower than the Pocketsphinx' error. It seems that our speaker adaptation is much more effective as the phonetic database for the SV (and the FD) is filled with the speaker's vowels and does not contain information from general acoustic model.

5   Conclusion and Future Work

In this paper we introduced a generalization of our FPD method for exponential family of distribution. As the synthesized equations (3), (7), (8) contain only evaluation of the KL divergence, we produced final expressions (10)-(12) to apply our method with an arbitrary distance. The experiment with Russian speech recognition showed the stability of the smoothing parameter's choice to a type of distance and object features. As a result, the usage of the FD method yields to the increase of the recognition accuracy (Tables 1,2) in comparison with conventional voting algorithm (5). The computing efficiency of the FD is obviously lower than for the SV technique due to calculation of the posterior probabilities (11). However, the phoneme recognition time is still reasonable even for real-time applications (Table 1).

The further research of the FD method may be continued in the following directions. First, it is an application to the continuous speech recognition [3] and aggregation of our solution with speaker-independent systems (e.g., Pocketsphinx [22]). Other possible direction is the application of our method with other set of objects recognition tasks, mainly, with still-to-still video-based face recognition [5, 14].

Acknowledgements. This study was carried out within "The National Research University Higher School of Economics Academic Fund Program in 2013-2014, research grant No. 12-01-0003".

References

1. Savchenko, A.V.: Probabilistic neural network with homogeneity testing in recognition of discrete patterns set. Neural Networks 46, 227–241 (2013)
2. Theodoridis, S., Koutroumbas, K.: Pattern Recognition, 4th edn. Elsevier Inc. (2009)
3. Benesty, J., Sondh, M., Huang, Y. (eds.): Springer Handbook of Speech Recognition. Springer (2008)
4. Savchenko, L.V., Savchenko, A.V.: Fuzzy Phonetic Decoding Method in a Phoneme Recognition Problem. In: Drugman, T., Dutoit, T. (eds.) NOLISP 2013. LNCS, vol. 7911, pp. 176–183. Springer, Heidelberg (2013)
5. Wang, H., Wang, Y., Cao, Y.: Video-based face recognition: a survey. World Academy of Science. Engineering and Technologies 60, 293–302 (2009)
6. Zadeh, L.A.: Fuzzy Sets. Information Control 8, 338–353 (1965)
7. Sarkar, M.: Fuzzy-rough nearest neighbor algorithms in classification. Fuzzy Sets and Systems 158(19), 2134–2152 (2007)
8. Kullback, S.: Information Theory and Statistics. Dover Pub. (1997)
9. Anusuya, M.A., Katti, S.K.: Speech recognition by Machine: A Review. International Journal of Computer Science and Information Security 6(3), 181–205 (2009)
10. Kipyatkova, I.S., Karpov, A.A.: An Analytical Survey of Large Vocabulary Russian Speech Recognition Systems. SPIIRAS Proceedings 12, 7–20 (2010)
11. Keener, R.W.: Theoretical Statistics: Topics for a Core Course. Springer, New York (2010)
12. Reddy, D.R.: Speech recognition by machine: a review. Proceedings of the IEEE 64(4), 501–531 (1976)
13. Hill, J.E.: The minimum of n independent normal distributions, http://www.untruth.org/~josh/math/normal-min.pdf
14. Savchenko, A.V.: Adaptive Video Image image Recognition recognition System Using using a Committee committee Machinemachine. Optical Memory and Neural Networks (Information Optics) 21(4), 219–226 (2012)
15. Specht, D.F.: Probabilistic neural networks. Neural Networks 3(1), 109–118 (1990)
16. Itakura, F., Saito, S.: An analysis–synthesis telephony based on the maximum likelihood method. In: Proc. of International Congress on Acoustics c-5-5, vol. 5, pp. 17–20 (1968)
17. Basseville, M.: Distance measures for signal processing and pattern recognition. Signal Processing 18, 349–369 (1989)
18. Mérialdo, B.: Multilevel Decoding for Very-Large-Size-Dictionary Speech Recognition. IBM Journal of Research and Development 32(2), 227–237 (1988)
19. Sirigos, J., Fakotakis, N., Kokkinakis, G.: A hybrid syllable recognition system based on vowel spotting. Speech Communication 38, 427–440 (2002)
20. Savchenko, A.V.: Phonetic words decoding software in the problem of Russian speech recognition. Automation and Remote Control 74(7), 1225–1232 (2013)
21. Savchenko, A.V.: Phonetic encoding method in the isolated words recognition problem. Journal of Communications Technology and Electronics 59(4), 310–315 (2014)
22. CMU Sphinx, http://cmusphinx.sourceforge.net/

Dynamic Filtering of Ranked Answers When Evaluating Fuzzy XPath Queries

Jesús M. Almendros-Jiménez[1], Alejandro Luna Tedesqui[2], and Ginés Moreno[2]

[1] Dep. of Languages and Computation, University of Almería, Spain
jalmen@ual.es
[2] Dep. of Computing Systems, University of Castilla-La Mancha, Spain
Gines.Moreno@uclm.es, Alejandro.Luna@alu.uclm.es

Abstract. We have recently designed an extension of the XPath language which provides ranked answers to flexible queries taking profit of fuzzy variants of *and*, *or* and *avg* operators for XPath conditions, as well as two structural constraints, called *down* and *deep*, for which a certain degree of relevance is associated. In practice, this degree is very low for some answers weakly accomplishing with the original query, and hence, they should not be computed in order to alleviate the computational complexity of the information retrieval process. In this work we focus on the scalability of our interpreter for dealing with massive XML files by making use of its ability for prematurely disregarding those computations leading to non significant solutions (i.e., with a poor degree of relevance according the preferences expressed by users when using the new command FILTER). Since our proposal has been implemented with a fuzzy logic language, here we exploit the high expressive resources of this declarative paradigm for performing "dynamic thresholding" in a very natural and efficient way, thus connecting with the so-called *top-k answering problem*, which is very well-known in the fuzzy logic and soft computing arena.

Keywords: Information Retrieval Systems, Fuzzy XPath, Information Filtering Systems, Fuzzy Filtering & Thresholding, Fuzzy Logic Programming.

1 Introduction

The *XPath* language [7] has been proposed as a standard for XML querying and it is based on the description of the path in the XML tree to be retrieved. XPath allows to specify the name of nodes (i.e., tags) and attributes to be present in the XML tree together with boolean conditions about the content of nodes and attributes. XPath querying mechanism is based on a boolean logic: the nodes retrieved from an XPath expression are those matching the path of the XML tree. Therefore, the user should know the *XML schema* in order to specify queries. However, even when the XML schema exists, it can not be available for users. Moreover, XML documents with the same XML schema can be very different in structure. Let us suppose the case of XML documents containing the

C. Cornelis et al. (eds.): RSCTC 2014, LNAI 8536, pp. 319–330, 2014.
© Springer International Publishing Switzerland 2014

curriculum vitae of a certain group of persons. Although they can share the same schema, each one can decide to include studies, jobs, training, etc. organized in several ways: by year, by relevance, and with different nesting degree.

Therefore, in the context of semi-structured databases, *flexible query languages* arise for allowing the formulation of queries without taking into account a rigid database schema, usually including too mechanisms for obtaining a certain *ranked list* of answers. The ranking of answers can provide *satisfaction degree* depending on several factors. In a structural XPath-based query, the main criteria to provide a certain degree of satisfaction depends on the *hierarchical deepness* and *document order*. Therefore the query language should provide mechanisms for giving *priority* to answers when they occur in different parts of the document. In this sense, the need for providing flexibility to XPath has recently motivated the investigation of extensions of the XPath language. We can distinguish those in which the main goal is the introduction of fuzzy information in data (similarity, proximity, vagueness, etc) [9,16,30,26] and the proposals in which the main goal is the handling of crisp information by fuzzy concepts [10,13,15,14,20]. Our work focuses on the second line of research.

In [10,13] authors introduce in XPath flexible matching by means of fuzzy constraints called *close* and *similar* for node content, together with *below* and *near* for path structure. In addition, they have studied *deep-similar* notion for tree matching. In order to provide ranked answers they adopt a *Fuzzy set theory*-based approach in which each answer has an associated numeric value (the membership degree). The numeric value represents the *Retrieval Status Value (RSV)* of the associated item. In the work of [15], they propose a satisfaction degree for XPath expressions based on associating a degree of importance to XPath nodes, and they study how to compute the best k answers. In both cases, authors allow the user to specify in the query the degree in which the answers will be penalized. On the other hand, in [14], they have studied how to relax XPath queries by means of rewriting in order to improve information retrieval in the presence of heterogeneous data resources. Our proposal also connects with the recent approaches of [27,28] but, as we are going to see, it is important to note that many of our fuzzy commands are directly inspired by the powerful expressive resources of the underlying fuzzy logic language used for implementing our tool.

As we will resume in Section 2, in [3,4,5,6] we have presented both an interpreter and a debugger coping with an extension of the XPath query language for managing flexible queries in a very natural way (the tool can be tested on-line via http://dectau.uclm.es/fuzzyXPath/). Our approach proposes two structural constraints called *down* and *deep* for which a certain degree of relevance can be associated. In such a way that *down* provides a ranked set of answers depending on the path is found from "top to down" in the XML document, and *deep* provides a set of answers depending on the path is found from "left to right" in the XML document. Both structural constraints can be combined. In addition, we provide fuzzy operators *and, or* and *avg* for XPath conditions. In this way, users can express the priority they give to answers. Such fuzzy operators can be combined to provide ranked answers. Our approach has been implemented with the so-called

«*Multi-Adjoint Logic Programming*» language (MALP in brief) [22] by using our «*Fuzzy LOgic Programming Environment for Research*» $\mathcal{FLOPER}$ [23,24,25], which can be freely downloaded from http://dectau.uclm.es/floper/.

We wish to remark now that our proposal is an extension of previous works about the implementation of XPath by means of logic programming [2], which has been extended to XQuery in [1]. The new extension follows the same encoding proposed in [1] in which a predicate called *xpath* is defined by means of PROLOG rules, which basically traverse the PROLOG representation of the XML tree by means of a PROLOG list. In order to implement *Fuzzy-XPath* by means of $\mathcal{FLOPER}$ we proceed similarly to the PROLOG implementation of XPath, but proposing a new (fuzzy) predicate called *fuzzyXPath* implemented in MALP. The new query language returns a set of ranked answers each one with an associated RSV. Such RSV is computed by easily using MALP rules (thus exploiting the correspondences between the languages *for-being* and *to-be* implemented), where the notion of *RSV* is modeled inside a multi-adjoint lattice, and usual fuzzy connectives of the MALP language act as ideal resources to represent new flexible XPath operators.

As we will see in Section 3, the main goal of this paper consists in the introduction of a new fuzzy command inside *Fuzzy-XPath* which comfortably relies on our implementation based on fuzzy logic programming. So, when «[FILTER=r]» precedes a fuzzy query, the interpreter *lazily* explores an input XML document for dynamically disregarding as soon as possible those branchs of the XML tree leading to irrelevant solutions with an RSV degraded below r, thus allowing the possibility of efficiently managing large files without reducing the set of answers for which users are mainly interested in.

2   A Fuzzy Extension of XPath

Following [3,4,5,6], our flexible XPath is defined by means of the following rules:

```
    xpath := [deepdown]path
     path := literal | text() | node | @att |
             node/path | node//path
     node := QName | QName[cond]
     cond := path op path
 deepdown := DEEP=degree, DOWN=degree
       op := > | = | < | and | or | avg
```

Basically, our fuzzy proposal extends XPath as follows:

- A given XPath expression can be adorned with «[DEEP $= r_1$, DOWN $= r_2$]» which means that the *deepness* of elements is penalized by r_1 and that the *order* of elements is penalized by r_2, and such penalization is proportional to the distance. In particular, «[DEEP $= 1$, DOWN $= r_2$]» can be used for penalizing only w.r.t. document order. *DEEP* works for //, and *DOWN* works for / and //.

```
<bib>
   <book year="2001" price="45.95">
      <title>Don Quijote de la Mancha</title>
      <author>Miguel de Cervantes Saavedra</author>
      <publications> <book year="1997" price="35.99">
                        <title>La Galatea</title>
                        <author>Miguel de Cervantes Saavedra</author>
                        <publications>
                              <book year="1994" price="25.99">
                              <title>Los trabajos de Persiles y Segismunda</title>
                              <author>Miguel de Cervantes Saavedra</author></book>
                        </publications></book>
      </publications></book>
   <book year="1999" price="25.65">
      <title>La Celestina</title>
      <author>Fernando de Rojas</author></book>
   <book year="2005" price="29.95">
      <title>Hamlet</title>
      <author>William Shakespeare</author>
      <publications>
         <book year="2000" price="22.5">
            <title>Romeo y Julieta</title>
            <author>William Shakespeare</author></book>
      </publications></book>
   <book year="2007" price="22.95">
      <title>Las ferias de Madrid</title>
      <author>Felix Lope de Vega y Carpio</author>
      <publications>
         <book year="1996" price="27.5">
            <title>El remedio en la desdicha</title>
            <author>Felix Lope de Vega y Carpio</author> </book>
         <book year="1998" price="12.5">
            <title>La Dragontea</title>
            <author>Felix Lope de Vega y Carpio</author></book>
      </publications></book>
</bib>
```

Fig. 1. Input XML document in our examples

- Moreover, the classical *and* and *or* connectives admit here a fuzzy behavior based on fuzzy logic, i.e., assuming two given RSV's r_1 and r_2, operator *and* is defined as $r_3 = r_1 * r_2$ and operator *or* returns $r_3 = r_1 + r_2 - (r_1 * r_2)$. In addition, the *avg* operator is defined as $r_3 = (r_1 + r_2)/2$.

In general, an extended XPath expression defines, w.r.t. a XML document, a sequence of subtrees of the XML document where each subtree has an associated RSV. XPath conditions, which are defined as fuzzy operators applied to XPath expressions, compute a new RSV from the RSVs of the involved XPath expressions, which at the same time, provides a RSV to the node. In order to illustrate these explanations, let us see some examples of our proposed fuzzy version of XPath according to the XML document shown in Figure 1 whose *skeleton* is depicted in Figure 2.

Example 1. Suppose the XPath query: « [DEEP=0.9,DOWN=0.8]//title », that requests *title*'s penalizing the occurrences from the document root by a proportion of 0.9 and 0.8 by nesting and ordering, respectively, and for which we obtain

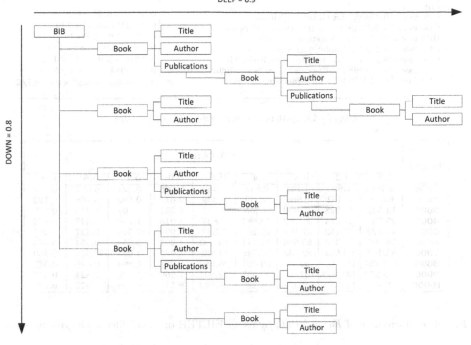

Fig. 2. XML skeleton represented as a tree

Document	RSV computation

<result>
   <title **rsv**="0.81">Don Quijote de la Mancha</title>
   <title **rsv**="0.6561">La Galatea</title>
   <title **rsv**="0.531441">Los trabajos de Persiles y ...</title>
   <title **rsv**="0.648">La Celestina</title>
   <title **rsv**="0.5184">Hamlet</title>
   <title **rsv**="0.419904">Romeo y Julieta</title>
   <title **rsv**="0.41472">Las ferias de Madrid</title>
   <title **rsv**="0.3359232">El remedio en la desdicha</title>
   <title **rsv**="0.26873856">La Dragontea</title>
 </result>
	$0.81 = 0.9^2$
	 $0.6561 = 0.9^4$
	 $0.531441 = 0.9^6$
	 $0.648 = 0.9^2 * 0.8$
	 $0.5184 = 0.9^2 * 0.8^2$
	 $0.419904 = 0.9^4 * 0.8^2$
	 $0.41472 = 0.9^2 * 0.8^3$
	 $0.3359232 = 0.9^4 * 0.8^3$
	 $0.26873856 = 0.9^4 * 0.8^4$

Fig. 3. Output of a query using *DEEP/DOWN*

Document	RSV computation

<result>
   <book **rsv**="0.5" ...> <title>Don Quijote ...</title> ...</book>
   <book **rsv**="1.0"...><title>La Celestina</title> ...</book>
   <book **rsv**="1.0" ...><title>Hamlet</title> ...</book>
   <book **rsv**="0.5" ...><title>Las ferias de Madrid</title> ...</book>
 </result>
	$0.5 = 0 + 1/2$
	 $1 = 1 + 1/2$
	 $1 = 1 + 1/2$
	 $0.5 = 1 + 0/2$

Fig. 4. Output of a query using *AVG*

Document	RSV computation

\<result\>
   \<title **rsv**="0.3645"\>La Galatea\</title\>
   \<title **rsv**="0.295245"\>Los trabajos de Persiles y... \</title\>
   \<title **rsv**="0.72"\>La Celestina\</title\>
   \<title **rsv**="0.288"\>Hamlet\</title\>
   \<title **rsv**="0.2304"\>Las ferias de Madrid\</title\>
   \<title **rsv**="0.2985984"\>El remedio en la desdicha\</title\>
   \<title **rsv**="0.11943936"\>La Dragontea\</title\>
 \</result\>
	$0.3645 = 0.9^3 * 1/2$
	 $0.295245 = 0.9^5 * 1/2$
	 $0.72 = 0.9 * 0.8 * 1$
	 $0.288 = 0.9 * 0.8^2 * 1/2$
	 $0.2304 = 0.9 * 0.8^3 * 1/2$
	 $0.2985984 = 0.9^3 * 0.8^4 * 1$
	 $0.11943936 = 0.9^3 * 0.8^5 * 1/2$

Fig. 5. Output of a query using all operators

Records	FILTER								
	0.1	0.2	0.3	0.4	0.5	0.6	0.7	0.8	0.9
1000	1.766	1.696	1.734	0.842	0.469	0.268	0.221	0.087	0.056
2000	6.628	6.432	6.998	3.242	1.439	0.677	0.599	0.168	0.122
3000	14.532	14.023	14.059	6.306	2.831	1.257	1.101	0.253	0.179
4000	25.535	24.684	24.722	10.883	4.827	1.918	1.794	0.345	0.242
5000	41.522	37.782	37.166	16.201	7.242	2.993	2.516	0.427	0.281
6000	58.905	55.354	55.596	24.411	10.993	4.207	3.554	0.554	0.373
7000	85.167	85.652	82.733	37.748	14.436	5.083	4.653	0.649	0.460
8000	137.737	102.816	102.763	69.401	26.680	8.273	5.894	0.690	0.481
9000	175.272	131.828	131.021	56.937	22.601	7.869	7.329	0.824	0.549
10000	195.613	185.201	167.676	95.286	26.649	9.516	9.595	0.973	0.742

Fig. 6. Performance of *Fuzzy-XPath* by using FILTER on XML files with growing sizes

the file listed in Figure 3. In such document we have included as attribute of each subtree, its corresponding RSV. The highest RSVs correspond the main *book*'s of the document, and the lowest RSVs represent the *book*'s occurring in nested positions (those annotated as related *publication*'s).

Example 2. Figure 4 shows the answer associated to the XPath expression: « /bib/book[@price<30 avg @year<2006] ». Here we show that books satisfying a *price* under 30 and a *year* before 2006 have the highest RSV.

Example 3. Finally, combining all operators «[DEEP=0.9,DOWN=0.8] //book [(@price>25 and @price<30) avg (@year<2000 or @year>2006)]/title», the RSV values are more scattered, as shown in Figure 5.

3   Using Filters for the Dynamic Thresholding of Queries

In [19,18] we have reported some *thresholding* techniques specially tailored for the MALP language, where the main idea consists in to dynamically create and evaluate filters for prematurely disregarding those superfluous computations leading to non-significant solutions. Somehow inspired by the same guidelines, we have recently equipped our fuzzyXPath interpreter with a new command with syntax «[FILTER=r]» (being r a real number between 0 and 1) which can be used just at the beginning of a query for indicating that only those answers with RSV greater of equal than r must be generated and reported.

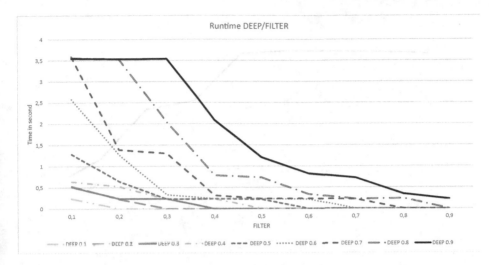

Fig. 7. Runtime for several *Fuzzy-XPath* queries varying DEEP and FILTER

So, if we consider a *Fuzzy-XPath* query with the following form «[FILTER=0.4]//book[@year<2000 avg @price<50]/title», we obtain nine answers, but only five if we fix «[FILTER=0.8]». Obviously, we would hope that the runtime of the second case should be lower than the first one since, as our approach does, there is no need for computing all solutions and then filtering the best ones. This desired dynamic behaviour when avoiding useless computations is reflected in Figure 6 which considers the effort needed for executing (excluding parsing/compiling time) a query like «[FILTER=r]//book[(@price>25 and @price<30) avg (@year<2000 or @year>2006)]»where each row represents the size of several XML files accomplishing with the same structure of our running example (but considering different nesting levels of tags book, title, author and publications), and each column refers to a different degree of the FILTER command. Here, the runtime is measured in seconds (the benchmarks have been performed using a computer with processor Intel Core Duo, with 2 GB RAM and Windows Vista) and each record in the input file refers to a different book (that is, the number of records coincides with the number of occurrences of tag book) which might contain other books inside its publications tag.

Moreover, in Figure 7 we continue with a similar query to the previous one, but also considering the DEEP command[1]. Here, for a large XML document with a fixed size, we express the number of seconds needed for executing such query when varying FILTER and DEEP, where it is easy to see that the behaviour is more and more improved whenever FILTER grows and DEEP decreases, as wanted. Note that the previous query makes use of the *avg* command and remember that its behaviour is defined, for two given RSV's r_1 and r_2, as

---

[1] This kind of statistics can be produced on-line for several XML files and *Fuzzy-XPath* queries via the following URL that we have just prepared for the interested reader: http://dectau.uclm.es/fuzzyXPath/fuzzyXPathEstatistic2.php#testing.

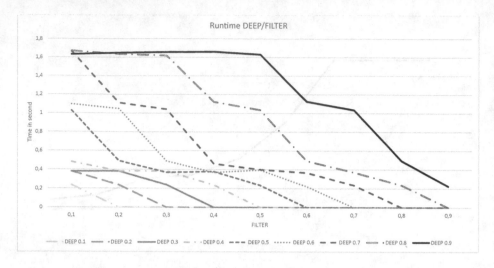

Fig. 8. Varying DEEP and FILTER in a query using $avg\{30, 1\}$

$r_3 = (r_1 + r_2)/2$. We have recently conceived a *priorized version* of such operator which let us to give different degrees of importance to its arguments. In general, $avg\{p_1, p_2\}$ is computed by $r_3 = (r_1 * p_1 + r_2 * p_2)/(p_1 + p_2)$ and hence, if in the previous query we use $avg\{30, 1\}$ instead of standard avg, we indicate that the first sub-condition (i.e., @price>25 and @price<30) is 30 times more important than the second one (i.e., @year<2000 or @year>2006), whereas $avg\{1, 30\}$ represent the inverse criterium. In Figures 8 and 9 we provide statistics in the same way than in Figure 7, but using now *average with priorities* 30-1 and 1-30, respectively.

Although the core of our application is written with (fuzzy) MALP rules, our implementation is based on the following items:

(1) We have reused/adapted several modules of our previous PROLOG-based implementation of (crisp) XPath described in [1,2].
(2) We have used the SWI-PROLOG library for loading XML files, in order to represent a XML document by means of a PROLOG term[2].
(3) The parser of XPath has been extended to recognize the new keywords FILTER, DEEP, DOWN, avg, etc... with their proper arguments.
(4) Each tag is represented as a data-term of the form: element(Tag, Attributes, Subelements), where Tag is the name of the XML tag, Attributes is a PROLOG list containing the attributes, and Subelements is a PROLOG list containing the sub-elements (i.e. sub-trees) of the tag. For instance, the document of Figure 1 is represented in SWI-PROLOG like in Figure 10. Loading of documents is achieved by predicate load_xml(+File,-Term) and writing by predicate write_xml(+File,+Term).
(5) Predicate fuzzyXPath(+ListXPath,+Tree,+Deep,+Down,+Filter,+Accum) receives six arguments: (1) ListXPath is the PROLOG representation of a

---

[2] The notion of *term* (i.e., data structure) is just the same in MALP and PROLOG.

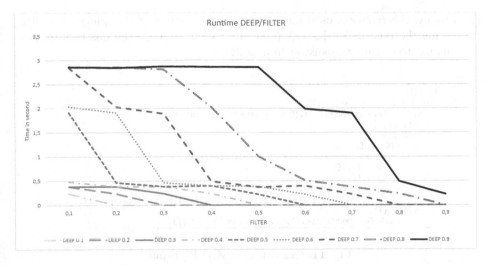

Fig. 9. Varying DEEP and FILTER in a query using $avg\{1,30\}$

```
[element(bib,[],
    [element(book,[year=2001,price=45.95],
        [element(title,[],[Don Quijote de la Mancha]),
        element(author,[],[Miguel de Cervantes Saavedra]),
        element(publications,[],
            [element(book,[year=1997,price=35.99],
                [element(title,[],[La Galatea]),
                element(author,[],[Miguel de Cervantes Saavedra]),
                element(publications,[],...])...]),1)])
```

Fig. 10. A data-term representing a XML document

XPath expression; (2) `Tree` is the term representing an input XML document; (3) `Deep/Down/Filter` have the obvious meaning, and finally (4) the last argument `Accum` (which is appropriately updated -maybe decreased- when going deeper in the exploration of the file) accumulates the sequence of penalties produced till reaching a concrete node, and it is very useful for deciding when performing a recursive call to the children of such node whenever the value of `Accum` is better than the one fixed by `Filter`.

(6) The evaluation of the query generates a *truth value* which has the form of a tree, called *tv tree*. For instance, the query shown in Example 1, generates the one illustrated in Figure 11. The main power of a fuzzy logic programming language like MALP w.r.t. PROLOG, is that instead of answering questions with a simple *true/false* value, solutions are reported in a much more tinged, documented way. Basically, the `fuzzyXPath` predicate traverses the PROLOG tree representing a XML document annotating into the *tv tree* the corresponding *deep/down* values according to the movements performed in the horizontal and vertical axis, respectively. In addition, the *tv tree* is annotated with the values of *and, or* and *avg* operators in each node.

(7) Finally, the *tv tree* is used for computing the output of the query, by multiplying the recorded values. A predicate called `tv_to_elem` has been implemented to output the answer in a pretty way.

```
tv(1, [],
   tv(0.9,[[],
      tv(0.9,[element(title,[],[Don Quijote de la Mancha]),[],
      tv(1,[[],[],
      tv(1,[[],
         tv(0.9,[[],
            tv(0.9,[element(title,[],[La Galatea]),[],
            tv(1,[[],[],
            tv(1,[[],
               tv(0.9,[[],
                  tv(0.9,[element(title,[],[Los trabajos de Persiles..]),...]),
      tv(0.8,[[],
         tv(0.9,[element(title,[],[La Celestina]),[],[]]),...
```

Fig. 11. Example of a MALP output

4   Conclusions and Future Work

In [3,4,5,6] we have recently enriched XPath with new constructs (both structural -*deep* and *down*- and constraints -*avg* and fuzzy versions of classical *or/and* operators-) in order to flexibly query XML documents. This paper has highlighted the benefits of using a new fuzzy command for filtering the set of ranked answers in a dynamic way, in order to reduce the runtime and complexity of computations when dealing with large files. Our approach represents the first real-world application developed with the fuzzy logic language MALP, for which we have recently developed some thresholding tabulation techniques[3] [19,18]. All these actions will be very useful for addressing in our framework the well-known "top-k ranking problem" (i.e. determining the top k answers to a query without computing the -usually wider, possibly infinite- whole set of solutions, which is strongly related with the FILTER command reported along this paper) inspired by [8,11,12,21,29,17].

Acknowledgements. We are grateful to anonymous reviewers for providing us valuable suggestions which have been used for improving the material compiled so far. This work was supported by the EU (FEDER), and the Spanish MINECO Ministry (*Ministerio de Economía y Competitividad*) under grant TIN2013-45732-C4-2-P, as well as by the Andalusian Regional Government (Spain) under Project P10-TIC-6114.

---

[3] We are now implementing into $\mathcal{FLOPER}$ this highly efficient procedural mechanism.

References

1. Almendros-Jiménez, J.M.: An Encoding of XQuery in Prolog. In: Bellahsène, Z., Hunt, E., Rys, M., Unland, R. (eds.) XSym 2009. LNCS, vol. 5679, pp. 145–155. Springer, Heidelberg (2009)
2. Almendros-Jiménez, J.M., Becerra-Terón, A., Enciso-Baños, F.J.: Querying XML documents in logic programming. Theory and Practice of Logic Programming 8(3), 323–361 (2008)
3. Almendros-Jiménez, J.M., Luna Tedesqui, A., Moreno, G.: A Flexible XPath-based Query Language Implemented with Fuzzy Logic Programming. In: Bassiliades, N., Governatori, G., Paschke, A. (eds.) RuleML 2011 - Europe. LNCS, vol. 6826, pp. 186–193. Springer, Heidelberg (2011)
4. Almendros-Jiménez, J.M., Luna, A., Moreno, G.: Fuzzy Logic Programming for Implementing a Flexible XPath-based Query Language. Electronic Notes on Theoretical Computer Science, ENTCS 282, 3–18 (2012)
5. Almendros-Jiménez, J.M., Luna Tedesqui, A., Moreno, G.: Annotating "Fuzzy Chance Degrees" When Debugging XPath Queries. In: Rojas, I., Joya, G., Cabestany, J. (eds.) IWANN 2013, Part II. LNCS, vol. 7903, pp. 300–311. Springer, Heidelberg (2013)
6. Almendros-Jiménez, J.M., Luna, A., Moreno, G., Vázquez, C.: Analyzing Fuzzy Logic Computations with Fuzzy XPath. In: Proc. of PROLE 2013, pp. 136–150. Universidad Complutense de Madrid (2013) ISBN: 978-84-695-8331-9; ("work in progress" track, extended version submitted to ECEASST)
7. Berglund, A., Boag, S., Chamberlin, D., Fernandez, M.F., Kay, M., Robie, J., Siméon, J.: XML path language (XPath) 2.0. In: W3C (2007)
8. Bruno, N., Chaudhuri, S., Gravano, L.: Top-k selection queries over relational databases: Mapping strategies and performance evaluation. ACM Trans. Database Syst. 27(2), 153–187 (2002)
9. Buche, P., Dibie-Barthélemy, J., Haemmerlé, O., Hignette, G.: Fuzzy semantic tagging and flexible querying of XML documents extracted from the Web. Journal of Intelligent Information Systems 26(1), 25–40 (2006)
10. Campi, A., Damiani, E., Guinea, S., Marrara, S., Pasi, G., Spoletini, P.: A fuzzy extension of the XPath query language. Journal of Intelligent Information Systems 33(3), 285–305 (2009)
11. Chang, K.C.-C., Hwang, S.W.: Minimal probing: supporting expensive predicates for top-k queries. In: Franklin, M.J., Moon, B., Ailamaki, A. (eds.) SIGMOD Conference, pp. 346–357. ACM (2002)
12. Chaudhuri, S., Gravano, L., Marian, A.: Optimizing top-k selection queries over multimedia repositories. IEEE Trans. Knowl. Data Eng. 16(8), 992–1009 (2004)
13. Damiani, E., Marrara, S., Pasi, G.: FuzzyXPath: Using fuzzy logic an IR features to approximately query XML documents. In: Melin, P., Castillo, O., Aguilar, L.T., Kacprzyk, J., Pedrycz, W. (eds.) IFSA 2007. LNCS (LNAI), vol. 4529, pp. 199–208. Springer, Heidelberg (2007)
14. Fazzinga, B., Flesca, S., Furfaro, F.: On the expressiveness of generalization rules for XPath query relaxation. In: Proceedings of the Fourteenth International Database Engineering & Applications Symposium, pp. 157–168. ACM (2010)
15. Bosc, P., Pivert, O., Mokhtari, A.: Top-k Answers to Fuzzy XPath Queries. In: Bhowmick, S.S., Küng, J., Wagner, R. (eds.) DEXA 2009. LNCS, vol. 5690, pp. 847–854. Springer, Heidelberg (2009)

16. Gaurav, A., Alhajj, R.: Incorporating fuzziness in XML and mapping fuzzy relational data into fuzzy XML. In: Proceedings of the 2006 ACM Symposium on Applied Computing, pp. 456–460. ACM (2006)

17. Ilyas, I.F., Beskales, G., Soliman, M.A.: A survey of top-k query processing techniques in relational database systems. ACM Comput. Surv. 40(4) (2008)

18. Julián-Iranzo, P., Medina-Moreno, J., Morcillo, P.J., Moreno, G., Ojeda-Aciego, M.: An unfolding-based preprocess for reinforcing thresholds in fuzzy tabulation. In: Rojas, I., Joya, G., Gabestany, J. (eds.) IWANN 2013, Part I. LNCS, vol. 7902, pp. 647–655. Springer, Heidelberg (2013)

19. Julián, P., Medina, J., Moreno, G., Ojeda-Aciego, M.: Efficient thresholded tabulation for fuzzy query answering. In: Bouchon-Meunier, B., Magdalena, L., Ojeda-Aciego, M., Verdegay, J.-L., Yager, R.R. (eds.) Foundations of Reasoning under Uncertainty. STUDFUZZ, vol. 249, pp. 125–141. Springer, Heidelberg (2010)

20. Li, H.G., Aghili, S.A., Agrawal, D., El Abbadi, A.: FLUX: fuzzy content and structure matching of XML range queries. In: Proceedings of the 15th International Conference on World Wide Web, pp. 1081–1082. ACM (2006)

21. Marian, A., Bruno, N., Gravano, L.: Evaluating top-k queries over web-accessible databases. ACM Trans. Database Syst. 29(2), 319–362 (2004)

22. Medina, J., Ojeda-Aciego, M., Vojtáš, P.: Similarity-based Unification: a multi-adjoint approach. Fuzzy Sets and Systems 146, 43–62 (2004)

23. Morcillo, P.J., Moreno, G.: Programming with Fuzzy Logic Rules by using the FLOPER Tool. In: Bassiliades, N., Governatori, G., Paschke, A. (eds.) RuleML 2008. LNCS, vol. 5321, pp. 119–126. Springer, Heidelberg (2008)

24. Morcillo, P.J., Moreno, G., Penabad, J., Vázquez, C.: A Practical Management of Fuzzy Truth Degrees using FLOPER. In: Dean, M., Hall, J., Rotolo, A., Tabet, S. (eds.) RuleML 2010. LNCS, vol. 6403, pp. 20–34. Springer, Heidelberg (2010)

25. Moreno, G., Vázquez, C.: Fuzzy logic programming in action with floper. Journal of Software Engineering and Applications 7, 237–298 (2014)

26. Oliboni, B., Pozzani, G.: An XML schema for managing fuzzy documents. In: Ma, Z., Yan, L. (eds.) Soft Computing in XML Data Management. STUDFUZZ, vol. 255, pp. 3–34. Springer, Heidelberg (2010)

27. Panzeri, E., Pasi, G.: An approach to define flexible structural constraints in xquery. In: Huang, R., Ghorbani, A.A., Pasi, G., Yamaguchi, T., Yen, N.Y., Jin, B. (eds.) AMT 2012. LNCS, vol. 7669, pp. 307–317. Springer, Heidelberg (2012)

28. Panzeri, E., Pasi, G.: Flex-basex: an xml engine with a flexible extension of xquery full-text. In: Proc. of the 36th International ACM SIGIR Conference on Research and Development in Information Retrieval, SIGIR 2013, pp. 1038–1084. ACM (2013), http://doi.acm.org/10.1145/2484028.248421

29. Re, C., Dalvi, N.N., Suciu, D.: Efficient top-k query evaluation on probabilistic data. In: Chirkova, R., Dogac, A., Özsu, M.T., Sellis, T.K. (eds.) ICDE, pp. 886–895. IEEE (2007)

30. Yan, L., Ma, Z.M., Liu, J.: Fuzzy data modeling based on XML schema. In: Proceedings of the 2009 ACM symposium on Applied Computing, pp. 1563–1567. ACM (2009)

Individual Decisions under Group Consensus

Tomoe Entani

Graduate School of Applied Informatics, University of Hyogo, Kobe, Hyogo 650-0047, Japan
entani@ai.u-hyogo.ac.jp

Abstract. In the group decision of this paper, it is assumed that the practice is entrusted to each decision maker. In such a decision problem, it is not necessary for a decision maker to obey the group decision completely, but necessary to consider it into his/her final decision. In this paper, when a group of decision makers give the comparisons of alternatives, their individual decisions are obtained as the interval weights of alternatives so as to have a common weight. The problem is formulated based on Interval AHP. By relaxing two conditions of the individual decisions for a consensus, a decision maker has to admit the modification of his/her initial judgments and/or the enlargement of his/her individual decision.

Keywords: Group decision making, Consensus, Interval analysis, Analytic hierarchy process.

1 Introduction

The decision problem in this paper follows AHP (Analytic Hierarchy Process). It is an approach to multi-criteria decision making problems and induces the preference of a decision maker from his/her judgments [1]. In the setting of AHP, a decision maker gives the comparisons of all pairs of alternatives intuitively, and his/her preference denoted as the weights of the alternatives is obtained from them.

Most decisions, at least in organizations and society, are the responsibility of groups rather than individuals [2]. In a group decision making problem, a group of decision makers try to reach a consensus based on their individual opinions. There are two possible goals of the group decision making. One is to induce the group decision which all the decision makers practice together [3]. The advantages of using AHP for this group decision making, such as showing the decision process to the decision makers, are explained in [4,5]. The popular technique to aggregate individuals into a group is geometric mean [6] and several other techniques are proposed [7,8]. In the cases where the group decision is practiced by all decision makers together, it is not necessary to show each decision maker his/her individual decisions. On the other hand, there are cases where the practice is entrusted to each decision maker in a group. S/he decides the individual decision under the group consensus at his/her discretion and practices it independently. The other possible goal is to induce such a final individual decision reflecting the group decision. For instance, a group of the managers of several branch offices are representatives of making the business policy of their company. Each manager has own opinion based on the peculiar situation of his/her branch office so that the initial individual opinions are not always the same. However, with discussion, they

C. Cornelis et al. (eds.): RSCTC 2014, LNAI 8536, pp. 331–338, 2014.

decide the company's policy. They reconsider and decide their branches' individual policies taking it into consideration. The final individual policies of all branches may not be exactly the same but should reflect the company's policy. In this example, the individual decision is up to each decision maker to some extent and may be different from the independently obtained decision from his/her initial judgments. Two possible goals of group decision making are to induce the group decision and the individual decisions, depending on group and individual practice, respectively. This paper focuses on the latter goal which tries to induce the individual decisions under the group consensus.

In group AHP, the group decision is obtained from the aggregated initial judgments or it is the aggregation of the independently obtained individual decisions [9]. As is mentioned before, AHP requires a decision maker to give the pairwise comparisons of alternatives, instead of his/her preference of alternatives directly. Therefore, both the group decision and the individual decisions are unknown in a decision making process. It is not necessary to aggregate the individual judgments beforehand and/or the individual decisions afterward. In this paper, all the individual decisions are obtained simultaneously from the initial individual judgments under the condition that there exists a group decision included in all of them. The method to induce the final individual decisions under the group consensus is proposed.

2   Preliminary

2.1   Individually Given Comparisons

There are m members in a group and they make decision on n alternatives. Each member of the group, decision maker k, gives his/her intuitive judgments on alternatives independently as the following pairwise comparison matrix A_k. S/he gives comparison a_{kij} of pair of alternatives i and j, without caring for the other pairs of alternatives $(i', j') \neq (i, j)$ and the other members $k' \neq k$.

$$A_k = \begin{bmatrix} 1 & \cdots & a_{k1n} \\ \vdots & a_{kij} & \vdots \\ a_{kn1} & \cdots & 1 \end{bmatrix} \forall k, \qquad (1)$$

whose element a_{kij} represents the importance ratio of alternative i to alternative j by decision maker k and is between 1/9 to 9; $a_{kij} \in \{1/9, 1/7, \ldots, 1, 3, \ldots, 9\}$, generally in AHP [1]. The comparisons are identical $a_{kii} = 1$ and reciprocal $a_{kij} = 1/a_{kji}$ so that s/he has to compare $n(n-1)/2$ pairs of alternatives. The comparison matrix is consistent, if and only if the following transitivity relations for all pairs of alternatives are satisfied.

$$a_{kij} = a_{kil}a_{klj} \ \forall (i, j, l). \qquad (2)$$

Decision maker k compares alternatives i and j and gives comparison a_{kij} in (1) without fixing their weights, w_{ki} and w_{kj} beforehand. That is, the weights of alternative i, implicitly used for giving comparisons a_{kij} and a_{kil} may not be equal; $w_{ki}^j \neq w_{ki}^l$. In this way, the weight of an alternative based on the given comparisons is uncertain. In order to reflect such uncertainty, the weight of an alternative is denoted as an interval in

Interval AHP [10,11]. Assuming the weight of alternative i in his/her mind as interval $W_{ki} = [\underline{w}_{ki}, \overline{w}_{ki}]$, s/he uses a real value in the interval $w_{ki}^j \in W_{ki} \leftrightarrow \underline{w}_{ki} \leq w_{ki}^j \leq \overline{w}_{ki}$ in giving comparison a_{kij}. Then, the individual interval weights of alternatives of decision maker k, W_k, is denoted as $W_k = (W_{k1}, W_{k2}, \ldots, W_{kn})$, where $W_{ki} = [\underline{w}_{ki}, \overline{w}_{ki}]$. The sum of the widths represents uncertainty in the decision of decision maker k.

2.2 Conditions of Individual Decision

The individual preference on alternatives is obtained from the given comparisons following three conditions. The 1st condition of the individual decision is the relation to the given comparison. The given comparison is included in the ratio of the corresponding interval weights

$$a_{kij} \in \frac{W_{ki}}{W_{kj}} = \frac{[\underline{w}_{ki}, \overline{w}_{ki}]}{[\underline{w}_{kj}, \overline{w}_{kj}]} \leftrightarrow \begin{cases} \dfrac{\underline{w}_{ki}}{\overline{w}_{kj}} \leq a_{kij} \leq \dfrac{\overline{w}_{ki}}{\underline{w}_{kj}} \ \forall i, j, \\ \epsilon \leq \underline{w}_{ki} \leq \overline{w}_{ki} \ \forall i, \end{cases} \tag{3}$$

where ϵ is a small positive number and the fraction of intervals is defined as its maximum range.

The 2nd condition is to normalize the interval weights by interval probability [12,13], since the comparison is denoted as a ratio as in (3).

$$\sum_{i \neq j} \overline{w}_{ki} + \underline{w}_{kj} \geq 1, \ \sum_{i \neq j} \underline{w}_{ki} + \overline{w}_{kj} \leq 1 \ \forall j, \tag{4}$$

where the redundancy of the intervals to make their sum be 1 is excluded. For instance, the 1st condition requires $\underline{w}_{kj}$ not to be too small. When the weights are real values as $\overline{w}_{ki} = \underline{w}_{ki} = w_{ki} \forall i$, two inequalities are replaced into $\sum_i w_{ki} = 1$ as ordinal probability.

The last condition is to make the interval weights be as close as possible to the given comparison. Because of the inclusion relation (3), the widths of interval weights are minimized as

$$\min \ \sum_i (\overline{w}_{ki} - \underline{w}_{ki}), \tag{5}$$

which also minimizes uncertainty in the decision of decision maker k. In case of the consistent comparisons which satisfy transitivity (2), it reaches the minimum 0, i.e., $\underline{w}_{ki} = \overline{w}_{ki} = w_{ki}$, where the weight is fixed regardless of compared alternatives. In the other cases, the weight of at least one alternative is an interval so that (5) is over 0.

Then, the individual interval weights are obtained from the individually given comparisons by the following linear programing (LP) problem in Interval AHP,

$$\begin{aligned} &\min \ \sum_{ki} (\overline{w}_{ki} - \underline{w}_{ki}), \\ \text{s. t. } &\frac{\underline{w}_{ki}}{\overline{w}_{kj}} \leq a_{kij} \leq \frac{\overline{w}_{ki}}{\underline{w}_{kj}} \ \forall i, j, k, \\ &\sum_{i \neq j} \overline{w}_{ki} + \underline{w}_{kj} \geq 1 \ \sum_{i \neq j} \underline{w}_{ki} + \overline{w}_{kj} \leq 1 \ \forall j, k, \\ &\epsilon \leq \underline{w}_{ki} \leq \overline{w}_{ki} \ \forall i, k, \end{aligned} \tag{6}$$

where the variables are the upper and lower bounds of the interval weights. Since three conditions for each decision maker are independent from the other decision makers', (6) for m decision makers simultaneously is divided into m LP problems for each decision maker.

3 Individual Decision under Group Consensus

3.1 Group Decision as Consensus

For a consensus, it is natural that the group decision $W_i = [\underline{w}_i, \overline{w}_i]$ should be included in any of the individual decisions $W_i = [\underline{w}_{ki}, \overline{w}_{ki}]$ so that $\underline{w}_{ki} \leq \overline{w}_i$ and $\underline{w}_i \leq \overline{w}_{ki}$. If the independently obtained individual interval weights satisfy

$$\max_k \underline{w}_{ki} \leq \min_k \overline{w}_{ki}, \tag{7}$$

then the group interval weight is defined as $W_i = [\max_k \underline{w}_{ki}, \min_k \overline{w}_{ki}]$. On the other hand, if $\max_k \underline{w}_{ki} > \min_k \overline{w}_{ki}$, the group decision as a consensus of all decision makers cannot be found. In this case, the independently obtained decisions may be improved by reflecting the group decision. It is one of the advantages of being a group to reflect each other for the better individual decisions.

3.2 Relaxing Conditions of Individual Decision

In order to reflect group decision into the final individual decisions, it is assumed that there is a consensus among. The final individual decisions are obtained so as to satisfy (7), which is rewritten as

$$\underline{w}_{li} \leq \overline{w}_{ki} \ \forall k, l, k \neq l \tag{8}$$

where for a pair of decision makers their lower bounds are smaller than the upper bounds of the others.

Instead of adding consensus constraint (8) into (6), where the individual decisions are independently obtained, its constraints are relaxed. They are the inclusion relation (3) and the widths of interval weights (5). That is, the initial individual comparisons are modified and/or the final individual decisions are enlarged.

As for the inclusion relation (3), the initial comparison a_{kij} is modified with the positive and negative excess into $a'_{kij} = a_{kij} + p_{kij} - n_{kij}$, where $0 \leq p_{kij}, n_{kij}$.

$$\frac{\underline{w}_{ki}}{\overline{w}_{kj}} \leq a'_{kij} = a_{kij} + p_{kij} - n_{kij} \leq \frac{\overline{w}_{ki}}{\underline{w}_{kj}} \ \forall i, j, k, \tag{9}$$

where a_{kij} can be too small or large considering the others so that $n_{kij} p_{kij} = 0$. The inclusion relation (9) is rewritten as follows.

$$\begin{aligned}
&\frac{\underline{w}_{ki}}{\overline{w}_{kj}} \leq a_{kij} + p_{kij}, \quad a_{kij} - n_{kij} \leq \frac{\overline{w}_{ki}}{\underline{w}_{kj}} \ \forall i, j, k, \\
&\leftrightarrow \underline{w}_{kj} \leq a_{kij}\overline{w}_{ki} + p_{kij}\overline{w}_{ki}, \quad \underline{w}_{kj} - n_{kij}\underline{w}_{kj} \leq \overline{w}_{ki} \ \forall i, j, k,
\end{aligned} \tag{10}$$

When one of the excesses is positive, the initial comparison a_{kij} is not included in the ratio of the final individual interval weights; $a_{kij} \notin W_{ki}/W_{kj}$. In order to include a_{kij} as much as possible, the sum of excesses is minimized.

$$\min \ \textstyle\sum_{kij}(n_{kij} + p_{kij}), \tag{11}$$

which, in other words, minimizes the modifications of the initial comparisons. Such a modified comparisons considering the group consensus may be more reliable than the initial ones because of the following two reasons. One is that the comparisons may not be perfect and have some errors, since they are given intuitively. The other is that a_{kij} may be a little different from what is in the decision maker's mind, since s/he chooses one value in $\{1/9, 1/7, ..., 1, ..., 7, 9\}$. The advantage and disadvantage of these values are discussed from the viewpoint of AHP [14].

As for the widths (5), if the widths of the individual interval weights are enlarged enough, it is possible to find their core as a group decision, such that $W_{ki} = [0,1] \forall k, i$ and $W_i = [0,1] \forall i$. In order for the final individual decisions to be as certain as possible, their widths are minimized as

$$\min \ \textstyle\sum_{ki}(\overline{w}_{ki} - \underline{w}_{ki}), \tag{12}$$

whose optimal value cannot be smaller than that of (6) because of adding the consensus constraint (8).

When a group of decision makers are required to reach a consensus, they have to admit that their final decisions are obtained from the modified judgments and/or enlarged from their independently obtained decisions.

3.3 Final Individual Decisions

By relaxing the conditions of the individual decision, the individual interval weights of all decision makers are obtained simultaneously so as to have a common weight. By adding consensus condition (8) to independent Interval AHP (6), the problem is formulated as follows.

$$\begin{aligned}
&\min \textstyle\sum_{ki}(\overline{w}_{ki} - \underline{w}_{ki}), \\
&\min \textstyle\sum_{kij}(n'_{kij} + p'_{kij}), \\
&\text{s.t.} \ \ \underline{w}_{ki} \leq a_{kij}\overline{w}_{kj} + p'_{kij}, \ \ a_{kij}\underline{w}_{kj} - n'_{kij} \leq \overline{w}_{ki} \ \forall i,j,k, \\
&\qquad 0 \leq n'_{kij}, p'_{kij} \forall i,j,k, \\
&\qquad \textstyle\sum_{i\neq j}\overline{w}_{ki} + \underline{w}_{kj} \geq 1, \ \ \textstyle\sum_{i\neq j}\underline{w}_{ki} + \overline{w}_{kj} \leq 1 \ \forall k,j, \\
&\qquad \epsilon \leq \underline{w}_{ki} \leq \overline{w}_{ki} \ \forall k,i, \\
&\qquad \underline{w}_{li} \leq \overline{w}_{ki} \ \forall i,k,l, k \neq l,
\end{aligned} \tag{13}$$

where $n'_{kij} = n_{kij}\underline{w}_{kj}$ and $p'_{kij} = p_{kij}\overline{w}_{kj}$ for calculation. The variables are the individual interval weights $W_{ki} = [\underline{w}_{ki}, \overline{w}_{ki}]$ and the excesses n'_{kij} and p'_{kij} in addition.

There are two objective functions on the enlargement of the individual decisions (12) and modifications of the comparisons (11). When the decision makers do not care the modifications of their initial comparisons, such that they are not confident of their judgments, it is enough to minimize the 1st objective function (12). As a result the initial judgments are not always included in the final individual decision. While, when the decision makers are confident of their judgments, they may focus on whether their initial judgments are reflected to their final decisions or not. The 2nd objective function (11) is primary minimized. As a result, the individual interval weights have wide ranges so that the final individual decisions are vague. These are the extreme cases by ignoring one of the objective functions. In practice, two independent objective functions in (13) are treated in various ways following multiobjective programming techniques. One of the simplest treatments is the application of weighting approach as min $(1-\lambda)\sum_{ki}(\overline{w}_{ki}-\underline{w}_{ki}) + \lambda\sum_{kij}(n_{kij} + p_{kij})$, where $\lambda \in [\epsilon, 1 - \epsilon]$. λ is given based on the confidence of the decision makers in their initial judgments.

4 Numerical Example

There are 3 decision makers who give the comparisons on 4 alternatives as in Table 1. By (6), the individual interval weights are independently obtained and shown next to each matrix. As for A3 and A4, all individual decisions have common weights, although, as for A1 and A2, they do not. Therefore, in this example, in order to reach a consensus, they need to admit the modifications of their initial comparisons and/or the enlargement of their final decisions.

Table 1. Initial comparisons and independent individual decisions

DM1	A1	A2	A3	A4	Interval weight	DM2	A1	A2	A3	A4	Interval weight
A1	1	2	3	4	0.500	A1	1	3	3	4	0.571
A2	1/2	1	2	3	0.250	A2	1/3	1	3	3	[0.190, 0.214]
A3	1/3	1/2	1	2	[0.125, 0.167]	A3	1/3	1/3	1	4	[0.071, 0.190]
A4	1/4	1/3	1/2	1	[0.083, 0.125]	A4	1/4	1/3	1/4	1	[0.048, 0.143]

DM3	A1	A2	A3	A4	Interval weight
A1	1	1	4	6	0.390
A2	1	1	1	2	[0.244, 0.390]
A3	1/4	1	1	3	[0.098, 0.244]
A4	1/6	1/2	1/3	1	[0.065, 0.122]

By (13), the individual interval weights with $\lambda = 0.2, 0.3$ and 0.6 are obtained simultaneously and shown in Table 2. It has two objective functions to minimize the enlargement of individual decisions and the modifications of judgments. The widths of the final individual interval weights and the modified comparisons are shown at the bottom two lines of Table 2. The less the confidence of the decision makers in their judgments, the smaller λ is given. Then, the widths of the individual interval weights become smaller, as well as the number of the modified comparisons increases. With $\lambda = 0.2$, the widths of the individual decisions are primary minimized so that all of them are obtained as the same real values. From the viewpoint that their initial judgments are flexible, it is reasonable for all decision makers to treat the certain group decision as their own decisions. While, with $\lambda = 0.6$, instead that no comparisons are modified, the widths are wider than those in Table 1 and the individual decisions are vague. From the viewpoint that the decision makers stick to their initial judgments, it is reasonable for all decision makers to accept the enlargement of their decisions as far as their initial judgments are reflected into their final decisions.

With $\lambda = 0.3$, DM1 and DM2 reach a consensus without modifying their initial comparisons, though, DM3 has to modify two comparisons, a_{312} and a_{323}. This may be because two comparisons by DM3 are unreliable comparing to the others'. In this way, the individual decision considering the group decision can be obtained by revising the suspicious initial comparisons.

All the independent decisions of A3 and A4 in Table 1 include [0.125, 0.167] and [0.083, 0.122], respectively, and such cores include the group decisions with all λs. With $\lambda = 0.6$, the group interval weights are (0.5, 0.25, [0.125, 0.167], [0.083, 0.1]), which don't satisfy (4), i.e., they are not normalized. By reducing the redundancy in them, the normalized intervals (0.5, 0.25, [0.15, 0.167], [0.083, 0.1]) are easily found. The detailed method to obtain the normalized intervals in a set of intervals is not discussed here, since the group decision in this study is necessary to ensure a consensus and is not used for a practice.

Table 2. Individual and group interval weights with $\lambda = 0.2, 0.3$ and 0.6

$\lambda = 0.2$	DM1	DM2	DM3	Group
A1	0.500	0.500	0.500	0.500
A2	0.250	0.250	0.250	0.250
A3	0.167	0.167	0.167	0.167
A4	0.083	0.083	0.083	0.083
Width	0	0	0	0
Modified comparison	$\{14, 23\}$	$\{12, 14, 23, 34\}$	$\{12, 13, 23, 24, 34\}$	

$\lambda = 0.3$	DM1	DM2	DM3	Group
A1	0.500	[0.500, 0.529]	0.500	0.500
A2	0.250	[0.176, 0.259]	0.250	0.250
A3	[0.125, 0.167]	[0.086, 0.169]	[0.125, 0.188]	[0.125, 0.167]
A4	[0.083, 0.125]	[0.042, 0.125]	[0.062, 0.125]	[0.083, 0.125]
Width	0.084	0.278	0.126	0.084
Modified comparison	-	-	$\{12, 23\}$	

$\lambda = 0.6$	DM1	DM2	DM3	Group
A1	0.500	[0.500, 0.529]	[0.404, 0.500]	0.500
A2	0.250	[0.176, 0.259]	[0.200, 0.404]	0.250
A3	[0.125, 0.167]	[0.086, 0.169]	[0.125, 0.200]	[0.125, 0.167]
A4	[0.083, 0.125]	[0.042, 0.125]	[0.067, 0.100]	[0.083, 0.100]
Width	0.084	0.278	0.408	0.059
Modified comparison	-	-	-	

5   Conclusion

The individual decisions of all decision makers are obtained simultaneously from their initially given comparisons under the group consensus. It is suitable when the practice is entrusted to each decision maker and the group decision itself is not used for a practice. Assuming the group decision as the core of individual decisions, all the individual interval weights have to have a common weight so that their maximum lower bound should be less than their minimum upper bound. Instead of adding the consensus constraint into independent Interval AHP, two conditions of the individual decisions are relaxed. As a result, the decision maker may admit the modifications of his/her initial judgments and/or the enlargement of his/her final decision. The problem is transformed into LP problem for calculation. Its two objective functions of minimizing the excess from the initial comparisons and the enlargement of the widths from the independently obtained ones. They are aggregated into one by weighting approach. When the decision makers stick to their initial judgments, the weight for minimizing the modifications is assumed to be greater than that for minimizing the widths. And vice versa, when they are not confident of their initial judgments. The modification of the initial comparisons is also reasonable since the initially given comparisons are not always perfect.

References

1. Saaty, T.L.: The Analytic Hierarchy Process. McGraw-Hill, New York (1980)
2. French, S., Maule, J., Papamichail, N.: Decision Behaviour, Analysis and Support. Cambridge University Press (2009)
3. Wenger, E.: Communities of Practice: Learning, Meaning, and Identity. Cambridge University Press (1999)
4. Dyer, R.F., Forman, E.H.: Group decision support with the Analytic Hierarchy Process. Decision Support Systems 8, 94–124 (1992)
5. Basak, I., Saaty, T.: Group decision making using Analytic Hierarchy Process. Mathematical and Computer Modelling 17(4/5), 101–109 (1993)
6. Aczel, J., Saaty, T.L.: Procedure for synthesizing ratio judgements. Journal of Mathematical Psychology 27, 93–102 (1983)
7. Altuzarra, A., Moreno-Jimenez, J.M., Salvador, M.: A Bayesian priorization procedure for AHP-group decision making. European Journal of Operational Research 182(1), 364–382 (2007)
8. Yeh, C.H., Chang, Y.H.: Modeling subjective evaluation for fuzzy group multicriteria decision making. European Journal of Operational Research 194(2), 464–473 (2009)
9. Forman, E., Peniwati, K.: Aggregating individual judgments and priorities with the Analytic Hierarchy Process. European Journal of Operational Research 108(1), 165–169 (1998)
10. Sugihara, K., Tanaka, H.: Interval evaluations in the Analytic Hierarchy Process by possibilistic analysis. Computational Intelligence 17(3), 567–579 (2001)
11. Sugihara, K., Ishii, H., Tanaka, H.: Interval priorities in AHP by interval regression analysis. European Journal of Operational Research 158(3), 745–754 (2004)
12. de Campos, L.M., Huete, J.F., Moral, S.: Probability intervals: a tool for uncertain reasoning. International Journal of Uncertainty 2(2), 167–196 (1994)
13. Tanaka, H., Sugihara, K., Maeda, Y.: Non-additive measures by interval probability functions. Information Sciences 164, 209–227 (2004)
14. Saaty, T.L., Ozdemir, M.S.: Why the magic number seven plus or minus two. Mathematical and Computer Modelling 38(3-4), 233–244 (2003)

Fuzzy Ontology-Based Approach for Automatic Construction of User Profiles

Mateus Ferreira-Satler[1], Francisco P. Romero[2], Jose A. Olivas[2], and Jesus Serrano-Guerrero[2]

[1] Universidade Federal de Ouro Preto
João Monlevade, Minas Gerais, Brasil
mateus@decea.ufop.br
[2] Deparment of Information Systems and Technologies
University of Castilla La Mancha
Ciudad Real, Spain
{FranciscoP.Romero,JoseAngel.Olivas,Jesus.Serrano}@uclm.es

Abstract. This paper shows a fuzzy ontology based approach to automatically build user profiles from a collection of user interest documents. The ontological representation of the user profile enhances the performance in tasks such as filtering, categorization and information retrieval. The proposed technique takes advantage of relevance measures to generate semantic representations of user context. The proposed work also presents a strategy for automatic generation of fuzzy ontologies to support user profile modeling. The experiments performed confirm that the automatically obtained fuzzy ontologies are good representation of the user's preferences. In order to test the applicability of the obtained ontologies, a text categorization experiment has been proposed and the obtained results indicate that the approach can be applied with satisfactory results and warrants further research.

Keywords: User Profile, Fuzzy Ontology, User Modeling, Text Mining.

1 Introduction

The World Wide Web presents new challenges to information retrieval [10]. The rapid growth of digital libraries, such as Internet, makes it difficult to human beings to access useful information conveniently and effectively. This is due to the fact that most of information is embedded in a non-structured or semi-structured way, which makes the search of a particular content a daunting and time consuming task. Traditional search engines usually use techniques that match the words in the query with the document content, and display as result thousands of pages of which only a few ones are really useful and relevant. In addition, search engines are not able to identify the nature and interests of a user in such dynamic environment, for that reason a user profile could be required to present the information in a manner that truly reflects user's needs. [5].

Some of the most important issues to take advantage in the process of constructing user profiles are the notions of the Semantic Web and Personalized

C. Cornelis et al. (eds.): RSCTC 2014, LNAI 8536, pp. 339–346, 2014.

Information Management, which use the semantic context of the presented information and the user preferences to facilitate the information storage and retrieval process [4]. On the other hand, ontologies have proven to be successful in handling a machine processable representation of information and have been used to model user context in various applications [5]. In [2] it is shown how a fuzzy ontology-based approach can improve semantic documents retrieval. The proposal is illustrated using an information retrieval algorithm based on an object-fuzzy concept network. The proposed fuzzy ontology, based on the semantic correlation between concepts, is capable to represent a dynamic knowledge of a domain adapting itself to the context. On the other hand, in [1] a personal ontology is defined using a knowledge extraction process from the general purpose ontology YAGO guided by a set of keywords.

A bottleneck in developing ontology-based systems stems from the fact that the conceptual formalism that supports by typical ontologies may not be sufficient to represent uncertainty that is commonly found in many application domains [11]. Moreover, most of ontologies modeling user preferences are application specific and its construction can be a long and costly task. The use of conceptual structures to provide users access useful information requires that there is something bridging the gap between the conceptual and real world.

This paper presents an approach that takes advantage of Fuzzy Logic [14] to automatically build the so-called fuzzy ontology of concept relations from a collection of documents. The ontology generated is then used for modeling a user profile. The idea is to provide a semantic representation of the user context based on characterizations represented by weighted concepts that are sense-related to the context itself. The ontology together with the concept characterizations are referred to as an *ontological user profile* of the document collection supplied by a particular user. The main contributions of this research work are the development of an automatic fuzzy ontology extraction strategy, based on co-occurrence and frequency of concepts in a document collection and a new ontology-based approach to represent user's preferences. Automatic ontology generation alleviates the knowledge acquisition bottleneck of manually constructing domain ontology.

The rest of the paper is organized as follows. Section 2 provides a detailed description of all stages needed to automatically construct an ontological user profile. The results of the experiments carried out are presented in section 3 and finally, some conclusions and future works are pointed out in section 4.

2   User Profile Generation Process

The proposal presented in this paper includes several phases of data processing to automatically generate fuzzy ontology-based user profiles from a previously selected document collection. This set of documents is provided with consideration to the user's role on the system, i.e. a *"content generator"* user will select the documents created by him/her, but, a *"content consumer or distributor"* user will select the documents retrieved and selected by him/her.

2.1 Linguistic Preprocessing

The building process is the following. First, the document set is preprocessed in order to characterize texts by their topically significant words, for this purpose, all non-textual information like digits, dates and punctuation marks is removed from the documents (lexical analysis). Next, collocations were extracted according to the method described in [3]. Finally, three techniques are used to reduce the vocabulary and make the representation of texts more meaningful: stop lists and stemming and zipf law. Language detection and spelling correction processes are also included in this stage.

2.2 Indexing

This stage aims to provide an index structure, called pre-ontology, which contains information about all terms generated in the previous stage. By analyzing the treated documents, the following term features are recovered: Term ID and List of Documents Features where the concept appears. For each document, the stored attributes are the following: ID, number of occurrences of the most frequent term in the document contents and a list of terms' positions in the document. Each term position is a tuple (p, s) where s represents the section or paragraph where the term is located and p the index of the term into the corresponding section.

2.3 User Relevant Concepts Identification

A set of chosen concepts that represent the user preferences are built using the pre–ontology as source data. Each concept has a weight on each document according to FIS-CRM Model $(fis - crm(c_i, d))$. The fundamental basis of FIS-CRM [8] is to "share" the occurrences of a contained word among the fuzzy synonyms that represent the same concept, and to "give" a fuzzy weight to the words that represent a more general concept that the contained one . In this way, a word may have a fuzzy weight in the new vector even if it is not contained in it, as long as the referenced concept underlies the document.

We distinguished three levels of concept relevance (relevant, sub-relevant and other), which will help to generate a semantic representation of the user context. Therefore, it is necessary to establish a method to quantify the value or usefulness of a concept in a documents set to identify a user. In this way, the weight $kmod_i^u$ of a concept c_i within a user profile u can be calculated as proposed in the equation 1. According to this formula a concept is regarded as relevant in a user profile if it occurs more frequently than other concepts in a certain user document set, but occasionally elsewhere. A concept that occurs frequently in several document set could be a relevant domain concept but it is not useful to represent the user.

$$kmod_i^u = \sum_{j \in u} w_{ij} \times \left(1 + \frac{docs\,(c_i, u)}{|D|}\right) \times Ln\left(\frac{|U|}{U\,(c_i)} + 1\right) \tag{1}$$

where w_{ij} represents the relevance degree of the concept c_i in the document d_j (d_j belongs to the user profile u) using the FIS-CRM model ($w_{ij} = fis-crm(c_i, d_j)$), $docs(c_i, u)$ is the number of documents in the user profile u in which the concept c_i occurs, $|D|$ is the total number of documents considered in the user profile, $|U|$ is the total number of user profiles in the context and $U(c_i)$ represents the number of user profiles in which the concept c_i has a positive membership degree.

Once the concepts weights are calculated, it is possible to identify the relevance distribution of all the concepts and use them to construct the groups according to the following statements: *Relevant concepts* (concepts with higher relevance degrees), *Sub-Relevant concepts* (concepts with relevance degrees higher than the average) and *Other concepts* concepts with relevance degree lower than the average. The user profile is defined by the most relevant concepts from the Relevant concepts set (75% to 100%). These concepts have an associated weight proportional to their importance. In this way, we can build a joint vector representation of the documents selected by the user. This representation does not correspond to the theoretical geometric center of the profile, neither in dimensions nor in values. It is used as a representative semantic description of user preferences.

2.4 User Ontology Generation

In our research, the final aim is to build a valued network of relations between the user relevant concepts indexed by the pre-ontology. A fuzzy ontology, in this context, may be considered as a set of directed graphs where each node represents an item and the edges denote that a concept "is related with" other concept. A relatedness degree (RD) is associated with each edge to represent the strength of the "is related with" association.

Our approach of generating a fuzzy ontology is based on the algorithm presented in [13]. This algorithm allows us to get a fuzzy measure of the generality degree (GD) between pair of words contained in the document collection. Formally, it defines $C = (d_1, d_2, \ldots, d_n)$ as a set of documents, where each document $D = (t_1, t_2, \ldots, t_m)$ is represented by a set of terms t_i. The generality degree (GD) between two terms t_i and t_j is defined as (Eq. 2):

$$GD(t_i, t_j) = \frac{\sum_{d \in D} occur(t_i, d) \otimes occur(t_j, d)}{\sum_{d \in D} occur(t_i, d)} \tag{2}$$

where $\otimes$ denotes a fuzzy conjunction operator, and the value of the function $ocurr(t_i, d)$, referring to the occurrence of term t_i in document d, is 1 if t_1 appears in d and 0 otherwise.

In this work, we have extended the idea of term occurrence to incorporate the concept frequency, and used it to calculate the relatedness degree (RD). Given a concept t_i, the weight of c_i in document d is represented by $fis - crm(c_i, d)$ and its membership value is defined according to the FIS-CRM model $fis-crm(c_i, d)$ mentioned above (Eq. 3).

$$RD\left(c_i, c_j\right) = \frac{\sum_{d \in D} \overline{fis - crm\left(c_i, d\right)} \otimes \overline{fis - crm\left(c_j, d\right)}}{\sum_{d \in D} \overline{fis - crm}\left(c_i, d\right)} \tag{3}$$

where $\overline{fis - crm\left(c_i, d\right)}$ is the normalized relevance of c_i in document d. In this work, it is assumed the association strength between concepts is asymmetric, because the proposed relationship is a hierarchical semantic connection. Regarding "the fuzziness" of the proposed approach, a ontology could considered a set of fuzzy sets, we will analyze the effect of six different fuzzy conjunction operators $(x \otimes y)$, defined by the following t-norms[7]: Minimum (Mi), Hamacher Product (HP), Algebraic Product (AP), Einstein Product (EP), Bounded Difference (BD) and Drastic Product (DP).

The fuzzy ontology is constructed by first calculating the relatedness degree (RD) for each pair of two distinct concepts from the pre-ontology (i.e. $RD(c_i, c_j)$ and $RD(c_j, c_i)$). Then, two tests are applied to select the relation that will be incorporated into the ontology. The first test is performed to eliminate redundant concept associations. For each triple $(c_i, c_j, RD(c_i, c_j))$ and $(c_j, c_i, RD(c_j, c_i))$, the one with higher RD value is added as an ontology association and the other is discarded. This decision strategy will choose a positive concept instance if one of the RD values is far from the other, or the strategy will choose a stronger association if the two RD values are close to each other. Finally, in the second test, less meaningful information is eliminated by removing from the ontology the associations that have a RD value lower than an $\alpha - cut$ defined by the user. Unrelated concepts $(RD = 0)$ are automatically excluded in this phase. The fuzzy ontology contains a description of the RD associations in the form of directed graphs.

2.5 User Profile Update

Additional documents of interest can be selected or created by users, and the user profile needs to be updated to incorporate this relevant information source. The update procedure should produce the same result as if the new documents had been available at the beginning of the profile generating process. To perform this task, we first need to carry out the linguistic *pre-processing* and the concept indexing for all new documents. As a result, another version of the *pre-ontology* is generated. Next, for each concept from this pre-ontology, the relevance weight is calculated and the user relevant concepts set is updated, incorporating new concepts or updating concept weights. Then, relevant, sub-relevant and other concepts groups are restructured by reordering each of them. Finally, the user ontology is updated. For each concept from the new pre-ontology, the relatedness degree (RD) is calculated according to all concepts from user *pre-ontology*. After applying redundant concept exclusion and $\alpha - cut$ elimination the new user profile is generated.

3   Experiments

The study group was shaped by 12 teachers from different countries. A total number of 135 resources was analyzed to generate the profiles of the users, i.e. each user selected around 11 resources as relevant. Resources are basically text based presentations or documents related to topics of research or teaching activities of the participant. User profile ontological definition were created for each participant, applying the methodology and algorithms described previously.

In order to choose the evaluate the method to measure the relevance of a concept in a user context, an evaluative analysis of the relevant concept set in terms of the standard definition of recall, precision, and F-measure (the harmonic mean between precision and recall) [12] have been carried out.

In order to evaluate the proposed method, we first analyzed the two sets of weighted concepts, according to the $kmod$ measure and a classic $tf - idf$ approach. The analysis was performed by comparing the well-ranked generated concepts with a set of concepts afforded by users. These concepts are keywords and a short description of users' interest areas. The higher the ranking of a term in a weighted concepts set that appears in the user's keyword list, the better the semantic representation of the user context and, consequently, the better the weighting method.

The $kmod$ measure gives us statistically better results($F = 0.35$) in user relevant concepts extraction. Analyzing the experiment results, it is important to highlight two aspects. Firstly, keywords and short descriptions given by users are not a standard de facto, which means that some relevant terms founded by the algorithm that do not appear in that context may still be important to determine the user profile. Secondly, the documents made available by users may include extended information about the treated area or information about topics that are not mentioned in keywords and short descriptions

On the other hand, the ontology statistics for the number of extracted relations in accordance with different fuzzy conjunction: the Einstein Product (EP), the Bounded Difference (BD) and the Drastic Product (DP). Some tests were carried out to evaluate the generated ontologies. In the first test, the hierarchical classification of the terms was analyzed by users. According to users, the ontology that best represents user preferences is that generated by using Einstein Product (EP), followed by Bounded Difference (BD) and Drastic Product (DP), respectively.

In addition, in this work we used WordNet [6] as golden standard to contrast them with our generated ontologies.. For this analysis, we compared all relationships from generated ontologies with four WordNet semantic relations (synonyms, coordinates, hyponyms and hypernyms). We obtain low precision and recall values and, consequently, low F-Measure values are not surprising ($F = 0.02$) . This may occur for several reasons. Firstly, the kind of non-taxonomic relationship generated by fuzzy ontology construction methods is not contained in a golden standard, which is often organized on syntactic and taxonomic levels. Secondly, due to the large amount of information comprised in golden standards, we were only able to make a first level analysis (i.e., relationships were compared with synonyms, but not with synonyms of synonyms). Thirdly, we generated on-tologies that specifically comprise the users' basic features and try to describe his/her particular domain. On the other hand, golden standards are very generic and focus on the description of a large portion of the world, making it quite dif-ficult to cover certain individual aspects, which are highly important in user ontologies.

Finally, we also performed some tests by contrasting users' ontologies rela-tionships with standard measures implemented in WordNet Similarity [9]. The precision is improved considerably if compared with previous analysis. However, the results still depends highly on the golden standard used, in this case, Word-Net.

On the other hand, we can observe that, similarly to the previous experiment, the ontology that presents better results, in terms of goodness, is that one gen-erated by using Einstein Product (EP), followed by Bounded Difference (BD) and Drastic Product (DP), respectively.

4 Conclusions

The use of fuzzy ontologies to represent user profiles has been proposed in this work. We discussed the shortcomings found in ontology and user profile construc-tion and presented an approach that takes advantage of text mining techniques to automatically extract relevant information from a collection of user documents, and use it to represent the user context in form of a fuzzy ontology-based user profile. We implemented and evaluated a concept weight measure, and several fuzzy conjunction operators used in the process.

One of the main strengths of this approach is the possibility of capturing relevant user information automatically, and representing it in a way that can be easily recovered and used by any application. In this way, starting from a document collection provided by the user, an automatic mechanism could be set out without requiring anything else from the user's point of view. The im-plementation and subsequent analysis of the strategy have showed the viability of automatic generation of fuzzy ontology-based user profiles. The experiment results indicated that user profiles generated are a good representation of the user context. Most of the information recovered, processed and then represented can be considered relevant for the user and characterize, in an organized way, the treated environment. Nevertheless, certain undesirable results strengthen the need of new techniques to improve the result obtained from this kind of structures.

Future work includes working with a larger number of users and a greater volume of documents in order to verify the quality of the model and to learn the subjectivity of users, points of view. Further research is directed towards the task of improving the user profile quality, using a pruning process to avoid con-cepts which have no significance. It is also necessary to consider the information provided a priori by the user.

References

1. Calegari, S., Pasi, G.: Personal ontologies: Generation of user profiles based on the YAGO ontology. Information Processing & Management 49(3), 640–658 (2013); personalization and Recommendation in Information Access
2. Calegari, S., Sanchez, E.: Object-fuzzy concept network: An enrichment of ontologies in semantic information retrieval. Journal of the American Society for Information Science and Technology 59(13), 153–2890 (2008)
3. Church, K., Hanks, P.: Word association norms, mutual information, and lexicography. Computational Linguistics 16(1), 22–29 (1990)
4. Golemati, M., Katifori, A., Vassilakis, C., Lepouras, G., Halatsis, C.: Creating an ontology for the user profile: Method and applications. In: Proceedings of the First International Conference on Research Challenges in Information Science, RCIS (2007)
5. Han, L., Chen, G.: A fuzzy clustering method of construction of ontology-based user profiles. Advances in Engineering Software 40(7), 535–540 (2009)
6. Miller, G.A.: WordNet: a lexical database for English. Commun. ACM 38(11), 39–41 (1995)
7. Novak, V., Mockor, J., Perfilieva, I.: Mathematical principles of fuzzy logic. Kluwer international series in engineering and computing science. Kluwer, Boston (1999)
8. Olivas, J.A., Garcés, P.J., Romero, F.P.: An application of the fis-crm model to the fiss metasearcher: Using fuzzy synonymy and fuzzy generality for representing concepts in documents. Int. J. Approx. Reasoning 34(2-3), 201–219 (2003)
9. Pedersen, T., Patwardhan, S., Michelizzi, J.: Wordnet:similarity: Measuring the relatedness of concepts. In: Demonstration Papers at HLT-NAACL 2004. HLT–NAACL–Demonstrations, pp. 38–41. Association for Computational Linguistics, Stroudsburg (2004)
10. Sendhilkumar, S., Geetha, T.V.: Personalized ontology for web search personalization. In: COMPUTE 2008: Proceedings of the 1st Bangalore Annual Compute Conference, pp. 1–7. ACM, New York (2008)
11. Tho, Q.T., Hui, S.C., Cao, T.H.: FOGA: A Fuzzy Ontology Generation Framework for Scholarly Semantic Web. In: Proceedings of the 2004 Knowledge Discovery and Ontologies Workshop (KDO 2004), Pisa, Italy (2004)
12. Van Rijsbergen, C.: Information Retrieval. Butterworth, London (1979)
13. Widyantoro, D.H., Yen, J.: Incorporating fuzzy ontology of term relations in a search engine. In: Proceedings of the BISC Int. Workshop on Fuzzy Logic and the Internet 2001, pp. 155–160. University of California, Berkeley (2001)
14. Zadeh, L.A.: Fuzzy logic and approximate reasoning. Synthese 30(3), 407–428 (1975)

Hierarchically Structured Recommender System for Improving NPS of a Company

Jieyan Kuang[1], Albert Daniel[1], Jill Johnston[1], and Zbigniew W. Raś[1,2]

[1] Univ. of North Carolina, College of Computing and Informatics, KDD Laboratory,
Charlotte, NC, 28223, USA
[2] Warsaw Univ. of Technology, Inst. of Computer Science, 00-665 Warsaw, Poland
{jkuang1,ras}@uncc.edu

Abstract. The paper presents a description of a hierarchically structured recommender system for improving the efficiency of a company's growth engine. Our dataset (NPS dataset) contains answers to a set of queries (called questionnaire) sent to a randomly chosen groups of customers. It covers 34 companies called clients. The purpose of the questionnaire is to check customer satisfaction in using services of these companies which have repair shops all involved in a similar type of business (fixing heavy equipment). These shops are located in 29 states in the US and Canada. Some of the companies have their shops located in more than one state. They can compete with each other only if they target the same group of customers. The performance of a company is evaluated using the Net Promoter System (NPS). For that purpose, the data from the completed questionnaires are stored in NPS datasets. We have 34 such datasets, one for each company. Knowledge extracted from them, especially action rules and their triggers, can be used to build recommender systems giving hints to companies how to improve their NPS ratings. Larger the datasets, our believe in the knowledge extracted from them is higher. We introduce the concept of semantic similarity between companies. More semantically similar the companies are, the knowledge extracted from their joined NPS datasets has higher accuracy and coverage. Our hierarchically structured recommender system is a collection of recommender systems organized as a tree. Lower the nodes in the tree, more specialized the recommender systems are and the same the classifiers and action rules used to build their recommendation engines have higher precision and accuracy.

1 Introduction

Net Promoter Score (NPS) is used to measure a customer's loyalty to a product or service provider [11], [12]. It is based on the response to a 1 to 10 scale with 0 being very unlikely to recommend the provider and 10 being very likely to recommend. Net Promoter System is based on the fundamental assumption that customers can be divided into three categories: *promoters*, *passives*, and *detractors*. Promoters are loyal enthusiasts who are buying from a company and urge their friends to do the same. Passives are satisfied but unenthusiastic customers who can be easily taken by the competition. Detractors are less than loyal customers who may urge their friends to avoid that company [10]. Customers are categorized based on their answers to the likelihood to recommend question. The figure below explains how these three categories are computed.

C. Cornelis et al. (eds.): RSCTC 2014, LNAI 8536, pp. 347–357, 2014.

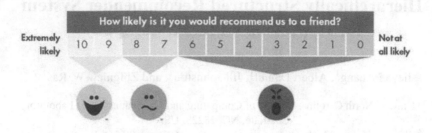

Fig. 1. Net Promoter Score (NPS)

Customers falling into interval 9-10 are seen as promoters, into 7-8 as passives, and into 0-6 as detractors. The partition into these three categories is widely accepted by business organizations but still other discretizations of NPS can be taken into consideration especially when classifiers extracted from NPS datasets do not have acceptable precission/recall for one of these three categories. The classical way to evaluate the efficiency of a company's growth engine is to compute NPS efficiency rating which is defined as the percentage of customers who are promoters minus the percentage who are detractors. Companies with the most efficient growth engines such as Amazon, Costco, Vanguard, or Dell have NPS efficiency ratings between 50 to 80 percent. But even these companies still have room for improvement.

In order to formulate actions for improving the performance of a company, we need to know why a customer is or is not likely to recommend the company to their colleagues/friends. This is why customers are asked to complete a questionnaire which gives us personal information about them and about their satisfaction with the services provided by a company. Examples of questions included in the questionnaire are give below:

- name of the customer
- name of the organization (client)
- invoice amount
- internal contact name (person with whom you deal in a company)
- how many days was needed to repair the equipment
- when the equipment was delivered to repair shop
- any disagreements?
- name/type of the equipment to be repaired
- are you satisfied with the job

More questionnaires are completed by customers, larger dataset for mining is available. Classifiers extracted from this dataset, defining each of the NPS categories, are evaluated using confusion matrix. If their accuracy and coverage is high then the action rules built from them will have high confidence as well. Different types of classifiers have been tested with a goal to identify which one has the highest accuracy/coverage.

Our dataset (NPS dataset) contains answers to a set of queries (called questionnaire) sent to a randomly chosen groups of customers. The purpose of the questionnaire is to check customer satisfaction in using services of 34 repair shops (clients) involved in a similar type of business (fixing heavy equipment). These shops are located in 34 states in the US and Canada and they can compete with each other only if they are geographically closely located. For each shop, we extracted a number of classifiers from its NPS dataset. We also extended the original NPS dataset by developing and adding new groups of attributes, including temporal attributes. The accuracy/coverage of several classifiers for the category promoter become very high for all 34 shops. However, the accuracy and coverage of classifiers for the categories passive and detractor still remains low. So, action rules can not be built from pairs of classification rules [5] unless we are interested to build recommender system for each shop (or groups of shops) which only will target customers who completed the questionnaire. To apply action rules successfully for other customers, we have to extract these rules either directly from the dataset or built them from action reducts [2],[5].

The concept of an action rule was proposed by Ras and Wieczorkowska in [7] and investigated further in [1], [3], [9], [14]. Action rules describe possible transitions of objects from one state to another with respect to a distinguished attribute called the decision [7]. In our application domain, we are only interested in transitions from detractors and passives to promoters. We assume that attributes used to describe customers are partitioned into stable and flexible. Values of flexible attributes can be changed. In our domain, invoice amount or name/type of equipment to be repaired are examples of stable attributes. "Client name", "how many days are needed by the shop to fix the equipment" are examples of flexible attributes. "Client name" is a flexible attribute because customers may decide to change shops for fixing their equipment. In early papers, action rules have been constructed from two classification rules $[(\omega \wedge \alpha) \rightarrow \phi]$ and $[(\omega \wedge \beta) \rightarrow \psi]$, where ω is a stable part for both rules. Action rule was defined as the term $[(\omega) \wedge (\alpha \rightarrow \beta)] \Rightarrow (\phi \rightarrow \psi)$, where ω is the description of clients for whom the rule can be applied, $(\alpha \rightarrow \beta)$ shows what changes in values of flexible attributes are required, and $(\phi \rightarrow \psi)$ gives the expected effect of the action. Let us assume that ϕ means *detractors* and ψ means *promoters*. Then, the discovered knowledge shows how values of flexible attributes need to be changed so the customers classified as detractors will become promoters.

2 Action Rules

In this section we recall the definition of an information system (also called a dataset), action set, and also recall the classical strategy of constructing action rules from action sets.

By an information system [4] we mean a triple $S = (X, A, V)$, where:

1. X is a nonempty, finite set of objects
2. A is a nonempty, finite set of attributes, i.e.
   $a : U \longrightarrow V_a$ is a function (can be partial function) for any $a \in A$, where V_a is called the domain of a
3. $V = \bigcup \{V_a : a \in A\}$.

For example, Table 1 shows an information system S with a set of objects $X = \{x_1, x_2, x_3, x_4, x_5, x_6, x_7, x_8\}$, set of attributes $A = \{a, b, c, d\}$, and the set of their values $V = \{a_1, a_2, b_1, b_2, c_1, c_2, d_1, d_2\}$.

Table 1. Information System S

	a	b	c	d
x_1	a_1	b_1	c_1	d_1
x_2	a_2	b_1	c_1	d_1
x_3	a_2	b_2	c_1	d_2
x_4	a_2	b_2	c_2	d_2
x_5	a_2	b_1	c_1	d_1
x_6	a_2	b_2	c_1	d_2
x_7	a_2	b_1	c_2	d_2
x_8	a_1	b_2	c_2	d_1

Additionally, we assume that $A = A_{St} \cup A_{Fl}$, where attributes in A_{St} are called *stable* and attributes in A_{Fl} are called *flexible*. "Customer name" is an example of a stable attribute. "Interest rate" for each customer account is an example of a flexible attribute.

Let $S = (X, A, V)$ is an information system, where $V = \bigcup \{V_a : a \in A\}$.

By an *atomic action set* we mean a singleton set containing an expression $(a, a_1 \rightarrow a_2)$ called atomic action, where a is an attribute and $a_1, a_2 \in V_a$. If $a_1 = a_2$, then a is called stable on a_1. Instead of $(a, a_1 \rightarrow a_1)$, we usually write (a, a_1) for any $a_1 \in V_a$.

By *Action Sets* we mean a smallest collection of sets such that:

1. If t is an atomic action set, then t is an action set.
2. If t_1, t_2 are action sets, then $t_1 \cup t_2$ is a candidate action set.
3. If t is a candidate action set and for any two atomic actions $(a, a_1 \rightarrow a_2)$, $(b, b_1 \rightarrow b_2)$ contained in t we have $a \neq b$, then t is an action set.

By the domain of an action set t, denoted by $Dom(t)$, we mean the set of all attribute names listed in t.

By an *action rule* we mean any expression $r = [t_1 \Rightarrow t_2]$, where t_1 and t_2 are action sets. Additionally, we assume that $Dom(t_2) \cup Dom(t_1) \subseteq A$ and $Dom(t_2) \cap Dom(t_1) = \emptyset$. The domain of action rule r is defined as $Dom(t_1) \cup Dom(t_2)$.

Now, we give an example of an action rule assuming that our information system S is represented by Table 1, a, c are stable and b, d are flexible attributes. Expressions (a, a_2), $(b, b_1 \rightarrow b_2)$, (c, c_2), $(d, d_1 \rightarrow d_2)$ are examples of atomic actions. Expression $(b, b_1 \rightarrow b_2)$ means that the value of attribute b is changed from b_1 to b_2. Expression (c, c_2) means that the value c_2 of attribute c remains unchanged. Expression $r = [\{(a, a_2), (b, b_1 \rightarrow b_2)\} \Rightarrow \{(d, d_1 \rightarrow d_2)\}]$ is an example of an action rule. The rule says that if value a_2 remains unchanged and value b will change from b_1 to b_2, then it is expected that the value d will change its value from d_1 to d_2.

3 Extracting Classifiers and Action Rules from NPS Datasets

In this section, we present a brief discussion on some classifiers built from NPS datasets representing answers to the customer satisfaction questionnaire completed by about 50,000 customers

The first result concerns classifiers built from the union of all NPS datasets which contains the answers collected from the customers for all the clients. We tested the classifiers available in WEKA by using the average confusion matrix for 10 random samplings, extracted from the union of all NPS datasets, each one covering about 1200 customers. J48 gave the best results which are presented in Table 2.

Table 2. Confusion Matrix for J48 and the original NPS dataset covering all clients

	Promoter	Passive	Detractor
Promoter	87	407	0
Passive	77	422	1
Detractor	29	170	0

We can see that Table 2 presents confusion matrix for the dataset with 494 promoters, 500 passives, and 199 detractors. So the average, total number of customers is 1193. NPS score is 494/1193 - 199/1193 = 0.41 - 0.17 = 0.24 (24 percent). The large number of *Passive* customers (almost 1/2) is the main reason of this very low NPS score. To improve that score, action rules need to be extracted from NPS dataset. The first step to achieve our goal is to improve the classifiers extracted from that dataset. A number of new attributes have been constructed using text mining methods and added to the original NPS dataset. These attributes include: *Overall satisfaction, Likelihood to be a repeated customer, Technician arrived when promised,* and *Repair completed correctly*. The new average confusion matrix obtained from 10 random samplings using J48 classifier for the extended NPS dataset is shown in Table 3. Obviously the NPS score did not change because the decision column is the same.

Table 3. Confusion Matrix for J48 and NPS dataset with new features

	Promoter	Passive	Detractor
Promoter	407	80	7
Passive	123	327	50
Detractor	23	77	99

Much better results we received for the extended NPS datasets covering certain combinations of clients, especially single clients. Table 4 shows confusion matrix for the Tree Classifier using Rough Set Exploration System ($RSES$) for a dataset which represents two clients.

Assume now that we use $RSES$ Tree Classifier to construct action rules by pairing classification rules describing *Detractor* with classification rules describing *Promoter*. The goal is to reclassify as many *Detractors* as possible to *Promoters*. The average confidence of action rules will be $0.993 \cdot 0.849 = 0.84$ (see the Accuracy column in Figure 2). Our action rules can target only 4.2 (out of 10.2) detractors. So, we can expect $4.2 \cdot 0.84 = 3.52$ detractors moving to the promoter status. The NPS score for the initial dataset covering two clients was 0.80. After applying our action rules we get:

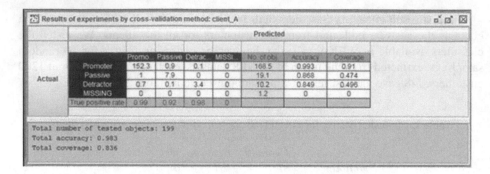

Fig. 2. Two-Clients Confusion Matrix for $RSES$ Tree Classifier

- Number of $Promoters = 168.5 + 3.52 = 172.2$
- Number of $Passives = 19.1$
- Number of $Detractors = 10.2$ - $3.52 = 6.68$

Total number of customers = 197.8. So, the new NPS = $\left[\frac{172.2}{197.8} - \frac{6.7}{197.8}\right] = 0.87 - 0.03 = 0.84$, which means 4 percent of improvement in NPS is expected.

4   Clustering of Clients Based on Semantic Similarity in NPS Datasets

In this section, we introduce the notion of semantic similarity between clients. Informally speaking, we say that two clients are semantically similar if they agree on the knowledge concerning $Promoter$, $Passive$, and $Detractor$ which is hidden in their NPS datasets. Stronger is the agreement, semantically more similar they are. NPS Datasets of two or more clients semantically similar can be joined together giving us larger NPS datasets for mining. Larger the datasets, our own confidence in the results shown in confusion matrices is higher.

Assume now that $RC[1]$, $RC[2]$ are the sets of classification rules extracted from the questionnaire-type datasets (NPS datasets) collected for clients $C1$, $C2$. Also, we assume that

$RC[1] = RC[1, Promoter] \cup RC[1, Passive] \cup RC[1, Detractor]$,
where $RC[1, Promoter] = \{r[1, Promoter, i] : i \in I_{Pr}\}$, $RC[1, Passive] = \{r[1, Passive, i] : i \in I_{Ps}\}$, $RC[1, Detractor] = \{r[1, Detractor, i] : i \in I_{Dr}\}$, where $\{r[1, Promoter, i] : i \in I_{Pr}\}$ is a collection of classification rules defining "Promoter", $\{r[1, Passive, i] : i \in I_{Ps}\}$ is a collection of classification rules defining "Passive", and $\{r[1, Detractor, i] : i \in I_{Dr}\}$ is a collection of classification rules defining "Detractor".

In a similar way, we define
$RC[2] = RC[2, Promoter] \cup RC[2, Passive] \cup RC[2, Detractor]$,
where $RC[2, Promoter] = \{r[2, Promoter, i] : i \in J_{Pr}\}$, $RC[2, Passive] = \{r[2, Passive, i] : i \in J_{Ps}\}$, $RC[2, Detractor] = \{r[2, Detractor, i] : i \in J_{Dr}\}$.

By $C1[1, Promoter, i]$, $C1[1, Passive, i]$, $C1[1, Detractor, i]$ we mean confidence of $r[1, Promoter, i]$, $r[1, Passive, i]$, and $r[1, Detractor, i]$ in a dataset for client $C1$, respectively.

By $C2[1, Promoter, i]$, $C2[1, Passive, i]$, $C2[1, Detractor, i]$ we mean confidence of $r[1, Promoter, i]$, $r[1, Passive, i]$, and $r[1, Detractor, i]$ in a dataset for client $C2$, respectively.

By $C2[2, Promoter, i]$, $C2[2, Passive, i]$, $C2[2, Detractor, i]$ we mean confidence of $r[2, Promoter, i]$, $r[2, Passive, i]$, and $r[2, Detractor, i]$ in a dataset for client $C2$, respectively.

By $C1[2, Promoter, i]$, $C1[2, Passive, i]$, $C1[2, Detractor, i]$ we mean confidence of $r[2, Promoter, i]$, $r[2, Passive, i]$, and $r[2, Detractor, i]$ in a dataset for client $C1$, respectively.

Now, we can introduce the concept of semantic similarity between clients $C1$, $C2$ denoted by $SemSim(C1, C2)$.

$$SemSim(C1, C2) =$$

$$\frac{\Sigma\{|C1[1,Promoter,k]-C2[1,Promoter,k]|:k\in I_{Pr}\}}{card(I_{Pr})} + \frac{\Sigma\{|C1[1,Passive,k]-C2[1,Passive,k]|:k\in I_{Ps}\}}{card(I_{Ps})} +$$
$$\frac{\Sigma\{|C1[1,Detractor,k]-C2[1,Detractor,k]|:k\in I_{Dr}\}}{card(I_{Dr})} + \frac{\Sigma\{|C2[2,Promoter,k]-C1[2,Promoter,k]|:k\in J_{Pr}\}}{card(J_{Pr})}$$
$$+ \frac{\Sigma\{|C2[2,Passive,k]-C1[2,Passive,k]|:k\in J_{Ps}\}}{card(J_{Ps})} + \frac{\Sigma\{|C2[2,Detractor,k]-C1[2,Detractor,k]|:k\in J_{Dr}\}}{card(J_{Dr})} .$$

Figure 3 shows the hierarchical clustering of 34 clients with respect to their semantic similarity. Clients which are semantically and geographically close to each other can have their datasets merged and the same considered as a single client from the business perspective (customers have similar opinion about them). For each state (its abbreviation is given), we list clients operating in that state with their respective NPS values:

AB- $\{(clint - 9, NPS = 0.503)\}$, AZ- $\{(client - 8, NPS = 0.802)\}$, CA- $\{(client - 13, NPS = 0.777), (client - 16, NPS = 0.767), (client - 17, NPS = 0.848), (client - 24, NPS = 0.724)\}$, ID- $\{(client - 30, NPS = 725)\}$, IL- $\{(client - 1, NPS = 0.836)\}$, KS- $\{(client - 7, NPS = 0.771)\}$, KY- $\{(client - 31, NPS = 0.804)\}$, LA- $\{(client - 19, NPS = 0.705)\}$, MN- $\{(client - 3, NPS = 0.823)\}$, MO- $\{(client - 7, NPS = 0.771), (client - 10, NPS = 0.788)\}$, MS- $\{(client - 23, NPS = 0.860), (client - 26, NPS = 0.828)\}$, NC- $\{(client - 4, NPS = 0.803), (client - 11, NPS = 0.797), (client - 12, NPS = 0.722), (client - 27, NPS = 0.760), (client - 32, NPS = 0.740)\}$, ND- $\{(client - 3, NPS = 0.823)\}$, NE- $\{(client - 21, NPS = 0.732)\}$, NV- $\{(client - 5, NPS = 0.771)\}$, OH- $\{(client - 22, NPS = 0.820), (client - 28, NPS = 0.779)\}$, OK- $\{(client - 29, NPS = 0.710)\}$, PA- $\{(client - 6, NPS = 0.788), (client - 25, NPS = 0.800)\}$, QC- $\{(client - 14, NPS = 0.721)\}$, SC- $\{(client - 2, NPS = 765), (client - 11, NPS = 0.797)\}$, SD- $\{(client - 2, NPS = 0.765), (client - 3, NPS = 0.823)\}$, SK- $\{(client - 18, NPS = 0.636)\}$, TN- $\{(client - 26, NPS = 0.828)\}$, TX- $\{(client - 15, NPS =$

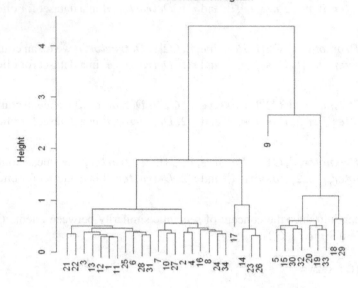

Fig. 3. Hierarchical clustering of 34 clients

0.762), $(client - 20, NPS = 0.675), (client - 29, NPS = 0.710)\}$, UT- $\{client - 5, NPS = 0.771)\}$, WV- $\{(client - 28, NPS = 0.779)\}$; WY- $\{(client - 30, NPS = 0.725), (client - 33, NPS = 0.772)\}$, VA- $\{(client - 11, NPS = 0.800)\}$.

From the Figure 2, we can observe that $\{23, 26\}$, $\{24, 34\}$, $\{1, 11\}$, $\{21, 22\}$, $\{28, 31\}$, $\{19, 33\}$ are examples of six clusters of semantically similar clients. Clients in the cluster $\{23, 26\}$ are both located in Mississippi so they are geographically close as well. Their NPS ratings are 0.860 and 0.828, respectively. Since they both target the same group of customers, Client 26 can improve its ratings by using recommendations based on action rules extracted from the NPS dataset covering both clients. Clients in the cluster $\{24, 34\}$ are far away from each other. One is in California and the other in Georgia. NPS rating for Client 24 is 0.724 and for Client 34 is equal to 0.80. They do not compete for the same group of customers, so the strategy for improvement of the NPS rating for Client 24 is more challenging. In this particular case, Client 24 has the worst NPS ratings among clients in California (Client 13, 16, 17). Client 16 is semantically the closest one to Client 24 and also its NPS rating is the closest to the NPS rating of Client 24. So, we have two options. We can merge NPS datasets of clients 16 and 24, and next extract action rules from the joined dataset to get improvement of the NPS score for Client 24 or we can merge the NPS datasets of Clients 24 and 34, and next extract action rules from the joined dataset. Following the first option we are targeting the same group of customers whereas in the second, the customers are from two different states and geographically far away from each other. It is quite possible that customers from two different states evaluating their local clients may follow different criteria when they answer the questionnaire. The optimal solution to solve this problem is to test both options and chose the one which is giving us better improvement in NPS score.

5   NPS-Based Recommender System

In this section, we present the methodology which can be followed to build a hierarchically structured recommender system with nodes being recommender systems of different generalization levels. Leaves of the tree represent personalized recommender systems built from classifiers and action rules extracted from NPS dataset of a single client (single company). The root of the tree represents the most generalized recommender system which is built from classifiers extracted from the union of all NPS datasets covering all 34 clients. The internal nodes of the tree are built following the dendrogram presented in Figure 3. Lower the nodes in the tree, more specialized the recommender systems are. Also, lower the nodes in the tree, the accuracy and precision of the classifiers assigned to them is higher.

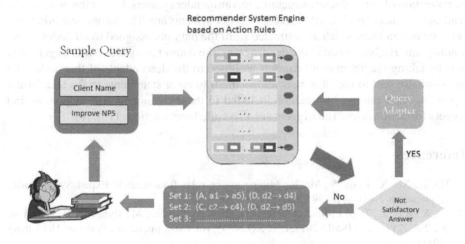

Fig. 4. Recommender System

Figure 4 presents Flexible Query Answering System (FQAS) built on the top of a hierarchically structured recommender system. Client (company) which is not satisfied with its current NPS ratings can submit a query to FQAS asking what can be done to improve its NPS. The easiest answer can be obtained from the recommender system constructed from the NPS dataset of that client but to get better recommendation we have to use recommender systems which are not personalized. Namely, we should consider getting help from recommender systems assigned to the nodes forming the path which starts from the leaf (our client) in the dendrogram and leads to the root. So, the process is bottom-up. For instance, if the Client 24 wants to improve its ratings, action rules from its NPS dataset are extracted. If the hints from the recommender system based on these rules are not satisfactory, Client 24 is clustered with Client 34 (see the dendrogram) since these two clients are semantically the most similar. If the NPS rating of Client 34 is higher than the rating of Client 24, action rules are extracted from the union of the corresponding two NPS datasets and hints based on them are given to the client. Otherwise, we should check all leaves in the dendrogram leading us to the parent of Client 24 and 34. We should chose that leaf which represents a client with NPS rating higher than the rating of Client 24 and also which is semantically the closest one to Client 24.

6 Conclusion

The paper presents preliminary results which finally will lead us to the construction of a flexible hierarchically structured recommender system for improving NPS of a company in a global competitive market. Thirty four companies (clients) form the domain for the agglomerative clustering algorithm based on their semantic distance. Clients are compared in terms of the similarity of their knowledge concerning the meaning of three concepts: promoter, passive, and detractor. The resulting dendrogram is a skeleton for the collection of hierarchically structured recommender systems. Lower the nodes in the dendrogram, more specialized the recommender systems are. The recommendations are based on action rules which are extracted from the datasets assigned to all nodes of the dendrogram. Higher a node in the dendrogram, the dataset assigned to it is larger - it is built by taking the union of all datasets assigned to the descendants of that node. The questionnaire sent to the customers allows them to enter statements in the text format explaining their ratings. Information included in these statements will help us to find triggers for action rules. The triggers are also called meta-actions [13], [6].

References

1. He, Z., Xu, X., Deng, S., Ma, R.: Mining action rules from scratch. Expert Systems with Applications 29(3), 691–699 (2005)
2. Im, S., Ras, Z., Tsay, L.-S.: Action reducts. In: Kryszkiewicz, M., Rybinski, H., Skowron, A., Raś, Z.W. (eds.) ISMIS 2011. LNCS (LNAI), vol. 6804, pp. 62–69. Springer, Heidelberg (2011)
3. Paul, R., Hoque, A.S.: Mining irregular association rules based on action and non-action type data. In: Proceedings of the Fifth International Conference on Digital Information Management (ICDIM), pp. 63–68 (2010)
4. Pawlak, Z.: Information systems - theoretical foundations. Information Systems Journal 6, 205–218 (1981)
5. Ras, Z., Dardzinska, A.: From Data to Classification Rules and Actions. In: Proceedings of the Joint Rough Sets Symposium (JRS 2007). LNCS (LNAI), vol. 4482, pp. 322–329. Springer (2011)
6. Raś, Z.W., Dardzińska, A.: Action Rules Discovery based on Tree Classifiers and Meta-Actions. In: Rauch, J., Raś, Z.W., Berka, P., Elomaa, T. (eds.) ISMIS 2009. LNCS (LNAI), vol. 5722, pp. 66–75. Springer, Heidelberg (2009)
7. Raś, Z.W., Wieczorkowska, A.A.: Action-rules: how to increase profit of a company. In: Zighed, D.A., Komorowski, J., Żytkow, J.M. (eds.) PKDD 2000. LNCS (LNAI), vol. 1910, pp. 587–592. Springer, Heidelberg (2000)

8. Raś, Z.W., Wyrzykowska, E.z., Wasyluk, H.: ARAS: Action rules discovery based on agglomerative strategy. In: Raś, Z.W., Tsumoto, S., Zighed, D.A. (eds.) MCD 2007. LNCS (LNAI), vol. 4944, pp. 196–208. Springer, Heidelberg (2008)
9. Rauch, J., Šimůnek, M.: Action rules and the GUHA method: Preliminary considerations and results. In: Rauch, J., Raś, Z.W., Berka, P., Elomaa, T. (eds.) ISMIS 2009. LNCS (LNAI), vol. 5722, pp. 76–87. Springer, Heidelberg (2009)
10. Reichheld, F.F.: The one number you need to grow. Harvard Business Review, 1–8 (December 2003)
11. SATMETRIX, NET Promoter (2014),
    http://www.satmetrix.com/net-promoter/
12. SATMETRIX, Improving your net promoter scores through strategic account management, white paper (2012)
13. Tzacheva, A., Ras, Z.W.: Association action rules and action paths triggered by meta-actions. In: Proceedings of 2010 IEEE Conference on Granular Computing, Silicon Valley, CA, pp. 772–776. IEEE Computer Society (2010)
14. Qiao, Y., Zhong, K., Wang, H.-A., Li, X.: Developing event-condition-action rules in real time active database. In: Proceedings of the 2007 ACM Symposium on Applied Computing, Seoul, pp. 511–516 (2007)

Building Contextual Student Group Recommendations with Fuzzy Logic

Krzysztof Myszkorowski and Danuta Zakrzewska

Institute of Information Technology, Lodz University of Technology,
Wolczanska 215, 90-924 Lodz, Poland
{kamysz,dzakrz}@ics.p.lodz.pl

Abstract. Groups of learners of similar features are often created in order to diversify the environment accordingly. However student preferences may differ depending on the context of the system usage. Each new student, who intends to join the community, should obtain context-aware recommendation of the group of colleagues matching his needs. In the paper, using fuzzy logic for modeling student groups is considered. We propose to build the possibility-based representation of each group. We assume that context can be modeled by a vector of weights. Then recommendations for new students are determined taking into account a degree of possibility of matching together with the respective context parameters. We examine the presented approach by taking into account learning style dimensions as attributes which characterize student preferences. The method is evaluated on the basis of experimental results obtained for data of different groups of real students.

Keywords: recommender systems, fuzzy logic, group modeling.

1 Introduction

Grouping students of similar characteristics, who should learn together enables to adjust learning resources appropriately. However different student traits may be important taking into account course requirements as well as the context of the system use. Accordingly, building context-aware student group recommendations can help students to join the suitable groups of colleagues while enrolling on different courses.

The group assignment of each new learner should guarantee his similarity to the group members. Effectiveness of recommendations depends on accuracy of group modeling. In the paper [1] group representation in the probabilistic form was proposed and accordingly modified Bayes classifier was used to build context aware group recommendations. In [2] fuzzy group representation has been defined with the use of linguistic terms corresponding to attribute values. The linguistic variables were associated with fuzzy numbers defined over domains of attributes. The recommendation process was based on the cardinality of the defined fuzzy sets. In the current research, we propose to build the possibility-based representation of each group. Attribute values of the group are represented by

C. Cornelis et al. (eds.): RSCTC 2014, LNAI 8536, pp. 358–365, 2014.

means of possibility distributions. In the recommendation process we apply the fuzzy pattern matching technique [3]. We assume that context can be modeled by a vector of weights. Then recommendations for new students are determined taking into account a degree of possibility of matching together with the respective context parameters. The proposed method is examined for student traits based on their learning style dimensions. It is validated, on the basis of experiments, done for real students' clusters.

The paper is organized as follows. The related work is described in the next section. Then the proposed methodology for building recommendations, including group modeling and context usage is depicted. In Section 4 the case study of students described by dominant learning style dimensions is presented. Next, results of experiments carried out on real students' data are displayed and discussed. Finally, concluding remarks and future research are outlined.

2 Related Work

Context-awareness was very often considered for recommendation purposes. Modeling context was discussed by Adomavicius & Tuzhilin, who introduced algorithmic paradigms for incorporating contextual information into the recommendation process [4]. Zheng et al. [5] considered differential context weighting approach, in which the contribution of each contextual variable is weighted. The broad review of context parameters as well as context aware e-learning systems was presented in [6]. Rosaci and Sarne [7], considered both aspects: student's profile and an exploited device. Their recommendations were built on the basis of the time spent by student on the particular Web site, taking into account type of a device used for navigating. Yang et al. [8] proposed learning resources recommendation system based on connecting similar students into small communities, where they can share resources and communicate with each other. Christodoulopoulos & Papanikolaou [9] discussed several factors that should be considered while assigning learners into groups. A survey of the state-of-the art in group recommendation was presented in [10]. Masthoff [11] described using of group modeling and recommendation techniques for recommending to individual users. A collaborative Bayesian network-based group recommender system has been proposed in [12]. Fuzzy Bayesian network was considered to build user group recommendations based on context of scenario analysis and inference [13].

Using fuzzy logic for student modeling was examined by several researchers. Authors used fuzzy sets to describe reality by means of linguistic terms, which are close to human nature. The intelligent system, using fuzzy logic for evaluation and classification of student performance on the basis of the structure of the observed learning outcome, was presented in [14]. In [15] for evaluation of intelligent learning systems the authors considered the use of fuzzy sets to specify the relevance or learning intensity of cognitive elements and fuzzy rules that establish or modify those fuzzy sets.

In several works researchers considered fuzzy approach to adapting and personalizing of the learning process to the students' needs. In the paper [16] authors applied fuzzy logic for defining a fuzzy ontology in which relationships between objects, attributes and classes were described by means of fuzzy relations. They elaborated a fuzzy ontology discovery algorithm to extract the concept maps representing students' knowledge structure. Fazlollahtabar and Mahdavi applied a neuro-fuzzy approach for obtaining an optimal learning path [17]. Characteristics of students were inferring on the base of teachers' opinions expressed by means of linguistic terms.

Fuzzy logic was included into recommendation techniques in e-learning by several researchers. Ferreira-Satler et al.[18] built a fuzzy user profiles-based recommendation engine by including fuzzy ontology obtained automatically taking into account learning objects published by users and used as the representation of their preferences. Serrano-Guerrero et al.[19], in turn, applied a fuzzy linguistic approach. Their algorithm suggested an adequate, depending on student profile, set of activities which aimed at strenghtening students' competences.

3   Recommender System

Let us assume that students are described by N attributes of nominal types A_i, $i = 1,..., N$. A tuple ST representing a student is of the form:

$$ST = (st_1, st_2, ..., st_N), \quad st_i \in DOM(A_i) , \tag{1}$$

where $DOM(A_i)$ stands for the domain of A_i. Attribute A_i may take on m_i nominal values $a_{i,j}$, $i = 1,..., N$, $j = 1,..., m_i$ where m_i stands for cardinality of $DOM(A_i)$: $m_i = card(DOM(A_i))$.

In e-learning, as the main factors which decide on context parameters, there should be mentioned course requirements. They determine student features together with their priorities, which should be taken into account during the process of course materials designing. Let $w_i, 0 \leq w_i \leq 1, i = 1, ..., N$ denote weights connected with attributes' priorities. They will be further used in the matching stage, which aims at suggesting the best choice of group of peers to learn together during the considered course.

Let us assume that there exist groups of students with attributes A_i, $i = 1,..., N$. Representative values of attributes A_i for groups are not determined uniquely. For creation of their representation one can apply tools for describing uncertain or imprecise information. One of them is the fuzzy set theory. Based on the concept of fuzzy sets there has been introduced the concept of possibility distributions [20].

Definition 1. *Let U be the universe of discourse, X be a variable on U and F be a fuzzy set with $\mu_F(u)$. The possibility distribution of X with respect to F is defined as*

$$\Pi_X = \{\pi_X(u)/u : u \in U, \quad \pi_X(u) = \mu_F(u)\}. \tag{2}$$

Each element u of U is assigned with a number $\pi_X(u) \in [0, 1]$ which is a possibility measure of $X = u$. If there exists $u \in U$ such that $\pi_X(u) = 1$ the possibility distribution Π_X is called normalized.

Let GS_k, $k = 1, 2, ..., NG$ be a group of students ST described by N attributes of nominal type (1). Let us denote by $c_{k,i}$ cardinality of the most frequent value of the attribute A_i in the group GS_k:

$$c_{k,i} = \max_j card\{ST \in GS_k : ST(A_i) = a_{i,j}\}. \tag{3}$$

As the representation of the attribute A_i for the group GS_k we will consider a fuzzy set $FS_{k,i}$ with the following membership function:

$$\mu_{FS_{k,i}}(a_{i,j}) = card\{ST \in GS_k : ST(A_i) = a_{i,j}\}/c_{k,i}. \tag{4}$$

Let a group GS_k be represented by a tuple $(gs_{k,1}, gs_{k,2}, ..., gs_{k,n})$. The membership function $\mu_{FS_{k,i}}(u)$ may be interpreted as a measure of possibility that objects from GS_k are characterized by a certain attribute value [21]. Thus each value $gs_{k,i}$ is represented by a normal possibility distribution:

$$gs_{k,i} = \{\pi_{k,i}(a_{i,j})/a_{i,j} : a_{i,j} \in DOM(A_i), j = 1, 2, ..., m_i\}. \tag{5}$$

The biggest values of $\pi_{k,i}(a_{i,j})$ indicate dominant attribute values in groups. Possibilistic form of the group representation allows to determine matching degrees of attributes of new students and classify them to appropriate groups.

Let a tuple $NST = (ns_1, ns_2, ..., ns_N)$, $ns_i \in DOM(A_i)$, represent a new student. Let us consider a group of students GS_k represented by a tuple $(gs_{k,1}, gs_{k,2}, ..., gs_{k,n})$. The compatibility degree between NST and GS_k can be estimated with the use of possibility measure, denoted by $Pos(NST, GS_k)$. This measure expresses the extent to which the considered values satisfy a comparison relation.

According to [3] a degree of possibility of matching NST to GS_k for the attribute A_i equals:

$$Pos(NST(A_i), GS_k(A_i)) = \pi_{k,i}(ns_i). \tag{6}$$

The total matching degree $Pos(NST, GS)$ for the group is a minimal value of $Pos(NST(A_i), GS_k(A_i))$, $i = 1, ..., N$:

$$Pos(NST, GS_k) = \min_i(Pos(NST(A_i), GS_k(A_i))). \tag{7}$$

Maximal value of $Pos(NST, GS_k)$ indicates the group that should be recommended for students.

The described way of recommendation assumes that all attributes are of equal importance. However, if certain attribute A_i is less important for the choice of the group and the matching degree $Pos(NST(A_i), GS_k(A_i))$ is low, then the group may be rejected regardless of matching degrees of other attributes. The problem may be resolved by introduction of weights as it was proposed in [3].

Let each attribute A_i be assigned with a number $w_i \in [0, 1]$. Let $w_i \in [0, 1]$ be the weight of importance of the attribute A_i. For the most important attributes $w_i = 1$. For attributes which are not considered during the recommendation process $w_i = 0$. A degree of possibility of matching NST to GS_k for A_i equals:

$$Pos(NST(A_i), GS_k(A_i)) = \max(\pi_{k,i}(ns_i), 1 - w_i). \tag{8}$$

The total matching degree $Pos(NST, GS_k)$ is expressed by the formula:

$$Pos(NST, GS_k) = \min_i \max(\pi_{k,i}(ns_i), 1 - w_i). \qquad (9)$$

Let us assume, that there are NG student groups, then the whole process of recommendation building will take place in the following way:

Algorithm of group recommendation building

Input   A set of NG groups GS_k, of students of N nominal attributes; a tuple NST representing a new student; a set of weights assigned with attributes;

Step 1: For each group $GS_k, k = 1, 2, ..., NG$ find its possibility-based representation according to (5)

Step 2: For the student NST find the group GL_{rec} with the maximal value of the matching degree (9)

Step 3: Recommend GL_{rec} to the student.

---

4   Students Characterized by Learning Styles

As an example for building contextual group recommendations, we will consider student models based on learning styles. Das et al. [6] mentioned learning styles as context parameters corresponding to media used by the learner. For the purpose of the evaluation of the proposed methodology, we will apply Felder & Silverman [22] model, where learning styles are described by means of 4 attributes which indicate preferences for 4 dimensions from among excluding pairs: active vs. reflective (L_1), sensing vs. intuitive (L_2), visual vs. verbal (L_3), and sequential vs. global (L_4) or balanced if the student has no dominant preferences. Attribute values belong to the set of odd integers from the interval [-11, 11]. These numbers describe scores for features represented by respective attributes. Each student can be modeled by a vector SL of 4 integer attributes:

$$SL = (sl_1, sl_2, sl_3, sl_4),\ \ sl_i \in \{-11, -9, -7, -5, -3, -1, 1, 3, 5, 7, 9, 11\} \ . \quad (10)$$

Let $jmax_i$ denotes the index of the most frequent value of the attribute L_i in the group. As the fuzzy group representative we will consider the following sets:

$$Rep_i = \{l_{i,j} : |jmax_i - j| \leq 2\}, 1 \leq i \leq 4 \ . \qquad (11)$$

For the new student $NSL = (nsl_1, nsl_2, nsl_3, nsl_4)$, and each group GL_k, $k =1,..., NG$ we can define a weighted recommendation error Err_k as follows:

$$err_{k,i} = \begin{cases} 1 & \text{if } nsl_i \notin Rep_i \\ 0 & otherwise \end{cases}, \qquad (12)$$

$$Err_k = \sum_{i=1}^{4} w_i * err_{k,i}. \qquad (13)$$

5   Experiments

The goal of the experiments was to evaluate the performance of the proposed recommendation technique taking into account different weight parameters. The evaluation was done by comparison of recommendation results obtained by using the proposed method, with the ones which match students the best according to the recommendation error defined by (13). The tests were carried out in the context of different courses represented by the set of weight parameters.

The experiments were done for two different datasets of real students' attributes representing their dominant learning styles as was presented in SL model (10). Students filled self-scoring questionnaire Index of Learning Styles (ILS) [23] for assessing preferences on 4 dimensions of the Felder & Silverman model. After selecting answers to 44 questions concerning student preferences during learning process, learners obtained scores for the respective dimensions. The first dataset contains data of 194 Computer Science students from different levels and years of studies, including part-time and evening courses. This data was further prepared by building groups of similar students. The second set contains data of students, who were to learn together with their peers from the first dataset and whose data was used for testing the recommendation efficiency. The set consists of 31 data of students studying the same master's course of Information Systems in Management.

The groups were created by clustering taking into account techniques, which are easy to understand for educators and for which input parameters can be easily determined. There were used: partitioning - K-means (KM), statistical - EM and hierarchical Farthest First Traversal (FFT). Clusters were created by using Open Source Weka software $http://www.cs.waikato.ac.nz/ml/weka$. As the number of data was not very big, clustering into 3,4,5,6 and 7 clusters were considered. Such approach allowed to check the proposed technique for groups of different qualities, similarity degrees and structures and to enable comparison of the method performance depending on the number of considered groups. Recommendations were built in the context of five different courses, where scores for student attributes are represented by vectors: $W1 = (1,1,1,1); W2 = (1,0,1,0); W3 = (0,1,1,0); W4 = (0,1,0,1); W5 = (1,1,1,0)$. Quantitative analysis of the results showed that the majority of the students obtained the best recommendations. The results did not show dependency between clustering schema and the percentage of properly assigned recommendations. The detailed results of quantitative analysis are presented in Table 1. The first two columns contain clustering method and the number of clusters. Next columns show percentage of students who obtained the best recommendations for different weight vectors.

Qualitative analysis showed the big influence of the weight parameters as well as group sizes on recommendations. In most of the cases, the recommendation errors took on the largest values when all student attributes were of the same importance. When the differences between cluster sizes were big, usually the larger group was suggested. The biggest error values can be notified when there exist students, whose profiles do not match any of existing groups. Such situation took place in the case of six clusters created by EM algorithm.

Table 1. Quantitative analysis depending on clustering schema

Schema	Cl. no	W1	W2	W3	W4	W5
EM	3	97.77%	90.13%	100%	100%	100%
	4	87.10%	80.65%	93.55%	87.10%	93.55 %
	5	83.87%	77.42%	97.77%	90.13%	90.13%
	6	58.06%	77.42%	70.97%	70.97%	64.52%
	7	74.19%	67.71%	83.87%	77.42%	93.55%
FFT	3	90.13%	87.10%	97.77%	97.77%	90.13%
	4	83.87%	80.65%	93.55 %	90.13%	87.10%
	5	80.65%	83.87%	87.10%	87.10%	87.10%
	6	77.42%	83.87%	77.42%	83.87%	77.42%
	7	80.65%	80.65%	70.97%	77.42%	74.19%
KM	3	90.13%	93.55%	97.77%	90.13%	90.13%
	4	77.42%	90.13%	90.13%	80.65%	83.87%
	5	87.10%	87.10%	77.42%	70.97%	80.65%
	6	83.87%	93.55%	97.77%	90.13%	87.10%
	7	70.97%	97.77%	87.10%	90.13%	80.65%

6   Concluding Remarks

In the paper, fuzzy logic for building student group recommendations in the context of different courses was pondered. There was proposed a possibilistic representation of the group. The considered technique was examined in the case of students described by dominant learning styles and the course context represented by the weight vector. Experiments done for datasets of real students and different group structures showed that for the majority of the students the system indicated the best possible choice of colleagues to learn together. However during experiments, situations when students did not match any of the existing groups occurred. In such cases contacting tutors seems to be the best solution.

The proposed method of recommendation building can be applied by educators during the process of course management as well as organization of joint activities for student groups of similar characteristics.

Future research will consist in further development of the recommendation tool, examination of attributes of different types and taking into account dynamical changes in group representations each time the recommendation is accepted.

References

1. Zakrzewska, D.: Building Context-Aware Group Recommendations in E-learning Systems. In: Jędrzejowicz, P., Nguyen, N.T., Hoang, K. (eds.) ICCCI 2011, Part I. LNCS, vol. 6922, pp. 132–141. Springer, Heidelberg (2011)
2. Myszkorowski, K., Zakrzewska, D.: Using Fuzzy Logic for Recommending Groups in E-learning Systems. In: Bădică, C., Nguyen, N.T., Brezovan, M. (eds.) ICCCI 2013. LNCS (LNAI), vol. 8083, pp. 671–680. Springer, Heidelberg (2013)

3. Dubois, D., Prade, H., Testemale, C.: Weighted fuzzy pattern matching. Fuzzy Set Syst. 28, 313–331 (1988)
4. Adomavicius, G., Tuzhilin, A.: Context-Aware Recommender Systems. In: Ricci, F., et al. (eds.) Recommender Systems Handbook, pp. 217–253. Springer (2011)
5. Zheng, Y., Burke, R., Mobasher, B.: Recommendation with Differential Context Weighting. In: Carberry, S., Weibelzahl, S., Micarelli, A., Semeraro, G. (eds.) UMAP 2013. LNCS, vol. 7899, pp. 152–164. Springer, Heidelberg (2013)
6. Das, M.M., Chithralekha, T., SivaSathya, S.: Static context model for context aware e-learning. Int. J. En. Sci. Tech. 2, 2337–2346 (2010)
7. Rosaci, D., Sarne, G.: Efficient personalization of e-learning activities using a multi-device decentralized recommender System. Comput. Intell. 26, 121–141 (2010)
8. Yang, F., Han, P., Shen, R.-M., Hu, Z.: A novel resource recommendation system based on connecting to similar e-learners. In: Lau, R., Li, Q., Cheung, R., Liu, W. (eds.) ICWL 2005. LNCS, vol. 3583, pp. 122–130. Springer, Heidelberg (2005)
9. Christodoulopoulos, C., Papanikolaou, K.: A group formation tool in an e-learning context. In: 19th IEEE ICTAI 2007, vol. 2, pp. 117–123 (2007)
10. Boratto, L., Carta, S.: State-of-the-Art in Group Recommendation and New Approaches for Automatic Identification of Groups. In: Soro, A., Vargiu, E., Armano, G., Paddeu, G. (eds.) Information Retrieval and Mining in Distributed Environments. SCI, vol. 324, pp. 1–20. Springer, Heidelberg (2010)
11. Masthoff, J.: Group Recommender Systems: Combining Individual Models. In: Ricci, F., et al. (eds.) Recommender Systems Handbook, pp. 677–702. Springer (2011)
12. Campos, L., Fernández-Luna, J., Huete, J., Rueda-Morales, M.: Managing uncertainty in group recommending processes. User Model. User-Adap. 19, 207–242 (2009)
13. Wang, J., Li, H., Zhao, H.: The contextual group recommendation. In: 5th Int. Conf. on Intelligent Networking and Collaborative Systems, pp. 127–131 (2013)
14. Vrettaros, J., Vouros, G.A., Drigas, A.S.: Development of an intelligent assessment system for solo taxonomies using fuzzy logic. In: Mellouli, K. (ed.) ECSQARU 2007. LNCS (LNAI), vol. 4724, pp. 901–911. Springer, Heidelberg (2007)
15. de Arriaga, F., El Alami, M.: Arriaga: Evaluation of Fuzzy Intelligent Learning Systems. In: Méndez-Vilas, A., et al. (eds.) Recent Research Developments in Learning Technologies. FORMATEX, Badajoz (2005)
16. Lau, R., Song, D., Li, Y., Cheung, T., Hao, J.: Towards A Fuzzy Domain Ontology Extraction. IEEE T. Knowl. Data En. 21, 800–813 (2009)
17. Faziolahtabar, H., Mahdavi, I.: User/tutor optimal learning path in e-learning using comprehensive neuro-fuzzy approach. Educ. Res. Rev. 4, 142–155 (2009)
18. Ferreira-Satler, M., Romero, F., Menéndez-Domínguez, V., Zapata, A., Prieto, M.: Fuzzy ontologies-based user profiles applied to enhance e-learning activities. Soft Comput. 16, 1129–1141 (2012)
19. Serrano-Guerrero, J., Romero, F., Olivas, J.: Hiperion: A fuzzy approach for recommending educational activities based on the acquisition of competences. Information Sciences 248, 114–129 (2013)
20. Zadeh, L.: Fuzzy sets as a basis for a theory of possibility. Fuzzy Set Syst. 1, 3–28 (1978)
21. Dubois, D., Prade, H.: The three semantics of fuzzy sets. Fuzzy Set Syst. 90, 141–150 (1997)
22. Felder, R., Silverman, L.: Learning and teaching styles in engineering education. Eng. Educ. 78, 674–681 (1988)
23. ILS Questionnaire, http://www.engr.ncsu.edu/learningstyles/ilsweb.html

Architecture and Message Protocol Proposal for Robot's Integration in Multi-Agent Surveillance System

Bruno Dias[1], Bruno Rodrigues[1], Jorge Claro[1], João Paulo Pimentão[1], Pedro Sousa[1], and Sérgio Onofre[2]

[1] Faculdade de Ciências e Tecnologia - UNL, Caparica, Portugal
{bm.dias,bm.rodrigues,j.claro}@campus.fct.unl.pt,
{pim,pas}@fct.unl.pt
[2] Holos SA, Caparica, Portugal
onofre@holos.pt

Abstract. Nowadays, surveillance systems are less dependent of humans' interactions, mostly in event's detection. However, there still are tasks to be performed by humans that could be delegated to autonomous robots. The integration of an autonomous service robot, as an agent, in a multi-agent surveillance system can reduce even more humans' dependency. This paper proposes a new architecture and a communication protocol that integrates ServRobot's autonomous robot into DVA's surveillance system. This integration allows the DVA's system to use the robot as a mobile unit for validation of triggered events, to perform surveillance missions and to gather sensors information.

Keywords: Software agents, surveillance, sensors, distributed systems, architecture, protocol, collective behavior, autonomous robot.

1 Introduction

DVA project[1] (partially sponsored by the European Regional Development Fund and the Portuguese Government) developed a surveillance intelligent system based on a multi-agent platform. DVA's system supports different sensors types and implements mechanisms of geo-referencing sensor's data and events. This distributed system implements different agent's types, such as: Sensor Agent, Processor Agent, Inference Agent, Action Agent, Mobile Agent, Backup Agent, Interface Agent and Monitor Agent.

ServRobot[2] (partially sponsored by the European Regional Development Fund and the Portuguese Government), is an autonomous robot capable of: following people or lines, teleoperation, execute predefined missions (such as: go to a specific GPS position), and obtain sensor data.

---

[1] http://dva.holos.pt
[2] http://servrobot.holos.pt

C. Cornelis et al. (eds.): RSCTC 2014, LNAI 8536, pp. 366–373, 2014.

The integration of ServRobot in DVA's system, as a Mobile Agent and Sensor Agent, could significantly improve DVA's performance, using it in the event's detection, fusing its sensor data with DVA traditional sensors, improving their inference engine; event's validation and event's handling using robot mobility and its environment perception. This paper proposes an architecture and message protocol to enable the implementation of this integration, allowing DVA to use ServRobot in different scenarios.

It is expected that this proposed integration will improve intelligent surveillance system, reducing false positives, using robot's sensors to complement DVA static sensors improving the reliability of events triggered and the need for human's interactions using the robot mobility to confirm and handling events.

In the reminder of this paper, Distributed Systems are described in Section 2; In Section 3 the proposed architecture is presented, followed by a description of the message protocol. This paper will conclude with an analysis of the work developed and definition of the research steps.

2   Distributed Systems

With the computing capabilities and network services evolution, distributed systems have widely sought to solve certain system's problems, since with a decentralized structure, it is possible to design a more robust and fault tolerance system [1–4].

Multi-Agent Systems (MAS) allow distributed, flexible, robust, scalable and reconfigurable systems, constituted by a team of multiple agents. These agents can be positioned at different locations and perform a wide-ranged monitoring. With these characteristics, it is possible to design systems with objectives such as protection of an international border against trespassing, timely detection of a bushfire, accurate analysis of the traffic state of a city, and so forth [2].

The development of MAS's in robotic applications has been widely used in various applications, often adding more intelligence and autonomy to systems. Since Robotic Systems are evolving from industrial robots that are only responsible for one task, (performing it automatically), to autonomous and mobile robots that can collaborate among themselves and use sensors to understand their context, it was necessary to develop new systems to take advantage of all these new capabilities [3]. Taking this into account, it is possible to manage all these new features and requirements with a MAS. This paradigm can be applied to Robotics System in several forms: Heterogeneous mobile robots [4]; Robots working in ambient intelligence environments [5], [6]; Collective robot swarms [7]; Mobile sensor robotics networks [8] and Multi-agent control systems [9].

In general, robotic systems are controlled by Robotic Software Frameworks. These frameworks are focused on providing scalability, reusability, deployment and debugging of the software developed in the system. There are many Open-Source Frameworks available for the development of Robotic Systems such as: Player, OROCOS, ROS, YARP, OpenRave, OpenRTM, and others. However, these frameworks don't provide specific services oriented to MAS because they were developed focusing in Robotics and Hardware Systems [9].

Multi-Agent Software Frameworks, are frameworks that achieve the necessary conditions of a MAS, mainly in the communication mechanisms such as messages, ports, topics, naming and lookup services, agent mobility, development and introspection tools. Examples of these frameworks are: JADE, Mobile-C, Zeus, and others [9].

3   Architecture

The described work (section 2), in MAS's, assumes that all the agents are developed in the same framework. Nevertheless, there are systems that could improve their features and performance by interacting with "agents" from other systems.

DVA is a Geo-referenced multi-agent surveillance system, composed by several agents: Sensor agent – provides sensor information; Processor agent – transforms sensor information into parameters; Inference agent – uses parameters in rules for event detection; Action agent – executes predefined actions for each event; Backup agent - stores all the system information; Interface agent – shows (in maps) the values of the sensors, events, actions and system status; Mobile agent – Associated with a human, equipped with a mobile device who is responsible to perform events' actions, such as confirming the event or handling the event; Monitor agent – monitors all system's agents, ensuring correct system performance.

Furthermore, ServRobot is a server/client system based on Player/Stage [9] framework, where different software modules are implemented allowing inclusion of new features as an autonomous service robot.

Both projects are operating independently of each other and it is not possible to include ServRobot as agent of DVA's system. In the next section a new architecture is proposed that extends DVA's architecture to allow ServRobot integration without compromising the operation of these separately. This way, ServRobot and DVA can continue to function as it is currently implemented differing only in ServRobot role, this being an agent available at DVA.

3.1   Proposed Architecture

The proposed architecture is represented at Fig. 1, which shows the current operational DVA's architecture elements in solid line. In dashed line, in the Fig. 1, are represented the proposed new elements used to extend the architecture to include ServRobot in the DVA system, using it as a mobile agent or as a mobile sensor.

The "Robot Mobile" agent handles the registration process of robots and manages their availability and their skills. This registration allows, for example, that when the Action Agent has to execute an action to confirm an event, it can use the robot (as an additional Mobile agent) to check event's location and to confirm event's occurrence. With this architecture, the Action agent has the possibility to delegate tasks to a robot, reducing the dependence of human resources connected to the system as Mobile agents.

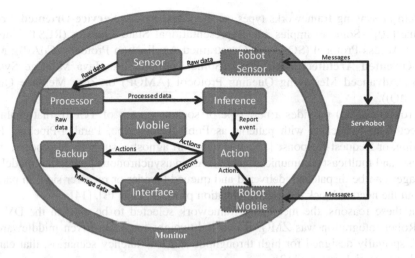

Fig. 1. Architecture proposal

The "Robot Sensor" agent will handle the reception of all sensory information sensed by the robot, integrating it as other standard DVA sensors. All the data collected by the Robot Sensor agent will be sent to a Processor agent, just as traditional DVA sensors. This agent can get single sensor values (following a request/reply pattern) or can subscribe to a sensor and receive its values periodically.

Details of implementation of these two new agents are described in the next section.

In summary, through this extended architecture, the DVA system will be capable of: sending a robot agent to execute a mission; get robot's sensors data; ensure the mobile surveillance of different areas; send a robot to confirm events, getting feedback from the environment. This architecture also makes possible for new devices to enter in the system as client devices, capable to teleoperate the robot or to get feedback from their sensors.

4   Architecture's Proposed Implementation

The architecture's proposed implementation requires a message protocol and a middleware framework to guarantee that all the functionalities reported before are available. Next sections describe details of the middleware framework proposed and of the messages designed to establish the communication between the DVA system and ServRobot.

4.1   Middleware Framework

There are several middleware messaging frameworks available to use as base of the proposed architecture. Middleware frameworks also allow the programmer to integrate applications developed for different executions contexts and in different times [10].

Main messaging frameworks types can be classified as: **Service-Oriented Architecture** [10], (Some examples are: Representational State Transfer (REST), Simple Object Access Protocol (SOAP) or Constrained Application Protocol (CoAP)); **Message-Oriented Architectures** (MOA) (Some examples are: Java Message System (JMS), Advanced Messaging Queuing Protocol (AMQP) or Zero Message Queue (ZeroMQ)).

ZeroMQ (ZMQ) provides a new type of sockets on top of TCP/IP that enable to connect N-to-N sockets with patterns as Publish-Subscribe, Parallel-Pipeline, Fair-Queuing, or Request-Response [11]. It is transport agnostic, supports in-process, inter-process, and multicast communication [11]. As an asynchronous processing model, the messages can be dispatched, delivered and queued (sender or receiver side) in parallel without the need to block the main application process [12], [13], [14].

For these reasons, the messaging framework selected to be used in the DVA to ServRobot integration was ZMQ. It is a lightweight message-driven middleware library, specially designed for high throughput and low latency scenarios, that can be found in financial systems [12].

4.2   Message Patterns in Use

The proposed architecture uses two of ZMQ message patterns: Request-Reply (REQ/REP) and Publish-Subscriber (PUB/SUB). Essentially, REQ/REP is used when an acknowledge is expected, for example on registration messages, direct orders; or one time sensor output requests. The concept of PUB/SUB is used when it's needed a periodically updates from a sensor. It's also used to send heartbeats to the connected clients as "I'm alive" messages.

4.3   Message Language

The language proposed for the exchanged messages, is the Extended Markup Language (XML). XML was adopted taking into account its advantages to: modulate the concepts of the scenario in study (instead a byte codification); make changes in the message protocol by modeling new objects and data types [15]; develop in different platforms, debug problems and validate the messages' composition [16, 17].

Also, with XML an important issue to this architecture is guaranteed: communication interface does not contain limitations, so in future, new functionalities can be added easily with scalability, for new sensor/modules in the autonomous robot or new devices in the system [9].

4.4   Message Types

Messages are divided into different types, being used each one for distinct purposes. The messages types used are:

- **Emergency Command** - Used to abort an activity that is being undertaken. Independently of the current status of the robot, when it receives an emergency command, the robot should stop immediately.

- **Heartbeat Message** - The heartbeat message is used as an "I'm alive." type of message. It is sent using a PUB/SUB messaging pattern and is present in all devices in the architecture.
- **Registration Message** - To join the system, the devices have to send a registration request to the "Robot Mobile" agent at DVA.
- **Simple Message** - Simple Messages are a regular messages defined in this protocol. These messages have a *DataType* attribute that specifies the type of data contained on the message to send and can take values such as: *Robot Status*, *Mission*, *Sensor* and *Reply*.

Reply
Reply type is used as an "answer" to every request made from "Robot Mobile" agent.

Mission
Mission messages are used to request execution of missions to the robot and get feedback from the result of an executed mission. These messages will be used by the "Robot Mobile" agent when it receives a notification from Action agent, using a REQ/REP pattern. Different mission types are available, taking into account the robot's capabilities. The mission arguments can be already stored at robot memory or sent in the message, as example on **Fig. 2.**

It is also possible get missions list available at the robot memory or get a more detailed data about one specific mission. This way "Robot Mobile" agent knows the robot capabilities available. It is possible to upload and remove missions on the robot's memory.

```
<msg type="SimpleMsg">
  <Header>... </Header>
  <DataFrame DataType="Mission">
    <Command>DoNewMission</Command>
    <Speed>10</Speed>
    <MissionType>GoToPtsAbsXY</MissionType>
    <XY>1;1</XY>
    <XY>1;2</XY>
  </DataFrame>
</msg>
```

Fig. 2. DoNewMission message example

The robot also deals with the concept of Reference Point. A Reference Point is a known location (by the Robot and "Robot Mobile" agent) that is associated with a label. This way it is possible to execute a mission at 'Room01' instead using its coordinates. There are messages that manipulate and get these Reference Points.

Sensors
Using these messages it is possible to get information about the robot's sensors, for example: get the list of all the sensors or get the detailed information of a specific sensor. This way, the "Robot Sensor" agent knows what kind of data it can gather from robot.

With these messages it is also possible for the "Robot Sensor" agent to subscribe or unsubscribe to specific sensor values that are published by the device. Those values are published with a certain Update Interval requested by the device when it subscribes it. There are other methods to get values from sensors following a REQ/REP, obtaining the sensors information in terms of a specific parameter (for example a temperature of the left motor) or type of parameter (for example all the temperatures from the robot's sensors).

Robot Status

With this message, the "Robot Mobile" agent will be able know the robot's availability and the operation mode (Idle, Mission Mode or Teleoperation Mode).

5   Conclusion

Given the developments of MAS's and robotic systems, it was shown that in certain applications, there are significant improvements in the integration of robotic agents in pre-existing multi-agent applications. But often, taking into account the differences between robotic frameworks and multi-agent frameworks, this process of integration is not trivial, it is necessary to implement an architecture and protocol for communication between systems.

The proposed architecture, presented in this paper, will allow the integration of ServRobot's robot in the DVA system without affecting the operation of the current implementation. Thus, the DVA system has the possibility of: using the robot as an agent to acknowledge events that are reported in the system, execute surveillance missions, replacing humans that could be not available to that task.

The use of robot's sensors to get feedback from the environment is another advantage, fusing the data perceived by the robot with the data from DVA sensors, will expectably improve the detection of events in the rules of inference currently in use in DVA. The mobility of the robot is a positive factor taking into account that the remaining DVA's system sensors are fixed. This way, the dependence on human actions can be decreased taking advantage of developments in robotics, integrating in the DVA's system the autonomous robot.

The next steps of the described work will be the development of the communication modules for the ServRobot and the DVA. These communication modules will include all the necessary functions that will be used to create, send, receive, interpret and debug the message flow.

Also we will be integrating mobile devices as Clients and make robot missions available in DVA's web interface.

References

1. Balasubramanian, B., Garg, V.K.: Fault Tolerance in Distributed Systems Using Fused Data Structures. IEEE Transactions on Parallel and Distributed Systems 24, 701–715 (2013)
2. Baig, Z.A.: Multi-agent systems for protecting critical infrastructures: A survey. Journal of Network and Computer Applications 35, 1151–1161 (2012)

3. Vidoni, R., García-Sánchez, F., Gasparetto, A., Martínez-Béjar, R.: An intelligent framework to manage robotic autonomous agents. Expert Systems with Applications 38, 7430–7439 (2011)
4. Group, I.C., Control, A.: A Cooperative Perception System for Multiple UAVs: Application to Automatic Detection of Forest Fires 23, 165–184 (2006)
5. Coradeschi, S., Saffiotti, A.: The Future of AI Symbiotic Robotic Systems: Humans, Robots, and Smart Environments, pp. 20–22 (2006)
6. Lin, F., Norrie, D.H.: Schema-based conversation modeling for agent-oriented manufacturing systems 46, 259–274 (2001)
7. Mondada, F., Gambardella, L.M.: Swarm-Bot: A New Distributed Robotic Concept, 193–221 (2004)
8. Howard, A., Matarić, M.J.: An Incremental Self-Deployment Algorithm for Mobile Sensor Networks, 113–126 (2002)
9. Iñigo-Blasco, P., Diaz-del-Rio, F., Romero-Ternero, M.C., Cagigas-Muñiz, D., Vicente-Diaz, S.: Robotics software frameworks for multi-agent robotic systems development. Robotics and Autonomous Systems 60, 803–821 (2012)
10. Frank, B., Kevlin, H., Douglas, C.: Pattern Oriented Software Architecture "A pattern Language for Distributed Computing". Pattern-Oriented Software Architecture, A Pattern Language for Distributed Computing (2007)
11. Piël, N.: ZeroMQ an introduction,
http://nichol.as/zeromq-an-introduction
12. Hintjens, P.: Code Connected, vol. 1. iMatix Corporation (2013)
13. Hintjens, P.: Multithreading Magic,
http://zeromq.org/blog:multithreading-magic
14. Hintjens, P.: ZeroMQ: The Guide, http://zguide.zeromq.org/page:all
15. Bray, T., Paoli, J.: Extensible markup language (XML). W3C 1 (2006)
16. Kostoulas, M., Matsa, M.: XML screamer: an integrated approach to high performance XML parsing, validation and deserialization. In: World Wide Web Conference Committee, pp. 93–102 (2006)
17. Wang, F., Li, J., Homayounfar, H.: A space efficient XML DOM parser. Data & Knowledge Engineering 60, 185–207 (2007)

Image Coverage Segmentation
Based on Soft Boundaries

Jiuzhen Liang, Yudi Gu, Lan Di, and Qin Wu

Department of Computer Science, Jiangnan University, Wuxi, Jiangsu, 214122, China
jzliang@jiangnan.edu.cn

Abstract. This paper studies image object coverage segmentation by introducing soft boundaries. By using soft boundaries, fuzzy image can be segmented into several classes with a sharing boundary which is called a soft boundary. In this paper, several concepts of boundaries are defined, namely, hard boundary, inner boundary and outer boundary. Soft boundary is defined by the subtraction between inner boundary and outer boundary of a set. Coverage segmentation algorithm and optimization method are proposed in this paper. Meanwhile, neighbor decision rules are used in classification of pixels to filter noise or outliers. Experiments and comparison with classical coverage segmentation methods are presented, including noise test on the proposed method with four kinds of boundaries and neighbor decision rules.

Keywords: image classification, coverage segmentation, fuzzy c-means, spectral projected gradient optimization, soft boundary.

1 Introduction

Image segmentation is a challenging problem which has been addressed more frequently than any other problems in image processing[1]. It aims at partitioning an image into a number of components constructed by a certain intra-component homogeneity and inter-component discontinuity[2]. This is generally considered to be both the most important and the most challenging task in image processing[3].

Recently, some researches focused on one specific type of fuzzy discrete object representations[1]. Lindblad et al[4][5], have utilized the coverage model to improve the estimation precision. They found that a possible lack of precision resulting from limited spatial resolution may be overcome by properly utilizing grey-level value contained in the images when estimating relevant features of the objects. If the criterion for membership of a pixel to an object is the coverage of the pixel by the continuous imaged object, and the assigned pixel value corresponds to the relative coverage of the pixel, the resulting membership is referred to as a coverage representation.

However, there are relatively few generally applicable methods which explicitly result in a coverage representation[1]. In [2], Sladoje presented a method that, based on any existing crisp segmentation, enhances it to a coverage segmentation by identifying boundary pixels and suitable re-evaluating their coverage values[3]. Compared to crisp segmentation, precision and accuracy of feature estimations are increased[4][5]. Compared to other fuzzy representations, advantages of the coverage model come from the

C. Cornelis et al. (eds.): RSCTC 2014, LNAI 8536, pp. 374–381, 2014.

knowledge of a particular and clearly defined membership function; bounds of estimation errors are derived utilizing these assumptions[1].

In this paper, we introduce some concepts of different boundaries and their usefulness in coverage segmentation. Then we provide an algorithm for coverage segmentation. Experiments are presented to compare the performance of the proposed model with different boundaries.

2   Related Work and Motivations

Models based on linear unmixing of image intensities are common in the field of image processing. They are frequently used, such as in remote sensing[6][7] and fluorescence microscopy[8][9]. Per-pixel linear unmixing has been shown to be highly noise sensitive[10]. Lindblad and Sladoje[1] addressed the general problem of estimating coverage values by combining intensity information with spatial smoothness criteria, to improve estimation accuracy. It can also process a multi-band input image containing several objects, and it does not require crisp segmentation as an input. For each object, pure class representatives are used. The coverage segmentation process is then based on energy function minimization with several regularization terms. However, this method is more computationally demanding than the method proposed in [2][11].

The work mentioned above is aimed to segment image with a membership matrix, but there is few work on how the class representatives influence segmentation and how to choose them. Fig1 shows image segmentations by different boundaries. In case (a), the representatives of the object are too high light. Many pixels are classified into background. In this case, the boundary is too close to the high light. In case (b), the representatives of the object are middle and adopt. This makes most of the object pixels to be classified. In this case, the boundary is not too clear to see in the original image, but it is useful for classification. In case (c), the representatives of the object are lower a little. This makes many of the background pixels classified into the object. In this case, the boundary is too near to the background, and a lot of pixels are misclassified.

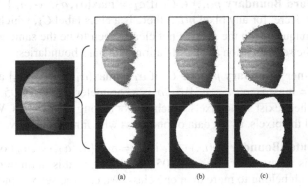

Fig. 1. Image segmentation by different boundaries with different representatives. (a)High grey averaged representatives; (b)Middle grey averaged representative; (c)low grey averaged representatives.

Fig.1 shows an example which needs to choose adaptive class representatives to separate the object from its background. However, how to choose the adaptive class representatives may depend on different images under different conditions. Further more, image segmentation depends not only on class representatives but also the value of the membership matrix for each pixel. In the following sections of this paper, we study the issues by introducing the soft boundary for image segmentation.

3   Models

Given a multi-band digital image I of size $N = width \times height$, on a discrete domain I_D, $I : I_D \to \mathbb{R}^b$, where b denotes the number of bands of the image. We consider I to be given as matrix of size $N \times b$, $I = [p_{i,k}]^{N \times b}$. In this representation, a row contains intensities of one pixel in each of the observed bands, and a column represents the pixel intensities of the image in one band. For convenience, denote the i-th pixel in the image I as $p(i, .)$. Let $\mathbb{W}_m$ denote the set of m-component (fuzzy) segmentation vectors[2],

$$\mathbb{W}_m = \{w = (w_1, w_2, \cdots, w_m) \in [0, 1]^m | \sum_{k=1}^{m} w_k = 1\} \tag{1}$$

A coverage segmentation of an image I into m components is a set of ordered pairs

$$\{((i, .), \alpha(i)) | (i, .) \in I_D, \alpha(i) \in \mathbb{W}_m\}, \alpha(i) \approx \frac{|p(i, .) \cap S_k|}{|p(i, .)|} \tag{2}$$

where S_k is the region of the k-th (out of m) image segment and $I_D \subseteq \mathbb{Z}^2$ is the discrete image domain. Generally, the set S_k is unknown. Therefore, the value of $\alpha(i)$ has to be estimated from the image data. We denote an arbitrary m component coverage segmentation with N elements (coverage vectors) by $\mathbb{A}_{N \times m}$.

Denote the set of pixels belong to the k-th class as C_k. We introduce the following definitions of a boundary used in image segmentation.

Definition 1. Hard Boundary $p(i, .) \in C_k$ if $\alpha_k = \max\{\alpha_1, \alpha_2, \cdots, \alpha_m\}$, where $\alpha_{(i,.)} = (\alpha_1, \alpha_2, \cdots, \alpha_m)$. Here, for any pixel $p(i, .)$ there is a class label C_k, which it belongs to without cross conquer, even the pixel is not clearly seen to be the same as other pixels in C_k. We call the segmentation as classification with hard boundaries.

Definition 2. Inner Boundary $p(i, .) \in C_k$ if $\alpha_k = \max\{\alpha_1, \alpha_2, \cdots, \alpha_m\}$ and $\alpha_k - \alpha_s > \sigma/m$, where $\alpha_s = \max\{\alpha_1, \alpha_2, \cdots, \alpha_m\} \backslash \alpha_k$, σ is a parameter, such as 0.05. According to this definition, there exist some pixels which do not belong to any class. We call the set S_I, including all the pixels, as the data of one class with inner boundary.

Definition 3. Outer Boundary $p(i, .) \in C_k$ if $\alpha_k = \max\{\alpha_1, \alpha_2, \cdots, \alpha_m\}$ or $\alpha_{max} - \alpha_k < \sigma/m$, where σ is a parameter, such as 0.05. According to this definition, there exists some pixels which belong to more than one class. We call the set S_O, including all the pixels, as the data of one class with outer boundary.

Definition 4. Soft Boundary Suppose S_I and S_O are the sets for the data of the same class C, define $S_B = S_O - S_I$ as the soft boundary of the data set of class C.

Fig.2(a) is an example of a Hard Boundary. Fig.2(b) is an example of an Inner Boundary. Fig.2(c) is an example of an Outer Boundary. Fig.2 is an example of Soft Boundary example.

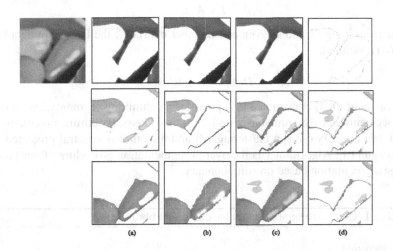

Fig. 2. Image segmentation by different boundary, (a) segmentation by hard boundaries; (b) segmentation by inner boundaries; (c) segmentation by outer boundaries; (d) soft boundaries between two classes

Generally, it holds that $S_I \in S_O$ and $S_I \cap S_O \neq \emptyset$. This means inner boundary and outer boundary can both used to separate one object from others, but with different standards. For an object in an image, if its soft boundary is null, i.e. $S_B = \emptyset$ and $S_I = S_O$, we call the soft boundary vanish. In this case, usually the boundary of the object is sharp. For an object in an image, its hard boundary is a special case of soft boundary with $\sigma = 0$. The hard boundary is defined the same as classical boundary used in [2]. When an object has a soft boundary, the outer boundary is usually much larger than the inner boundary. When the soft boundary is vanished, the object has a sharp boundary.

4 Algorithm

For any image I defined in Section 3, our goal is to obtain a coverage segmentation of I corresponding to m classes (objects). Each pixel is assigned a vector of length m whose components give the relative area of the pixel covered by each of the m classes. A coverage segmentation of the image I is, in accordance with (2) and the following notation introduced for I. Denoted $A = [\alpha_{i,j}]_{N \times m}$, where $\alpha_{i,j} \in [0, 1]$ is the coverage of the pixel with index $i(i = 1, 2, \cdots, N)$ by a class (object) S_j. Assuming spatially non-overlapping classes S_j which partition the image, each row of A sums up to one.

The proposed segmentation method models the image intensities I as a non-negative linear mixture (i.e., a convex combination) of the pure class representatives. The pure class representatives (often referred to as end-members) can be represented by a matrix

$C = [c_{j,k}]_{m \times b}$; where $c_{j,k}$ corresponds to the image value of a class j in the band k. Using the introduced notation, we can, conveniently, express that I is approximately a linear mixture of the end-members as follows

$$I \approx A \cdot C \tag{3}$$

For a given image I and a given end-member matrix C, the following data fidelity term $D(A)$ is defined,

$$D(A) = \| I - AC \|^2 \tag{4}$$

Minimization of $D(A)$ constrained to A provides an unmixing segmentation. In the following segmentation algorithm based on Eq.(4), two classical optimization methods are referred, one is fuzzy c-mean clustering[12] and the other is spectral projected gradient algorithm[13]. Algorithm 1 is the overall segmentation procedure of our proposed coverage segmentation based on soft boundary.

Algorithm 1. Coverage segmentation based on soft boundary

---

begin
  Data preparing
    Image data
    Class number
  Initializing samples C and membership matrix A
    Select C mutually or automatically
    Random initialize A
  Update and optimize A
    Update A by FCM
    Optimize A by the SPG algorithm
  Classification
    Compute inner boundary, outer boundary and soft boundary
    Classification by A and neighbor decision
end

---

5   Experiments and Discussion

In this section, we first present a qualitative comparison with crisp and fuzzy pixel-wise classification of the "peppers" image. Then we evaluate the sensitivity of the proposed method to noise on synthetic data, and compare the results with the method in [5]. Finally, we present a coverage segmentation on an image of astronautics with rough boundaries, and compare the performance with the method presented in [1].

The results of a crisp segmentation for the "peppers" image based on a Fuzzy clustering (FCM) and SPG algorithm, are shown in Fig.3. Fig.3(a),(b),(c),(d),(e) show classification results of the segmentation based on Hard Boundary, Inner Boundary, Outer Boundary, Soft Boundary and Neighbor decision, respectively. From Fig.3, we find that by using soft boundary, some overlapping parts of the object are detected and departed from other parts. We call these parts of pixels as debated region, which is error possible if trying to determine their classes. Moreover, Fig.3(e) shows that, by using neighbor decision rule, some parts of noise and boundaries are clearly classified and connected comparing with Fig.3(a),(b),(c). Especially, for the middle of Fig.3(e), there are some yellow parts of pixels from the right(red) now classified into the yellow, which is it should be from the point view of color construction theory.

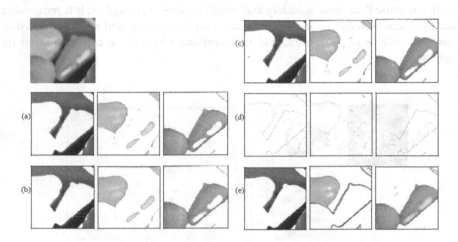

Fig. 3. Segmentation with different boundaries (a) hard boundary. (b)inner boundary. (c) outer boundary. (d) soft boundary. (e) neighbor decision.

In order to test how optimization of the objection function Eq.(4) may improve the segmentation in detail, we provide the comparison of the segmentation results in Fig.4. It shows the coverage segmentation results of two cases, i.e. optimization with/without SPG algorithm. Experiments show that some image segmentation results are not sensitive to SPG optimization algorithm, but most image samples, e.g. Fig.4, are very sensitive to SPG optimization because of the tiny difference between classes in gray value of each bands.

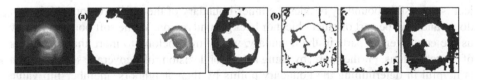

Fig. 4. Coverage segmentation results for optimization with/without SPG algorithm. (a) optimization with SPG algorithm. (b) optimization without SPG algorithm.

Fig.5 shows the coverage segmentation results of "Nebula", with noise addition. It is clearly visible that the image noise has a rather strong negative impact on the result. In addition, the obtained segmentation is crisp and does not provide coverage information.

We explore how the methods perform for increasing levels of noise. As a reference, coverage values for a hard boundary, inner boundary, outer boundary and neighbor decision boundary, are presented. Coverage values are computed by 256×256 times super-sampling (the object is digitized at 256 higher resolution and coverage values are approximated as the relative number of covered pixels for each square of 256×256 pixels)[1]. From Fig.5, one may find that outer boundary and hard boundary are more sensitive to noise than inner boundary and neighbor decision boundary. It is reasonable since the inner boundary is the smallest set in all boundaries, and neighbor decision boundary is robust to noise because of its smoothness by a linear combination of its neighbors.

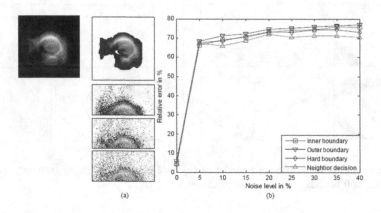

Fig. 5. (a): (top) Test objects for one class of "Nebula". (second) Part of object with 5% noise added. (third) Coverage segmentation result with 10% noise. (bottom) Coverage segmentation result with 15% noise. (b): Average absolute error of coverage values of object border pixels with different noise levels.

Experiments show that the proposed model has advantage in describing the over-lapping region of objects. The inner boundary of the object can be considered as the determined boundary of an object region with no doubt. The outer boundary of the object is a possible boundary besides the inner boundary of the object region which is in doubt and may be part of other object or background region. While soft boundary show us the controversial parts in the boundary, which may belong to more than one class of object regions. Furthermore, by using neighbor decision rule, coverage segmentation is helpful in determining the boundary points by their neighbors' membership value besides by their own membership value.

6   Conclusion

We proposed a method for coverage segmentation of fuzzy images based on information obtained from soft boundaries and neighbor decision. Inner boundaries and outer boundaries are applied to describe the determined and possible regions of objects, respectively. The performance of the experiments show that the proposed method improves the coverage segmentation of color objects with fuzzy boundaries and noise. For fuzzy color objects, we demonstrated that the proposed method can compete with traditional coverage segmentation methods, especially when the images contain noise. In case of noise free images, the proposed method dropped a little bit behind, which we believe is caused by the grey-level quantization during the down-sampling. The proposed method provides an extended research on the high-resolution of fuzzy images.

Acknowledgement. This work is supported by Blue Project of Universities in Jiangsu Province Training Young Academic Leaders Object, the six talent peaks project of Jiangsu Province (No. DZXX-028) and National Natural Science Foundation of China (No.61170121).

References

1. Lindblad, J., Sladoje, N.: Coverage segmentation based on linear unmixing and minimization of perimeter and boundary thickness. Pattern Recognition Letters 33(6), 728–738 (2012)
2. Sladoje, N., Lindblad, J.: Pixel coverage segmentation for improved feature estimation. In: Foggia, P., Sansone, C., Vento, M. (eds.) ICIAP 2009. LNCS, vol. 5716, pp. 929–938. Springer, Heidelberg (2009)
3. Sladoje, J., Lindblad, N.: The coverage model and its use in image processing, mathematics subject classification. In: Mathematics Subject Classification, pp. 68U10, 65D18 (2010)
4. Sladoje, N., Nystr, I., Saha, P.K.: Measurements of digitized objects with fuzzy borders in 2d and 3d. Image Vision Comput. 23(2), 123–132 (2005)
5. Sladoje, N., Lindblad, J.: High-precision boundary length estimation by utilizing gray-level information. IEEE Trans. Pattern Anal. Mach. Intell. 31(2), 357–363 (2009)
6. Keshava, N., Mustard, J.: Spectral unmixing. Signal Process. 19(1), 44–57 (2002)
7. Villa, A., Chanussot, J., Benediktsson, J.A., Jutten, C.: Spectral unmixing for the classification of hyperspectral images at a finer spatial resolution. J. Sel. Topics Signal Processing 5(3), 521–533 (2011)
8. Dickinson, M.E., Bearman, G., Tille, S., Lansford, R., Fraser, S.E.: Multi-spectral imaging and linear unmixing add a whole new dimension to laser scanning fluorescence microscopy. Biotechniques 31(6), 1272–1278 (2001)
9. Haraguchi, T., Shimi, T., Koujin, T., Hashiguchi, N., Hiraoka, Y.: Spectral imaging fluorescence microscopy. Genes to Cells 7(9), 881–887 (2002)
10. Gong, P., Zhang, A.: Noise effect on linear spectral unmixing. Annals of GIS 5(1), 52–57 (1999)
11. Lidayova, K., Lindblad, J., Sladoje, N., Frimmel, H.: Coverage segmentation of thin structures by linear unmixing and local centre of gravity attraction. In: Proc. 8th International Symposium on Image and Signal Processing and Analysis, pp. 83–88 (2013)
12. Nock, R., Nielsen, F.: On weighting clustering. IEEE Trans. Pattern Anal. Mach. Intell. 28(8), 1223–1235 (2006)
13. Birgin, E.G., Martínez, J.M., Raydan, M.: Nonmonotone spectral projected gradient methods on convex sets. SIAM Journal on Optimization 10(4), 1196–1211 (2000)

A Modified Fuzzy Color Histogram Using Vision Perception Difference of Pixels Location

Jie Zhao[1,2], Gang Xie[1,*], and Wenjing Zhao[1]

[1] College of Information Engineering , Taiyuan University of Technology,
No.79 West Yingze Street, Taiyuan 030024, Shanxi, China
[2] Department of Computer Engineering, Taiyuan University, No.18 South Dachang Road,
Taiyuan 030012, Shanxi, China
tydxcomputer@163.com, {xiegang,zhaowenjing}@tyut.edu.cn

Abstract. Conventional color histogram is sensitive to noisy interference such as illumination and quantization errors. Furthermore, small changes in the conventional color histogram might result great changes due to large dimension or histogram bins. This paper presents a modified fuzzy color histogram. Firstly, it considers each pixel's color associated to all the histogram bins according to fuzzy logic and provides a histogram with single-dimension based on the CIELAB color space. Thus, it has the capacity to tolerate noisy interference and reduce computational complexity. Then, it combines the information about the location of pixels in an image to record the human vision perception variation in different spatial positions within an image. The proposed histogram is further exploit in the application of image indexing and retrieval. Experimental results show that the proposed histogram is more accurate and efficient in retrieving the user-interested images.

Keywords: Fuzzy color histogram, Fuzzy logic, Spatial position, Vision perception.

1   Introduction

Among the characteristics of color images, the color of an image constitutes a powerful visual cue and the most robust characteristic[13,14]. Therefore, many researchers have recently used the color property to state the characteristic of an image and developed more accurate image retrieval methods. The color histogram is one of the common methods using color property[1-3]. With the advantages of simplicity and invariance to translation and rotation of the image, the color histogram is widely used in image indexing for image retrieval, video retrieval and object tracking[16-18].

The primary weakness of the conventional color histogram is sensitive to noisy interference, since it cannot consider the color similarity across different bins and the color dissimilarity in the same bin. Furthermore, large dimension or histogram bins can result in extensive computation. To address these issues, the concept called fuzzy

---

* Corresponding author.

C. Cornelis et al. (eds.): RSCTC 2014, LNAI 8536, pp. 382–389, 2014.

color histogram (FCH) has been proposed[4-9]. FCH considers the color similarity of each pixel's color associated to all the histogram bins through fuzzy-set membership function. Two main methods are used to FCH production for content-based image retrieval. One method is fuzzy c-means algorithm [15] that is an unsupervised classification algorithm. It is a useful method to produce the fuzzy color histogram. For example, Ju et al. [6] determined the bin number of FCH and computed the bins' membership values by using fuzzy c-means method. Since fuzzy c-means algorithm is very sensitive to the initialization condition, Khang et al.[10] split the color image hierarchically into multiple homogeneous regions, and merged those regions to obtain the initialization condition for fuzzy c-means algorithm. The other method is fuzzy inference system according to Fuzzy logic introduced by Zadeh[11]. Konstantinidis et al.[4] proposed to link the three elements into a single histogram by means of a fuzzy inference system, and the created FCH was used for image retrieval. The experimental results prove that the method is flexible and tolerant to imprecise data, and the expertise can be incorporated easily and efficiently. Maryam et al.[5] used OWA (ordered weighted average) aggregation operator in fuzzy inference system. The OWA fuzzy linking color histogram can reveal more detailed image information to description particular significance of the image.

Generally speaking, FCH created by fuzzy inference system can contain only one dimension to reduce computational complexity, and the method is relatively easy to implement. However, the existing FCHs are all accompanied with some drawbacks. For example, they do not take into account the spatial distribution of image color and the change degree of pixel colors. This paper proposes a spatial fuzzy linking color histogram (SFLCH) to address the problems and make the image retrieval more accurate. SFLCH integrates the fuzzy linking color histogram with the spatial information to depict the color distribution and the spatial information.

The paper is organized as follows: Section 2 defines the modified fuzzy color histogram, color space selection and the structure of fuzzy inference system; Section 3 illustrates the experiments and the analysis of the results; and the conclusion is stated in Section 4.

2   The Proposed Fuzzy Color Histogram

2.1   Fuzzy Linking Color Histogram Creation

Color Space Transformation

In order to create fuzzy linking color histogram, we selected CIELAB color space which is a perceptually uniform color space and approximates the way that humans perceive color. CIELAB color space is nearly linear with visual perception, or at least as close as any color space is expected to sensibly get. Since they are based on CIE system of color measurement, which is itself based on human vision, CIELAB is device independent and coloring information is referred to the color of the white point of the system. CIELAB is intended to mimic the logarithmic response of the eye. CIELAB was found to perform better than other color spaces (such as RGB, HSV, LCH) in various retrieval tests performed in the laboratory for this exact purpose that the histogram creation method is assessed to retrieve similar images from a widely diverse image collection [19].

The color space transformation from RGB to CIELAB needs to be operated pixel by pixel. RGB value cannot be transformed directly to the CIELAB color space. The transformation process requires two steps[12]. Firstly, RGB values are transformed to XYZ tristimulus values as follows:

$$X = 0.4339R + 0.3762G + 0.1899B$$

$$Y = 0.2126R + 0.7152G + 0.07218B \qquad (1)$$

$$Z = 0.01776R + 0.1095G + 0.8729B$$

The CIELAB equation is then applied, which involves the evaluation of cube roots as follows:

$$L^* = 116f(\frac{Y}{Y_0}) - 16$$

$$A^* = 500[f(\frac{X}{X_0}) - f(\frac{Y}{Y_0})] \qquad (2)$$

$$B^* = 200[f(\frac{Y}{Y_0}) - f(\frac{Z}{Z_0})]$$

where

$$f(q) = \sqrt[3]{q} \qquad q > 0.008856$$

$$f(q) = 7.787q + \frac{16}{116} \qquad q \leq 0.008856$$

X_0, Y_0, and Z_0 are tristimulus values of the nominally white object-color stimulus.

In CIELAB, the luminance is represented as L^*, the relative greenness-redness as A^*, and the relative blueness-yellowness as B^*. The range of CIELAB color space can be calculated through Equations (1) and (2).

Fuzzy Inference System
After color space transformation, the fuzzy linking color histogram is created by a fuzzy inference system. The structure of the proposed fuzzy inference system is shown in Fig.1.

Fuzzification of input variables is achieved by using triangular member function. L^* is subdivided into only three regions. A^* and B^* components have more weight than L^* component. A^* is subdivided into five regions: green, greenish, middle, reddish and red. B^* is subdivided into five regions: blue, bluish, middle, yellowish

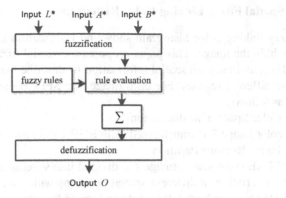

Fig. 1. Structure of the proposed fuzzy inference system

and yellow. Fig.2(a)-(c) show membership functions of the inputs. The output of the system is shown in Fig.2(d). It has only 10 equally divided trapezoidal membership functions.

Three components (L^*, A^* and B^*) are linked to lead to the one output of the system in Mamdani type of fuzzy inference according to 27 fuzzy rules which were established through the expert experience.

The defuzzification phase is performed by using the largest of maximum (LOM) method, and the resulting fuzzy set is defuzzified to produce output variable O which is a crisp number.

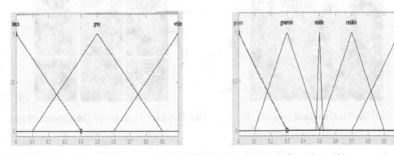

(a) Membership function of input variable "L^*"(b) Membership function of input variable "A^*"

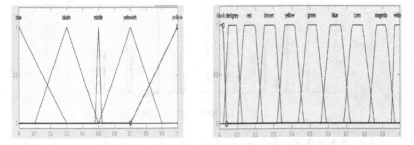

(c) Membership function of input variable "B^*"(d) Membership function of output variable "O"

Fig. 2. Membership function of the inputs (L^*, A^* and B^*) and the output ("O")

2.2   Spatial Fuzzy Linking Color Histogram

The fuzzy linking color histogram shows no information about the location of these pixels within the image. This paper proposes the spatial fuzzy linking color histogram (SFLCH) combined with spatial information to describe the color distribution of pixel colors in different regions. For each image, the SFLCH creation process can be illustrated as follows:

Step 1: Color Space Transformation

The color image T is transformed from RGB color space to CIELAB color space.

Step 2: Image Regions Partition

In CIELAB color space, image T is divided into several regions using human vision perception variation in different spatial positions within an image. This paper selects nine divided regions from left to right and top to bottom.

Step 3: SFLCH Creation

SFLCH describes the location and the color distribution of pixels in each divided region. The fuzzy histogram of each region consists of 10 bins representing black, dark grey, red, brown, yellow, green, blue, cyan, magenta and white. So the image T has a SFLCH totaling 9×10 bins. Fig.3 shows the images and SFLCH of the flower.

3   Experimental Results

(a) The original image

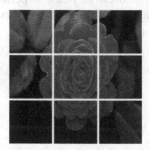
(b) The nine regions of the flower image

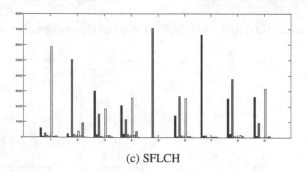
(c) SFLCH

Fig. 3. The images and fuzzy histogram of the flower

The experiments were all run on MATLAB. To check the retrieval efficiency of the modified fuzzy color histogram, we have to test the performance of retrieving the user-interested images using three methods from the Corel Image Gallery. Method 1 is SFLCH; Method 2 is the conventional histogram; Method 3 is the fuzzy linking color histogram. 20 similar images are retrieved for each query image. The average precision of each kind of query results were obtained, as shown in Fig.4. As a result, the proposed method is proved to be more accurate than other methods. SFLCH combines the spatial information into FLCH and SFLCH is superior to FLCH.

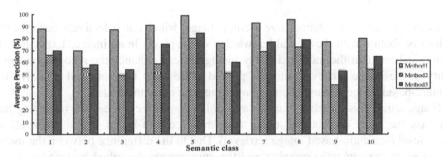

Fig. 4. The average precision of 4 retrieval image methods. Method 1 is SFLCH; Method 2 is the conventional histogram; Method 3 is the fuzzy linking color histogram.

(a) The classic fuzzy linking color histogram

(b) SFLCH

Fig. 5. The retrieval results of query image "horse"

Fig.5 shows the retrieval results of image "horse", which use the classic fuzzy linking color histogram and the modified fuzzy color histogram respectively. The query image is at the upper-left corner, and other 20 images are the retrieval similar images sorted by the similarity ratio. The experimental result clearly reveals that the proposed histogram is much more accurate than other methods for the first 20 returned images of the test image database.

4   Conclusion

SFLCH is presented to characterize a color image, which can effectively describe the color distribution with the spatial knowledge of the pixels in an image. Experimental results indicate that the modified fuzzy histogram can enhance the recognition capability of the image retrieval, and has proved to be more accurate and effective for content-based image retrieval being comparing with other methods.

Image retrieval is generally known as a collection of techniques for retrieving images on the basis of features, either low-level (content-based image retrieval) or high- level (semantic-based image retrieval). Due to the difference between the users' concerns on the semantic meaning and the appearances described by low-level features, many researchers began to find the way how to bridge the semantic gap. They provide many state-of-the-art achievements in filling the semantic gap[20-22]. For examples, literature [20] proposed the supervised multiclass labeling that leads to optimal annotation and retrieval. Images are represented as bags of localized feature vectors and joint modeling of semantic label and visual feature distributions. Literature [21] proposed a new dimensionality reduction algorithm for relevance feedback to enhance the performance of image retrieval.   Images are represented by three types of popular global features: color, texture and shape. Because the visual features cannot well describe semantic contents, the system involves the user's performance in the loop to bridge the semantic gap.

Since semantic-based features rely on low-level ones, in this paper we modified the most widely used low-level feature "color histogram" that is a statistical tool to bear concrete information. We would like to find an efficient method of low-level features extraction in order to establish a good foundation for the semantic retrieval. That is our next research direction to bridge the gap between the low-level visual features and the high-level semantic meanings by the modified color histogram.

Acknowledgements. This work was supported by Talent Special Foundation of Taiyuan Science and Technology Project (120247-28).

References

1. Smeulders, A.W.M., Worring, M., Santini, S., et al.: Content-based image retrieval at the end of the early years. IEEE Transactions on Pattern Analysis and Machine 22(12), 1349–1380 (2000)

2. Datta, R., Li, J., Wang, J.: Content-based image retrieval: Approaches and trends of the new age. In: 7th ACM SIGMM International Workshop on Multimedia Information Retrieval, pp. 253–262 (2005)
3. David, F.: In Multimedia Information Retrieval and Management: Technological Fundamentals and Applications. Springer, Heidelberg (2010)
4. Konstantinidis, K., Gasteratos, A., Andreadis: Image retrieval based on fuzzy color histogram processing. Optics Communications 248(4), 375–386 (2005)
5. Mahmoudi, M.T., Beheshti, M., et al.: Content-based image retrieval using OWA fuzzy linking histogram. Journal of Intelligent and Fuzzy Systems 24(2), 333–346 (2013)
6. Han, J., Ma, K.-K.: Fuzzy color histogram and its use in color image retrieval. IEEE Transactions on Image Processing 11(8), 944–952 (2002)
7. Das, S., Sural, S., Majumdar, A.K.: Detection of hard cuts and gradual transitions from video using fuzzy logic. International Journal of Artificial Intelligence and Soft Computing 1(1), 77–98 (2008)
8. Tan, K.S., et al.: Novel initialization scheme for Fuzzy C-Means algorithm on color image segmentation. Applied Soft Computing 13(4), 1832–1852 (2013)
9. Kim, W., Kim, C.: Background subtraction for dynamic texture scenes using fuzzy color histograms. IEEE Signal Processing Letteers 19(3), 127–130 (2012)
10. Lo, C.-C., Wang, S.-J.: Video segmentation using a histogram-based fuzzy c-means clustering algorithm. Computer Standards and Interfaces 23(5), 429–438 (2001)
11. Zadeh, L.: Fuzzy sets and systems. In: Proc. Sympos. on System Theory, pp. 29–37 (1965)
12. Ford, A., Roberts, A.: Color space conversions. Westminster University, London (August 11, 1998)
13. Lee, H.-Y., Kang, I.-K., Lee, H.-K., Suh, Y.-H.: Evaluation of feature extraction techniques for robust watermarking. In: Barni, M., Cox, I., Kalker, T., Kim, H.-J. (eds.) IWDW 2005. LNCS, vol. 3710, pp. 418–431. Springer, Heidelberg (2005)
14. Theoharatos, C., Laskaris, N.A., et al.: A generic scheme for color image retrieval based on the multivariate Wald-Wolfowitz test. IEEE Transactions on Knowledge and Data Engineering 17(6), 808–819 (2005)
15. Bezdek, J.C.: Pattern Recognition with Fuzzy Objective Function Algorithms. Kluwer Academic Publishers, Norwell (1981)
16. Rasheed, W., et al.: Image retrieval using maximum frequency of local histogram based color correlogram. In: 2nd International Conference on Multimedia and Ubiquitous Engineering, pp. 322–326 (2008)
17. Leichter, I., Lindenbaum, M., Rivlin, E.: Mean Shift tracking with multiple reference color histograms. Computer Vision and Image Understanding 114(3), 400–408 (2010)
18. Yoo, H.-W., Cho, S.-B.: Video scene retrieval with interactive genetic algorithm. Multimedia Tools and Applications 34(3), 317–336 (2007)
19. Konstantinidis, K., Andreadis, I.: Performance and computational burden of histogram-based color image retrieval techniques. Journal of Computational Methods in Sciences and Engineering, 141–147 (2005)
20. Carneiro, G., Chan, A.B., Moreno, P.J., Vasconcelos, N.: Supervised Learning of Semantic Classes for Image Annotation and Retrieval. IEEE Transactions on Pattern Analysis and Machine Intelligence 29(3), 394–410 (2006)
21. Bian, W., Tao, D.: Biased Discriminant Euclidean Embedding for Content-Based Image Retrieval. IEEE Transactions on Image Processing 19(2), 545–554 (2010)
22. Vogel, J., Schiele, B.: Semantic Modeling of Natural Scenes for Content-Based Image Retrieval. Int. J. Comput. Vision 72(2) (2007)

Author Index